BASIC BIOCHEMISTRY

MAX E. RAFELSON, Jr., Ph.D.

Associate Dean, Biological and Behavioral Sciences and Services, and Professor of Biochemistry, Rush Medical College, Rush-Presbyterian-St. Luke's Medical Center, Chicago

STEPHEN B. BINKLEY, Ph.D.

Dean, Graduate College, and Professor of Biological Chemistry, University of Illinois Medical Center, Chicago

JAMES A. HAYASHI, Ph.D.

Professor of Biochemistry, Rush Medical College, Rush-Presbyterian-St. Luke's Medical Center, Chicago

Basic Biochemistry

Third Edition

THE MACMILLAN COMPANY
New York

COLLIER-MACMILLAN LIMITED
London

Printed in the United States of America

THE MACMILLAN COMPANY
866 THIRD AVENUE, NEW YORK, NEW YORK 10022

COLLIER-MACMILLAN CANADA, LTD., TORONTO, ONTARIO

Library of Congress catalog card number: 79-136264

PRINTING 3456789 YEAR 3456789

PREFACE TO THIRD EDITION

THE THIRD EDITION of this textbook again reaffirms the purpose of the first two editions, namely, to introduce the beginning student to the principles and viewpoints of biochemistry, and to a core of essential facts, without presenting an overwhelming amount of factual material. In order to accomplish this goal, it has been necessary to assume that the reader has an adequate background in biology and chemistry. As a consequence, the definitions of basic biologic terms and the mechanisms of organic reactions have not been emphasized. In limiting the size of the volume to less than 400 pages, the authors have purposely excluded material that some instructors might have wished included. However, we believe that sufficient areas have been presented in depth for the student to obtain a basic understanding of the functions of biologic systems at the molecular and cellular levels and to enable him to progress to other topics in physiological chemistry.

Several of the chapters have been either completely rewritten or extensively revised to include new material of importance and to delete sections of lesser significance. Some changes have been made in the general organization. These include the incorporation and integration of chemistry and metabolism into single chapters for carbohydrates (Chapter 5), lipids (Chapter 7), and nucleic acids (Chapter 8). Metabolic regulation has been expanded and is again integrated throughout the appropriate areas of metabolism.

water-soluble vitamins are presented as cofactors in the development of metabolic pathways, rather than as a separate chapter. Hormones that have well-established metabolic roles are integrated into the appropriate areas of metabolism. Fat-soluble vitamins and hormones for which clear-cut biochemical mechanisms may not be known are discussed in the sections on descriptive chemistry. Rather than starting with detailed mechanisms, and combining them into a total picture, metabolism is first presented from the overall point of view, after which detailed mechanisms are developed to fit the larger picture. Because of the current importance of control mechanisms and protein synthesis, the authors have included considerable material on these subjects.

References are limited to a select few that will lead the reader to other helpful sources of information. The authors are hopeful that the student will continue to use, for reference purposes, one or more of the excellent comprehensive texts* that have extensive bibliographies, and to which chapter references are given.

M. E. R.
S. B. B.
J. A. H.

* *Textbook of Biochemistry,* by West, Todd, Mason, and Van Bruggen (4th ed., 1966. The Macmillan Co.); *Principles of Biochemistry,* by White, Handler, and Smith (4th ed., 1968, McGraw-Hill); and *Biological Chemistry,* by Mahler and Cordes (1st ed., 1966, Harper & Row).

CONTENTS

BASIC BIOCHEMISTRY

Chapter 1

ACIDS, BASES, AND BUFFERS

BIOCHEMISTRY is a study of biology at the molecular level. It is concerned with the molecules that make up the structure of cells and organs, that is, molecular anatomy. Much is known about these molecules, and rapid progress is being made in working out the molecular organization of the cell. Biochemistry is also concerned with molecular physiology, that is, the function of molecules in carrying out the needs of the cells and organs. Again there is considerable information in this area, and the knowledge about the chemical reactions taking place in cells is increasing rapidly.

CELLULAR ENVIRONMENT

It is common knowledge that the function and activity of organisms depend on the environment and that relatively small changes in temperature, humidity, and pressure can be incompatible with life. In the same manner the function of molecular reactions depends on the environment in which the reaction takes place. A significant portion of human biochemistry is concerned with the chemical mechanisms for maintaining the cellular environment in a state that is optimal for proper function.

The overall function of the cell is to convert food, with the help of biologic catalysts called enzymes, into energy and protoplasm. In this process a number of byproducts are formed and must be eliminated. This activity is expressed by the equation

$$\text{Food} \xrightarrow{\text{enzymes}} \text{Energy} + \text{Protoplasm} + \text{Byproducts}$$

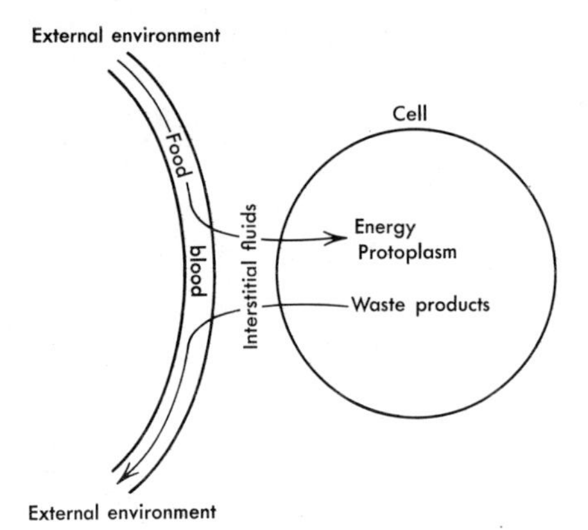

Figure 1.1. Diagram illustrating relation of cell to external environment.

In a complex multicellular animal, food must be transported to the cell, and the byproducts must be carried away from the cell. In man the cell is connected to the external environment by means of the circulatory system (Figure 1.1). Through this mechanism the cells are protected temporarily from large physical and chemical changes. Food can be supplied continuously and the byproducts removed continuously, and at the same time concentrations of many ions and chemical substances within the cell are controlled within narrow limits.

Since the reactions taking place in a cell are carried out in aqueous solution, one of the most important ions is the hydrogen ion. The ionic and tertiary structures of the protein molecules, important components of all protoplasm, are dependent on the hydrogen ion concentration.

Figure 1.2 shows the influence of hydrogen ion concentration on the ionic forms of a protein. The protein molecule has several carboxyl and amino groups in its structure. At an intermediate hydrogen ion concentration the net charge of the molecule will be zero. At a higher hydrogen ion concentration the molecule will possess a positive charge. At a lower hydrogen ion concentration the molecule will carry a negative charge. The physical and chemical properties of proteins depend on the charge on the molecule; for example, there is a hydrogen ion concentration at which a protein will have a minimum solubility. If cellular proteins are to function efficiently, they must have specific ionic forms.

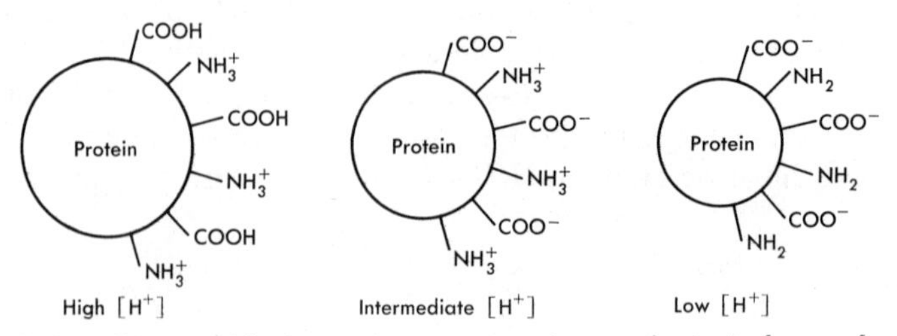

Figure 1.2. Influence of hydrogen ion concentration on the ionic forms of a protein.

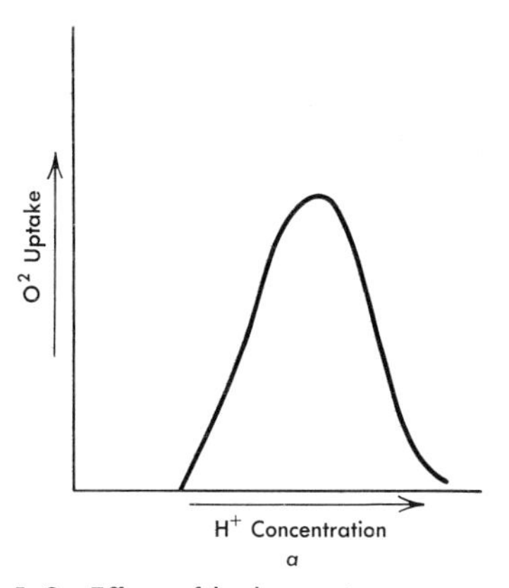

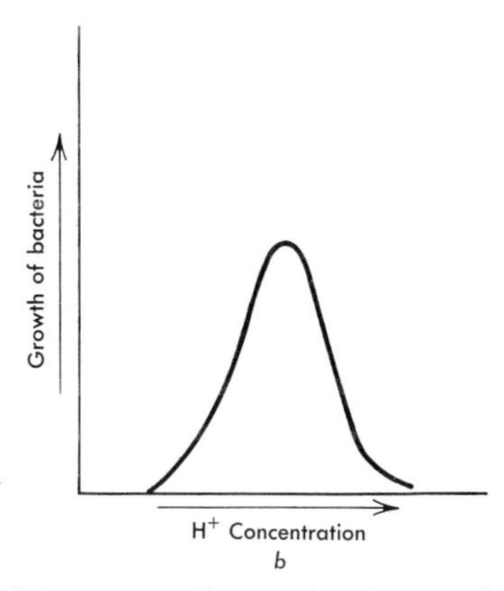

Figure 1.3. Effect of hydrogen ion concentration on (*a*) oxygen utilization by tissue and (*b*) rate of growth of bacteria.

It is possible to measure the amount of oxygen utilized by tissue and to show that the rate of uptake depends on the hydrogen ion concentration (see Figure 1.3a). Likewise, the rate of growth of bacteria depends on the hydrogen ion concentration (see Figure 1.3b).

ACIDS AND BASES

Brönsted Theory

According to the Brönsted theory an acid is a substance that supplies H^+ (protons) and a base is a substance that reacts with H^+ (protons); for example, in the equation $HB \rightleftharpoons H^+ + B^-$, HB is an acid and B^- is a base. Water reacts with protons, $H^+ + H_2O \rightleftharpoons H_3O^+$, to form the hydronium ion. In this case water is a base and the hydronium ion is an acid. In future discussions H^+ will be used, but it will be understood that it actually exists in aqueous solution as the hydronium ion.

Some examples of acids and bases are

Acids		Conjugate Bases	
HCl	$\rightleftharpoons$	Cl^-	$+ H^+$
H_2CO_3	$\rightleftharpoons$	HCO_3^-	$+ H^+$
HCO_3^-	$\rightleftharpoons$	$CO_3^=$	$+ H^+$
H_3PO_4	$\rightleftharpoons$	$H_2PO_4^-$	$+ H^+$
$H_2PO_4^-$	$\rightleftharpoons$	$HPO_4^=$	$+ H^+$
NH_4^+	$\rightleftharpoons$	NH_3	$+ H^+$
RNH_3^+	$\rightleftharpoons$	RNH_2	$+ H^+$
HOH	$\rightleftharpoons$	OH^-	$+ H^+$
H_3O^+	$\rightleftharpoons$	H_2O	$+ H^+$

It should be noted that one substance (HCO_3^-, for example) may react with a proton (base) or it may supply a proton (acid).

Methods of Expressing Concentrations of Acids

Acids may differ in strength, and this strength may be expressed in terms of hydrogen ion concentration as grams per liter.

0.1 *M* HCl has 0.1 g H^+ per liter
H_2O has 0.000,0001 g H^+ per liter (10^{-7})
0.1 *M* NaOH has 0.000,000,000,0001 g H^+ per liter (10^{-13})

It would be time consuming and complicated to make calculations with such unwieldy numbers. Therefore, it is common practice to use the convenient designation pH, the negative logarithm of the hydrogen ion concentration, which is expressed symbolically as $pH = -\log [H^+]$. For water, then

$$\begin{aligned} \log 10^{-7} &= -7 \\ -\log 10^{-7} &= 7 \\ pH &= 7 \end{aligned}$$

Intracellular pH has been determined by various methods. Depending on the method used and the condition of the cell, researchers obtained values ranging from about 5.2 (slightly acid) to 8.5 (slightly alkaline). The pH of fluids from most cells appears to be close to neutral—that is, pH 7.0.

Problem

Blood has a $[H^+] = 5 \times 10^{-8}$. Calculate its pH.

$$\begin{aligned} pH &= -\log [H^+] \\ &= -\log (5 \times 10^{-8}) \\ &= -(\log 5 + \log 10^{-8}) \\ &= -(0.7 - 8) \\ &= 7.3 \end{aligned}$$

IONIZATION OF WATER

The pH of water, 7, may be derived from the following relationships.

$$H_2O \rightleftharpoons H^+ + OH^-$$

$$\frac{[H^+][OH^-]}{[H_2O]} = K \qquad [H_2O] \text{ is constant,}$$

therefore $[H^+][OH^-] = K_w = 10^{-14}$

since $[H^+] = [OH^-]$

then $[H^+]^2 = 10^{-14}$

and $[H^+] = 10^{-7}$

therefore $pH = 7$

The pH values for some common fluids are

0.1 *N* NaOH	13.0	Urine (average)	4.8–7.5
0.1 *N* NH_4OH	11.1	0.1 *N* CH_3COOH	2.9
Pancreatic juice	7.5–8.0	Gastric juice	1.0
Bile	5.5–7.0	0.1 *N* HCl	1.0
Blood	7.35–7.45		

STRENGTH OF ACIDS

Strong acids are completely ionized; for example

$$HCl \longrightarrow H^+ + Cl^-$$

Weak acids, such as acetic, are partially ionized and are treated as equilibrium reactions; for example

$$HB \rightleftharpoons H^+ + B^-$$

$$\frac{[H^+][B^-]}{[HB]} = K_a$$

The ionization constant, K_a, for acetic acid $\cong 10^{-5}$.

A designation analogous to that for pH is the pK_a and is defined as the negative logarithm of the ionization constant

$$pK_a = -\log K_a$$

For acetic acid

$$pK_a = -\log 10^{-5}$$

$$pK_a = 5$$

Some examples of weak acids are

Strength of Acid	Acid	K_a	pK_a
Moderately strong	H_3PO_4	10^{-2}	2
Moderately weak	CH_3COOH	10^{-5}	5
Very weak	NH_4^+	10^{-9}	9
Extremely weak	glucose	10^{-13}	13

In the case of phosphoric acid, which can furnish three protons, there are an ionization constant and a pK_a for each.

$$H_3PO_4 \rightleftharpoons H_2PO_4^- + H^+ \; (pK_a = 2) \rightleftharpoons HPO_4^{=} + H^+ \; (pK_a = 7) \rightleftharpoons PO_4^{\equiv} + H^+ \; (pK_a = 12)$$

Problem

Calculate the pH of 0.15 *M* acetic acid.

$$K_a = 2 \times 10^{-5}$$

$$\text{HAc} \rightleftharpoons \text{H}^+ + \text{Ac}^-$$

$$(0.15 - x) \rightleftharpoons (x) + (x)$$

$$\frac{[x][x]}{[0.15 - x]} = 2 \times 10^{-5}$$

Neglect the concentration of x in the denominator since it is much smaller than 0.15. Then

$$[x]^2 = 0.3 \times 10^{-5} = 3 \times 10^{-6}$$

$$[x] = 1.73 \times 10^{-3} \text{ moles H}^+ \text{ liter}$$

$$\text{pH} = -\log [\text{H}^+] = -(\log 1.73 + \log 10^{-3}) = -(0.24 - 3.0) = 2.76$$

BUFFERS, WEAK ACIDS IN PRESENCE OF SALTS

Consider the ionization of a weak acid

$$\text{HB} \rightleftharpoons \text{H}^+ + \text{B}^-$$

$$\frac{[\text{H}^+][\text{B}^-]}{[\text{HB}]} = K_a$$

Solve for $[\text{H}^+]$

$$[\text{H}^+] = K_a \frac{[\text{HB}]}{[\text{B}^-]}$$

In a mixture of an acid and its salt most of the acid, HB, is not ionized. Therefore the concentration of acid [HB] is approximately the same as the original concentration of acid. The salt is completely ionized; therefore the value of $[\text{B}^-]$ is approximately the same as the salt concentration. Very little B^- is supplied by HB.

It is possible to write the equation

$$[\text{H}^+] = K_a \frac{[\text{acid}]}{[\text{base}]}$$

in the following form

$$-\log [\text{H}^+] = -\log \left(K_a \frac{[\text{acid}]}{[\text{base}]}\right) = -\log K_a - \log \frac{[\text{acid}]}{[\text{base}]}$$

or

$$\text{pH} = \text{p}K_a + \log \frac{[\text{base}]}{[\text{acid}]}$$

This is called the Henderson–Hasselbalch equation.

The pK_a is the negative logarithm of the ionization constant and is the pH at which the concentration of acid is equal to the concentration of the conjugate base. For example, to calculate the pH of a solution of acetic acid that is half neutralized

$$pK_a = 4.6$$

$$\text{pH} = 4.6 + \log \frac{0.5}{0.5} = 4.6$$

For each tenfold change in concentration there is a change of 1 in pH. For example, what is the pH of a solution containing 1 mole of sodium acetate and 0.1 mole acetic acid?

$$\text{pH} = 4.6 + \log \frac{1}{0.1} = 5.6$$

Problem

How many moles of sodium acetate and how many moles of acetic acid are needed to prepare 1 liter of a 0.1 *M* solution having a pH 5.0?

$$\text{pH} = 4.6 + \log \frac{[\text{base}]}{[\text{acid}]}$$

$$5.0 = 4.6 + \log \frac{[\text{B}]}{[\text{A}]}$$

$$\log \frac{[\text{B}]}{[\text{A}]} = 0.4$$

$$\frac{[\text{B}]}{[\text{A}]} = 2.5$$

Therefore 2.5 moles of salt would be required for each mole of acid. Since a 0.1 *M* solution is desired

$$\frac{2.5}{3.5} \times 0.1 = 0.071 \text{ mole sodium acetate}$$

$$\frac{1}{3.5} \times 0.1 = 0.029 \text{ mole acetic acid}$$

Dissolve 0.071 mole sodium acetate and 0.029 mole acetic acid in water and dilute to 1 liter.

TITRATABLE ACIDITY

Titratable acidity is total acidity or total potential acidity. The pH might be termed the true acidity. In the following example both solutions are 0.1 *M* but have different pH values.

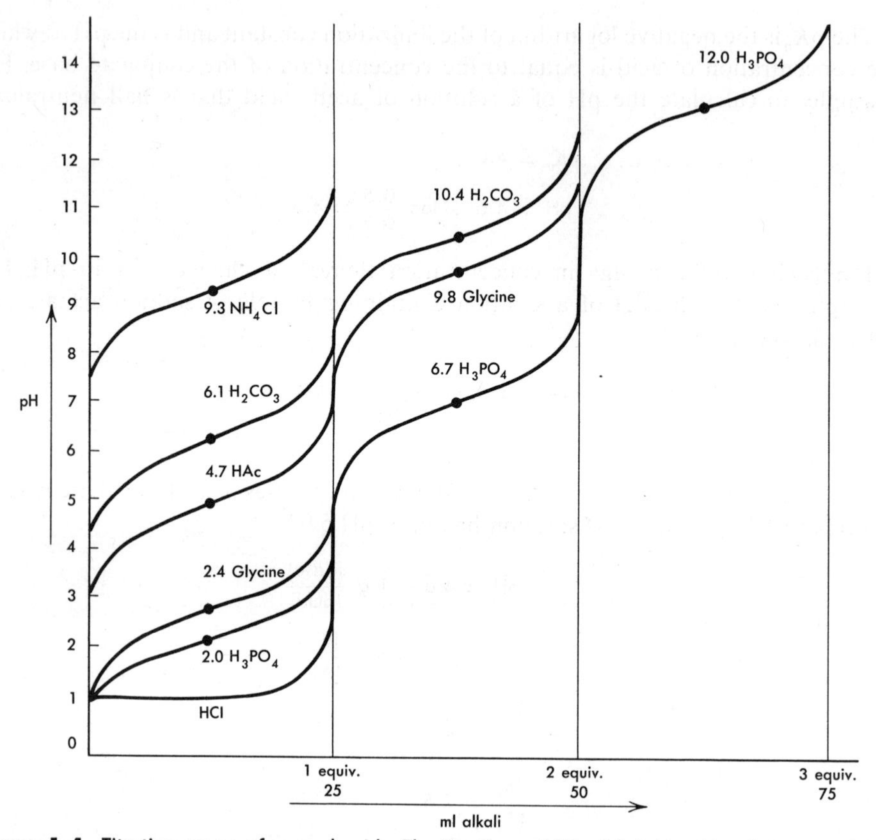

Figure 1.4. Titration curves of several acids. The titrations of 25 ml 0.1 *M* acid with 0.1 *M* NaOH.

$$HCl \rightleftharpoons H^+ + Cl^- \qquad 0.1\ M\ \text{pH}\ 1$$

$$HAc \rightleftharpoons H^+ + Ac^- \qquad 0.1\ M\ \text{pH}\ 3$$

When titrated, the same amount of alkali would be used for each. This is titratable acidity.

TITRATION CURVES

When the pH of a solution is followed during the titration of an acid, characteristic curves may be constructed by plotting pH against equivalents of alkali. In Figure 1.4 the titration curves for several acids are shown.

It should be noted that titration curves for weak acids constitute a family of curves, the pK_a for the particular acid determining the position of the curve. The curve is relatively flat at the pK_a and remains so for about 1 pH unit in either direction. The small change in pH on the addition of acid or base in this region of the curve is characteristic of a weak acid in the presence of its salt and is known as the buffering action. A solution of the weak acid and its salt is called a buffer solution.

During the titration of a weak acid—for example, acetic acid—salt will be formed, so that at any point there will be a definite ratio of salt (base) and acid. By substituting various ratios in the Henderson-Hasselbalch equation, it is possible to obtain theoretic curves that are very close to the experimental curves. At values greater that ± 1 pH unit from the pK_a of an acid the experimental curves may deviate from the calculated values. The Henderson–Hasselbalch equation cannot be applied to the stronger acids that have pK_a values below 2. HCl is completely ionized and has no pK_a.

It is possible to determine the pK_a of an unknown acid from a titration curve. Also, by knowing the pK_a of a weak acid, it is possible to prepare a buffer having a specific pH. A knowledge of the pK_a values for the acids in biologic systems is required for an understanding of the chemical mechanisms by which biologic systems control pH. The application of titration curves to amino acid and protein chemistry will be discussed in Chapter 2.

The buffer capacity is defined as the gram equivalents of hydrogen ion required to change 1 liter of 1 *M* buffer by 1 pH unit. The capacity of a buffer will depend on the concentration of acid and base present. The pH will depend on the ratio of base to acid and, for the biologic systems, will be independent of concentration. In biologic systems the normal range of concentration of buffers is 0.05–0.1 *M*.

REFERENCES

Christensen, H. N. L.: *pH and Dissociation: A Learning Program*. W. B. Saunders Co., Philadelphia, 1963.

Frisell, W. R.: *Acid–Base Chemistry in Medicine*. The Macmillan Co., New York, 1968.

Chapter 2

PROTEINS

PROTEINS, large complex molecules, serve as the basis of protoplasm. All living cells contain protein, and growth in cells is synonymous with protein synthesis. Proteins are important for many cellular functions, some of which are listed.

1. They are the chief structural units of protoplasm.
2. Protein in the diet serves as the primary source of amino acids, the building blocks for cellular proteins.
3. The biologic catalysts known as enzymes are proteins.
4. Proteins play an important role in the transport of water, inorganic ions, organic compounds, and oxygen.
5. Some of the hormones, the regulators of chemical reactions, are proteins or peptides.
6. Antibodies are complex proteins.
7. Viruses are nucleoproteins.

Proteins are large molecules varying in molecular weight from a few thousand to a few million. Some representative molecular weights of proteins are insulin, 5500; ribonuclease, 13,700; trypsin, 23,800; hemoglobin, 68,000; fibrinogen, 450,000; thyroglobulin, 630,000; and viruses, a few to several million.

The average composition of proteins is carbon, 50 per cent; hydrogen, 7 per cent; oxygen, 23 per cent; nitrogen, 16 per cent; sulfur, 0 to 3 per cent; phosphorus, 0 to 3 per cent. Elements such as iron, iodine, copper, manganese, and zinc are found in specific proteins. Since nitrogen may be determined easily by the Kjeldahl method, it has been common practice to determine nitrogen and multiply this value by 6.25 to obtain the amount of protein in a preparation. Although proteins may contain organic moieties such as carbohydrates and lipids, the primary building units are the amino acids.

AMINO ACIDS

The products of hydrolysis of simple proteins are α-amino acids.

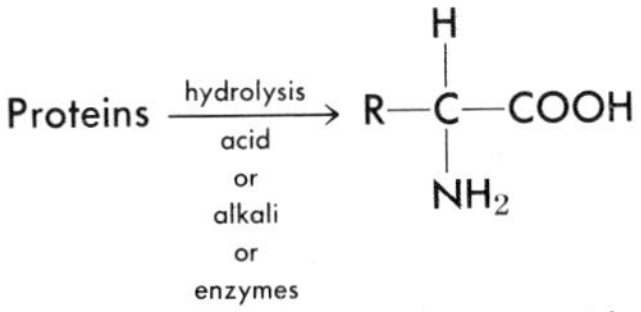

$$\text{Proteins} \xrightarrow[\text{acid or alkali or enzymes}]{\text{hydrolysis}} R{-}\underset{\underset{NH_2}{|}}{\overset{\overset{H}{|}}{C}}{-}COOH$$

α-Amino acids

Classification

The amino acids commonly found in proteins are classified as follows.

Aliphatic Monoamino Monocarboxylic Acids

Glycine	$H_2N{-}CH_2{-}COOH$
Alanine	$CH_3{-}\underset{\underset{NH_2}{|}}{\overset{\overset{H}{|}}{C}}{-}COOH$
Valine	$CH_3{-}\underset{\underset{CH_3}{|}}{CH}{-}\underset{\underset{NH_2}{|}}{CH}{-}COOH$
Leucine	$CH_3{-}\underset{\underset{CH_3}{|}}{CH}{-}CH_2{-}\underset{\underset{NH_2}{|}}{CH}{-}COOH$
Isoleucine	$CH_3{-}CH_2{-}\underset{\underset{CH_3}{|}}{CH}{-}\underset{\underset{NH_2}{|}}{CH}{-}COOH$
Serine	$\underset{\underset{OH}{|}}{CH_2}{-}\underset{\underset{NH_2}{|}}{CH}{-}COOH$
Threonine	$CH_3{-}\underset{\underset{OH}{|}}{CH}{-}\underset{\underset{NH_2}{|}}{CH}{-}COOH$

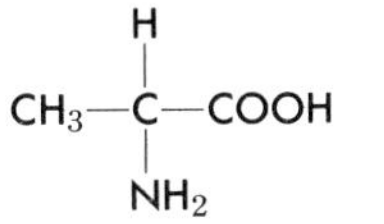

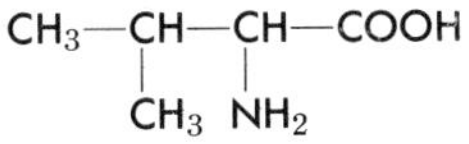

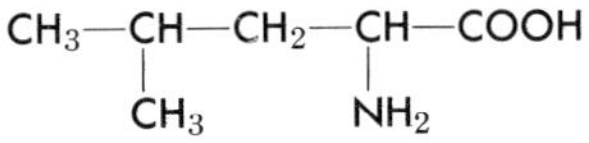

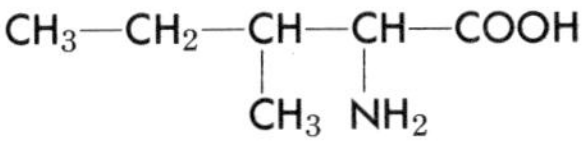

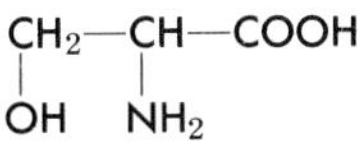

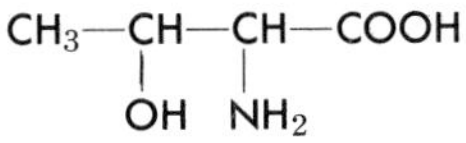

Amino Acids Containing Sulfur

Cysteine $CH_2(SH)—CH(NH_2)—COOH$

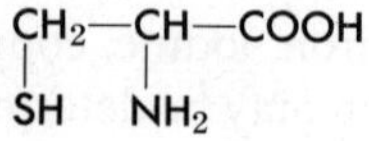

Cystine $HOOC—CH(NH_2)—CH_2—S—S—CH_2—CH(NH_2)—COOH$

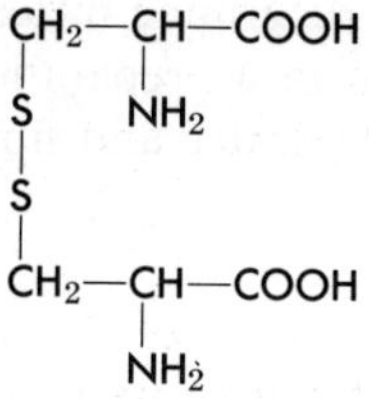

Methionine $CH_3—S—CH_2—CH_2—CH(NH_2)—COOH$

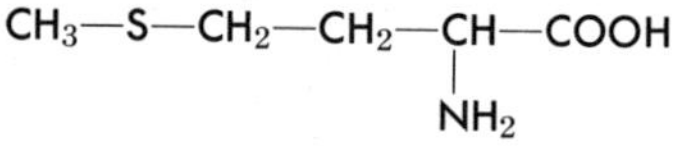

Aromatic Monoaminomonocarboxylic Acids

Phenylalanine $C_6H_5—CH_2—CH(NH_2)—COOH$

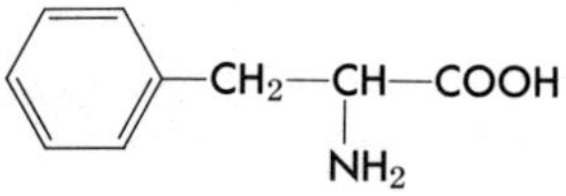

Tyrosine $HO—C_6H_4—CH_2—CH(NH_2)—COOH$

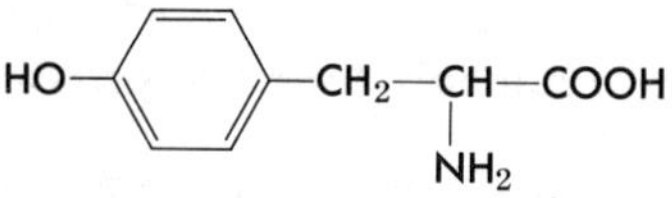

Monoaminodicarboxylic Acids

Aspartic acid $HOOC—CH_2—CH(NH_2)—COOH$

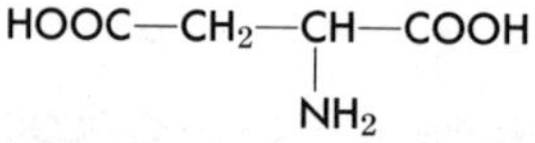

Glutamic acid $HOOC—CH_2—CH_2—CH(NH_2)—COOH$

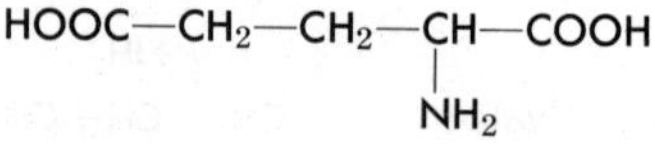

Diaminomonocarboxylic Acids

Lysine $CH_2(NH_2)—CH_2—CH_2—CH_2—CH(NH_2)—COOH$

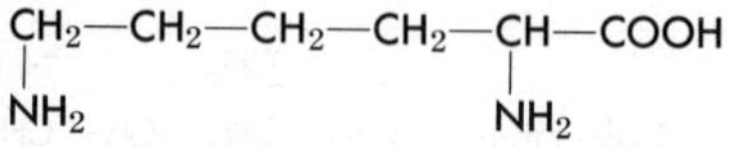

Arginine $H_2N—C(=NH)—NH—CH_2—CH_2—CH_2—CH(NH_2)—COOH$

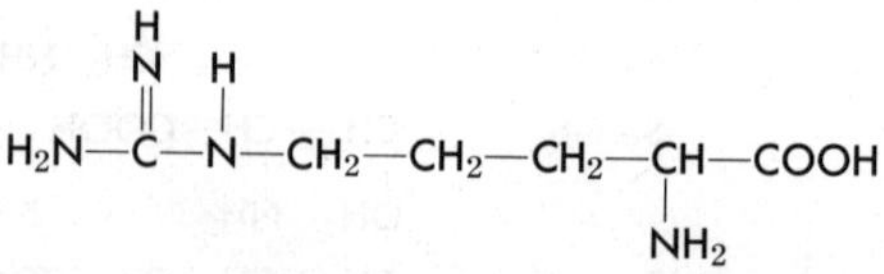

Hydroxylysine $H_2N—CH_2—CH(OH)—CH_2—CH_2—CH(NH_2)—COOH$

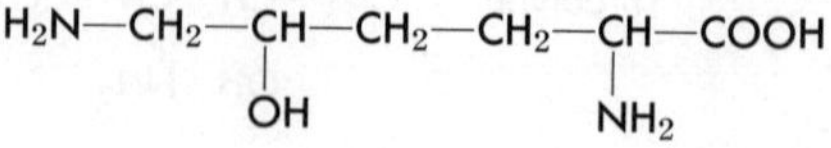

Heterocyclic Amino Acids

Tryptophan

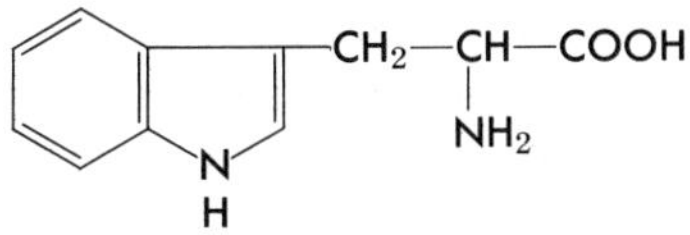

$$\text{(indole ring)}-CH_2-\underset{\displaystyle NH_2}{\underset{|}{CH}}-COOH$$

N
H

Histidine

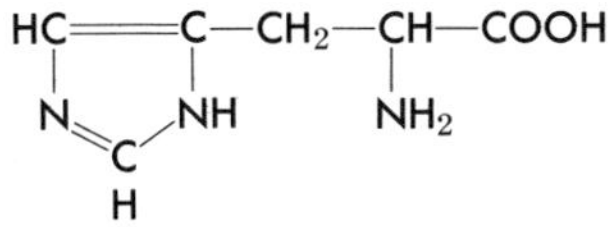

$$\underset{\displaystyle N\!=\!\underset{\displaystyle H}{C}-NH}{HC=C}-CH_2-\underset{\displaystyle NH_2}{\underset{|}{CH}}-COOH$$

Proline

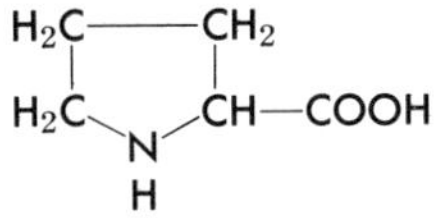

$$\begin{array}{l} H_2C\text{———}CH_2 \\ H_2C\diagdown\ \ \ \diagup CH-COOH \\ \qquad\ \ \underset{\displaystyle H}{N} \end{array}$$

Hydroxyproline

$$\begin{array}{l} HO-HC\text{———}CH_2 \\ \qquad\ H_2C\diagdown\ \ \ \diagup CH-COOH \\ \qquad\qquad\quad \underset{\displaystyle H}{N} \end{array}$$

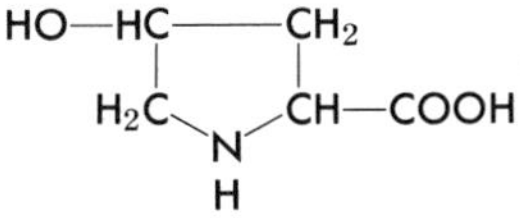

Other amino acids are found as minor constituents of proteins. Also, several important amino acids are not found in proteins but are of considerable importance in the operation of biologic systems. These will be studied at the appropriate time (see pages 223, 245).

Configuration

Except for glycine all the above listed amino acids contain an asymmetric carbon atom; and, as expected, the amino acids found in proteins are optically active, belonging to the L-series.

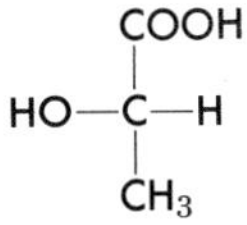

$$\begin{array}{c} COOH \\ | \\ HO-C-H \\ | \\ CH_3 \end{array} \qquad\qquad \begin{array}{c} COOH \\ | \\ H_2N-C-H \\ | \\ CH_3 \end{array}$$

L(+)-Lactic acid 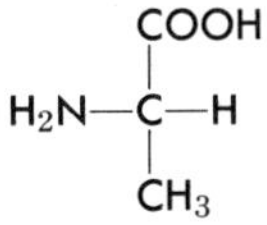 L(+)-Alanine

When amino acids are treated with alkali or certain chemical reagents, optical configuration is modified and a mixture of D and L forms is obtained. The amino acid is said to be racemized; a racemic mixture consists of equal amounts of the two forms.

Acid and Base Properties

Because the amino acids have high melting points, low solubility in organic solvents, low volatility, and high dipole moments, investigators have proposed an inner salt structure. This structure also accounts for other physical properties. Compounds that have this salt structure within the molecule are called *zwitterions.*

$$R{-}\underset{\underset{NH_3^+}{|}}{CH}{-}COO^-$$

Zwitterion structure

Salt formation within such a compound is strictly comparable to the neutralization of acetic acid with ammonia.

$$CH_3{-}COOH + NH_3 \rightleftharpoons CH_3{-}COO^- + NH_4^+$$

When an acid is added to a solution of an amino acid, protons are added to the carboxyl group. When base is added, protons are removed from the amino group; for example

$$\underset{A}{R{-}\underset{\underset{NH_3^+}{|}}{CH}{-}COOH} \underset{HCl}{\overset{NaOH}{\rightleftharpoons}} \underset{B}{R{-}\underset{\underset{NH_3^+}{|}}{CH}{-}COO^-} \underset{HCl}{\overset{NaOH}{\rightleftharpoons}} \underset{C}{R{-}\underset{\underset{NH_2}{|}}{CH}{-}COO^-}$$

The amino acid will have a positive charge (form A) in an acid solution and, if placed in an electrical field, will migrate to the cathode. The amino acid will have a negative charge (form C) in an alkaline solution and will migrate to the anode in an electrical field. There is a pH at which the amino acid will have a net charge of zero (form B) and will not migrate in an electrical field; this pH is called the *isoeletric* pH (IpH).

When titration curves are constructed, glycine is found to have two pK_a values, 2.4 and 9.8. Titration curves have been constructed for all the amino acids and the pK_a values determined. The values for three typical amino acids are as follows

Amino Acids	pK_{a1}	pK_{a2}	pK_{a3}	IpH
Glycine	2.4	9.8	—	6.1
Aspartic acid	2.0	4.0	9.8	3.0
Lysine	2.2	8.9	10.5	9.7

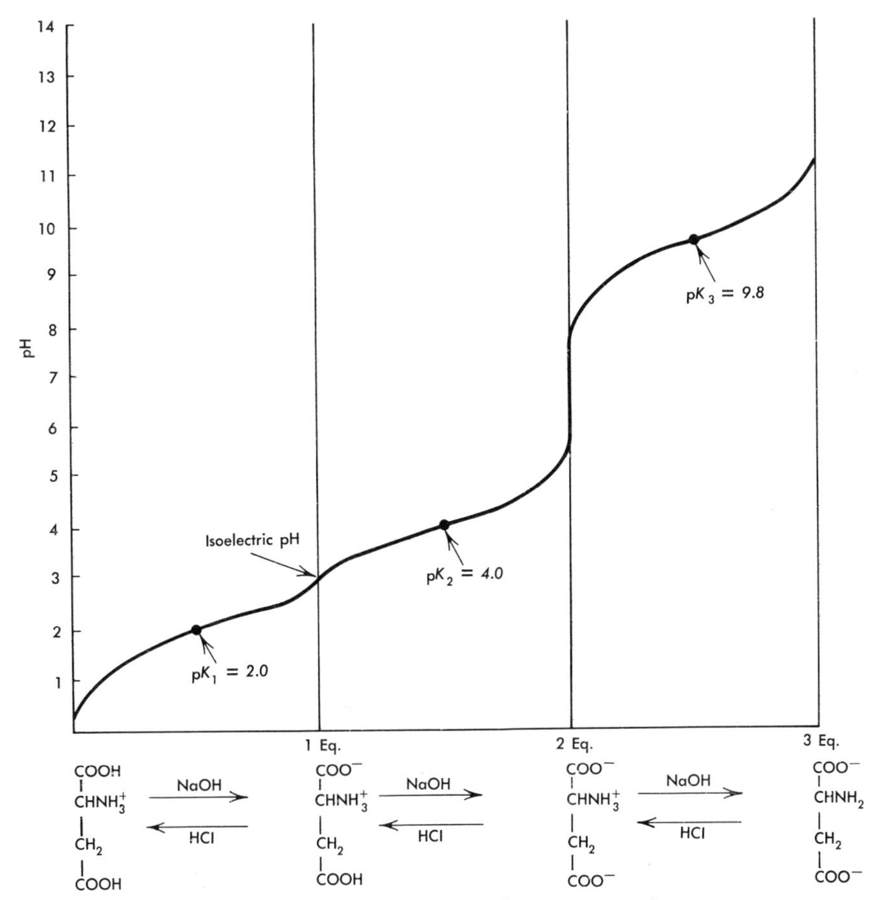

Figure 2.1. Titration curve for aspartic acid.

Figure 2.1 shows the titration curve for aspartic acid. Similar curves can be drawn for each of the amino acids. Students should draw curves for glutamic acid and lysine and determine the ionic species present at each pK_a. Also, they should determine the isoelectric pH.

Selected Chemical Reactions of α-Amino Acids

The following are only a few of the many reactions exhibited by amino acids. Those chosen illustrate a type similar to a known biochemical reaction.

1. Formation of Esters and Amides

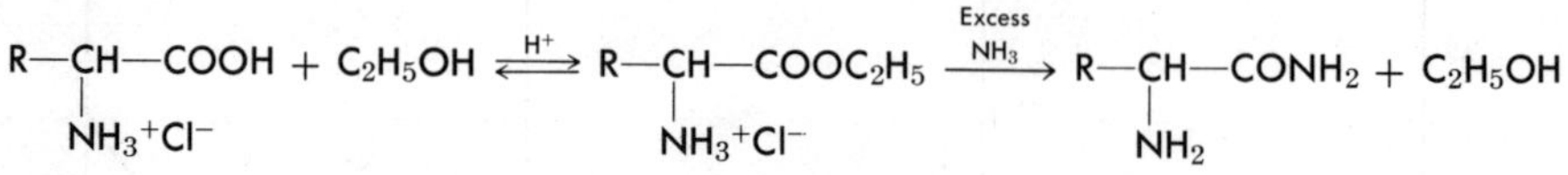

$$\underset{\displaystyle NH_3^+Cl^-}{R-CH-COOH} + C_2H_5OH \xrightleftharpoons{H^+} \underset{\displaystyle NH_3^+Cl^-}{R-CH-COOC_2H_5} \xrightarrow[]{\text{Excess } NH_3} \underset{\displaystyle NH_2}{R-CH-CONH_2} + C_2H_5OH$$

A special type of amide linkage is important in proteins (page 18).

2. Decarboxylation

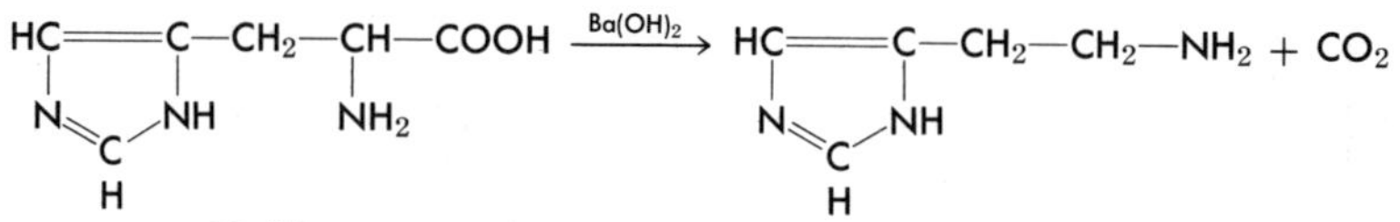

$$\text{Histidine} \xrightarrow{Ba(OH)_2} \text{Histamine} + CO_2$$

(Histidine: $HC{=}C(-CH_2-CH(NH_2)-COOH)$ imidazole ring $N{=}CH-NH$; Histamine: $HC{=}C(-CH_2-CH_2-NH_2)$ imidazole ring $N{=}CH-NH$)

Histidine Histamine

Histamine is formed in animals by the decarboxylation of histidine. It promotes secretion of acid by the stomach and is involved in allergic reactions.

3. Acylation

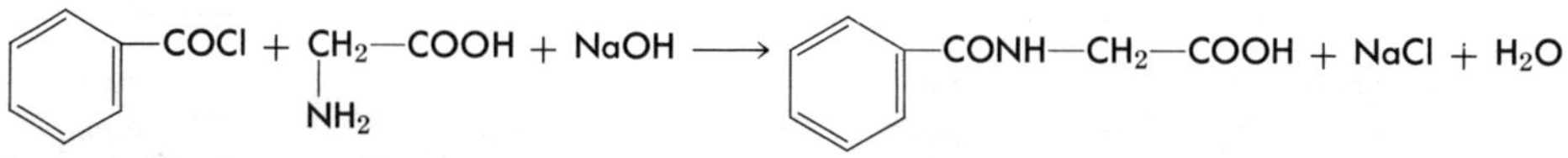

$$C_6H_5-COCl + \underset{\displaystyle NH_2}{CH_2-COOH} + NaOH \longrightarrow C_6H_5-CONH-CH_2-COOH + NaCl + H_2O$$

Benzoyl chloride Glycine Hippuric acid

Benzoic acid is excreted as hippuric acid when it is fed to animals.

4. Methylation

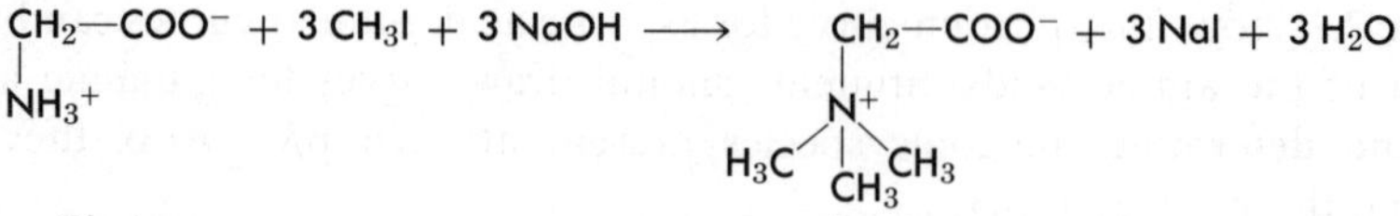

$$\underset{\displaystyle NH_3^+}{CH_2-COO^-} + 3\,CH_3I + 3\,NaOH \longrightarrow \underset{\displaystyle N^+(CH_3)_3}{CH_2-COO^-} + 3\,NaI + 3\,H_2O$$

Glycine Betaine

Betaine is an important compound for supplying methyl groups in biologic reactions.

5. Reactions with Nitrous Acid

$$\underset{\displaystyle NH_2}{\underset{|}{CH_2}}\text{—COOH} + \text{HONO} \longrightarrow \underset{\displaystyle OH}{\underset{|}{CH_2}}\text{—COOH} + N_2 + H_2O$$

The reaction is carried out under conditions such that only an amino group alpha to a carboxyl group reacts at a significant rate and the liberated gas is measured. This is the basis of the Van Slyke method for α-amino nitrogen.

6. Reaction with CO_2

$$\text{R—}\underset{\displaystyle NH_2}{\underset{|}{\text{CH}}}\text{—COO}^-\text{Na}^+ + CO_2 \xrightarrow{\text{NaOH}} \text{R—}\underset{\displaystyle \text{NH—COO}^-\text{Na}^+}{\underset{|}{\text{CH}}}\text{—COO}^-\text{Na}^+ + \text{HOH}$$

↑
Carbamino group

The reaction of carbon dioxide with the amino groups of hemoglobin to form carbamino groups is important for the transport of carbon dioxide from the tissues to the lungs.

7. Oxidative Deamination—Ninhydrin Reaction

The oxidation of an α-amino acid to ammonia, carbon dioxide, and an aldehyde takes place in the presence of mild oxidizing agents such as ninhydrin. Ninhydrindantin (reduced ninhydrin) reacts with another molecule of ninhydrin and the released ammonia to form a colored compound. The reactions may be formulated as follows

$$\text{R—}\underset{\displaystyle NH_2}{\underset{|}{\text{CH}}}\text{—COOH} \xrightarrow{(O)} \text{R—}\underset{\displaystyle \text{NH}}{\underset{\|}{\text{C}}}\text{—COOH} + (2\ \text{H})$$

$$\text{R—}\underset{\displaystyle \text{NH}}{\underset{\|}{\text{C}}}\text{—COOH} \xrightarrow{H_2O} \text{R—}\underset{\displaystyle \text{O}}{\underset{\|}{\text{C}}}\text{—COOH} + NH_3$$

$$\text{R—}\underset{\displaystyle \text{O}}{\underset{\|}{\text{C}}}\text{—COOH} \longrightarrow CO_2 + \text{RCHO}$$

$$2\ \text{Ninhydrin} + (2\ \text{H}) + NH_3 \longrightarrow \text{Colored compound}$$

The formation of the colored compound is used for the detection of amino acids and for their quantitative determination.

The oxidative removal of an α-amino group to furnish an α-keto acid represents an important pathway for metabolism of amino acids.

Peptide Formation

The peptide bond is a special case of an amide bond formed between an amino group of one amino acid and the carboxyl group of a second amino acid.

$$R\text{—}\underset{\underset{NH_2}{|}}{CH}\text{—}\overset{\overset{O}{\|}}{C}\text{—}\boxed{OH + H}\,\overset{H}{N}\text{—}\underset{\underset{R'}{|}}{CH}\text{—}COOH \longrightarrow$$

$$R\text{—}\underset{\underset{NH_2}{|}}{CH}\text{—}\overset{\overset{O}{\|}}{C}\text{—}\overset{H}{N}\text{—}\underset{\underset{R'}{|}}{CH}\text{—}COOH + R''\text{—}\underset{\underset{NH_2}{|}}{CH}\text{—}COOH \longrightarrow$$

Dipeptide

$$R\text{—}\underset{\underset{NH_2}{|}}{CH}\text{—}\overset{\overset{O}{\|}}{C}\text{—}\overset{H}{N}\text{—}\underset{\underset{R'}{|}}{CH}\text{—}\overset{\overset{O}{\|}}{C}\text{—}\overset{H}{N}\text{—}\underset{\underset{R''}{|}}{CH}\text{—}COOH$$

Tripeptide

By adding additional amino acids through peptide bonds, it is possible to form tetrapeptides, pentapeptides, and so on to polypeptides. It should be noted that polypeptides have a common chain or backbone made up of peptide bonds separated by $\text{—}\overset{\overset{R}{|}}{CH}\text{—}$ groups. They have a free amino group on one end of the chain and a free carboxyl group on the other. As is expected, the chemical, physical, and biologic properties of a peptide are determined by the nature of the R groupings.

Examples of R-groupings are

R =	pK_a =
$\text{—}CH_2\text{—}COOH$	4
$\text{—}CH_2\text{—}CH_2\text{—}COOH$	4
$\text{—}(CH_2)_3\text{—}CH_2\text{—}NH_2$	10
$\text{—}(CH_2)_3\text{—}NH\text{—}C(\text{=}NH)\text{—}NH_2$	12.5
$\text{—}CH_2\text{—}C_6H_4\text{—}OH$	10
$\text{—}CH_2\text{—}SH$	10
$\text{—}CH_2\text{—}OH$	>12

Compounds containing two or more peptide bonds react with $Cu(OH)_2$ to form a violet-colored complex. This reaction (biuret reaction) is the basis of a quantitative method for the determination of proteins.

Examples of Peptides of Biologic Origin

1. *Glutathione*

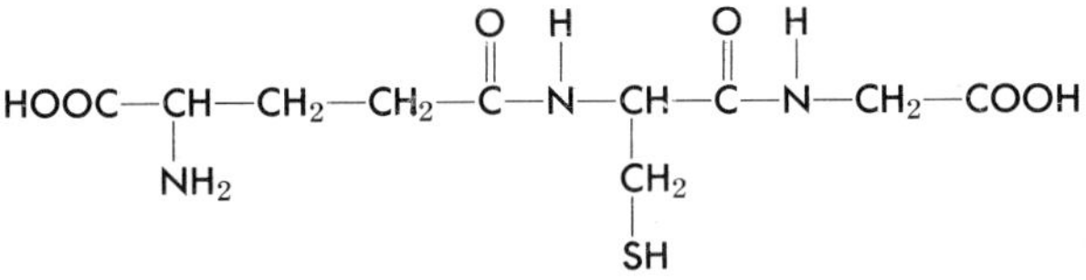

```
                               O  H       O  H
                               ‖  |       ‖  |
HOOC—CH—CH2—CH2—C—N—CH—C—N—CH2—COOH
     |                 |
     NH2               CH2
                       |
                       SH
```

γ-Glutamylcysteinylglycine

Glutathione exists in an oxidized form which has a disulfide bond analogous to cystine. The maintenance of the compound in the reduced form is important for the proper function of the red blood cell.

2. *Oxytocic Hormone*

The hormone isolated from the posterior lobe of the pituitary causes contraction of smooth muscle and has been used routinely in obstetrics to initiate labor. It is an octapeptide with the structure shown below.

3. *Vasopressin*

Another hormone isolated from the posterior lobe of the pituitary is vasopressin. This hormone causes a rise in blood pressure because of constriction of the peripheral vessels and has a profound effect upon salt and water balance. Loss of the ability to produce this hormone leads to the condition known as diabetes insipidus.

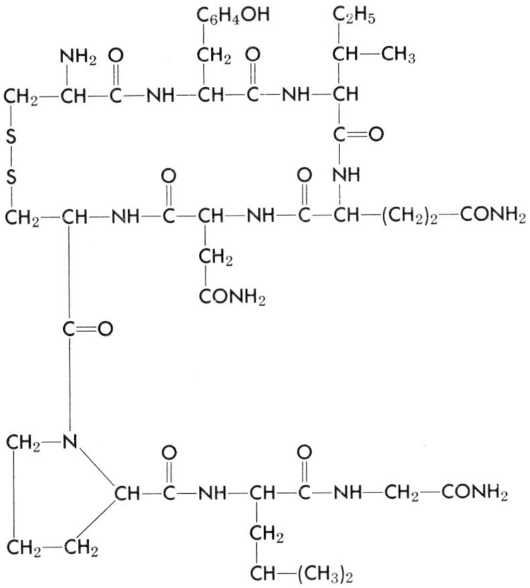

```
                    C6H4OH         C2H5
                    |              |
      NH2 O         CH2 O          CH—CH3
      |   ‖         |   ‖          |
CH2—CH—C—NH—CH—C—NH—CH
|                                  |
S                                  C=O
|                                  |
S               O              O   NH
|               ‖              ‖   |
CH2—CH—NH—C—CH—NH—C—CH—(CH2)2—CONH2
    |           |
    |           CH2
    |           |
    |           CONH2
    |
    C=O
    |
    |
CH2—N
|     \          O         O
|      \         ‖         ‖
|       CH—C—NH—CH—C—NH—CH2—CONH2
|      /              |
CH2—CH2               CH2
                      |
                      CH—(CH3)2
```

Oxytocin

Vasopressin (in beef) is an octopeptide having six of its amino acids the same as oxytocin. Its structure is

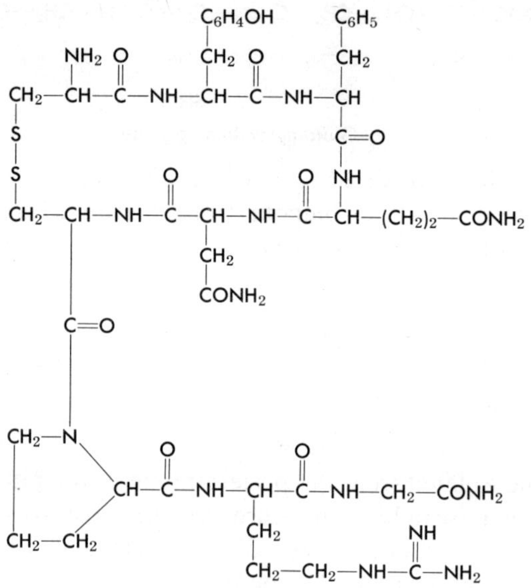

```
                         C6H4OH        C6H5
                         |             |
           NH2  O        CH2  O        CH2
           |    ||       |    ||       |
     CH2---CH---C---NH---CH---C---NH---CH
     |                                 |
     S                                 C=O
     |                                 |
     S               O            O    NH
     |               ||           ||   |
     CH2---CH---NH---C---CH---NH---C---CH---(CH2)2---CONH2
           |             |
           |             CH2
           |             |
           |             CONH2
           |
           C=O
           |
           |
     CH2---N
     |      \        O            O
     |       \       ||           ||
     |        CH-----C---NH---CH---C---NH---CH2---CONH2
     |       /                |
     |      /                 CH2                 NH
     CH2---CH2                |                   ||
                              CH2---CH2---NH------C---NH2
```

Vasopressin

Lysine replaces arginine in hog vasopressin. This is an example of the chemical basis of species specificity.

4. ACTH or Corticotropins

The terms *corticotropin* and *ACTH* are used to represent a group of substances which have been isolated from the pituitary gland. These hormones stimulate the adrenal cortex to produce steroid hormones. C-corticotropin from sheep and corticotropin-A from the pig each contain 39 amino acids. C-corticotropin has one more serine unit than corticotropin-A, whereas the latter has an additional leucine unit. These corticotropins exhibit melanotropic activity, which may be related to a common sequence of amino acids in these hormones and the melanocyte-stimulating hormone (MSH).

MSH	Pro.	Tyr.	Lys.	Met.	Glu.	His.	Phe.	Arg.	Tyr.	Gly	Ser.
	4	5	6	7	8	9	10	11	12	13	14
ACTH	1	2	3	4	5	6	7	8	9	10	11
	Ser.	Tyr.	Ser.	Met.	Glu.	His.	Phe.	Arg.	Tyr.	Gly.	Lys.

5. Antibiotics

Several antibiotics are peptides or contain peptides within the molecule. In many cases one or more of the amino acids belongs to the D series. An example of this class of substances is gramicidin-S. It possesses a cyclic structure and contains D-phenylalanine.

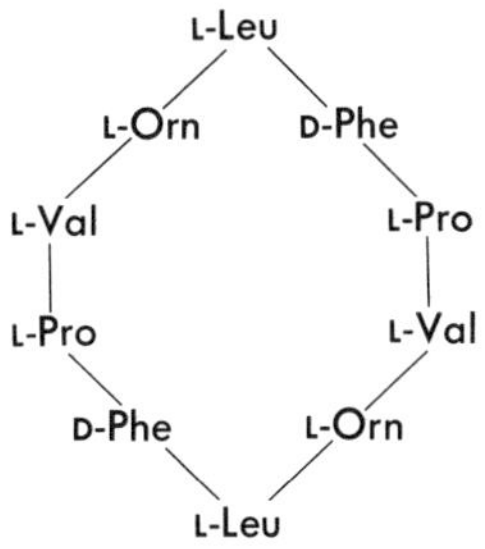

```
               L-Leu
              /     \
         L-Orn       D-Phe
        /                 \
   L-Val                   L-Pro
     |                       |
   L-Pro                   L-Val
        \                 /
         D-Phe       L-Orn
              \     /
               L-Leu
```

Gramicidin-S

Amino acids having unusual structures are present in several of the antibiotics. An example is penicillamine, a derivative of valine.

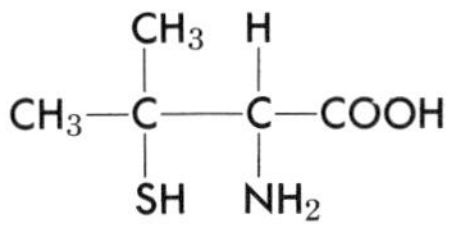

```
         CH3  H
          |   |
   CH3—C——C—COOH
          |   |
         SH  NH2
```

Penicillamine

Synthesis of Peptides

Vasopressin, MSH, gramicidin-S, ACTH, and a number of peptides of similar complexity have been synthesized. The development of methods for protecting reactive groups, that is, $—NH_2$, —SH, and —OH, and methods for establishing the peptide bond have received the attention of many investigators within the last few years. The following is an example from the many which are useful. In this case the amino group in alanine is prevented from entering into peptide bond formation by reacting with benzyloxycarbonyl chloride. After the peptide bond has been established, the benzyloxycarbonyl group is removed by reduction with hydrogen and a catalyst. This method leaves the peptide bond intact.

$$C_6H_5{-}CH_2{-}O{-}\overset{O}{\overset{\|}{C}}{-}Cl + H_2N{-}\underset{CH_3}{\underset{|}{CH}}{-}COOH \longrightarrow C_6H_5{-}CH_2{-}O{-}\overset{O}{\overset{\|}{C}}{-}\overset{H}{N}{-}\underset{CH_3}{\underset{|}{CH}}{-}COOH$$

$$C_6H_5{-}CH_2{-}O{-}\overset{O}{\overset{\|}{C}}{-}\overset{H}{N}{-}\underset{CH_3}{\underset{|}{CH}}{-}COOH + PCl_5 \longrightarrow C_6H_5{-}CH_2{-}O{-}\overset{O}{\overset{\|}{C}}{-}\overset{H}{N}{-}\underset{CH_3}{\underset{|}{CH}}{-}\overset{O}{\overset{\|}{C}}{-}Cl$$

$$C_6H_5{-}CH_2{-}O{-}\overset{O}{\overset{\|}{C}}{-}\overset{H}{N}{-}\underset{CH_3}{\underset{|}{CH}}{-}\overset{O}{\overset{\|}{C}}{-}Cl + H_2N{-}\underset{CH{-}(CH_3)_2}{\underset{|}{CH}}{-}COOH \longrightarrow$$

$$C_6H_5{-}CH_2{-}O{-}\overset{O}{\overset{\|}{C}}{-}\overset{H}{N}{-}\underset{CH_3}{\underset{|}{CH}}{-}\overset{O}{\overset{\|}{C}}{-}\overset{H}{N}{-}\underset{CH{-}(CH_3)_2}{\underset{|}{CH}}{-}COOH$$

$$C_6H_5{-}CH_2{-}O{-}\overset{O}{\overset{\|}{C}}{-}\overset{H}{N}{-}\underset{CH_3}{\underset{|}{CH}}{-}\overset{O}{\overset{\|}{C}}{-}\overset{H}{N}{-}\underset{CH{-}(CH_3)_2}{\underset{|}{CH}}{-}COOH \xrightarrow[\text{catalyst}]{H_2}$$

$$C_6H_5{-}CH_3 + CO_2 + H_2N{-}\underset{CH_3}{\underset{|}{CH}}{-}\overset{O}{\overset{\|}{C}}{-}\overset{H}{N}{-}\underset{CH{-}(CH_3)_2}{\underset{|}{CH}}{-}COOH$$

Alanylvaline

An ingenious method developed by Dr. R. B. Merrifield at the Rockefeller Institute for the synthesis of large peptides (simple protein) has been used successfully for the synthesis of ribonuclease, an enzyme containing 124 amino acid units in a single chain. The unique feature of this method is the fact that the carboxyl terminal amino acid is covalently attached to a resin and the amino acids are added sequentially to elongate the chain. After completion of the synthesis, the bond linking the peptide to the resin is cleaved. Near quantitative yields were obtained at each step, and all the amino acids were introduced without purification of the intermediate peptides. The sequence of reactions is as follows

$H_2N{-}CH(R){-}C(=O){-}ONa$ + $ClCH_2{-}C_6H_4{-}$(resin) ⟶ $H_2N{-}CH(R){-}C(=O){-}O{-}CH_2{-}C_6H_4{-}$(resin)

Resin matrix

Amino acid coupled to resin

$BOC{-}NH{-}CH(R'){-}C(=O){-}OH$

coupling reagent

$BOC{-}NH{-}CH(R'){-}C(=O){-}NH{-}CH(R){-}C(=O){-}O{-}CH_2{-}C_6H_4{-}$(resin)

BOC = butyloxycarbonyl group

The protecting group (BOC) is removed and the reaction sequence repeated with the next amino acid. After the last amino acid is added, the peptide is removed from the resin.

This method has also been used for the synthesis of insulin and a number of other biologically interesting molecules. The further development of this method should make the rapid synthesis of large molecules available to investigators. The ability to synthesize macromolecular polypeptides and to modify their structures is an important step in studying the relationship of the structure of the compounds to biological activity. It is possible that small peptides with interesting and useful biologic functions can be synthesized.

PROTEINS

Classification

Proteins are classified on the basis of both chemical and physical properties. In many cases it is difficult to place a single protein in a particular group; nevertheless, a uniform method for classifying these substances has been very useful. The system recommended by a joint committee of the American Society of Biological Chemists and the American Chemical Society places the proteins in three main groups: (1) simple, (2) conjugated, and (3) derived proteins.

1. Simple Proteins

a. Albumins: soluble in water and salt solutions (egg albumin, serum albumin)
b. Globulins: sparingly soluble in water, soluble in salt solutions (serum globulins, many globulins from seeds)
c. Prolamines: soluble in 70 to 80 per cent alcohol and insoluble in water or absolute alcohol (gliadin from wheat, zein from corn)
d. Glutelins: insoluble in neutral solvents but soluble in dilute acid or base (glutelin from wheat)
e. Scleroproteins: insoluble in aqueous solvents (collagens, elastins, keratins)

2. Conjugated Proteins

These are proteins that are combined with characteristic groups such as lipids, nucleic acids, carbohydrates, and other nonprotein substances.

a. Nucleoproteins: nucleic acids combined with basic proteins such as histones and protamines (found in the nucleus, microsome, and mitochondrion)
b. Mucoproteins: contain large amounts of carbohydrates and amino sugars (blood group substances, gonadotropin, and mucins are examples of such compounds)
c. Glycoproteins: contain smaller amounts of carbohydrates, particularly amino sugars (serum globulins are typical examples of this group of proteins)
d. Lipoproteins conjugated with lecithin, cholesterol, and other lipids (found primarily in brain, nerve tissue, and as structural components in all cells)
e. Chromoproteins: colored proteins (hemoglobin and the respiratory pigments are typical examples)
f. Metalloproteins: contain Mg, Mn, Fe, Co, Zn, Cu, and so forth
g. Phosphoproteins: proteins other than nucleoproteins that contain phosphorus (casein is a good example of this type)

3. Derived Proteins

Derived proteins are substances obtained when proteins are altered by chemical or physical methods. These include proteoses, coagulated proteins, and peptones.

Composition and Structure of Proteins

Since some 20 different amino acids are present in proteins, the number of variations is tremendous. It has been calculated that the total weight of one molecule of each of the possible isomers of insulin would be 10^{280} g. The mass of the earth is 10^{27} g. This must mean that proteins have specific structures that are genetically predetermined. In order to understand the biologic function of proteins, it is necessary to know the amino acid composition, the amino acid sequence, (primary structure), the coiling and folding of the peptide chain (secondary structure), the folding of the coiled structure (helix) or other organized structures into a higher state of organization (tertiary structure), and the organization of polypeptide subunits into a multiunit structure (quaternary structure).

Determination of Amino Acids

Partition Chromatography

In 1941, Martin and Synge introduced the technique of chromatography into the field of amino acid analysis. The first technique involved the establishment of equilibrium between a stationary liquid phase and a mobile liquid phase. The stationary phase was water bound in silica gel, and the mobile phase was chloroform. A solution of acetylated amino acids in chloroform was poured onto a column of silica gel, and the column was then washed with chloroform. The rate of movement of a compound depends on its partition coefficient; the greater the relative solubility in chloroform, the faster the rate of travel down the column. If sufficient solvent flows through the column, the substances can be removed one at a time. The progressive separation of three substances is illustrated in Figure 2.2.

This method is useful for a few of the amino acids, but more desirable modifications have been devised. Strips or sheets of filter paper have been used to hold the stationary phase, and a number of mobile solvents have been found that will give separation of the free amino acids. A small sample (5 to 15 μg of amino acid) is applied about 5 cm from the end of a strip of paper (1.5 by 50 cm) (see Figure 2.3). The strip is then supported with the end bearing the sample dipping into the solvent in a covered chamber. As the solvent moves up the paper, the amino acids will also move up the paper, each at a different rate. After a few hours the paper is dried and sprayed with a solution of ninhydrin which gives a color with the amino acids. An amino acid will be found localized in a spot on the paper. The ratio of the distance of the spot from point of application to the distance of the solvent front

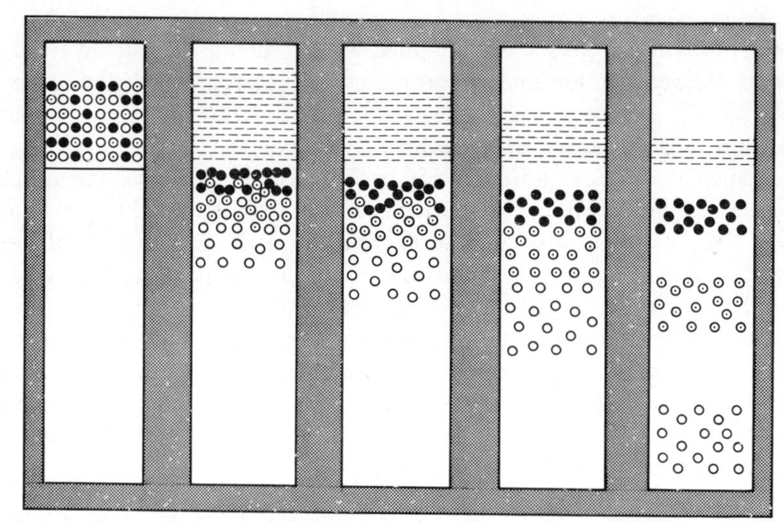

Figure 2.2. Partition column chromatography of amino acids. (From W. H. Stein and S. Moore: *Sci. Am.*, March, 1951.)

from the same point is the R_f value. Each amino acid has an R_f value characteristic for that amino acid in a given set of experimental conditions. When a sheet of filter paper is used instead of a strip, the sample is applied to one corner of the chromatogram developed. After several hours the paper is dried and developed in a second direction with a different solvent. In this way it is possible to separate all of the

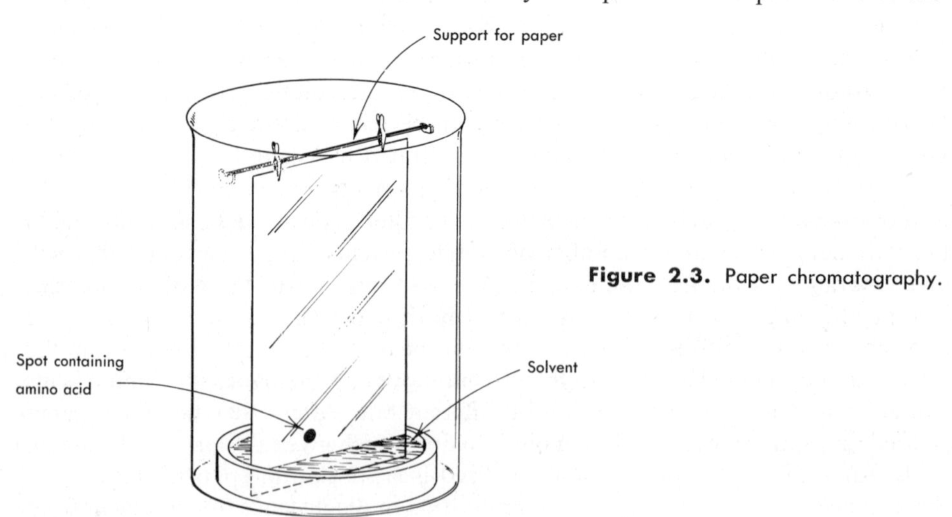

Figure 2.3. Paper chromatography.

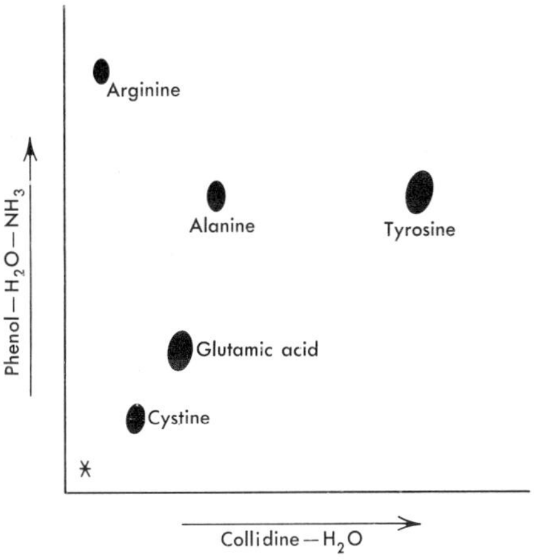

Arginine
Alanine
Tyrosine
Glutamic acid
Cystine
Phenol—H_2O—NH_3
Collidine—H_2O

Figure 2.4. Two-dimensional paper chromatography.

amino acids. The spots containing the amino acids can be cut out and determined quantitatively by micromethods. This process is called "two-dimensional paper chromatography" and is illustrated in Figure 2.4.

Ion Exchange Chromatography.

The separation of amino acids on columns of ion exchange resins is the method of choice for their quantitative estimation. Ion exchange resins are insoluble synthetic polymers containing acidic or basic groups. The acidic or cation exchange resins contain —SO_3H, —COOH, or phenol groups. The acidic groups on the surface of the resin dissociate, as do other acid groups, and can be titrated with bases to give salts. The surfaces will bind cations such as Ca^{++}, Na^+, NH_4^+ or organic bases, R—NH_3^+.

There is a definite force binding each cation to the resin. Among other factors the force of binding will depend on the number of charges, the size of the molecule, and the pK_a involved. If a cation exchange resin in the form of its sodium salt is treated with an organic base (R—NH_3^+ + Cl^-), the following reaction takes place.

$$\text{Resin—}SO_3^-Na^+ + \text{R—}NH_3^+Cl^- \rightleftharpoons \text{Resin—}SO_3^-NH_3^+\text{—R} + NA^+Cl^-$$

The reaction can be pushed either to the right or left, depending on the amounts of materials used. A small amount of the organic base can be exchanged on a relatively large amount of resin and then removed by treating with a large amount of Na^+Cl^-. Cations other than sodium may be used.

There is an analogous situation with the basic or anion exchange resins. The reaction of an organic acid with a basic resin is as follows

$$\text{Resin—}(NR_3^+)OH^- + RCOO^-Na^+ \rightleftharpoons \text{Resin—}(NR_3^+)^-OOC\text{—}R + Na^+OH^-$$

The organic anion can be removed by a large amount of another anion, such as —OH^-, —Cl^-, and so forth.

In practice the resins are placed in columns, the substances to be separated are dissolved in a small volume and placed on the column, and the column is washed with a buffer solution. Organic anions can be separated with an anion exchange resin by washing with the proper concentration of buffer. The anions will appear one after another in the effluent from the column.

The following diagram shows the sequence of the separation of three amino acids placed on a cation exchange column and eluted with a dilute acid.

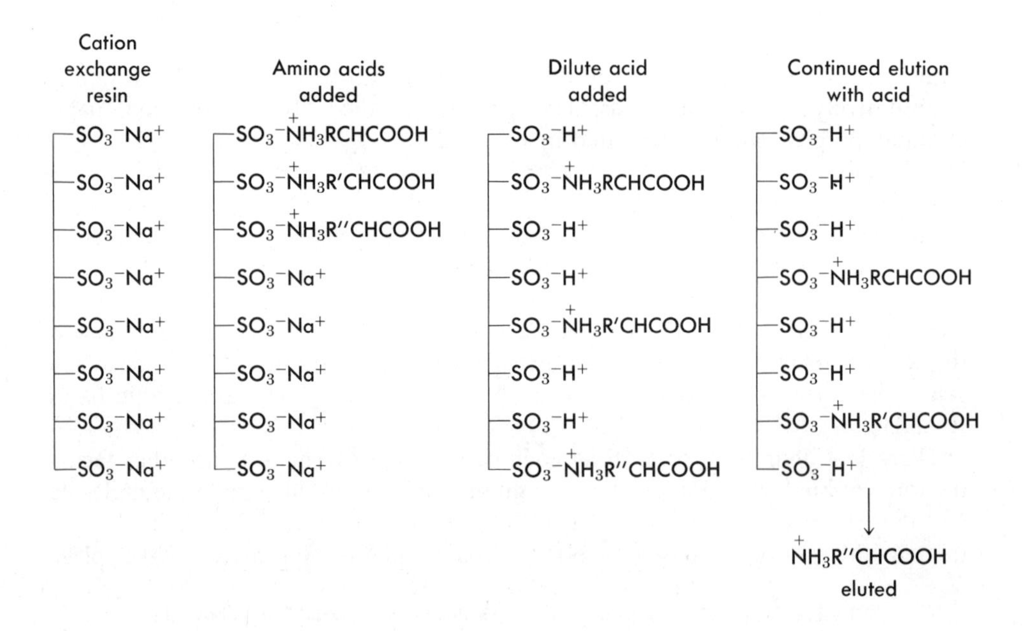

Figure 2.5 shows the results obtained from a sample of 2 mg of a mixture of 17 amino acids using a column of a cation exchange resin (Dowex 50). The mixture of amino acids was obtained from a hydrolysate of ribonuclease. Table 2.1 shows the results obtained from the analysis of the enzyme ribonuclease. The amount of each amino acid present was calculated from the area under the peaks of the curve.

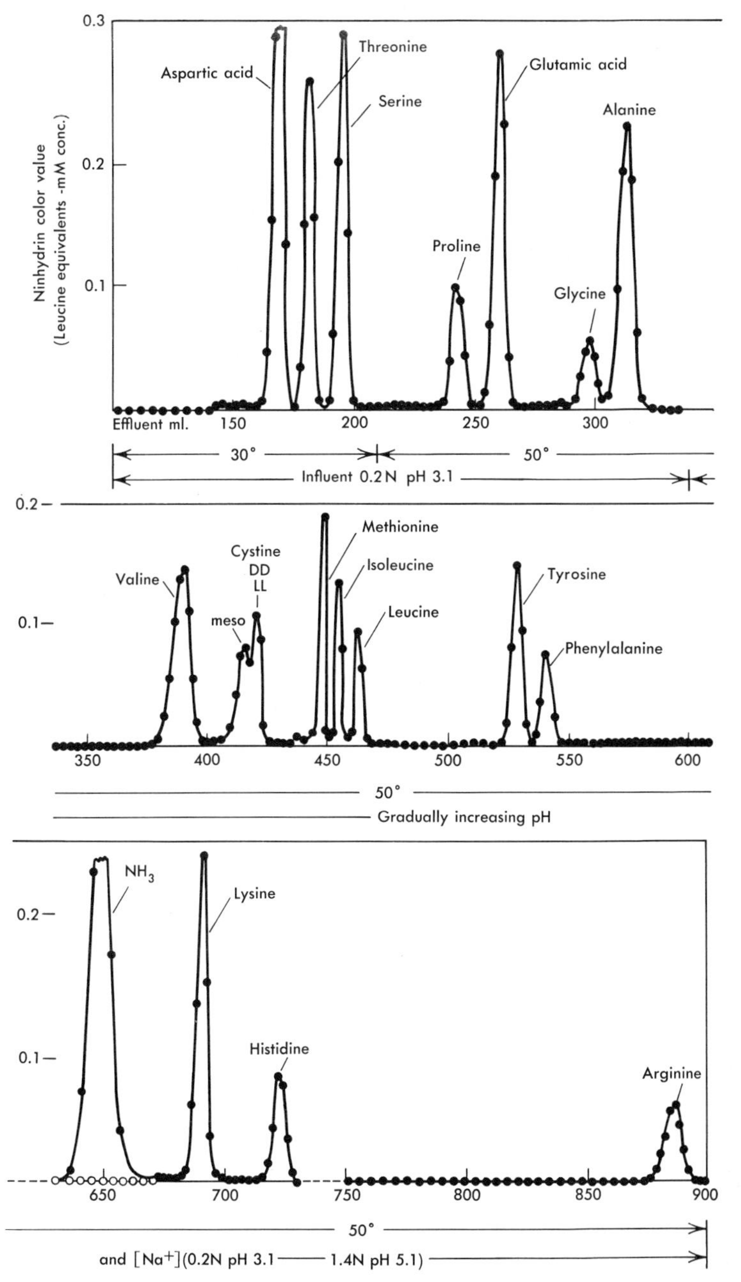

Figure 2.5. Amino acid analysis of a hydrolysate of the enzyme ribonuclease. (From C. H. W. Hirs, W. H. Stein, and S. Moore: *J. Biol. Chem.*, **211**:941, 1954.)

Table 2.1
Amino Acid Composition of Ribonuclease from Beef Pancreas

Amino Acid	No. Residues per Molecule M.W. 13, 683	Amino Acid	No. Residues per Molecule M.W. 13, 683
Aspartic acid	15	Methionine	4
Glutamic acid	12	Proline	4
Glycine	3	Phenylalanine	3
Alanine	12	Tyrosine	6
Valine	9	Histidine	4
Leucine	2	Lysine	10
Isoleucine	3	Arginine	4
Serine	15	Amide NH_3	(17)
Threonine	10		
Half-cystine	8	Total number of residues 124	

This method for the quantitative determination of amino acids has been completely automated, making it possible to find the amino acid composition of a peptide or protein in a few hours.

Amino Acid Sequence (Primary Structure)

The characteristic properties of a protein are determined chiefly by the number and sequence of amino acids in the polypeptide chain and the organization of the peptide chain. It is possible for a protein to contain more than one peptide chain; for the polypeptide chain to be cylic (gramicidin-S); for the chain to be branched through an —R group of lysine, aspartic acid, or glutamic acid; and for the chains to be branched or looped through bonds like the disulfide linkage (—S—S—). Information about such linkages is available for several simple proteins. The elucidation of the complete structure of insulin by Sanger is a classic development in this field (Figure 2.6).

Sanger introduced the reagent 1-fluoro-2,4-dinitrobenzene for the identification of the amino acid at the amino end of the peptide chain. The reagent reacts readily in mild alkaline solution with free amino groups to form a linkage which is stable to hydrolysis.

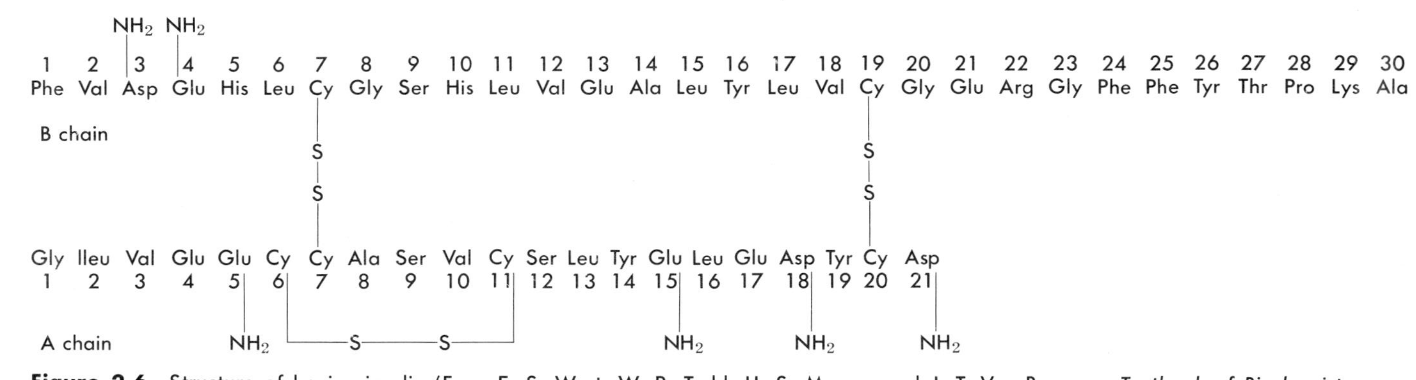

Figure 2.6. Structure of bovine insulin (From E. S. West, W. R. Todd, H. S. Mason, and J. T. Van Bruggen: *Textbook of Biochemistry*, 4th ed. The Macmillan Co., New York, 1966.)

$$O_2N\text{-}C_6H_3(NO_2)\text{-}F + H\text{-}N(H)\text{-}C(H)(R)\text{-Protein} \longrightarrow O_2N\text{-}C_6H_3(NO_2)\text{-}N(H)\text{-}C(H)(R)\text{-Protein} \xrightarrow{H^+}$$

1-Fluoro-2,4-dinitrobenzene

$$O_2N\text{-}C_6H_3(NO_2)\text{-}N(H)\text{-}C(H)(R)\text{-COOH} + \text{Amino acids}$$

Dinitrophenylamino acid

DNP-Amino acids are yellow, ether soluble, and easily identified. Sanger showed that insulin contained two peptide chains, one with phenylalanine and the other with glycine on the amino end. The two chains were separated by breaking the disulfide bonds and isolating each form. The glycyl chain (chain A) was found to contain 21 amino acid residues and the phenylalanine chain (chain B) 30 residues. The chains were broken into small peptides by hydrolysis with acids, and the peptide units were isolated and identified. By knowing the structure of the peptides, Sanger was able to determine the amino acid sequence in both chains. An example of the procedure is shown with an octapeptide from the A chain. From the structures of peptides numbered 2 through 8 below he was able to deduce the structure for peptide number 1.

gly.	ileu.	val.	glu.	glu.	cys.	cys.	ala.	1
	ileu.	val.	glu.	glu.				2
				glu.	cys.	cys.	ala.	3
gly.	ileu.	val.	glu.					4
	ileu.	val.	glu.		cys.	cys.	ala.	5
			glu.	glu.				6
		val.	glu.					7
	ileu.	val.		glu.	cys.			8

Enough peptides were identified in order to be able to draw the complete structure of the insulin molecule (Figure 2.6). The chains are held together by disulfide bonds.

The structures of pig, sheep, horse, and whale insulins have been determined and found to differ in the 8, 9, 10 amino acid sequence in chain A. In bovine insulin these are alanyl-seryl-valyl; in pig insulin, threonyl-seryl-isoleucyl; in sheep insulin, alanyl-glycyl-valyl; in horse insulin, threonyl-glycyl-isoleucyl; and in whale insulin, threonyl-seryl-isoleucyl.

Other Methods of Sequence Analysis

THE EDMAN PROCEDURE. The Edman procedure is the most useful method for determining the amino acid at the N terminal end of the chain. By this method the one amino acid at the end is removed, leaving the polypeptide chain intact. Repeating the procedure gives the second amino acid in the chain, and then the third, and so forth.

$$C_6H_5\text{—N=C=S} + NH_2\text{—}\overset{R'}{\overset{|}{CH}}\text{—}\overset{O}{\overset{\|}{C}}\text{—NH—}\overset{R''}{\overset{|}{CH}}\text{—}\overset{O}{\overset{\|}{C}}\text{—} \longrightarrow$$

Phenylisothiocyanate

$$C_6H_5\text{—NH—}\overset{S}{\overset{\|}{C}}\text{—NH—}\overset{R'}{\overset{|}{CH}}\text{—}\overset{O}{\overset{\|}{C}}\text{—NH—}\overset{R''}{\overset{|}{CH}}\text{—}\overset{O}{\overset{\|}{C}}\text{—} \xrightarrow{\text{Acid}} C_6H_5\text{—}\underset{\underset{O=C\text{—}CH(R')\text{—}NH}{|}}{N}\text{—}C\text{=S} + NH_2\text{—}\overset{R''}{\overset{|}{CH}}\text{—}\overset{O}{\overset{\|}{C}}\text{—}$$

Phenylthiohydantoin-yellow,
soluble in ether

This procedure has been automated making it possible to determine the amino acid sequence of a polypeptide in a relatively short time.

Carboxypeptidase

The enzyme carboxypeptidase splits peptide bonds adjacent to a free carboxyl group. As soon as the first amino acid is split from the chain, the enzyme will split the second peptide bond, and so on down the chain. Following the rate of release of each amino acid allows one to determine the sequence of three or four amino acids at the carboxyl end of the chain. Aminopeptidase may be used in a similar manner for determining amino acids at N end of the peptide.

Chymotrypsin, Trypsin, and Pepsin

Sanger was able to break insulin into small peptides by using weak acids at controlled temperatures for various periods of time. Enzymes have been used to break polypeptide chains at specific points.

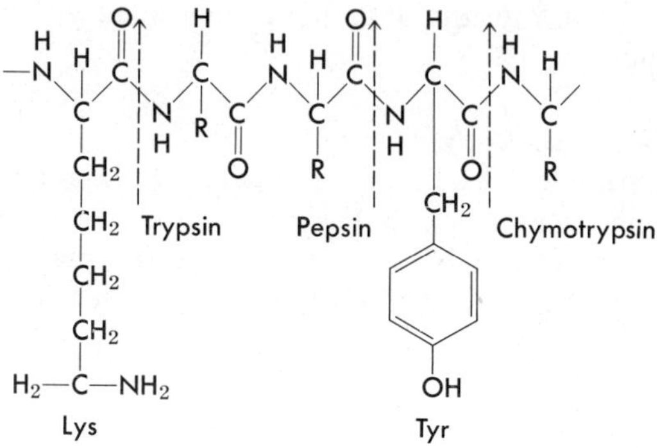

Trypsin Pepsin Chymotrypsin

Lys Tyr

Chymotrypsin attacks peptide bonds at the carboxyl side of tyrosine or phenylalanine; pepsin at the amino side of tyrosine or phenylalanine; and trypsin at the carboxyl side of peptide bonds containing lysine or arginine. Treatment of a protein with one or more of three enzymes will usually give peptides which can be isolated in pure form.

Other useful methods are available for sequence analysis. The methods given above, however, are the important ones and have enabled the biochemist to determine the sequence of the 124 amino acids in ribonuclease (Figure 2.7), the 158 amino acids in the protein unit of tobacco mosaic virus, and several other complex proteins.

Secondary Structure of Proteins

Fibrous Proteins

The fibrous proteins are represented by hair, silk fibroin, fibrin, collagen, and myosin of muscle. Evidence for the organization of peptide chains has been obtained from chemical and physical properties, but x-ray studies have yielded the most information.

Fibrous proteins consist of long peptide chains organized in bundles parallel to the fiber axis. A small portion of a molecule of silk fibroin is shown in the following structure

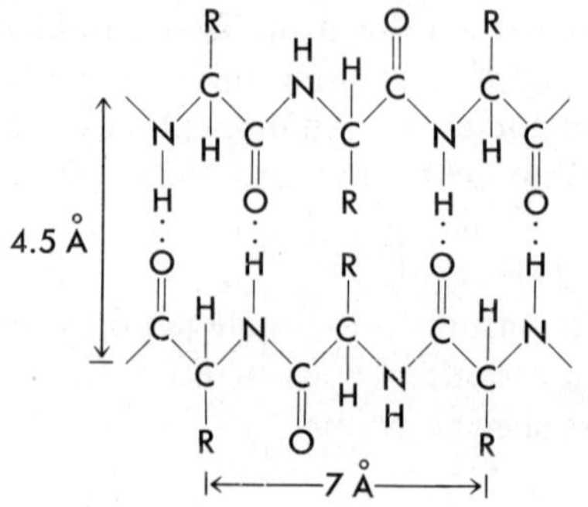

4.5 Å

7 Å

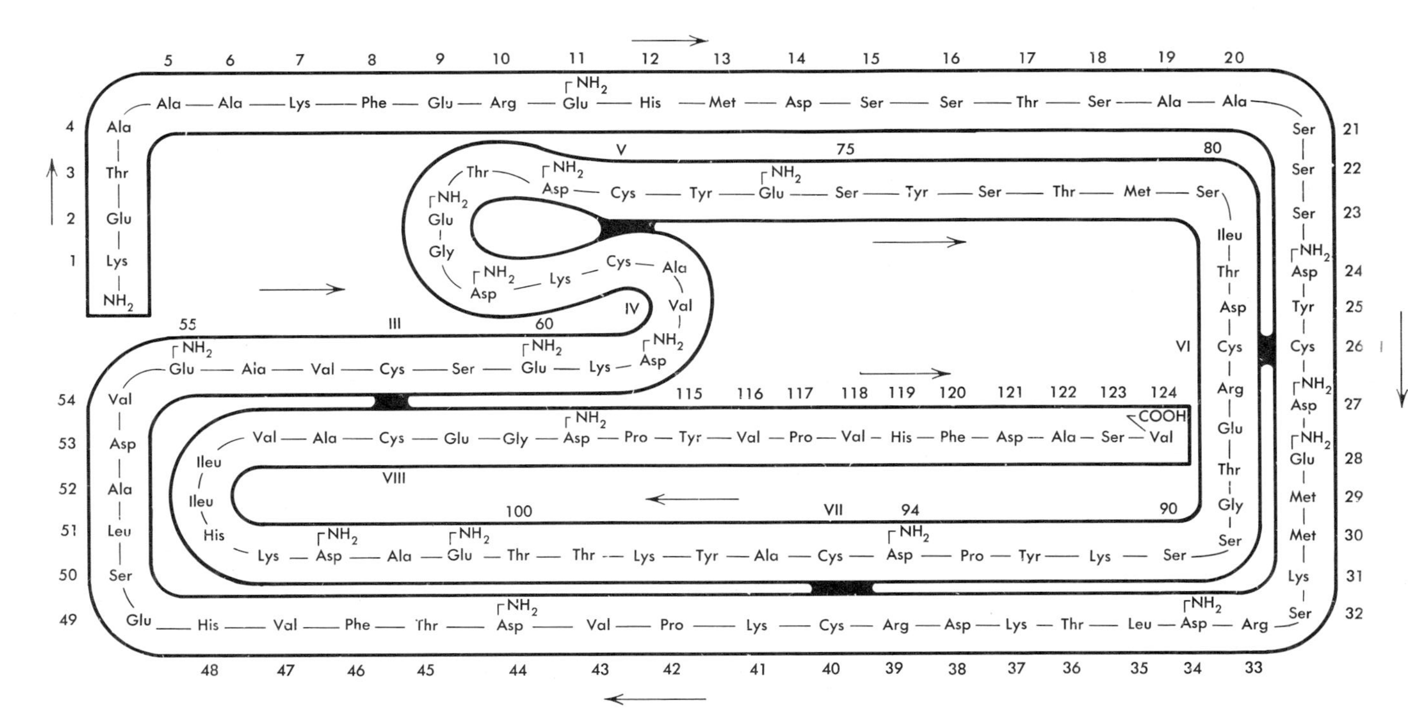

Figure 2.7. The two-dimensional structure of bovine pancreatic ribonuclease A. (From D. G. Smyth, W. H. Stein, and S. Moore: *J. Biol. Chem.*, **238**:227, 1963.)

The peptide linkages lie in the plane of the paper. The R groups lie above or below the plane and alternate along the chain. The peptide chains are held together by hydrogen bonds. It should be noted that the two peptide chains shown in the example are antiparallel. Parallel chains can also be held together by hydrogen bonding. Such an arrangement is called the "pleated sheet" structure.

A number of other types of cross linkages have been postulated for holding chains together. We have seen that the two chains in insulin are held together by two disulfide bonds. Salt linkages between the carboxyl of one R group and the amino of another R group are probably important, as are Van der Waals forces.

Other types of linkages are of questionable importance in the simple proteins. In conjugated proteins other bonds became important; for example, metal chelation.

Pauling and his coworkers made extensive studies of simple amides of amino acids and simple polypeptides. They concluded that all the atoms in the peptide linkage lie in one plane, that the two α-carbon atoms at each peptide bond have a *trans* configuration, and that the maximum hydrogen bonding is realized. Figure 2.8 shows the planar arrangement and bond angles in the peptide linkage. The "pleated sheet" structure illustrates hydrogen bonding between two polypeptide chains. Of course several peptides could be involved in this manner to form a macromolecular sheet.

Taking into account the above conclusions and assuming that the most stable arrangement would be a closely packed structure, Pauling proposed that a single peptide chain would form a helix through intramolecular hydrogen bonding. The most stable arrangement would be a repetitive unit having 3.7 amino acids per turn. Figure 2.9 shows a model of a β helix having a left hand turn. Figure 2.10 shows a simplified model of the same molecule with a right hand turn (α helix). The α helix is more stable and is the structure commonly found in proteins.

Tertiary Structure of Proteins

Most of our information about how the large polypeptides are rolled or coiled into definite structures has come from the x-ray analysis of crystalline proteins with biologic activity, for example, myoglobin and the enzymes. Kendrew and associates at Cambridge Univeristy were the first to complete such a detailed study. They showed that the three-dimensional structure of myoglobin is that shown in Figures 2.11 and 2.12. The segments of α-helix are represented by straight lines in Figure 2.12. The remainder of the polypeptide chain is random.

More recently, the three-dimensional structure of the enzyme lysozyme has been determined. It contains three separate sections of α helix and one region in which the chain makes a hair-pin turn to form a pleated sheet arrangement (Figure 2.13).

That the amino acid sequence in a protein must favor a certain configuration or conformation was shown by reducing the four disulfide linkages in ribonuclease to the —SH form with sodium borohydride. On careful oxidation with air the reduced

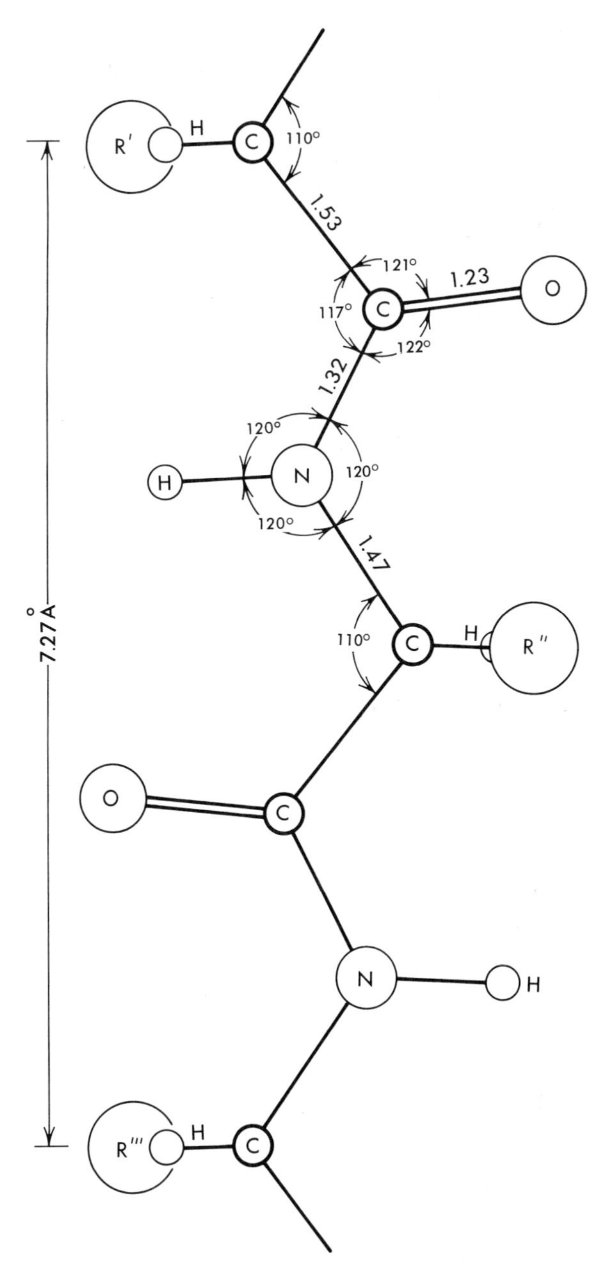

Figure 2.8. Dimensions of the polypeptide chain. (From L. Pauling, R. B. Corey, and H. R. Branson: *Proc. Nat. Acad. Sci.,* **37:**206, 1951.)

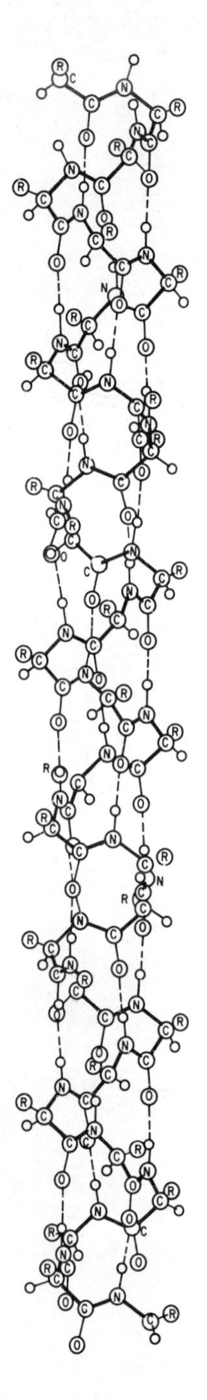

Figure 2.9. The helix with 3.7 residues per turn. (From L. Pauling, R. B. Corey, and H. R. Branson: *Proc. Nat. Acad. Sci.*, **37:**207, 1951.)

Figure 2.10. α Helix (C* = carbon atoms bearing R groups).

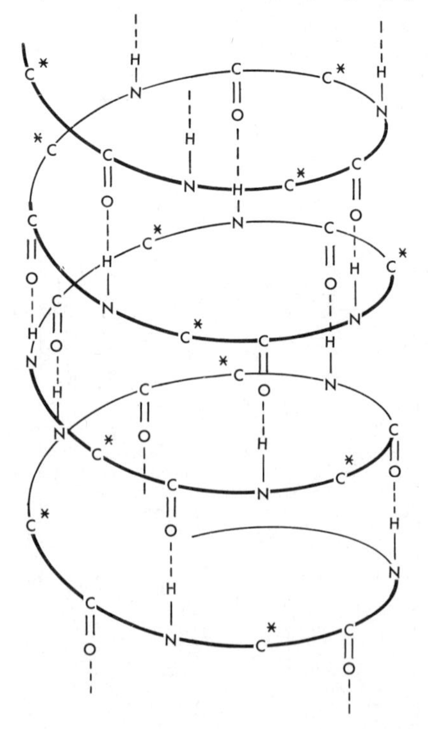

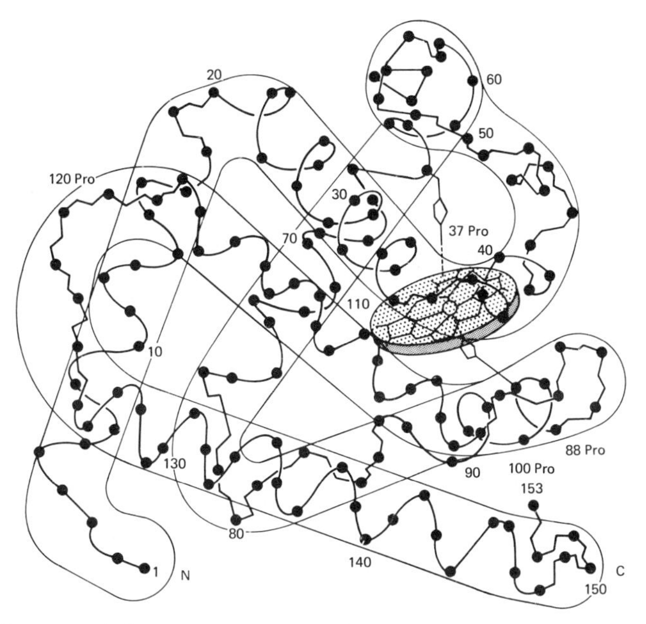

Figure 2.11. Drawing of the tertiary structure of myoglobin as deduced from x-ray analysis. (From "The Hemoglobin Molecule" by M. F. Perutz. Copyright © 1964 by Scientific American, Inc. All rights reserved.

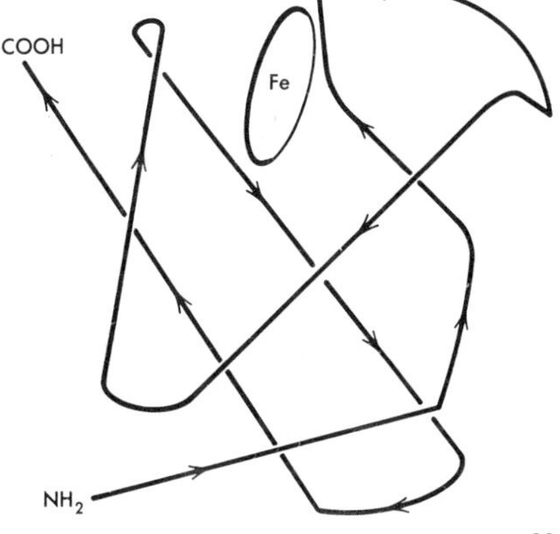

Figure 2.12. The course of the polypeptide chain deduced from the 2-Å Fourier synthesis. (From J. C. Kendrew, R. E. Dickerson, B. E. Strand, R. G. Hart, D. R. Davies, D. C. Phillips, and V. C. Shore: *Nature*, **185**:426, 1960.)

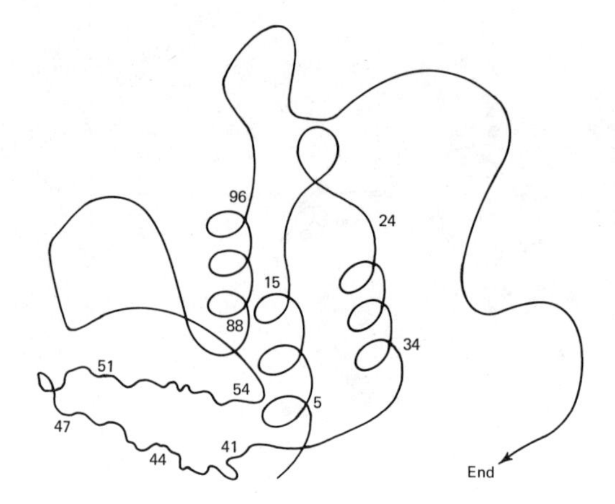

Figure 2.13. The structure of lysozyme, showing the three sections of α helix and the one section of pleated sheet structure (residues 41 to 54).

compound reformed the same —S—S— bonds. Coiling and folding in these proteins must occur in a definite manner, and the molecules must have a highly specific internal structure. This structure is maintained by hydrogen bonding and specific forces such as were discussed above.

Quaternary Structure of Proteins

Many proteins are composed of a number of polypeptide units. These units may be held together by disulfide linkages, as is the case with insulin, or by hydrogen bonding and Van der Waals forces. The subunits may be identical or composed of two or more different structures. Insulin possesses two different polypeptide chains. Hemoglobin is a tetramer having two α and two β chains; apoferritin, which combines with iron to form the iron storage protein ferritin, contains 20 subunits. The coat protein of a virus results from the aggregation of many subunits.

The hemoglobin molecule has served as an important model for studies on the relationship of the structure of a macromolecule to its function. It is interesting to note that, although the α and β chains differ considerably in their primary structures, they are very similar in both the secondary and tertiary structure (Figure 2.14). The chains do possess regions of similarity. On the other hand hemoglobin S from a patient with sickle cell anemia differs from normal hemoglobin in having glutamic acid at the 6 position in each of the β chains substituted by valine (page 330). Even though a small section of the protein may be able to carry out a specific biologic function, the remainder of the molecule may be important for conferring desirable physical properties such as solubility, viscosity, and so forth.

The conformation or arrangement of the subunits with respect to one another is important in the function of oligomers. As we shall see later the oxygenation of the heme in one subunit changes the affinity of the heme moieties in the remainder of the units for oxygen, resulting in the sigmoid nature of the oxygen dissociation

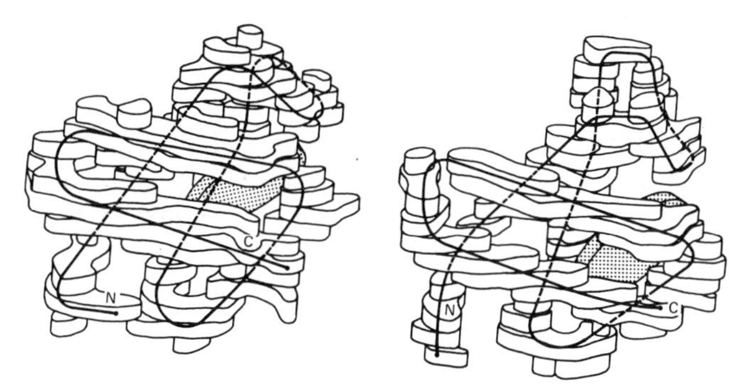

Figure 2.14. Illustration of α- hemoglobin (left) and β- hemoglobin (right). (From "The Hemoglobin Molecule" by M. F. Perutz. Copyright © 1964 by Scientific American, Inc. All rights reserved.

curve for hemoglobin (page 355). There are many examples of the effect of small molecules on enzyme activity, and some of these will be discussed in the next chapter.

Physical Properties of Proteins

Molecular Weight

FROM COMPOSITION. The minimum molecular weight of a protein can be determined by assuming that the molecule contains only one atom of an element. This method is useful for proteins that contain a metal. Hemoglobin contains 0.34 per cent iron (atomic weight = 55.8)

$$M_{\text{min}} = \frac{(55.8)(100)}{0.34} = 16{,}700$$

which indicates the presence of four atoms of iron in the hemoglobin molecule since the molecular weight by other methods is 64,500.

SEDIMENTATION. The most useful method for determining the molecular weight of proteins is based on the use of the ultracentrifuge. In this method the molecules are subjected to centrifugal forces up to 500,000 times gravity. The rate at which the protein will move depends on several factors including the force exerted, the shape, size, and density of the proteins, and the density and viscosity of the medium. The relationship of these factors is expressed quantitatively in the following equation

$$M = \frac{RTS}{D(1 - vP)}$$

where R is the gas constant, T the absolute temperature, D the diffusion coefficient, v the partial specific volume of the substance, P the density of the solution, and S the sedimentation coefficient.

The diffusion coefficient is defined as the quantity of material diffusing per second across a surface area of 1 cm^2 when the concentration of the solute is unity. If a fairly concentrated solution of a protein solution is separated from the pure solvent by a porous disc, the rate of diffusion can be measured by chemical means or by optical means that depend on measuring the change in refractive index as the solute diffuses into the pure solvent. v is the increment in volume when 1 g of the dry solute is dissolved. The sedimentation coefficient may be obtained by two general methods. If the ultracentrifuge is operated at relatively low speeds, the large molecules will move toward the periphery of the tube to a point at which sedimentation is exactly opposed by diffusion. Prolonged times are required for establishing equilibrium, and this factor limits the usefulness of the method. This is the *sedimentation equilibrium* method. Alternatively if the ultracentrifuge is operated at high speeds, the rate of sedimentation is in excess of the rate of diffusion, and the velocity of sedimentation can be measured. This is the *sedimentation velocity* method. Molecules of the same weight and shape will move through the solvent and form sharp moving boundaries that can be followed by changes in refractive index. The change in refractive index is observed by means of the Schlieren lens system. The sedimentation coefficient S can be calculated from the formula,

$$S = \frac{dx}{dt}\frac{1}{\omega^2 x}$$

where x is the distance from the center of rotation and ω the angular velocity in radians per second. The S value for proteins is between 1 and 200×10^{-13} sec. The Svedberg unit S is equal to a sedimentation coefficient of 10^{-13} sec, and a value of 5×10^{-13} would be $5S$.

Several modifications of the sedimentation method are useful, and the student should consult more advanced texts for these methods. The technique employing the *approach to sedimentation equilibrium* in particular is very useful.

DIFFUSION. The diffusion coefficient may be used to obtain information concerning the size and shape of protein molecules. The diffusion coefficient is related to the *molar frictional coefficient, F*, which is the force that acts on 1 gram-molecule of the diffusing substance to give it a velocity of 1 cm per sec and is expressed by the equation

$$D = \frac{RT}{F}$$

F is equal to Nf where N is Avogadro's number and f is the force per molecule. According to Stoke's Law

$$f = 6\pi\eta r$$

where η is the viscosity of the medium and r is the radius of a spherical molecule. Since the volume of a sphere is $\frac{4}{3}\pi r^3$ and the volume of a molecule is Mv/N, where v is the partial specific volume and M the molecular weight, the diffusion coefficient for a spherical molecule is

$$D_0 = \frac{RT}{Nf_0} = \frac{RT}{6\pi\eta N\left(\frac{3\,Mv}{4\pi N}\right)^{1/3}}$$

This enables one to determine M or, knowing M, to calculate a theoretical diffusion coefficient. The calculated values of D_0 are usually larger than the experimentally determined values because very few proteins are spheres. The ratio D_0/D, which is equal to f/f_0, is assumed to be a measure of the dissymmetry of the molecule and is related to the ratio of the major and minor axes of the molecule (see Fig. 2.15).

OSMOTIC PRESSURE. The osmotic pressure method has been used successfully for determining the molecular weight of proteins. According to the gas law

$$Pv = \frac{g}{M}RT$$

where P is the osmotic pressure in atmospheres, v is the volume of solution in liters, g is the weight of solute in grams per liter, M is molecular weight, R is the gas constant (0.082 liter atmosphere per mole per degree), and T the absolute temperature. The measures are made by placing the protein solution in a bag made from a semipermeable membrane, placing the bag in a buffered solution, and measuring the pressure developed in the bag by the rise in height of solution in a capillary tube. Difficulties encountered in this method are (1) equilibrium is attained slowly, (2) the gas law is valid only at low concentrations, (3) the observed osmotic pressure is small and difficult to measure, and (4) measurements must be made at the

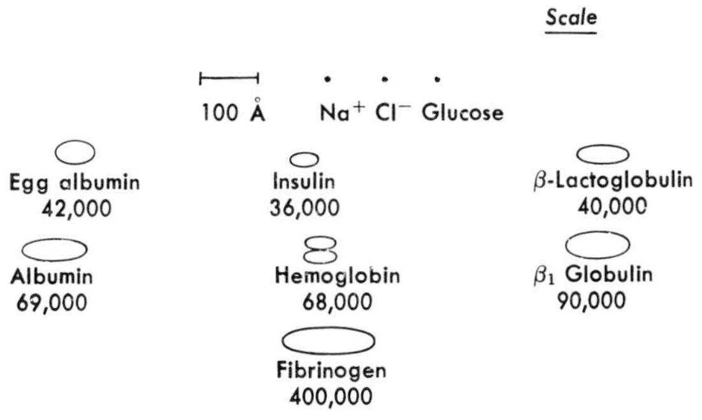

Figure 2.15. Relative size and shape of common proteins.

isoelectric point of the protein where the protein has no charge. The measurement of osmotic pressure at a pH at which a protein is charged leads to an increased osmotic pressure.

VISCOSITY. The measurement of intrinsic viscosity of large molecules has been helpful in obtaining information about the tertiary structure (folding) of large molecules. Compact globular proteins have a low intrinsic viscosity, randomly coiled chains an intermediate value, and coiled proteins a high value. Hemoglobin is compact and has a low intrinsic viscosity, whereas oxidized ribonuclease has an intermediate value and collagen a high value.

X-RAY DIFFRACTION. X-ray diffraction patterns using a finely powdered sample of crystalline protein or a single crystal have been useful for determining the dimensions of a unit cell and the number of molecules per unit. The method may be used to obtain an estimate of the molecular weight. It is one of the most potent methods for determining structures of large molecules and was the method used for elucidating the detailed structure of hemoglobin.

LIGHT SCATTERING. Light scattering is based on the principle that, when light is passed through a medium containing suspended particles, a portion of the light is scattered in various directions. The apparatus consists of a source of monochromatic light, a sample holder, and a phototube that can measure the intensity of the

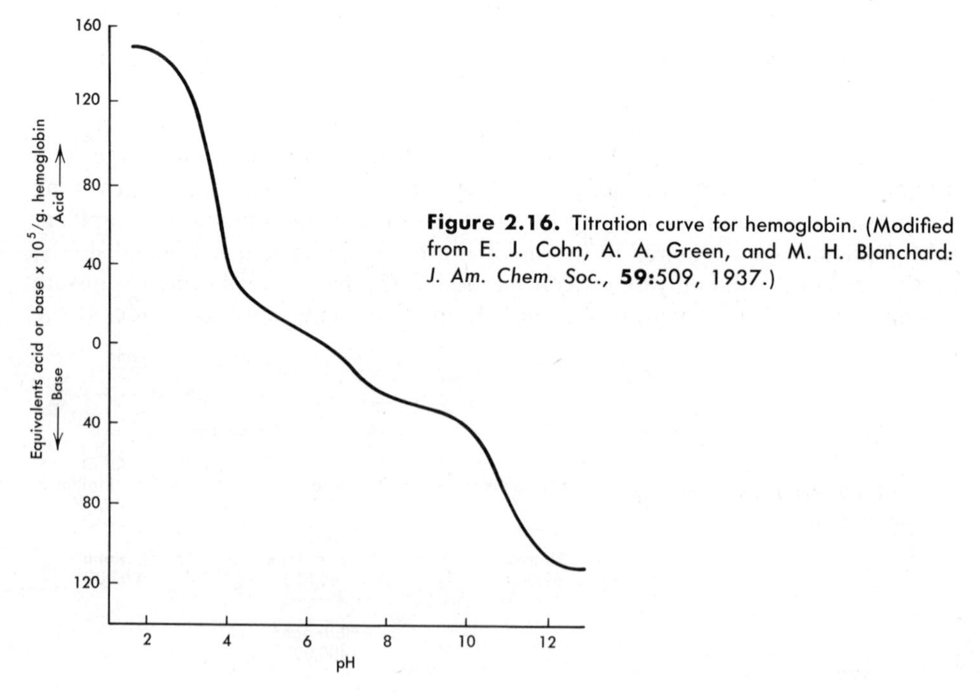

Figure 2.16. Titration curve for hemoglobin. (Modified from E. J. Cohn, A. A. Green, and M. H. Blanchard: *J. Am. Chem. Soc.*, **59**:509, 1937.)

scattered light at various angles with respect to the incident light. The scattering of the light is, under ideal conditions, related to the size of the particle and may be used to estimate the molecular weight of a protein. The dissymmetry of scattering at various angles has been used to estimate the shape of particles. Since measurements can be made very rapidly, the method is useful for following protein reactions and interactions.

OPTICAL ROTATORY DISPERSION. Optical rotatory dispersion has been used to determine the secondary structure (helix content) of proteins. The optical rotation of a protein will depend on the constituent amino acids and on helix content. If optical rotation is measured at several wavelengths and in solvents that either promote or disrupt helix formation, it is possible to estimate the amount of helix normally present in the protein.

INFRARED SPECTROSCOPY. A polypeptide chain exhibits characteristic absorption in the infrared spectrum due to the N—H and $>C=O$ stretching modes. The absorption bands will shift when a randomly coiled peptide changes to an α helix or interracts to form parallel chains. Such shifts give characteristic changes that can be identified. Additional information can be obtained by the use of plane polarized infrared light.

Acid and Base Properties

The properties of proteins as electrolytes are determined by the ionizable groups in the molecule. Since there is only one free α-amino and one free end-carboxyl group for each peptide chain, most of the ionizable groups must come from the R groups. It is estimated that β-lactoglobulin with a molecular weight of 40,000 has 4 imidazolium groups (histidine), 27 ammonium groups (lysine), 59 γ-carboxyl groups (glutamic acid), 7 guanidinium groups (arginine), and 30 amide groups. Proteins are complex cations in acid solution and when titrated with alkali show overlapping pK_a values for the hydrogen ion dissociation of the many groups. As alkali is added, zwitterions and then anions are formed. The situation is very similar to the titration of an amino acid. Figure 2.16 shows the titration curve for hemoglobin.

Even though there are no sharp characteristic breaks in the titration curve of a protein, it is possible to learn a great deal about the ionizable groups present. It should be noted that proteins are good buffers and play an important role in the control of pH in biologic systems.

The isoelectric pH of a protein is the pH at which the protein does not migrate in an electric field. At this pH the number of positive charges is equal to the negative charges, giving a net charge of zero. The total charge may be high. Proteins are cations at pH values lower than the isoelectric pH and anions at pH values higher than the isoelectric pH.

The physical properties of a protein are at a minimum at the isoelectric pH. At this pH, mobility in an electric field, osmotic pressure, swelling capacity, viscosity, and solubility are minimal. These minimal properties are useful for determining the isoelectric pH.

Electrophoresis

The electrophoretic method of analysis of proteins is based on the principle that proteins in solutions at a pH above their isoelectric points have a net negative charge and will move toward the anode when placed in an electric field; proteins in solution at a pH below their isoelectric points have a net positive charge and will move toward the cathode; and protein molecules of the same kind will move at the same rate. Tiselius designed a U-tube into which a solution of the protein in a buffer could be layered under the buffer, an electric current applied, and the movement of the protein measured by means of an optical device. At the beginning all of the proteins will be present at the boundary between the buffer and solution. Under the influence of the electric current the protein with the greatest charge will move faster than the other proteins and will form a moving boundary. The protein moving at the next faster rate will form a moving boundary behind the first one and so on. At each boundary there will be a difference in refractive index which will be related to the concentration of the protein forming the boundary, and this difference can be measured quantitatively (Figure 2.17).

Electrophoretic patterns have been studied for several disease states and are of importance in understanding and treating diseases (Chapter 10). Electrophoretic patterns for normal serum and plasma are shown in Figure 2.18.

A method of electrophoresis using simple equipment has been developed. A solution of the proteins is streaked or spotted on a strip of filter paper wet with a buffer of the proper pH. When an electric potential is applied across the paper, different protein molecules migrate at different rates and separate into zones. These zones can be identified by spraying with a solution of a dye or dipping the paper into a solution of a dye. The amount of protein can be determined by measuring the intensity of the dye. This method is described in more detail in Chapter 10.

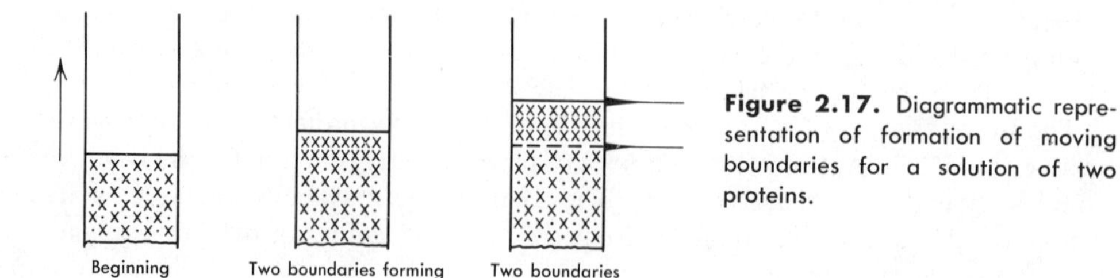

Figure 2.17. Diagrammatic representation of formation of moving boundaries for a solution of two proteins.

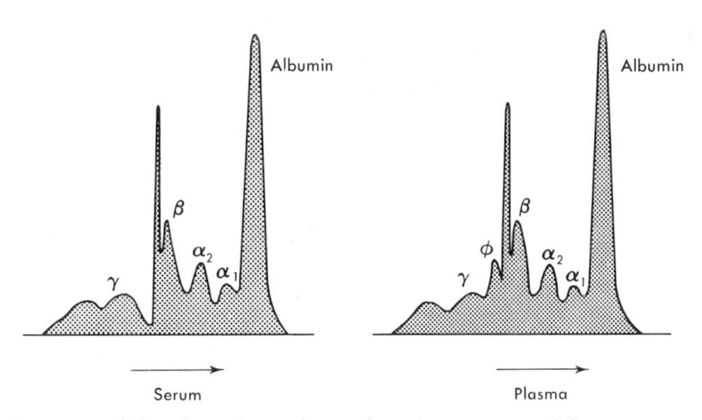

Figure 2.18. Tracings of free-boundary electrophoretic patterns of human serum and plasma.

Reaction of Proteins

With Water

Proteins have many polar groups which combine with water. These groups are —COOH, $—NH_2$, —OH, —NH—, —CO—, $—COO^-$, $—NH_3^+$, and so on. Proteins are more hydrated in acidic or basic solutions than at their isoelectric points. This is probably due to the greater attraction of ionized groups for water. Although most of the water is in equilibrium with the protein, much of it cannot be removed by ordinary methods. The amount of water held by a protein depends on concentration of the protein, competition for the water by other ions, and the pH of the solution.

The phenomenon of "salting in" of proteins is probably caused by forces of attraction between salt and protein at low salt concentration, leading to increased solubility. At high concentrations of salt there is competition for water between the protein and the salt, and the protein is "salted out." The salting out of proteins is an effective method of purification.

With Ions

Proteins are precipitated in solutions alkaline to the isoelectric pH by positive ions such as Zn^{++}, Cd^{++}, Hg^{++}, Fe^{+++}, Ca^{++}, and Pb^{++}. At this pH the protein has a net negative charge. The precipitated protein can be separated from the metal by acidifying the solution or precipitating the metal with H_2S. Protamine insulin is an example of the combination of a protein having a negative charge with a protein having a positive charge to form an insoluble form of the hormone. The isoelectric point of insulin is below pH 7 and that of protamine above 7.

Negative ions combine with proteins in solutions more acid than the isoelectric pH to form salts. The common precipitants in this group are trichloracetic acid, tungstic acid, phosphotungstic acid, sulfosalicylic acid, picric acid, and tannic acid.

Proteins will combine with dyes having positive or negative charges, the particular reaction depending on the pH of the solution. This is the basis of many of the staining procedures.

Reactions Caused by Specific Groups in Proteins

Proteins contain functional groups in the R side chains on the carbon α to the peptide linkage. These groups are contributed by the amino acids and may be the hydroxyl of serine, the imidazole group of histidine, the carboxyl group from aspartic acid or glutamic acid, and so forth. In general these groups give reactions characteristic of each. However, a particular group may be buried in the internal structure of the protein and not react or react very slowly. In addition to the general reactions used by the organic chemist (for example, acetylation of an amino or hydroxyl group), several reagents have been developed that give specific information about the protein. Much of the information we have concerning the active sites or enzymes comes from studies using these special reagents.

A number of color reactions that depend on specific groupings in amino acids and proteins are known. The biuret and ninhydrin reactions have been mentioned. The xanthoproteic test depends on nitration of an aromatic ring; Millon's test depends on the formation of a colored mercury complex with the tyrosine portion of a protein; the Hopkins-Cole test involves a reaction of tryptophan with glyoxylic acid; sodium nitroprusside reacts with —SH groups; the Sakaguchi reaction is specific for the guanido group; and the Sullivan test is used for the determination of cystine in proteins.

Most proteins contain aromatic amino acids which exhibit ultraviolet light absorption at 280 mμ. The measurement of light absorption at this wavelength is a convenient method for determining the amount of a protein in solution.

Denaturation

Denaturation of a protein involves changes in the molecule that leads to changes in physical properties such as decreased solubility. These changes involve alteration of the internal structure and in the arrangement of peptide chains but do not involve breaking of peptide bonds. Proteins are denatured by heating, by standing in acid or alkaline solution, by treatment with organic solvents such as ethanol or acetone, by high concentration of urea, by detergents, and by physical means such as x-ray, shaking, and light.

When globular proteins are denatured, they change to substances with properties of fibrous proteins. These changes are brought about by unfolding and uncoiling. Denaturation leads to an uncovering of active groups such as —SH or the phenolic —OH of tyrosine. The isoelectric pH of denatured proteins is higher than that of the parent substance; they have less capacity to combine with water, they are more easily digestible, and they are biologically inactive.

In most cases denaturation is not reversible, but there are some exceptions. Hemoglobin can be denatured in acid solution and the process reversed by neutralization under the correct conditions.

Isolation and Purification

Because of the ease with which proteins may be modified, the investigator has to control the conditions for the isolation and purification very carefully. The temperatures must be kept low; care must be exercised in the use of organic solvents, acids, bases, and heavy metals.

The first step in the purification of a protein is the preparation of a solution or extract. Cells may be broken by grinding with sand, sonic oscillations, release of pressure, or change of osmotic pressure. They are then extracted with water, dilute salt solutions, dilute acids, or dilute alkali. The previous removal of lipids by acetone and ether extraction often facilitates obtaining the protein in solution. Once in solution the protein may be precipitated by adjusting the pH to the isoelectric pH. The protein may be purified by salting-out of solution or precipitation with ethanol or acetone. Low molecular weight substances can be removed by dialysis. Cohn was able to separate many of the proteins from plasma by careful adjustment of pH, salt concentration, alcohol concentration, and temperature.

Chromatography on ion exchange substances, particularly those prepared from cellulose, has become a useful method for preparing purified proteins. DEAE cellulose, an anion-exchange substance containing a diethylaminoethyl function, and CM cellulose, having a cation exchange function, are most useful. The columns are developed by washing with buffers of increasing ionic strength and either increasing or decreasing the pH. Adsorption chromatography on hydroxylapatite, calcium phosphate gel, alumina gel, and Celite is often employed with good results. Proteins are also separated by chromatography on Sephadex, which is a dextrin and available in a number of pore sizes. The molecules are separated according to size, and the investigator can obtain an estimation of the molecular weight of a protein by choosing a Sephadex with the proper pore size. More recently diethylaminoethyl and carboxymethyl functions have been linked to Sephadex making it possible to carry out separations based on both charge and size.

The preparation of a purified protein usually involves the use of a number of these methods. The choice of procedure will depend on the type of impurities and the properties of the protein to be purified.

Once a protein has been prepared, it is important to determine its purity. Crystallization is not a reliable criterion. Several crystalline proteins have been found to be mixtures. Behavior in the ultracentrifuge is one of the more reliable methods of determining homogeneity. If the protein moves as a single boundary over a wide range of pH values in the electrophoresis apparatus, this is good evidence of purity.

Immunologic techniques are helpful in determining purity (Chapter 10).

Usually data obtained from sedimentation studies and in the electrophoresis apparatus are considered good evidence for purity.

REFERENCES

Anson, M. L., and Edsall, J. T. (eds.): *Advances in Protein Chemistry,* Vols. 1–24. Academic Press, New York, 1944–1970.

Fruton, J. S., and Simmonds, S.: *General Biochemistry,* 2nd ed. John Wiley & Sons, Inc., New York, 1958.

Mahler, R., and Cordes, H.: *Biological Chemistry,* Chaps. 2 and 3. Harper and Row, New York, 1966.

West, E. S.; Todd, W. R.; Mason, H. S.: and Van Bruggen, J. T.: *Textbook of Biochemistry,* 4th ed., Chap. 8. The Macmillan Co., New York, 1966.

White, A.; Handler, P.; and Smith, E. L.: *Principles of Biochemistry,* 4th ed., Chaps. 5, 6, 7, and 8. McGraw-Hill, New York, 1968.

Chapter 3

ENZYMES

PLANTS synthesize glucose, starch, protein, lipids, and other cellular constituents from CO_2, water, inorganic nitrogen, and other inorganic substances. The source of energy for doing this chemical work is sunlight. The plant is a chemical machine doing chemical work and in the process locks up the energy from the sun in the form of chemical compounds. Animals break down the starch, lipids, and protein of plants and use this locked up energy for the production of heat, locomotion, physical work, and the synthesis of proteins and chemical substances needed for their own use.

It is significant that most plants and animals carry out all the chemical reactions involved at temperatures of 20° to 40° C, and at a pH of about 7. In order for chemists to carry out similar reactions in the laboratory, high temperatures and strong acids or bases are required. The synthesis of relatively simple peptides may require several steps and many hours. Cells synthesize protein containing many peptide bonds with ease and under mild conditions. Certain cells can very easily convert glucose quantitatively to lactic acid. The chemist has difficulty preparing any lactic acid from glucose.

The cell contains biologic catalysts called enzymes. These are proteins and have many properties similar to chemical catalysts with which you are familiar. Their protein nature imparts both chemical and biologic properties peculiar to this class of substances. More than 100 enzymes have been prepared in crystalline form.

CATALYSTS

Before a discussion of the enzymes it is necessary to review the simple concepts of catalysis and energetics.

In order for a chemical reaction to take place, the reaction must be thermodynamically possible and the molecules must be activated. The change in free energy, ΔF, of a reaction is the work in calories that can be obtained from a reaction that proceeds spontaneously, or it is the amount of work that must be put into a reaction in order to make it go. When a chemical system proceeds from one energy state to another and in the process energy becomes available for work, this change in capacity to do work is designated as $-\Delta F$ (exergonic reaction). When it is necessary to put energy into a chemical system in order to have it change from one chemical state to a second, then ΔF is positive ($+\Delta F$) (endergonic reaction). When ΔF for a chemical reaction is negative, this energy may be used for running another chemical reaction that has a positive ΔF; the energy can be used as heat; the energy can be used to perform mechanical work, for example, contraction of muscle; or the energy can be converted to electric energy, for example, nerve transmission. ΔF is expressed as calories per mole of reactant transformed; for example, the conversion of 1 mole of glucose to CO_2 and H_2O makes 690,000 calories available for work.

The fact that a reaction can proceed with a $-\Delta F$ does not necessarily mean it will take place. Before a chemical reaction can take place, the molecules must be activated. Molecules become activated by collision with other molecules, by heating, or by receiving energy in some form. Heating a chemical reaction is a common method of activating the molecules. A 10°C rise in temperature about doubles the speed of a reaction. Once the molecules of the reaction are activated, they can change to products of the reaction and energy is made available for work. This is represented in Figure 3.1, where the energy of activation, E_a, is required to activate A to B. The activated B then changes to C, the energy of activation is recovered, and an amount of energy ΔF is made available.

If we assume that activities are equal to concentration, then for the equation

$$A + B \rightleftharpoons C + D$$

it can be shown that

$$\Delta F = \Delta F^\circ + RT \ln \frac{[C][D]}{[A][B]}$$

where ΔF° is the standard free energy change for the reaction, R the gas constant, and T the absolute temperature. At equilibrium $\Delta F = 0$ and

$$\Delta F^\circ = -2.3\ RT \log K$$

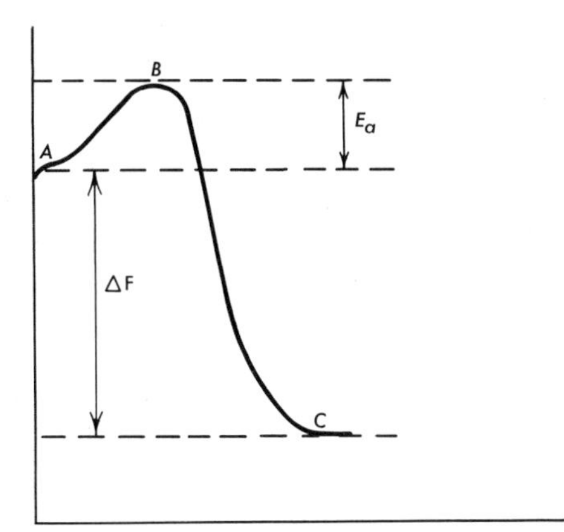

Figure 3.1. Schematic representation of activation energy and ΔF.

When the concentration of each of the reactants is unity, log 1 = 0 and $\Delta F = \Delta F^\circ$. In other words ΔF° is the energy available for chemical work when 1 mole of each of the reactants is converted to 1 mole of each of the products under conditions such that the ratio of products to reactants is maintained at unity.

Most biochemical reactions involve the gain or loss of a proton and take place at a pH of 7.0. It is a custom to express ΔF° as determined at a pH other than 0 as $\Delta F_0'$ and ΔF as $\Delta F'$.

Let us consider the effect of a catalyst on the following reaction

$$A + B \underset{V_1}{\overset{V_2}{\rightleftarrows}} C + D$$

At equilibrium the forward reaction, V_1, is proportional to the concentrations of A and B and the reverse reaction, V_2, is proportional to the concentration of C and D and $V_1 = V_2$. A catalyst only affects the rate at which equilibrium is attained. It does this by lowering the energy of activation and thereby affects the rate equally in both directions. A catalyst does not change the equilibrium constant or the ΔF. One molecule of the catalyst may be responsible for the activation of several thousand molecules of the substrate. The turnover number is defined as the number of moles of substrate converted per mole of catalyst per unit time.

The effect of the catalyst on a reaction is illustrated by the hydrolysis of ethyl acetate. This same equilibrium mixture will be obtained whether we start with 1 mole each of ethyl acetate and H_2O or 1 mole each of acetic acid and ethyl alcohol.

$$CH_3COOC_2H_5 + H_2O \rightleftarrows CH_3COOH + C_2H_5OH$$

The same equilibrium can be obtained much faster by adding traces of OH^- or H^+ or the enzyme lipase.

In biologic systems the concentrations of reactants and products are not unity and equilibrium is never obtained. However, there does exist a *steady state* in which reactants are added to the system continuously and the products removed continuously. Under these conditions the above energy relationships are only approximations. Nevertheless they have been very useful in helping to understand biochemical reactions.

Nomenclature and Classification

Enzymes are named and classified in terms of the reactions they catalyze. The individual enzymes are named by adding the suffix -ase to the name of the substrate acted upon, for example, urease, arginase, phosphatase, and so forth. Some of the names of enzymes which appeared in the early literature have been retained, for example, trypsin, pepsin, and the like.

The Commission on Enzymes of the International Union of Biochemistry has recently proposed that the enzymes be classified in six main divisions. The classification is based on the chemical reaction catalyzed by an enzyme. The divisions are similar to those previously used, but in many cases names are different. We will retain the names of the enzymes now in common usage and present a summary of the new classification more for the purpose of emphasizing the types of reactions than for introducing a new system of nomenclature. The six new names introduced should help the student who encounters the new nomenclature to recognize that he may be meeting a familiar enzyme with a new name, for example, an oxidoreductase may be a dehydrogenase.

I. Oxidoreductases

This group of enzymes includes the dehydrogenases, oxidases, and peroxidases.

II. Transferases

These enzymes catalyze the transfer of groups. Examples are the kinases which transfer phosphate from ATP to other compounds and the transaminases which transfer amino groups. Several enzymes in this class will be encountered.

III. Hydrolases

There are many enzymes in this group. As the name implies, they catalyze the hydrolysis of compounds. This group includes the proteolytic enzymes, lipases, amylases, and many others.

IV. Lyases

These enzymes catalyze the nonhydrolytic removal of groups. Examples are the decarboxylases and aldolases.

V. Isomerases

Enzymes in this class catalyze *cis-trans* isomerizations, epimerizations, racemizations, intramolecular transfers, and intramolecular isomerizations.

VI. Ligases

Ligases catalyze the coupling together of two molecules with the breaking of a pyrophosphate bond.

CHEMICAL PROPERTIES

A large number of enzymes have been prepared in crystalline form and found to be proteins. Many of the enzymes are conjugated proteins having a metal or some other nonprotein fragment attached to the protein. During the past few years there has been a rapid increase in the knowledge concerning the primary, secondary, tertiary, and quaternary structure of enzymes and the enzyme ribonuclease has been synthesized. This information has come from a study of the kinetics of enzymic reactions, the chemical determination of primary structure (page 30), by specifically modifying the protein with chemical reagents, and by physical chemical studies, with x-ray crystallographic analysis being particularly useful.

MEASUREMENT OF ACTIVITY (ENZYME KINETICS)

The activity of an enzyme may be measured by following the chemical change catalyzed by the enzyme. The enzyme is incubated under proper conditions with the substrate and samples withdrawn at short intervals and analyzed. The sample may be analyzed for the disappearance of substrate or the appearance of products. As an example, the action of invertase on sucrose can be followed by the appearance of glucose by means of its reducing properties. This same reaction could be followed by the change in optical rotation. The enzymatic hydrolysis of a protein could be followed by decrease in the biuret reaction or through the formation of amino groups (Van Slyke method) or the appearance of ninhydrin positive materials.

If we follow the progress of an enzymatic reaction, we obtain results shown in Figure 3.2. The reaction velocity soon begins to decrease; this may be due to disappearance of substrate and formation of products leading to equilibrium. This may also be due to changes in pH or inhibition of the enzyme by products of the reaction. Therefore, it is necessary to study the reaction under conditions such that these factors do not influence the results. The time required to produce a certain change may be determined. In this way separate enzyme preparations can be

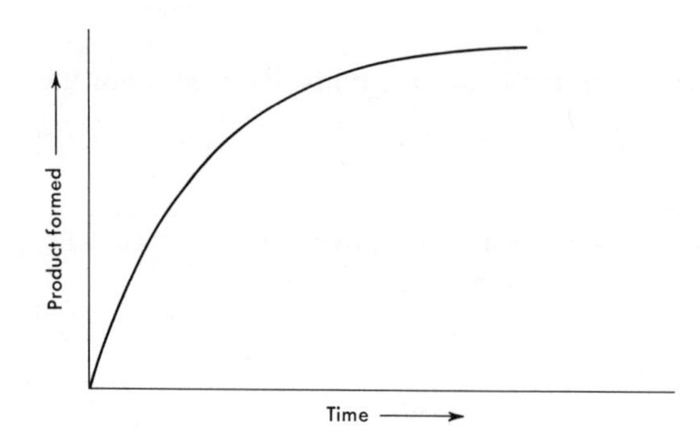

Figure 3.2. Time course of an enzyme-catalyzed reaction.

compared because the same change in pH, inhibition, and so forth, should occur in both preparations. A second method is to study the reaction over a very short period of time. In this procedure the change in pH, and so forth, is negligible. Such rates are called instantaneous velocities.

Influence of Temperature

A rise in temperature has a dual effect on an enzyme-catalyzed reaction. First, there is about a doubling of the reaction rate for each 10°C rise in temperature, the same as for most chemical reactions. However, as the temperature rises, the rate of inactivation of the catalyst by denaturation increases. In this situation the optimum temperature will depend on the reaction time. For short time intervals the optimum temperature may be much higher than for a longer time (see Figure 3.3).

Influence of pH

The catalytic activity of an enzyme is evident over a narrow range of pH and within this range has a maximum at a particular pH, known as the optimum pH. Generally the optimum pH is characteristic of the enzyme. If an enzyme acts on

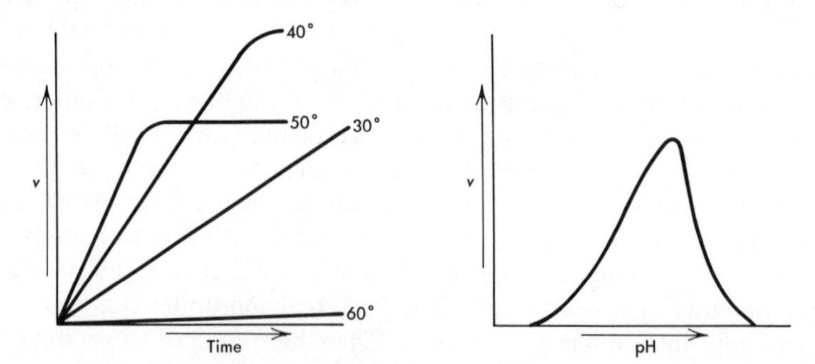

Figure 3.3. Effects of temperature and pH on enzyme action.

two different substrates, both of which may be ionic in nature, it may exhibit an optimum pH for each of these substrates. Such is the case with pepsin, which has an optimum pH between 1.5 and 2.5, depending on the particular protein being used as a substrate. Proteins have many charges on the molecule, and the charge will depend on the pH. The activity of the enzyme may require specific ionic structures in the substrate which are found in different proteins at different pH values.

Most enzymes have their optimum between pH 5 and 7 and are often most stable at these pH values. There are a number of exceptions.

Influence of Substrate

Under the proper conditions the rate of a reaction catalyzed by an enzyme will depend on the concentration of the enzyme and substrate. If the enzyme concentration is held constant, the initial reaction velocity increases with increasing substrate concentration until a limiting rate is reached. The results of such an experiment are shown in Figure 3.4.

In order to explain these events Michaelis and Menten assumed that the enzyme and substrate form a complex which is converted to products and enzyme.

$$\text{Enzyme (E)} + \text{Substrate (S)} \underset{K_2}{\overset{K_1}{\rightleftarrows}} \text{Enzyme-substrate complex (ES)}$$

$$\downarrow K_3$$

$$\text{Enzyme (E)} + \text{Products (P)}$$

It is further assumed that the rate of conversion of substrate to products is determined by the rate of conversion of ES to E and P.

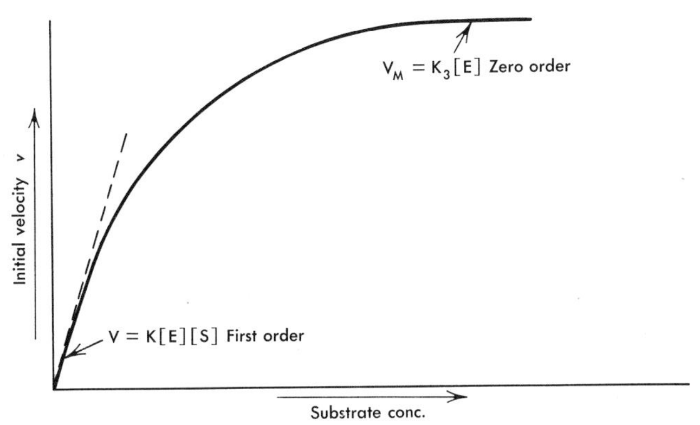

Figure 3.4. The effect of substrate concentration on the velocity of enzyme action.

In Figure 3.4 the portion of the curve at low substrate concentration represents a first-order reaction. The rate is dependent on both enzyme and substrate concentrations. As the substrate concentration is increased, the rate approaches a maximum rate V. In this portion of the curve the enzyme is completely bound in the form of ES. Under these conditions E = ES and the rate V is proportional to the concentration of E. This is represented diagramatically in Figure 3.5.

Michaelis-Menten Equation

In the equation

$$E + S \underset{k_2}{\overset{k_1}{\rightleftharpoons}} ES \xrightarrow{k_3} \text{Products} + E \tag{3.1}$$

the following symbols are used:

[E] = total concentration of enzyme
[S] = total concentration of substrate
[ES] = concentration of enzyme-substrate complex
[E] − [ES] = concentration of free enzyme
k_1, k_2, k_3 = respective velocity constants of the three reactions.

For the rate of formation of [ES] one obtains

$$\frac{[d\,ES]}{dt} = k_1([E] - [ES])[S] \tag{3.2}$$

since the rate of formation of [ES] is proportional to the concentration of substrate and free enzyme. The rate of disappearance of [ES] is given by

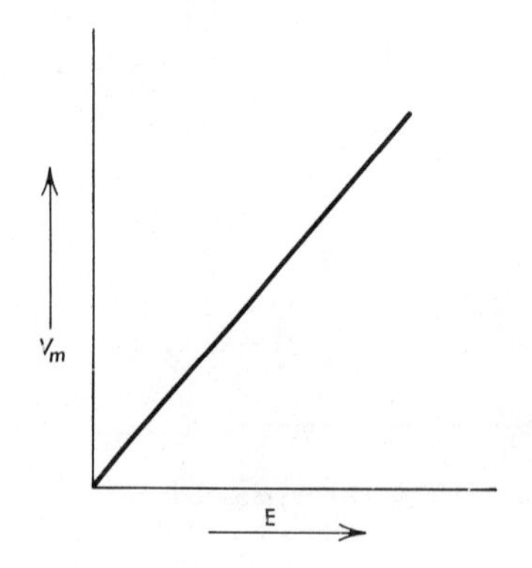

Figure 3.5. Effect of enzyme concentration on maximum velocity of the reaction.

$$-\frac{d\,[\mathrm{ES}]}{dt} = k_2[\mathrm{ES}] + k_3[\mathrm{ES}] \tag{3.3}$$

In the steady state when the rates of formation and disappearance of [ES] are equal

$$k_1([\mathrm{E}] - [\mathrm{ES}])[\mathrm{S}] = k_2[\mathrm{ES}] + k_3[\mathrm{ES}] \tag{3.4}$$

and upon rearrangement of terms

$$\frac{[\mathrm{S}]([\mathrm{E}] - [\mathrm{ES}])}{\mathrm{ES}} = \frac{k_2 + k_3}{k_1} = K_m \tag{3.5}$$

Employing K_m (the Michaelis constant) as equal to $(k_2 + k_3)/k_1$ and solving for [ES], one obtains the equation for the steady state of concentration of [ES].

$$[\mathrm{ES}] = \frac{[\mathrm{E}][\mathrm{S}]}{K_m + [\mathrm{S}]} \tag{3.6}$$

The observed initial velocity (v) of the reaction is proportional to [ES]. Thus

$$v = k_3[\mathrm{ES}] \tag{3.7}$$

When the substrate concentration is high in relation to the enzyme concentration, all the enzyme is present as [ES] and the velocity of the reaction is maximal (V_{max}).

$$V_{max} = k_3[\mathrm{E}] \tag{3.8}$$

Substituting the value of [ES] in equation (3.6) for [ES] in equation (3.7) and dividing by equation (3.8) gives the Michaelis-Menten equation

$$v = \frac{V_{max}[\mathrm{S}]}{K_m + [\mathrm{S}]} \tag{3.9}$$

When $v = V_{max}/2$, $K_m = [\mathrm{S}]$, and K_m is the substrate concentration (in moles per liter) which will give half maximum velocity. For any enzyme-substrate system the K_m will have a characteristic value independent of the concentration of the enzyme. If an enzyme acts on several substrates, it will have a characteristic K_m for each substrate.

For the determination of K_m one of several more convenient forms of the Michaelis-Menten equation is usually employed. The reciprocal of equation (3.9) is a commonly used linear form of the equation:

$$\frac{1}{v} = \frac{K_m + [S]}{V_{max}[S]} = \frac{K_m}{V_{max}[S]} + \frac{[S]}{V_{max}[S]} \tag{3.10}$$

$$\frac{1}{v} = \frac{K_m}{V_{max}}\left(\frac{1}{[S]}\right) + \frac{1}{V_{max}}$$

As shown in Figure 3.6, if one plots $1/v$ against $1/S$, the slope of the line is K_m/V_{max} and the intercept on the ordinate is $1/V_{max}$. Since V_{max} can be obtained from the intercept, it is possible to calculate K_m. In addition, the intercept on the abscissa is equal to $-1/K_m$.

SPECIFICITY

In general, enzymes are quite specific in their activity. This is a characteristic that clearly differentiates them from other catalysts. This specificity is probably an important property whereby the cell is able to control its chemical reactions.

The enzyme arginase will catalyze the decomposition of L-arginine to ornithine and urea but will not act on the D isomer.

$$\underset{\text{L-Arginine}}{H_2N{-}CH(COOH){-}(CH_2)_2{-}CH_2{-}NH{-}C({=}NH){-}NH_2} \xrightarrow[H_2O]{\text{arginase}} \underset{\text{L-Ornithine}}{H_2N{-}CH(COOH){-}(CH_2)_2{-}CH_2{-}NH_2} + \underset{\text{Urea}}{NH_2{-}C({=}O){-}NH_2}$$

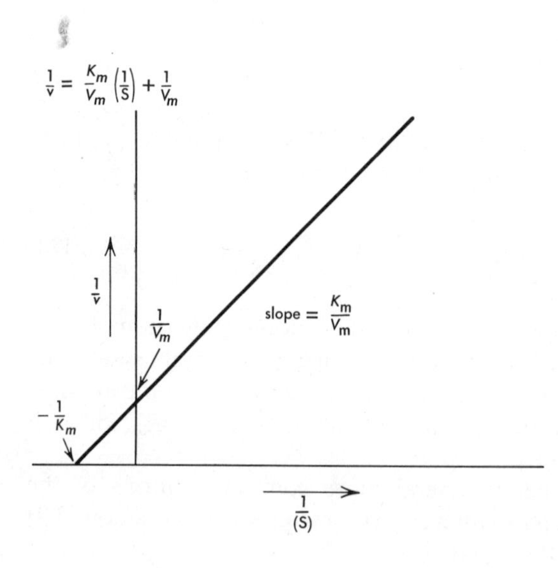

Figure 3.6. Lineweaver-Burk plot of the Michaelis-Menten equation.

A glycosidase may be specific for splitting an α or a β linkage to a glucose molecule but may not be specific for the aglycone portion of the molecule. In this particular class of enzymes various degrees of specificity can be found.

Lipases or esterases are known that will split most ester bonds. These probably represent the least specific enzymes.

The most significant work on enzyme specificity has been done with the proteolytic enzymes. These enzymes are the endopeptidases pepsin (stomach), trypsin (pancreas), chymotrypsin (pancreas), and the exopeptidases-aminopeptidase (intestine), carboxypeptidase (pancreas), and dipeptidases (many tissues). The endopeptidases catalyze splitting of peptide bonds within peptide chains; the exopeptidases act on the end of peptide chains.

Trypsin

Trypsin hydrolyzes compounds containing L-lysine or L-arginine in which the ϵ-amino or guanido grouping is free.

```
                              CH2NH2
                                |               ↑
     H   R     O    H   (CH2)3          O ! H   R    O
     |   |     ‖    |     |             ‖ ! |   |    ‖
   —N—CH—C—N—CH————————C+N—CH—C—
                                                !
                                            HO ! H
```

Trypsin will hydrolyze dipeptides containing arginine at a slow rate.

Chymotrypsin

Chymotrypsin is most active toward compounds containing an aromatic amino acid.

```
                        (C6H5)
                          |               ↑
     H   R     O        CH2  O ! H   R    O
     |   |     ‖    H     |   ‖ ! |   |    ‖
   —N—CH—C—N—CH—C+N—CH—C—
                                  !
                              HO ! H
```

Pepsin

Pepsin is most active on peptides having adjacent aromatic amino acids and is active on compounds having one aromatic amino acid. In the latter case the peptide bond is split on the amino side of the aromatic substituent.

Carboxypeptidase

Carboxypeptidase attacks many peptides that contain different terminal amino acids, splitting off the amino acid at the carboxyl end of the chain. The rate of splitting depends on the amino acid being released; aromatic amino acids are the most sensitive. This enzyme has been useful in the determination of the amino acid sequence in a peptide chain.

Aminopeptidase

This enzyme requires a free amino group for activity and catalyzes the splitting of a number of different types of side chains. Leucine aminopeptidase has been studied extensively and shown to require the presence of a metal (Mn^{++} or Mg^{++}) for activity.

Dipeptidase

This enzyme requires that both the amino and carboxyl groups of the dipeptide be free. A metal ion is required for activity; cobalt (Co^{++}) gives maximal effect.

The most plausible explanation for the specificity of the enzymes is the formation of an enzyme-substrate complex involving the attachment of specific groups of the substrate to specific sites on the enzyme.

INHIBITORS

Some of the best information in support of the formation of an enzyme-substrate complex through specific groupings on the enzyme has come from a study of specific inhibitors.

In simple competitive inhibition the inhibitor combines with the enzyme in a reversible manner at the same site that is used for the formation of ES. The competition is dependent on the relative concentrations of substrate and inhibitor.

$$E + S \rightleftharpoons ES \longrightarrow E + P$$
$$E + I \rightleftharpoons EI$$

One of the classic examples is the inhibition of succinic dehydrogenase by malonic acid.

$$\underset{\text{Succinic acid}}{\begin{array}{c} COOH \\ | \\ CH_2 \\ | \\ CH_2 \\ | \\ COOH \end{array}} + \underset{A}{\text{Acceptor}} \xrightarrow[\text{dehydrogenase}]{\text{Succinic}} \underset{\text{Fumaric acid}}{\begin{array}{c} HOOC{-}C{-}H \\ \| \\ H{-}C{-}COOH \end{array}} + AH_2$$

When the ratio of the concentration of malonic acid to succinic acid is 1:50, there is a 50 per cent inhibition of the reaction. The inhibition may be decreased by adding substrate or increased by adding more of the inhibitor.

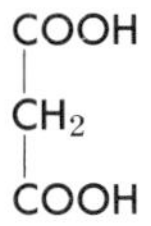

$$\begin{array}{c} COOH \\ | \\ CH_2 \\ | \\ COOH \end{array}$$

Malonic acid

In this example, malonic acid is the next lower homolog of succinic acid. It has two carboxyl groups separated by one less $—CH_2$ group. Because of this similarity in structure it can fit onto the active site of succinic dehydrogenase.

Substances that are similar in structure to the substrates with which the enzyme normally combines and that inhibit enzyme activity are called "metabolic antagonists" or "antimetabolites."

Another example of competitive inhibition is sulfanilamide and *p*-aminobenzoic acid.

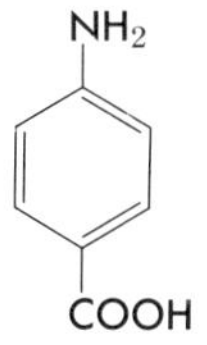

NH_2—C_6H_4—$COOH$
p-Aminobenzoic acid

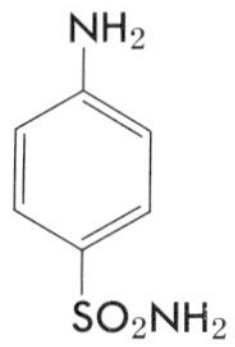

NH_2—C_6H_4—SO_2NH_2
Sulfanilamide

Sulfanilamide competes with *p*-aminobenzoic (PABA) for the site on an enzyme that incorporates PABA into folic acid which is required for the growth and function of certain microorganisms.

In the case of noncompetitive inhibitors there is no relationship between the amount of inhibition and substrate concentration. Examples of this type of inhibition are found with the heavy metals Ag^+, Hg^{++}, Pb^{++}, and so forth. A noncompetitive inhibitor used to determine the presence of —SH groups is iodoacetamide. The reaction is not reversible.

$$R—SH + ICH_2CONH_2 \longrightarrow R—S—CH_2CONH_2 + HI$$

p-Chloromercuribenzoate also reacts with —SH groups. Fluoride will combine with magnesium and inhibit enzymes that require Mg^{++}. The number of substances belonging to this class of inhibitors is quite large. They have been very useful in studying enzyme activity.

If data obtained by the effect of inhibitors on an enzymic reaction are treated by the Lineweaver-Burk type of equation, the type of plot shown in Figure 3.7 is obtained.

Note that the lines for competitive inhibition intercept the axis at the same point. Since 1/S is plotted against $1/v$, (S) would be infinitely high at this point and the inhibitor would have no effect. Since the velocity is slower, the plotting of $1/v$ results in a steeper slope. The effect of a noncompetitive inhibitor is to remove enzyme. Thus, the intercept on the vertical axis will be higher because V will be less.

ACTIVATORS

Many enzymes require the presence of metal ions in order to activate the enzyme. These metal ions are called activators. It has been postulated that the metal ion forms a coordination complex between the enzyme and the substrate. Leucine aminopeptidase is an example of an enzyme that requires a metal for activity. Its combination with the substrate has been formulated as follows

$CH_3—CH(CH_3)—CH_2—CH—N(H_2)$ and $C(=O)—N(H_2)$, both N coordinated to Mn^{++}, which is bound to Protein

Many enzymes have a metal ion firmly bound to the protein, and it is not easily separated. Examples are the copper-containing ascorbic acid oxidase and the zinc-containing carbonic anhydrase.

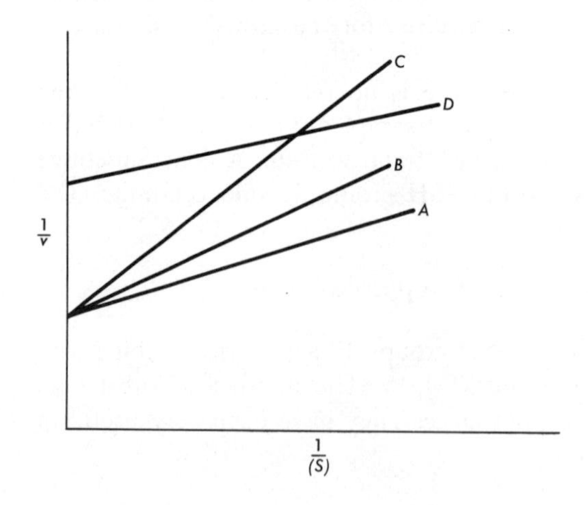

Figure 3.7. A, normal, uninhibited; B and C, two concentrations of competitive inhibitor; D, noncompetitive inhibitor. The lines for the competitive inhibitor intercept the vertical axis at the same point because (S) is very high.

The proteinases—pepsin, trypsin, and chymotrypsin—are found in inactive forms which have been called *zymogens* or *proenzymes*. At the acid pH of the gastric juice a polypeptide is liberated from pepsinogen and pepsin is released. The process is autocatalytic since pepsin will also split off the protecting polypeptide. There is an intestinal enzyme, enterokinase, which specifically catalyzes the conversion of trypsinogen to trypsin. Trypsin itself can act an an activator. The conversion of chymotrypsinogen to chymotrypsin is catalyzed by trypsin. Again the activation involves the release of a small peptide.

There are circumstances in which enzymes can be activated by removal of an inhibitor. Heavy metals may be removed by H_2S. An enzyme having a —SH group at the active site may easily be oxidized by air or by a mild oxidizing agent to the —S—S— form, which is inactive. Addition of a mild reducing agent may activate the enzyme.

THE PRIMARY STRUCTURE OF ENZYMES

The amino acid sequence has been determined for several enzymes, for example, ribonuclease, lysozyme, pepsin, trypsin, chymotrypsin, papain, carboxypepidase, and others. Information about the primary structure is important in relating particular peptide sequences to helix structure. It has been shown that alanine, glutamic acid, leucine, lysine, methionine and tyrosine promote helix formation, whereas valine, isoleucine, serine, threonine and proline are nonhelix forming. The determination of the amino acid sequence for an enzyme isolated from several species can give pertinent information about the active site of the protein as well as the evolutionary aspects. Margoliash has studied the homology of cytochrome c, a protein electron carrier, that is present in all eukaryotic organisms. He established the primary structures of 23 vertebrate, 4 invertebrate, 3 fungal, and 1 higher plant cytochrome c and concluded that all the cytochromes were ancestrally homologous; that is, at one time these proteins had a common ancestral gene.

THE NATURE OF THE ENZYME-SUBSTRATE COMPLEX

The side chain residues in the protein molecule contain several functional groups that can be involved in binding the substrate to the protein and in the catalytic process. Early attempts to find the specific groups responsible for the function of an enzyme involved reacting the protein with chemical reagents and determining the effect on activity. Usually, but not always, the active groups are exposed on the surface of the molecule and react rapidly. A study of change of activity with time gave valuable information. Recently, ingenious reagents and new methods have given more detailed information.

One approach has been to trap the intermediate complex, break the polypeptide chain, and determine the amino sequence of the peptide bearing the substrate. An example is the case of an aldolase in which the aldehyde function of substrate reacts with the ε-amino group of a lysine residue to form an anil. The anil is reduced by sodium borohydride to a stable amine, the protein hydrolyzed to peptides, and the amino-substituted peptide isolated and identified. The general reaction is as follows.

$$\text{—}NH_2 + \underset{\displaystyle R}{\overset{\displaystyle H}{C}}{=}O \longrightarrow \text{—}N{=}\underset{\displaystyle R}{\overset{\displaystyle H}{C}} \longrightarrow \text{—}\overset{\displaystyle H}{N}\text{—}\underset{\displaystyle R}{\overset{\displaystyle H}{C}}\text{—}H \longrightarrow \text{Peptides}$$

Stable

The same principle has been used to show that a serine residue is located at the active site of several hydrolytic enzymes. The use of radioactive substrates simplifies identification.

A number of specific analogs of the substrates have been devised that are bifunctional and can identify additional functional residues; for example, an analog of the substrate for chymotrypsin containing an alkylating group was used to show that a histidine residue was located near the active serine residue even though the two amino acids were separated by about 140 amino acids.

The information deduced by modification of enzymes with specific reagents has been verified by x-ray analysis in the cases where such information is available. At the present time x-ray analysis is the best technique available for studying detailed structure at the active site. Phillips was able to determine the functional residues in the active site of lysozyme from x-ray analysis and a knowledge of the amino acid sequence. In this case very little information was available from chemical modification of the molecule (see page 40). The molecule contains a cleft between the "pleated sheet" section and the helix sections into which the substrate fits.

Lysozyme hydrolyzes polymers of N-acetylglucosamine and N-acetylmuramic acid (page 113). Figure 3.8 is a schematic representation of such a six-unit polymer fitted into the cleft of lysozyme. The catalytic site lies between D and E and involves asparatic acid 52 and glutamic acid 35.

It is information like this, along with kinetic studies and stereochemical considerations, that has led to the conclusion that the active site of any one enzyme contains subsites made up of specific amino acid residues. These subsites are positioned in such a manner that the substrate binds through specific bonding. At this point in our discussion we could describe the active site as a template onto which a substrate will fit.

The template hypothesis is useful and serves to explain much of what is known about enzyme activity. However, the theory has been inadequate to explain many observations. When phosphoglucomutase, which catalyses the conversion of glucose-6-phosphate to glucose-1-phosphate, is incubated with glucose-6-phosphate,

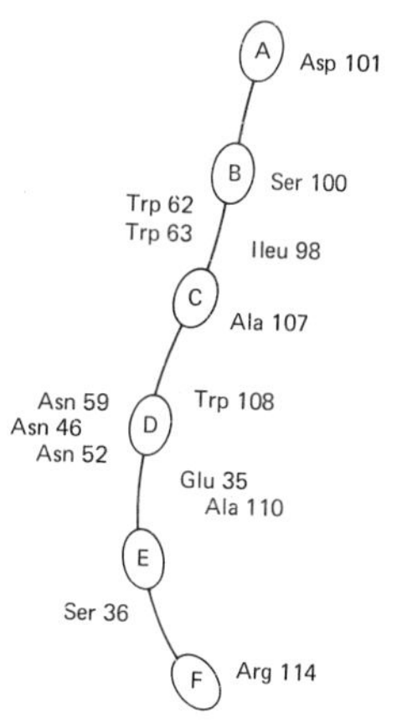

Figure 3.8. Schematic representation of the active site in the cleft region of lysozyme. (From D. E. Koshland and K. E. Neet, *Ann. Rev. Biochem.*, **37**:364, 1968.)

there is enhancement of absorption of ultraviolet light by tyrosine and tryptophan residues in the enzyme; at least one sulfhydryl group becomes more active; the activities of a cysteine, a lysine, and a methionine are reduced; and the phosphate group which is an ester on serine participates in the enzymic reaction. In order to explain these observations, Koshland has proposed that the enzyme has a flexible active site and that the combination with the substrate leads to a new conformation of the protein molecule. Active groups in the new form can become protected and less active. Groups that were in the center of the molecule and inactive become exposed and active. This concept can explain the change in activity of a sulfhydryl group or a tyrosine group that is far removed from the active site. This explanation is illustrated in the following diagram.

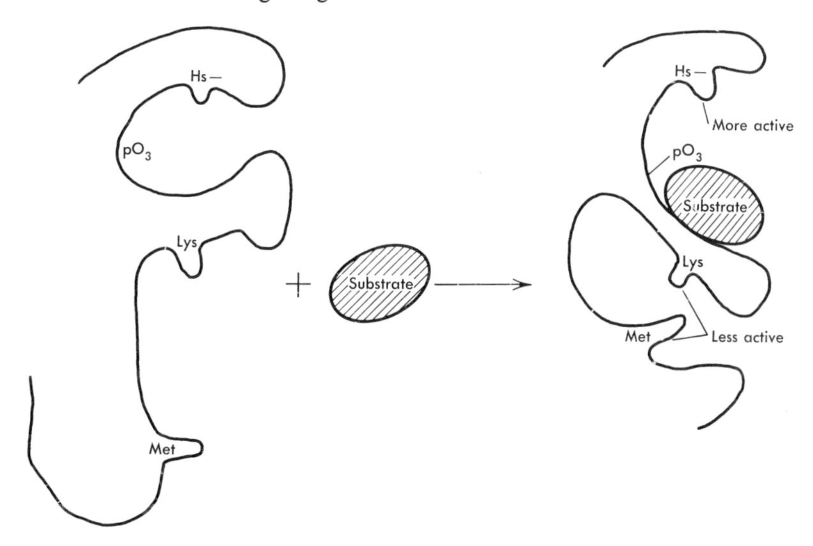

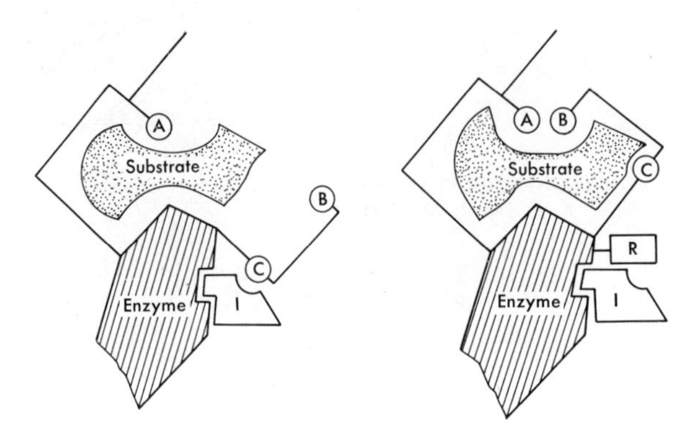

Figure 3.9. Effect of inhibitors and reagents on flexible active site enzyme. *Left:* inhibitor attracts group C and prevents proper alignment of catalytic group B. *Right:* reagent R prevents juxtaposition of group C with inhibitor I and effect of I nullified without changing binding affinity of I. (From D. E. Koshland, Jr.: *Fed. Proc.,* **23:**719, 1964.)

The flexible site theory helps to explain how compounds unrelated in structure to the normal substrate for an enzyme can increase or inhibit activity. In Figure 3.9 the inhibitor I could be bound to the molecule at a site other than the active site. In such a position it would have an affinity for C a grouping in the active site and prevent it from attaching to the substrate. In other words the attachment of I changes the conformation of the protein molecule. Such an inhibitor can act competitively or noncompetitively (page 62). Since the inhibitor is acting at a different site from the substrate, it can have a different structure and properties. Also a molecule different in structure from the substrate can activate or increase the activity of an enzyme. In this situation the compound can change the conformation to bring the attachment sites for the substrate into the correct position, or it might remove the effect of an inhibitor as shown in Figure 3.9. The attachment of R nullifies the inhibitory effect of I. It is possible that some of the hormones act in this manner.

Small molecules that change the affinity of an enzyme for a substrate have been termed effectors, modifiers, or modulators; positive effectors are substances that increase affinity, whereas negative effectors decrease affinity of the enzyme for substrates. Figure 3.10 shows the response of an enzyme to a negative effector and a positive effector. If a positive effector is added at some point on the control curve, the same rate of reaction can be maintained at a much lower substrate concentration c, or a much higher rate can be obtained if the substrate concentration remains constant, point b.

These considerations are extremely important for the regulation of metabolic reactions. For example, if the cell produces an amino acid by a number of steps each of which requires an enzyme, the amino acid acts as a negative effector on the enzyme catalyzing the first step and thereby regulates its own synthesis. This is termed a negative feedback system and is illustrated as follows

A ⟶ B ⟶ C ⟶ D ⟶ AA

Several examples of positive and negative effectors will be discussed in the following chapters.

Another important consideration in the activity of an enzyme is the quaternary structure, that is, the arrangement or conformation of the monomer units in the enzyme molecule. Most, if not all, enzymes subject to control by small molecules contain two or more monomer units. The effect is called an *allosteric effect* and is defined as the effect of a modifier (effector) bound at a site which is topographically distinct from the active site and influences the binding or activity at the active site. The hemoglobin molecule, which is composed of two α and two β chains, has served as a model. Each chain has one site for combination with oxygen. The combination of oxygen with one site can change the conformation of the quaternary structure and thereby the affinity of oxygen for the second site and so forth for the third and fourth sites. The oxygen dissociation curve for hemoglobin exhibits a sigmoid character similar to that shown in Figure 3.10. This is a special case in which binding at the active site in one of monomers changes the activity of the site in the other monomers. However, most enzymes operative in metabolic control have a site for the binding of ligands that is distinct from the active site. The schematic illustration in Figure 3.11 shows an enzyme containing subunits with allosteric sites.

Isozymes are enzymes from a single species having the same kind of activity but differing in chemical structure. They may exhibit different Michaelis constants or other quantitative characters. The most extensively studied system is lactic dehydrogenase. At least five enzymes have been identified by electrophoresis. Each enzyme is composed of four units. The units are of two types, H units found in heart lactic dehydrogenase and M units characteristic of the muscle dehydrogenase. It is possible

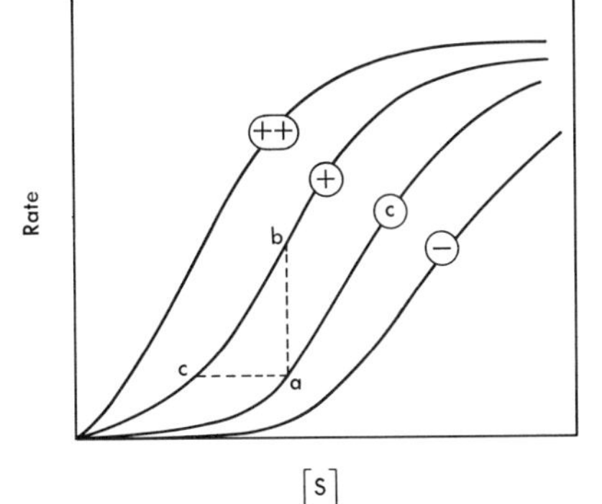

Figure 3.10. Generalized substrate response curves for a regulatory enzyme. The rate of the reaction catalyzed by the enzyme is plotted on the vertical coordinate, and substrate concentration on the horizontal. Curve identifications: c, control; −, negative effector added; +, positive effector added; ++, higher concentration of positive effector. (From D. E. Atkinson: *Science*, **150:**851, 1965.)

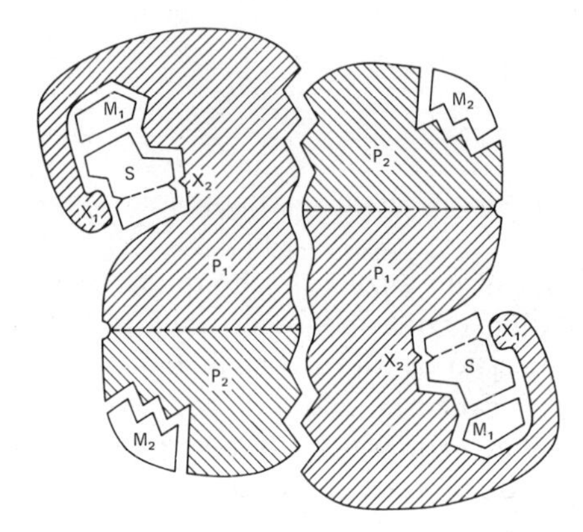

Figure 3.11. Schematic illustration of an enzyme containing identical subunits in which the bond to be broken in the substrate (dotted lines) is in contact with two catalytic residues, X_1 and X_2. Modifiers at the active site (M_1) and the allosteric site (M_2) aid in maintaining the proper conformation of the enzyme. P_1 and P_2 may represent two different polypeptide chains or parts of the same polypeptide chain. (From D. E. Koshland and K. E. Neet, *Ann. Rev. Biochem.*, **37**:364, 1968.)

then to have enzymes composed of HHHH, HHHM, HHMM, HMMM, and MMMM. All five have been found. The heart enzyme is composed of four H units, and the muscle enzyme is composed of four M units. The H_4 enzyme has a molecular weight of 151,000 and a K_m equal to 8.9×10^{-5}, whereas M_4 has a molecular weight of 140,000 and a K_m of 3.2×10^{-3}. The heart enzyme is inhibited by pyruvate. The muscle enzyme is less readily inhibited.

PROSTHETIC GROUPS AND COENZYMES

In many instances enzymes are conjugated proteins containing nonamino acid portions. These groups are frequently called prosthetic groups; they are important in bringing about the combination of the enzyme with the substrate, and they form a part of the active center on the enzyme. In many cases the prosthetic group is easily removed from the protein by dialysis. The enzyme carboxylase dissociates into a protein (apoenzyme) and thiamine pyrophosphate (coenzyme or cofactor). Neither the protein nor the coenzyme alone has activity. For the enzymatic reaction to take place, the coenzyme must be intimately attached to the protein and serve as at least a portion of the "active site" on the enzyme surface. There is no clear distinction between a prosthetic group and a coenzyme. A distinction is often made on the basis of the degree of binding to the protein molecule.

Many of the coenzymes are derivatives of the vitamins. A coenzyme function is known for each of the water-soluble vitamins except vitamin C (page 113), but the biochemical function of some of the fat-soluble vitamins is unknown. As more is learned about these substances such functions may be discovered.

In our discussion of coenzymes, pyridoxal phosphate will be used an an example of one that is involved in amino acid metabolism. The coenzymes involved in dehydrogenation reactions also will be discussed here. Others such as folic acid and vitamin B_{12} are discussed in the chapter on nitrogen metabolism (pages 229, 240).

Pyridoxal Phosphate

Pyridoxal, pyridoxine, and pyridoxamine constitute a family of water-soluble vitamins required for the growth of some microorganisms. There is no known nutritional requirement for this vitamin in man.

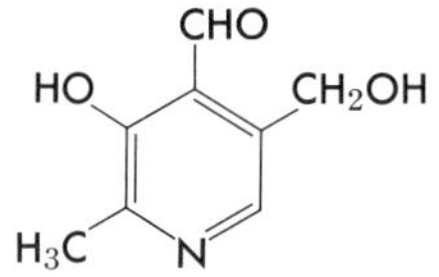

Pyridoxal

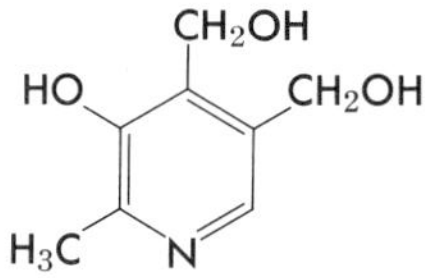

Pyridoxine

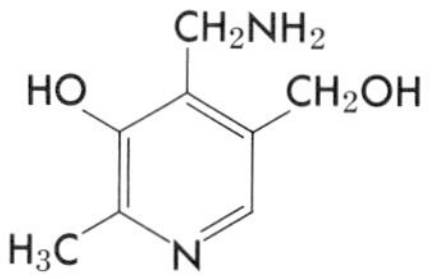

Pyridoxamine

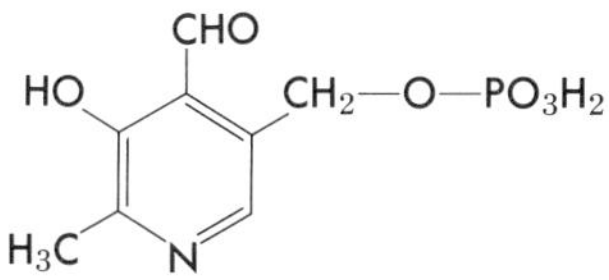

Pyridoxal phosphate

Pyridoxal phosphate functions as the coenzyme for several apoenzymes involved in amino acid metabolism. These enzymes may catalyze decarboxylation, transamination, elimination of H_2O or H_2S, and racemization. The key intermediate is probably the Schiff base, in which the cationic nitrogen atom of the pyridine ring acts as an electron sink to withdraw electrons.

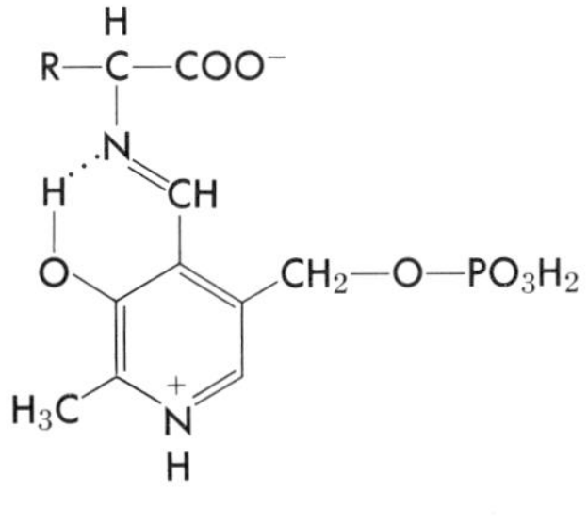

The reaction which takes place will depend on the nature of R and the apoenzyme to which pyridoxal phosphate is attached. This is another example which can be explained by assuming specific binding sites on the enzyme. In this case the specific binding is responsible for the specific activation of certain bonds in the amino acid. The enzyme bringing about the decarboxylation of tyrosine would cause a shift of electrons so that CO_2 would be eliminated

$$\text{HO}-\text{C}_6\text{H}_4-\text{CH}_2-\underset{\underset{\underset{\underset{\text{R}}{|}}{\text{CH}}}{\overset{\|}{\text{N}}}}{\overset{\text{H}}{\overset{|}{\text{C}}}}-\text{COOH} \xrightarrow[\text{decarboxylase}]{\text{tryosine}} \text{HO}-\text{C}_6\text{H}_4-\text{CH}_2\text{CH}_2\text{NH}_2 + \text{Pyridoxal phosphate} + \text{CO}_2$$

Tyrosine → Tyramine + Pyridoxal phosphate + CO_2

When serine or cysteine is involved, there is elimination of the β substituent along with a hydrogen from the α position.

$$\overset{\text{OH}}{\overset{|}{\text{CH}_2}}-\underset{\underset{\underset{\underset{\text{R}}{|}}{\text{CH}}}{\overset{\|}{\text{N}}}}{\overset{\text{H}}{\overset{|}{\text{C}}}}-\text{COOH} \xrightarrow[\text{dehydrase}]{\text{serine}} \underset{\underset{\text{CH}_3}{|}}{\overset{\overset{\text{COOH}}{|}}{\text{C=O}}} + \text{NH}_3 + \text{R}-\text{CHO}$$

Pyruvic acid; Pyridoxal phosphate

An extremely important reaction in which pyridoxal phosphate serves as the coenzyme is transamination.

$$\text{R}^1-\underset{\underset{\text{NH}_2}{|}}{\text{CH}}-\text{COOH} + \text{RCHO (Pyridoxal phosphate)} \rightleftharpoons \text{R}^1-\underset{\underset{\text{O}}{\|}}{\text{C}}-\text{COOH} + \text{RCH}_2\text{NH}_2 \text{ (Pyridoxamine phosphate)}$$

$$\text{RCH}_2\text{NH}_2 \text{ (Pyridoxamine phosphate)} + \text{R}^2-\underset{\underset{\text{O}}{\|}}{\text{C}}-\text{COOH} \rightleftharpoons \text{RCHO (Pyridoxal phosphate)} + \text{R}^2-\underset{\underset{\text{NH}_2}{|}}{\text{CH}}-\text{COOH}$$

The amino group is transferred from an amino acid to a keto acid. The reaction is fully reversible.

Oxidative Decarboxylation

At this point those coenzymes involved in the reaction of pyruvic acid to acetic acid and CO_2 will be considered. This is a key reaction in aerobic oxidative decarboxylation. The mechanism for this reaction is very similar to that for the oxidative decarboxylation of other α-keto acids; for example, α-ketoglutaric acid.

$$\underset{\underset{\text{CH}_3}{|}}{\overset{\overset{\text{COOH}}{|}}{\text{C=O}}} + \tfrac{1}{2}\text{O}_2 \longrightarrow \overset{\overset{\text{COOH}}{|}}{\text{CH}_3} + \text{CO}_2$$

Pyruvic acid; Acetic acid

Cocarboxylase

Thiamine is one of the water-soluble vitamins that is essential for man. A deficiency of this vitamin in the diet leads to the development of beriberi. Thiamine is converted in the liver by thiamine-kinase to thiamine pyrophosphate, the active form of the vitamin.

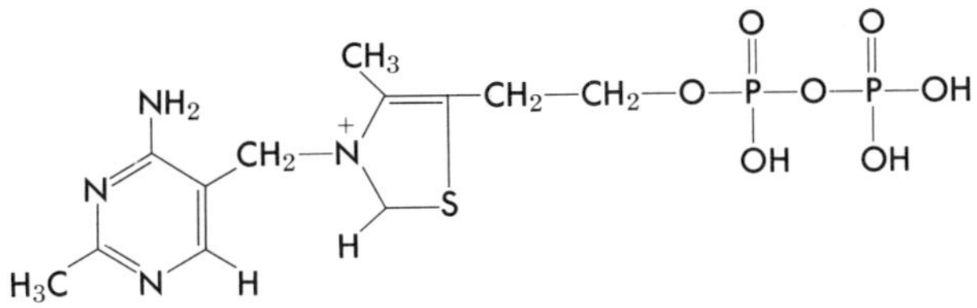

Thiamine pyrophosphate (cocarboxylase) (TPP)

Decarboxylases that catalyze the decarboxylation of pyruvic acid to acetaldehyde and CO_2 are known. In mammalian tissue, however, the decarboxylation requires an oxidative process with the participation of lipoic acid as well as thiamine pyrophosphate.

The following mechanism for participation of thiamine pyrophosphate as a coenzyme has been proposed.

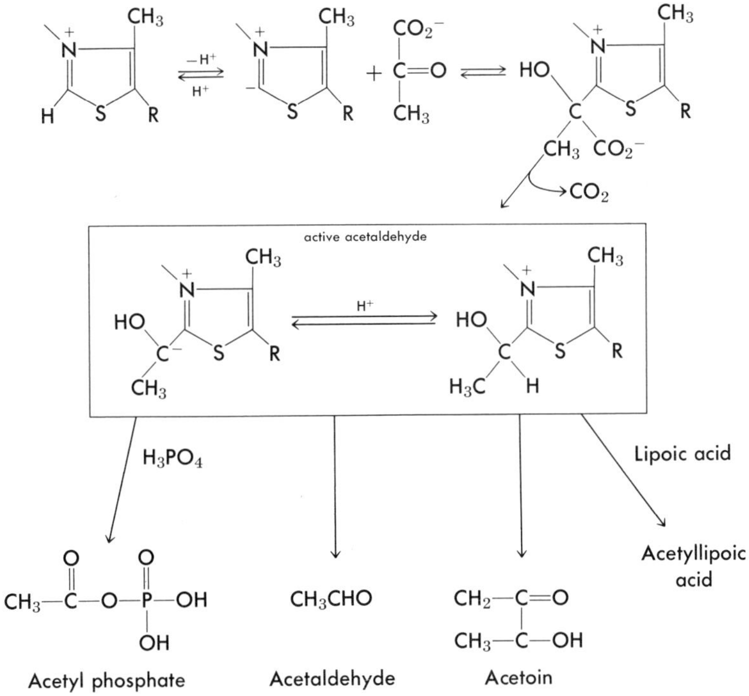

active acetaldehyde

H_3PO_4

Lipoic acid

Acetyllipoic acid

CH_3CHO

Acetyl phosphate

Acetaldehyde

Acetoin

Lipoic acid is 6,8-dithiooctanoic acid and has been called the "pyruvate oxidation factor," "protogen," and "acetate replacement factor." It is required for the oxidation

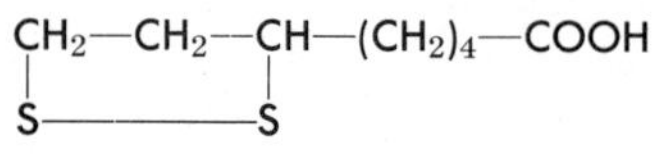

$$\begin{array}{l} CH_2-CH_2-CH-(CH_2)_4-COOH \\ |\qquad\qquad\quad | \\ S\text{———————}S \end{array}$$

of pyruvate by *S. faecalis* and can replace acetate for the growth of *L. casei*. In the oxidative decarboxylation of pyruvate in animal tissues, the disulfide linkage is reduced to form reduced lipoic acid, and the two-carbon unit representing the

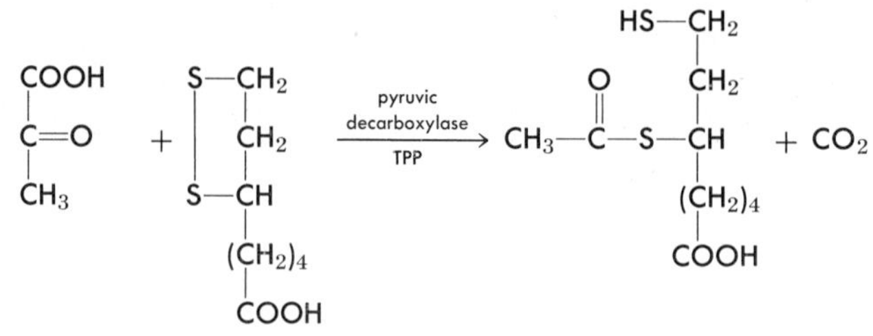

$$\begin{array}{c} COOH \\ | \\ C{=}O \\ | \\ CH_3 \\ \text{Pyruvic acid} \end{array} + \begin{array}{c} S-CH_2 \\ |\qquad | \\ |\quad CH_2 \\ |\qquad | \\ S-CH \\ | \\ (CH_2)_4 \\ | \\ COOH \\ \text{Lipoic acid} \end{array} \xrightarrow[\text{TPP}]{\text{pyruvic decarboxylase}} \begin{array}{c} HS-CH_2 \\ | \\ CH_2 \\ O\qquad | \\ \| \qquad | \\ CH_3-C-S-CH \\ | \\ (CH_2)_4 \\ | \\ COOH \\ \text{Acetyldihydrolipoic acid} \end{array} + CO_2$$

α- and β-carbons of pyruvate appear as the thioester of acetate. In other words, there has been an oxidation from the level of acetaldehyde to an ester of acetic acid.

If lipoic acid is to act catalytically in the overall reaction, it is necessary to remove the acetyl group and oxidize the reduced lipoic acid to the oxidized form. Two addition coenzymes are involved in these two transformations: one to accept the acetyl group and the other to accept the two electrons.

Coenzyme A

Pantothenic acid was first recognized as a growth factor for certain microorganisms. Coenzyme A is a complex nucleotide having pantothenic acid as a portion of its structure. Lipmann found it was required for a certain acetylation reaction. The structure is as follows

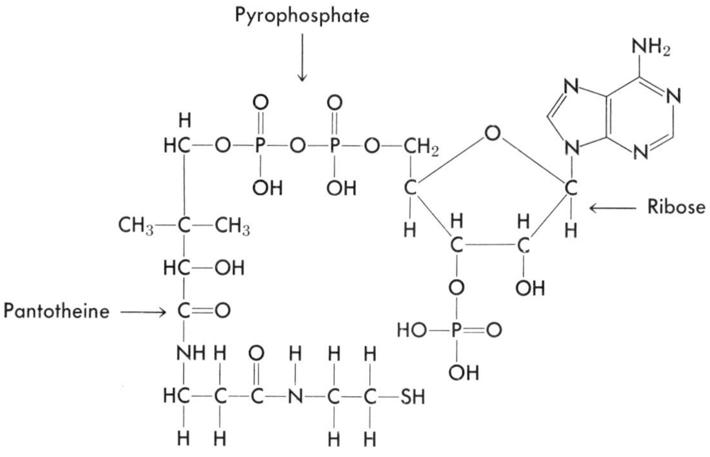

Pyrophosphate

Pantotheine

Ribose

Coenzyme A

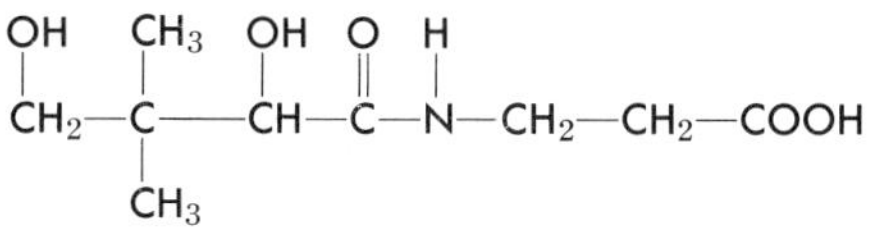

$HOCH_2-C(CH_3)_2-CH(OH)-CO-NH-CH_2-CH_2-COOH$

Pantothenic acid

The presence of a nucleotide in the structure is a common feature of many of the coenzymes. The functional group of coenzyme A is the —SH group, and the structure is abbreviated to CoASH. It reacts with acetyl dihydrolipoic acid as follows

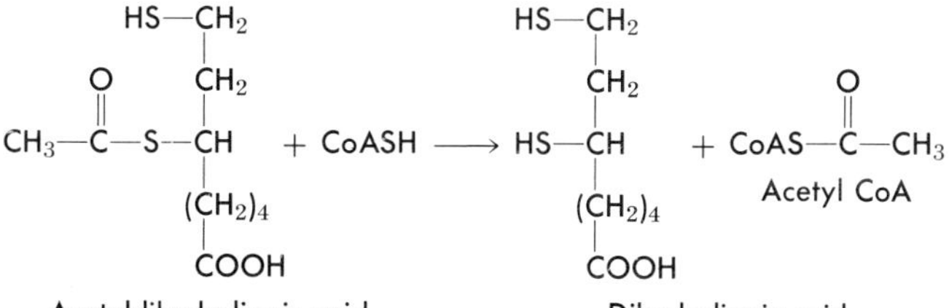

$$CH_3-CO-S-CH(CH_2CH_2SH)-(CH_2)_4-COOH + CoASH \longrightarrow HS-CH(CH_2CH_2SH)-(CH_2)_4-COOH + CoAS-CO-CH_3$$

Acetyldihydrolipoic acid · Dihydrolipoic acid · Acetyl CoA

Acetyl CoA enters into acetylation and condensation reactions. It is the key compound for introducing acetate into the tricarboxylic acid cycle, for the synthesis of fats, for acetylation, for synthesis of steroids, and for other important metabolic reactions which will be discussed.

The reduced form of lipoic acid is then oxidized to lipoic acid. The two electrons are transferred through a number of steps to molecular oxygen. The coenzymes involved in this series of reactions constitute the electron transport system.

Nicotinamide Adenine Dinucleotide

NAD^+ is the first in the series. Again, it is a nucleotide having the vitamin niacin as a part of its structure. A deficiency of niacin was recognized as being responsible for pellagra in man and blacktongue in dogs. The structure of NAD^+ is as follows

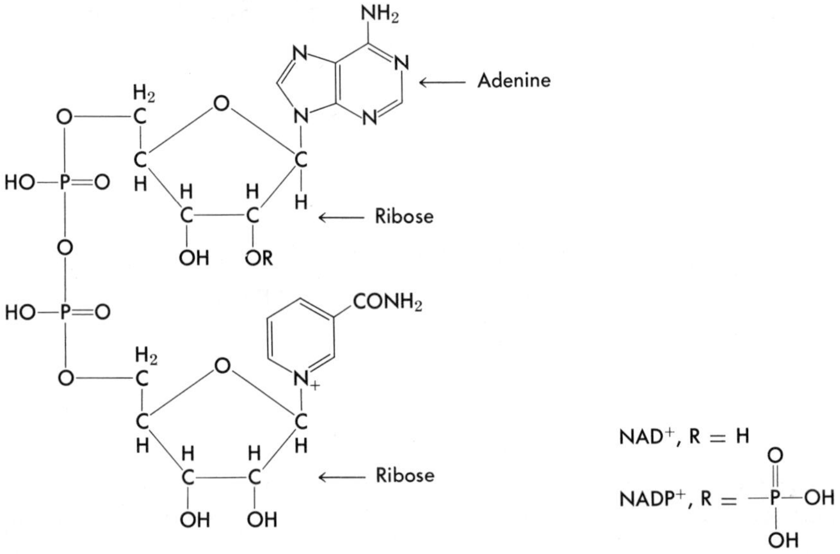

NH_2 ← Adenine

HO—P=O

HO—P=O

← Ribose

$CONH_2$

OH OR

OH OH

← Ribose

NAD^+, R = H

$NADP^+$, R = $-P(=O)(OH)-OH$

Nicotinamide adenine dinucleotide phosphate ($NADP^+$) has a third phosphate ester group, as shown. In the older literature NAD^+ was designated DPN^+ and $NADP^+$ was termed TPN^+. When NAD^+ and $NADP^+$ are used in a text discussion hereafter, the plus charges will be omitted and they will be shown as NAD and NADP; however, the plus charges will be retained when they appear in equations.

The function of NAD is illustrated in the oxidation of reduced lipoic acid.

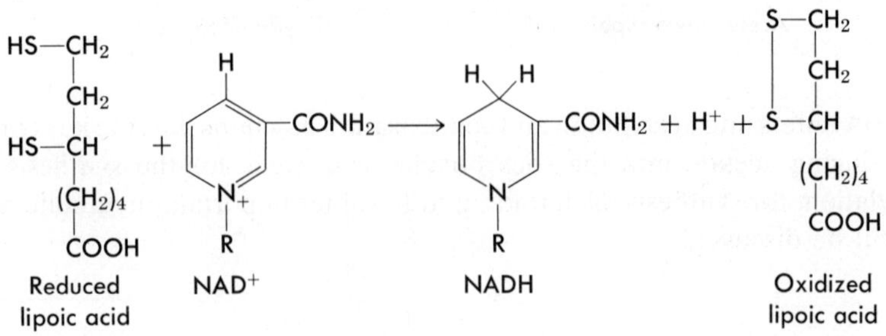

$$HS-CH_2-CH_2-CH(SH)-(CH_2)_4-COOH + NAD^+ \longrightarrow NADH + H^+ + \text{oxidized lipoic acid}$$

Reduced lipoic acid | NAD^+ | NADH | Oxidized lipoic acid

NAD, and in some important cases NADP, can accept electrons from many substrates. The enzymes involved are dehydrogenases and the specificity is due to the apoenzyme (protein) portion of the enzyme. The lactic dehydrogenase of muscle and other tissues is specific for L-lactic acid and has NAD as its coenzyme.

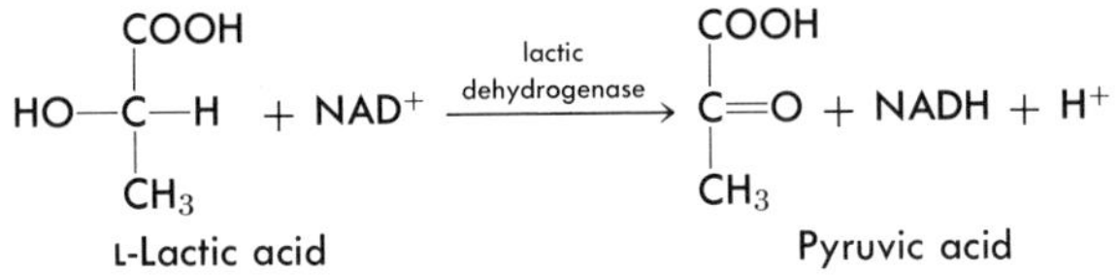

$$\underset{\text{L-Lactic acid}}{HO{-}\overset{COOH}{\underset{CH_3}{C}}{-}H} + NAD^+ \xrightarrow{\text{lactic dehydrogenase}} \underset{\text{Pyruvic acid}}{\overset{COOH}{\underset{CH_3}{C}}{=}O} + NADH + H^+$$

A system introduced by Baldwin will be used for writing such reactions.

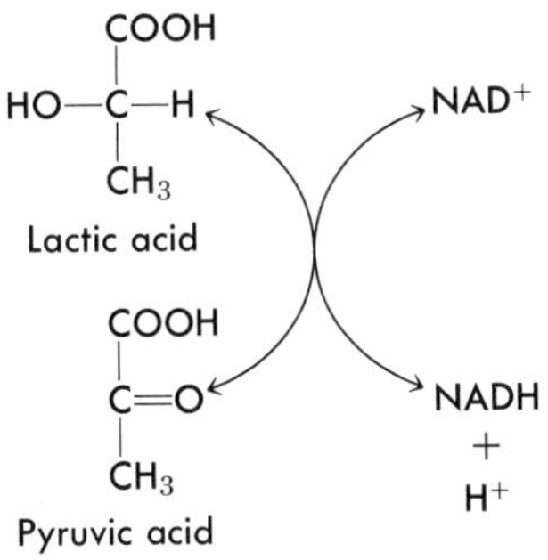

$$\underset{\text{Lactic acid}}{HO{-}\overset{COOH}{\underset{CH_3}{C}}{-}H} \rightarrow NAD^+$$

$$\underset{\text{Pyruvic acid}}{\overset{COOH}{\underset{CH_3}{C}}{=}O} \leftarrow NADH + H^+$$

Flavoproteins are electron acceptors that contain the vitamin riboflavin. Riboflavin is required for growth of the rat. It is the fluorescent material in whey and occurs in many foods. Pellagra and beriberi are usually complicated by a riboflavin deficiency. The functional coenzyme exists in two forms called flavin mononucleotide (FMN) and flavin dinucleotide (FAD). Although FMN is a phosphate ester, it is not a true nucleotide. Also, it should be noted that these compounds contain the sugar alcohol ribitol and not ribose.

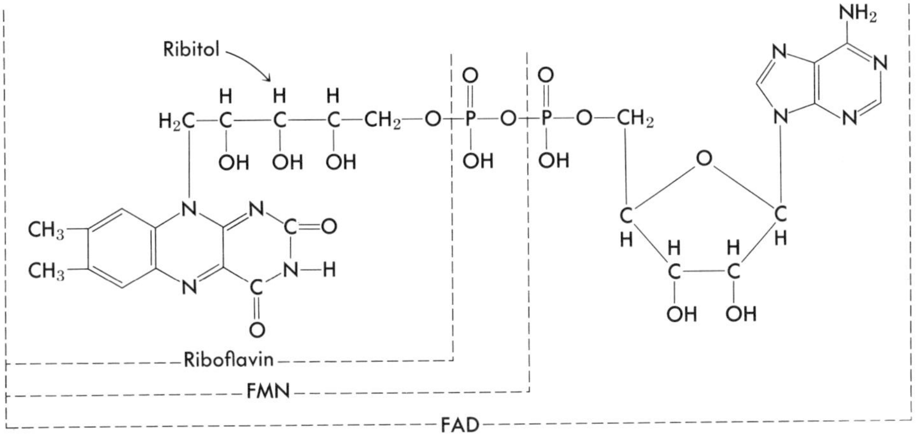

Ribitol

Riboflavin

FMN

FAD

The functional portion of the molecule is in the isoalloxazine ring structure and is illustrated by the oxidation of reduced NAD.

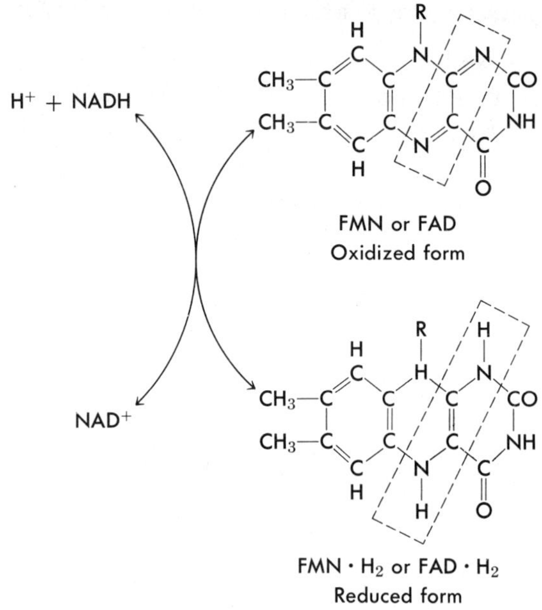

H^+ + NADH

NAD^+

FMN or FAD
Oxidized form

$FMN \cdot H_2$ or $FAD \cdot H_2$
Reduced form

Starting with dihydrolipoic acid, we can now write

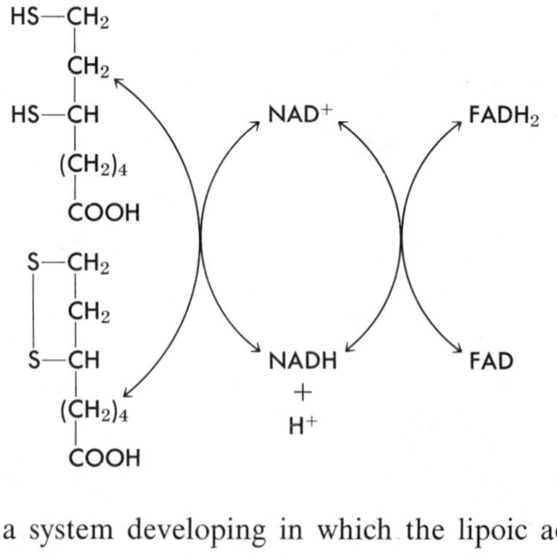

NAD^+ $FADH_2$

NADH + H^+ FAD

Here we can see a system developing in which the lipoic acid can be reduced by pyruvic acid and oxidized by NAD. As long as there is sufficient oxidized NAD in the system, one molecule of lipoic acid will be responsible for the oxidation of many molecules of pyruvic acid. The same argument can be applied to NAD. By keeping FAD in the oxidized form, NAD will perform in a catalytic manner. Actually cells have a system of cytochrome enzymes which accept the electrons from FAD and transfer them to molecular oxygen.

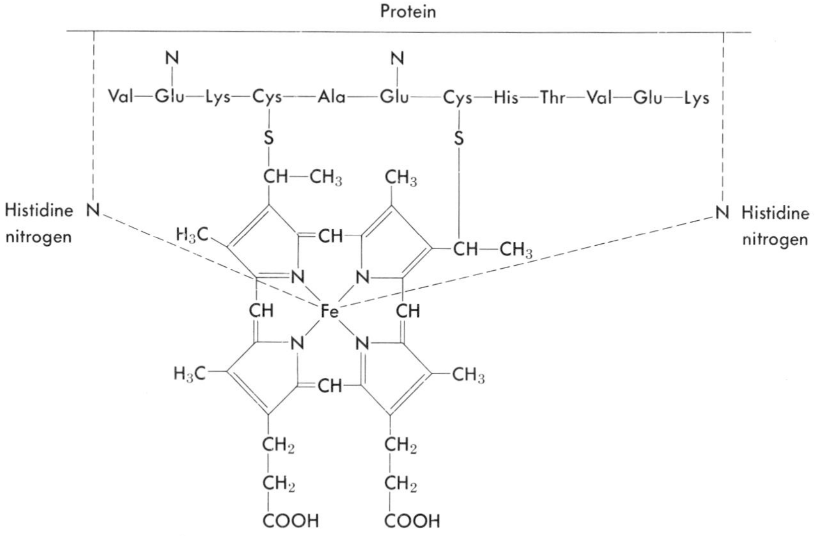

Cytochrome C

Cytochromes

The cytochromes are heme-containing proteins in which iron is the functional portion of the molecule. The structure of cytochrome c is shown. The reaction $Fe^{+++} + e^{-} \rightleftharpoons Fe^{++}$ is responsible for the transport of the electrons in these substances (see page 93).

The following represents a respiratory chain containing three cytochrome enzymes.

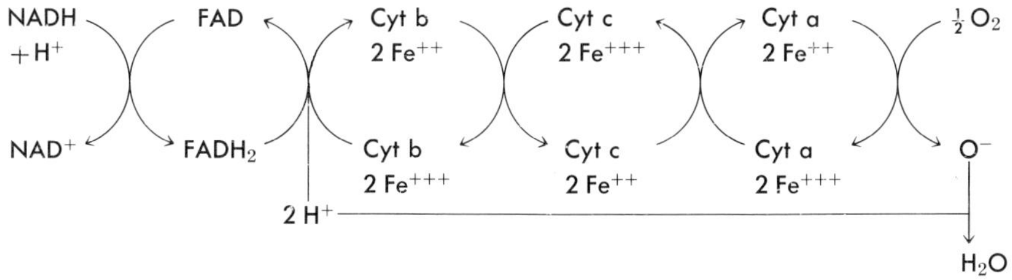

This system will function with catalytic quantities of all the enzymes as long as substrate and oxygen are supplied. This chain is very similar to the flow of electrons along a wire. There is a potential difference of over 1 volt between the substrate and oxygen. The transfer of the electrons by steps makes it possible for the cell to tap into the circuit and use the energy for its work (Chapter 4).

Returning to the oxidative decarboxylation of pyruvic acid, the following is a summary of these reactions.

$$\underset{\text{Pyruvate}}{CH_3-CO-COOH} + TPP + \underset{\text{Lipoic acid}}{\text{(S–S)R–COOH}} \longrightarrow \underset{\text{Acetyl lipoic acid}}{CH_3-\overset{O}{\overset{\|}{C}}-S\text{(HS)R–COOH}} + CO_2 + \underset{\text{Thiamine pyrophosphate}}{TPP}$$

$$\underset{\text{Acetyl lipoic acid}}{CH_3\overset{O}{\overset{\|}{C}}-S\text{(HS)R–COOH}} + \underset{\text{Coenzyme A}}{CoASH} \longrightarrow \underset{\text{Acetyl CoA}}{CH_3\overset{O}{\overset{\|}{C}}-SCoA} + \underset{\text{Reduced lipoic acid}}{\text{(HS)(HS)R–COOH}}$$

$$\underset{\text{Reduced lipoic acid}}{\text{(HS)(HS)R–COOH}} + NAD^+ \longrightarrow \underset{\text{Oxidized lipoic acid}}{\text{(S–S)R–COOH}} + NADH + H^+$$

$$NADH + H^+ + FAD \longrightarrow FADH_2 + NAD^+$$

$$FADH_2 + 2(Fe^{+++}Cyt\ b) \longrightarrow FAD + 2(Fe^{++}Cyt\ b) + 2\ H^+$$

$$2(Fe^{++}Cyt\ b) + 2(Fe^{+++}Cyt\ c) \longrightarrow 2(Fe^{+++}Cyt\ b) + 2(Fe^{++}Cyt\ c)$$

$$2(Fe^{++}Cyt\ c) + 2(Fe^{+++}Cyt\ a) \longrightarrow 2(Fe^{++}Cyt\ a) + 2(Fe^{+++}Cyt\ c)$$

$$2(Fe^{++}Cyt\ a) + \tfrac{1}{2}O_2 \longrightarrow 2(Fe^{+++}Cyt\ a) + O^-$$

$$O^- + 2\ H^+ \longrightarrow H_2O$$

Summation

$$CH_3-CO-COOH + CoASH + \tfrac{1}{2}O_2 \longrightarrow CH_3-\overset{O}{\overset{\|}{C}}-SCoA + CO_2 + H_2O$$

The coenzymes which were used are thiamine pyrophosphate, lipoic acid, NAD, FAD, and cytochromes.

CELLULAR ORGANIZATION OF ENZYMES

In discussing metabolism, we will have occasion to refer to enzyme activities in the cell and to locate the activity in special parts of the cell. Figure 3.12 shows the organization of the typical cell. Special note should be taken of the mitochondria, ribosomes, nucleus, membranes, and cytoplasm. Specific enzymatic reactions can be attributed to each of these structures. The electron transport system and most oxidative enzymes are found in the mitochondria. There are two pathways for the synthesis of fatty acids: one in the mitochondria and the other in the cytoplasm. Protein synthesis takes place on the ribosomes. Cells also contain small granules called lysosomes. These contain digestive or hydrolytic enzymes that can break down large molecules into their smaller constituents. Rupture of the lysosomes results in lysis of the cell.

The cell is a highly organized structure and cannot be considered as a bag of enzymes. It is known that structures such as the mitochondria are also highly organized and that the various enzymes are held in a spatial relationship to one another in such a manner that efficient operation and control are possible.

Cells can be broken by grinding with sand or other abrasives, by freezing and thawing, by rapid change in pressure, by high-frequency sound waves, by a change in osmotic pressure, and by enzymic treatment. The choice of method will depend on the type of cell to be broken. Red blood cells are readily broken by the change

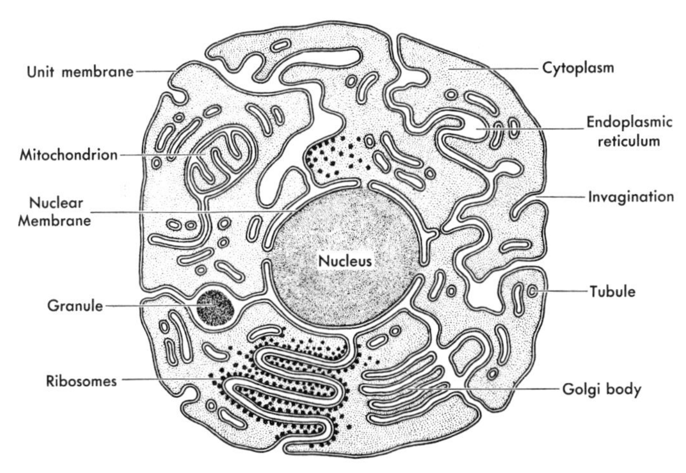

Figure 3.12. Schematic diagram of a cell. (From "The Membrane of the Living Cell" by J. David Robertson. Copyright © 1962 by Scientific American, Inc. All rights reserved.

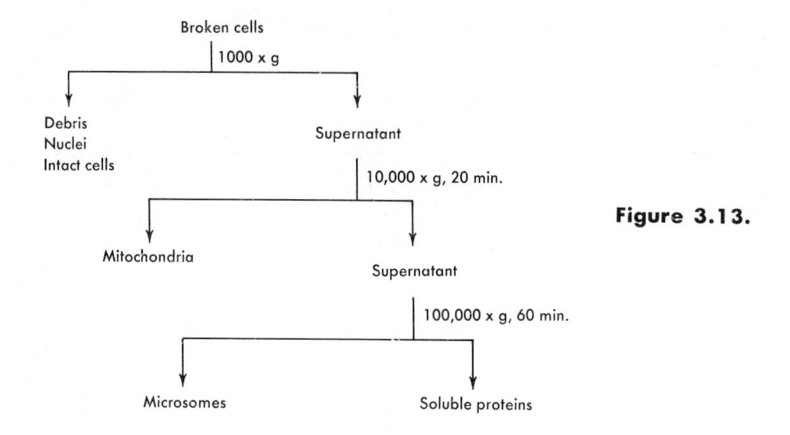

Figure 3.13.

in osmotic pressure when blood is diluted with water. The disruption of bacteria can be difficult and may require strong forces.

Broken cells can be fractionated by ultracentrifugation into various components. Figure 3.13 shows a scheme for such a separation.

REFERENCES

Colowick, S. P., and Kaplan, N. O. (eds.): *Methods in Enzymology,* Vols. 1–16. Academic Press, New York, 1955–1969.

Dixon, M., and Webb, E. C.: *Enzymes.* Academic Press, New York, 1964.

International Union of Biochemistry: *Report of the Commission on Enzymes.* Pergamon Press, Oxford; The Macmillan Co., New York, 1961.

Mahler, H. R., and Cordes, E. H.: *Biological Chemistry,* Chaps. 6, 7, 8, and 9. Harper and Row, New York, 1966.

Nord, F. F. and Werkman, C. H. (eds.): *Advances in Enzymology,* Vols. 1–31. Interscience Publishers, New York, London, 1941–1969.

Symposium on enzyme conformation in relation to controlled activity. *Fed. Proc.,* **23:**719–35, 1964.

Symposium on mechanism of action of coenzymes. *Fed. Proc.,* **20:**956–1021, 1961.

West, E. S.; Todd, W. R.; Mason, H. S.; and Van Bruggen, J. T.: *Textbook of Biochemistry,* 4th ed., Chap. 11. The Macmillan Co., New York, 1966.

Chapter 4

HIGH-ENERGY COMPOUNDS AND OXIDATIVE PHOSPHORYLATION

PLANTS manufacture glucose by using solar energy to combine CO_2 and H_2O molecules. The glucose molecule, then, possesses potential chemical energy in the same way as water behind a dam or a boulder poised on a hillside possess potential mechanical energy.

Oxidation of the glucose molecule, whether by combustion or by a series of physiological reactions at 37°C results in the liberation of energy. The means by which that energy is released determines whether that energy performs useful work or is wasted as light and heat resulting from combustion. The oxidative reactions which occur during the catabolism of glucose in animals release energy in a usable form for use in anabolic reactions vital to growth and reproduction.

In this chapter we shall consider the energy yields from metabolic oxidation reactions and the capture and storage of that energy in high-energy compounds.

THE FREE ENERGY OF OXIDATION-REDUCTION REACTIONS

In Chapter 3 we learned that the conversion of reactants to products could release energy for useful work if the change in free energy (ΔF) was negative. The same considerations hold for oxidation-reduction reactions in which there are electrons transferred and ΔF is directly related to the differences in potential, ΔE_0, of the reactants.

The standard reference electrode to which these reactants are compared is the hydrogen electrode which is arbitrarily assigned a value of $E_0 = 0.000$ volt at pH 0. Since a proton is formed in the reaction

$$\frac{1}{2} H_2 \rightleftharpoons H^+ + 1 \text{ electron}$$

the oxidation-reduction potential at pH 7.0 will be -0.420 volt and is designated as E_0'. This may be calculated from the relationship:

$$E' = E_0 + \frac{RT}{nf} \ln \frac{\text{(Oxidized)}}{\text{(Reduced)}} \tag{4.1}$$

where R is the gas constant, T the absolute temperature, n the number of electrons transferred, and f the Faraday. When n equals 1 and the temperature is 20°C, the expression becomes

$$E' = E_0 + 0.06 \log \frac{\text{(Oxidized)}}{\text{(Reduced)}} \tag{4.2}$$

and for the special case at pH 7.0 for H_2 at 1 atm

$$E_0' = E_0 + 0.06 \log \frac{(10^{-7})}{(1)} = -0.420$$

The observed potential for the system Fe^{+++} + 1 electron $\rightleftharpoons$ Fe^{++} compared to the hydrogen electrode (E_0') is +0.75 volt. This is E_0' and is the voltage obtained in a half-cell containing unit concentrations of Fe^{+++} and Fe^{++} compared to the standard hydrogen half-cell at pH 7.0. It may be calculated from equation 4.2 that, when the ratio of $(Fe^{+++})/(Fe^{++}) = 10/1$, the potential is +0.81 volt; and when the ratio is 1/10, the potential is 0.69 volt. In biologic systems the concentrations of the oxidized and reduced forms are usually maintained at ratios other than unity. In considering a table of oxidation-reduction potentials, such as Table 4.1, it should be realized that these are values for unit concentrations and that the steady-state situation in a cell will differ from these, the difference depending on concentrations.

When a proton enters into the oxidation-reduction reaction, the potential of the systems will depend on the concentration of H^+ as well as the concentrations of the reduced and oxidized forms of the substance. Equation 4.1 can be used to calculate this effect in the same manner as it was used to calculate the value of the H_2 potential at pH 7.0. In biologic systems the pH is usually near 7.0 and the values we will use are E_0' values.

Table 4.1
E_0' Values for Some Biologic Systems

O_2/H_2O	0.82 volt
O_2/H_2O_2	0.30 volt
Cytochrome a; Fe^{+++}/Fe^{++}	0.29 volt
Cytochrome c; Fe^{+++}/Fe^{++}	0.22 volt
Cytochrome b; Fe^{+++}/Fe^{++}	0.12 volt
Ubiquinone; oxid/red	0.10 volt
Yellow enzyme; $FMN/FMNH_2$	−0.12 volt
Oxalacetate/malate	−0.17 volt
Pyruvate/lactate	−0.19 volt
Lipoic acid; oxid/red	−0.29 volt
NAD/NADH	−0.32 volt
H^+/H_2	−0.42 volt

Those substances with the more positive potentials, for example, $Fe^{++}/Fe^{+++} = +0.75$, are good oxidizing agents and will accept electrons from substances of lower potential. The substances with low values are good reducing agents.

It is possible to derive the equation $\Delta F_0' = -nf\Delta E_0'$, where $\Delta F_0'$ is the standard free energy change of the reaction, n the number of electrons transferred, f the Faraday (23,000 cal per volt equivalent), and $\Delta E_0'$ the difference in potential between the oxidizing and reducing systems at pH 7.0. Let us consider the overall reaction of $NADH + H^+ + \frac{1}{2}O_2 \rightleftharpoons NAD^+ + H_2O$. The potential for NADH/NAD is -0.320 and for $H_2O/\frac{1}{2}O_2$ is 0.816. The difference $\Delta E_0'$ between the oxidizing agent and reducing agent, is $0.816 - (-0.320) = 1.136$. It follows that

$$\begin{aligned} \Delta F &= -(2 \times 23{,}000)(1.136) \\ &= -52{,}256 \text{ cal.} \end{aligned}$$

and in the reaction of

$$CoASH + NAD^+ + CH_3{-}\overset{\overset{\displaystyle O}{\|}}{C}{-}COOH \longrightarrow CH_3{-}\overset{\overset{\displaystyle O}{\|}}{C}{-}S{-}CoA + CO_2 + H_2O + NADH + H^+$$

it is possible for the cell to have available 52,256 calories from the oxidation of 1 mole of NADH. It should be emphasized that a knowledge of the free energy change tells us the amount of energy available if the transformation takes place. It does not give any information about the chemical mechanism of the conversion nor does it tell us anything about how the energy is utilized. NADH is stable in the presence of oxygen and is oxidized only when the catalysts and necessary chemical machinery are present for doing the job. We have seen how the electrons are transferred by the respiratory chain from NADH to the flavin system and through the cytochromes to oxygen.

As was pointed out earlier, the reaction of pyruvic acid to acetyl CoA and CO_2 represents one important step in the metabolism of glucose. In the oxidation of 1 mole of glucose to CO_2 and water there are 12 pairs of electrons available. A consideration of the chemical mechanisms by which the cell obtains energy from the transfer of these electrons as well as energy from nonoxidative reactions will constitute the section on carbohydrate metabolism (page 116). Before discussing this subject let us consider the question of how the organism can utilize the energy.

HIGH-ENERGY COMPOUNDS

Energy from physiological reactions is generally transferred and stored through the medium of high-energy compounds. Most of the known high-energy compounds contain phosphorous. It was the study of the role of phosphorous-containing compounds in the fermentation of glucose by yeast and in supplying energy for muscle contraction which led to an early appreciation of the significance of adenosine triphosphate and creatine phosphate as high-energy compounds. In the last few years, several other high-energy compounds of biologic importance have been found.

High-energy compounds are compounds which, on hydrolysis, release a large amount of free energy. In general, these compounds are unstable to acid and alkali. The key compound is adenosine triphosphate.

AMP
ADP
ATP

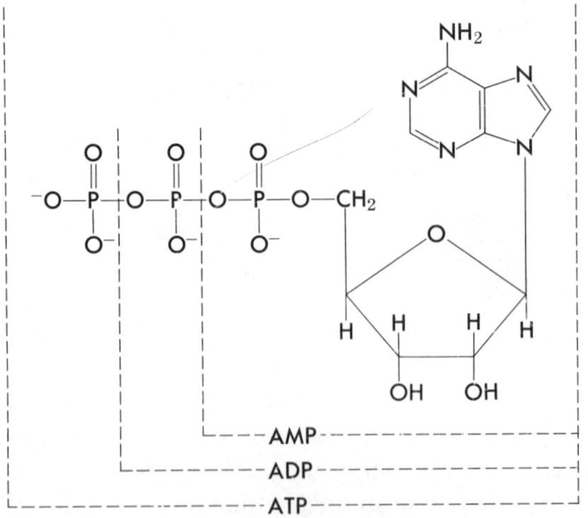

AMP—Adenosine monophosphate
ADP—Adenosine diphosphate
ATP—Adenosine triphosphate

In this compound the linked phosphate groups form acid anhydrides analogous to acetic anhydride

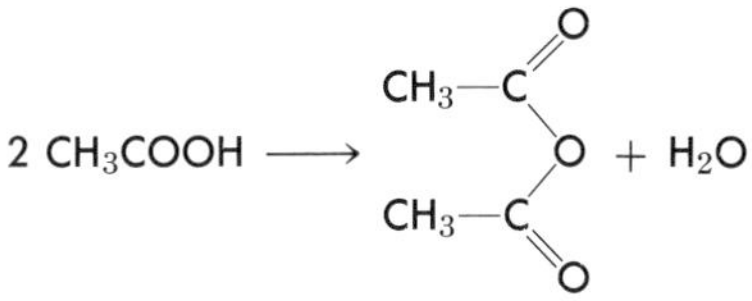

$$2\ CH_3COOH \longrightarrow \begin{matrix} CH_3-C(=O) \\ \quad\ \ \diagdown O \\ \quad\ \ \diagup \\ CH_3-C(=O) \end{matrix} + H_2O$$

Several others of the high-energy compounds described later contain anhydride structures, except that they are mixed anhydrides of phosphoric acid and a carboxylic acid (1,3-diphosphoglyceric acid, acetyl phosphate), or of phosphoric acid and carbamic acid (carbamyl phosphate).

Hydrolysis of ATP to ADP and inorganic phosphate liberates about 7000 calories.

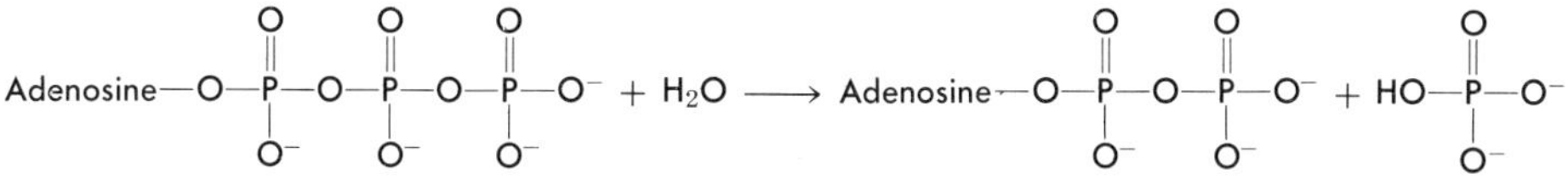

$$\text{Adenosine}-O-\underset{O^-}{\overset{O}{\overset{\|}{\underset{|}{P}}}}-O-\underset{O^-}{\overset{O}{\overset{\|}{\underset{|}{P}}}}-O-\underset{O^-}{\overset{O}{\overset{\|}{\underset{|}{P}}}}-O^- + H_2O \longrightarrow \text{Adenosine}-O-\underset{O^-}{\overset{O}{\overset{\|}{\underset{|}{P}}}}-O-\underset{O^-}{\overset{O}{\overset{\|}{\underset{|}{P}}}}-O^- + HO-\underset{O^-}{\overset{O}{\overset{\|}{\underset{|}{P}}}}-O^-$$

$$\Delta F_0' = -7000 \text{ calories (pH} = 7.0)$$

About the same amount of energy is released in the hydrolysis of ADP to AMP and inorganic phosphate (Pi) or the hydrolysis of ATP to AMP and pyrophosphate (PPi).

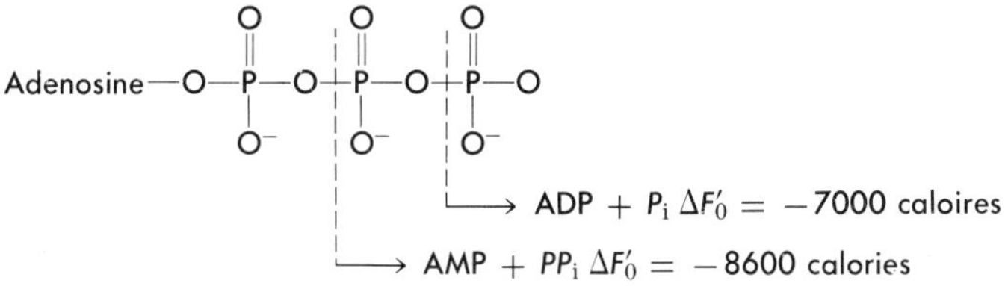

$$\text{Adenosine}-O-\underset{O^-}{\overset{O}{\overset{\|}{\underset{|}{P}}}}-O\,\vdots\,\underset{O^-}{\overset{O}{\overset{\|}{\underset{|}{P}}}}-O\,\vdots\,\underset{O^-}{\overset{O}{\overset{\|}{\underset{|}{P}}}}-O$$

$$\longrightarrow ADP + P_i\ \Delta F_0' = -7000 \text{ caloires}$$

$$\longrightarrow AMP + PP_i\ \Delta F_0' = -8600 \text{ calories}$$

Nucleotides of uridine, guanosine, cytidine, and inosine corresponding to ADP and ATP behave similarly and are also important in biologic systems (Chap. 8).

Other High-Energy Compounds

1. Other Phosphate-Containing Compounds

a. 1,3-Diphosphoglyceric acid

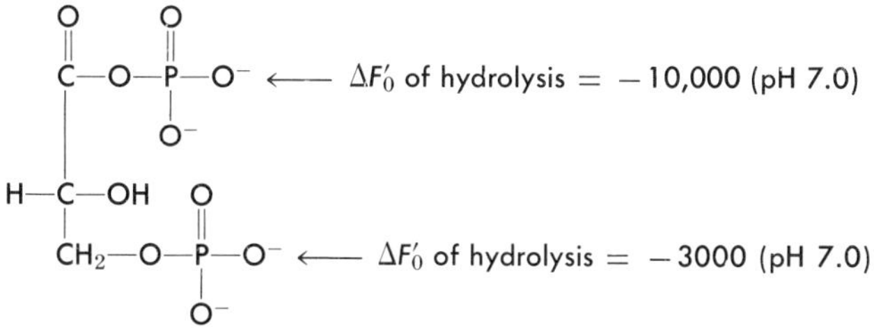

$$\begin{matrix} \overset{O}{\overset{\|}{C}}-O-\underset{O^-}{\overset{O}{\overset{\|}{\underset{|}{P}}}}-O^- & \longleftarrow \Delta F_0' \text{ of hydrolysis} = -10{,}000 \text{ (pH 7.0)} \\ | & \\ H-C-OH & \\ | & \\ CH_2-O-\underset{O^-}{\overset{O}{\overset{\|}{\underset{|}{P}}}}-O^- & \longleftarrow \Delta F_0' \text{ of hydrolysis} = -3000 \text{ (pH 7.0)} \end{matrix}$$

1,3-Diphosphoglyceric acid

In this compound the energy is released when cleavage takes place between the carboxyl group and the phosphate group. The energy of hydrolysis of the phosphate ester grouping in position 3 is about $\Delta F_0' = -3000$ calories. This is generally true for phosphate esters of alcohols; for example

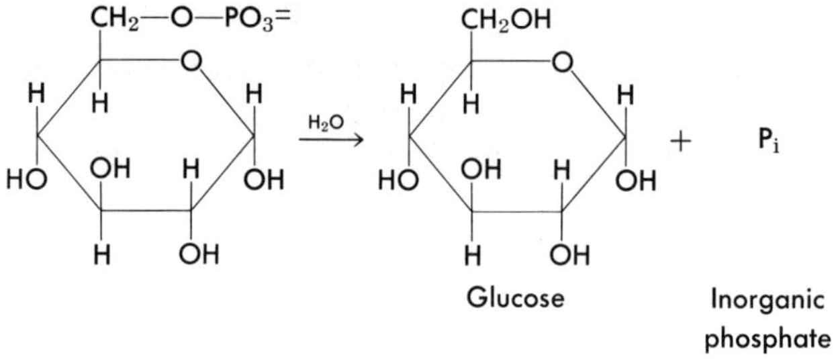

$$\text{Glucose 6-phosphate} \xrightarrow{H_2O} \text{Glucose} + P_i \text{ (Inorganic phosphate)}$$

$\Delta F_0' = -3000$ calories (pH = 7.0)

b. Phosphenolpyruvic acid

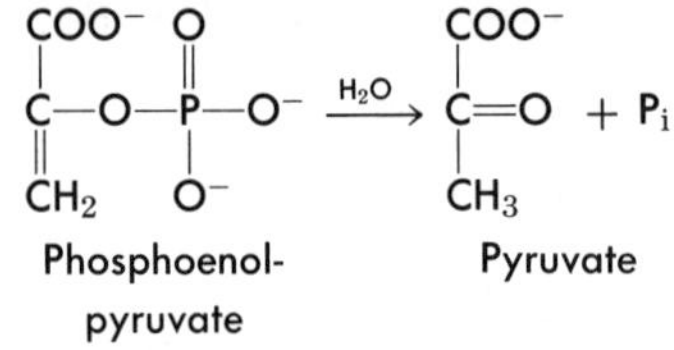

$$CH_2{=}C(COO^-){-}O{-}PO_3^{2-} \xrightarrow{H_2O} CH_3{-}C(=O){-}COO^- + P_i$$

Phosphoenolpyruvate → Pyruvate

$\Delta F_0' = -12{,}000$ calories (pH = 7.0)

This is a key intermediate in the metabolism of glucose. It is an enol phosphate, that is, a combination of phosphoric acid with an acidic enol.

c. Acetyl phosphate

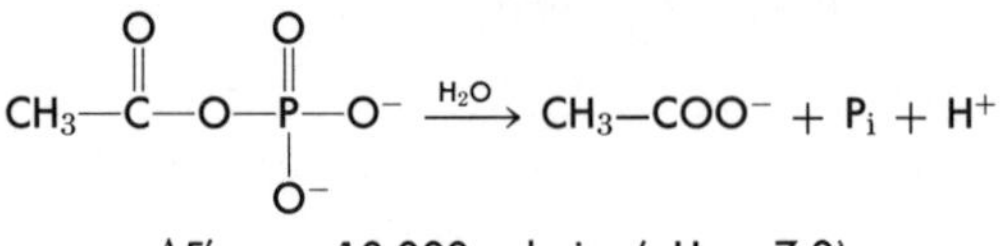

$$CH_3{-}C(=O){-}O{-}P(=O)(O^-){-}O^- \xrightarrow{H_2O} CH_3{-}COO^- + P_i + H^+$$

$\Delta F_0' = -10{,}000$ calories (pH = 7.0)

d. Carbamyl phosphate

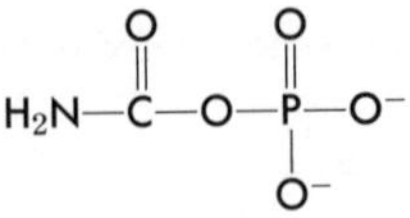

$$H_2N{-}C(=O){-}O{-}P(=O)(O^-){-}O^-$$

Carbamyl phosphate is an important compound in the synthesis of urea and pyrimidines.

2. Phosphorus-Nitrogen Compounds

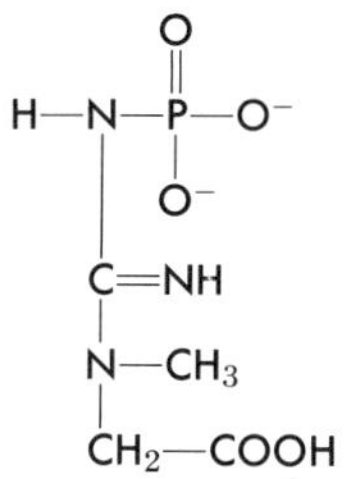

$$\begin{array}{l} \qquad\quad O \\ \qquad\quad \| \\ H{-}N{-}P{-}O^- \\ \quad\ | \quad | \\ \quad\ | \quad O^- \\ \quad\ C{=}NH \\ \quad\ | \\ \quad\ N{-}CH_3 \\ \quad\ | \\ \quad\ CH_2{-}COOH \end{array}$$

Creatine phosphate

$\Delta F'_0 = -10{,}000$ calories for hydrolysis at pH $= 7.0$

This compound represents an energy storage form which can be transferred to ADP to form ATP; or in a situation where ATP is being generated faster than it is being used, the muscle can store the energy in the form of creatine phosphate.

$$\text{ADP} + \text{Creatine phosphate} \rightleftharpoons \text{ATP} + \text{Creatine}$$

3. Thioesters

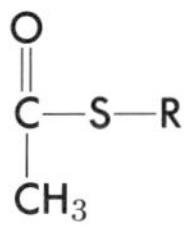

$$\begin{array}{l} O \\ \| \\ C{-}S{-}R \\ | \\ CH_3 \end{array} \qquad\qquad \begin{array}{l} O \\ \| \\ C{-}SCoA \\ | \\ CH_3 \end{array}$$

Thioester of acetic acid

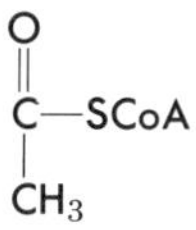

CoA = Coenzyme A

Free energy of hydrolysis = −8000 calories

Reasons for a Large $-\Delta F'_0$ of Hydrolysis

1. Increase in Resonance

There are more resonance forms for acetate and inorganic phosphate than there are for acetylphosphate.

2. Electrostatic Repulsion

The oxygen in the pyrophosphate-type linkage is electronegative and the phosphorus has a residual positive charge. The energy required to overcome the repulsion between the like charges is liberated on hydrolysis.

3. Free Energy of Ionization

When acetyl phosphate is hydrolyzed, an H^+ is produced. The $\Delta F_0'$ of ionization is about -3000 calories.

4. Change in Free Energy Due to Isomerization

Most of the energy change on hydrolysis of phosphoenolpyruvate is due to the isomerization of enolpyruvate to pyruvate.

More than one of these changes may be responsible for the energy change; for example, reasons one and three are operating in the hydrolysis of acetylphosphate, and the increase of resonance as well as a separation of charges is an important factor in the hydrolysis of P—O—P bonds.

Transformations at the Various Energy Levels

The free energy change in the transfer of phosphate from one high-energy compound to another would be small, and it would be expected that such reactions would take place. Examples are

a. Creatine phosphate + ADP ⇌ ATP + Creatine

b. ATP + UDP ⇌ ADP + UTP

c. $HOOC{-}C(O{-}Ⓟ){=}CH_2$ + ADP ⇌ $HOOC{-}C({=}O){-}CH_3$ + ATP

Phosphoenol-pyruvic acid ⇌ Pyruvic acid

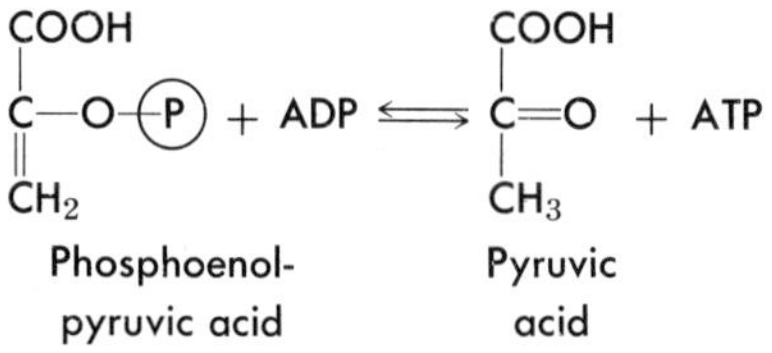

d. $HO{-}P(O)(OH){-}O{-}C({=}O){-}CHOH{-}CH_2O$Ⓟ + ADP ⇌ $COOH{-}CHOH{-}CH_2O$Ⓟ + ATP

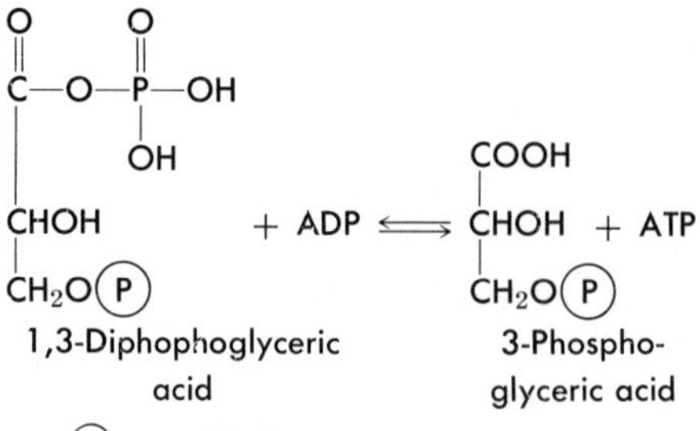

1,3-Diphophoglyceric acid ⇌ 3-Phospho-glyceric acid

Ⓟ = $-PO_3H_2$

Utilization of ATP

The high-energy compounds are used to drive endergonic reactions. The formation of glucose-6-phosphate from glucose and phosphate requires about 3000 calories. The reaction of glucose with ATP catalyzed by a kinase has a $\Delta F_0'$ of about -4000 calories

$$\text{Glucose} + \text{ATP} \xrightarrow{\text{hexokinase}} \text{Glucose-6-phosphate} + \text{ADP}$$

There are many reactions of this type in biologic systems.

So far in this chapter and in Chapter 3 we have seen that cells have enzymes which can catalyze specific exergonic reactions. We have seen that cells have the catalytic machinery for carrying out oxidative reactions and transferring the electrons to oxygen. We have also looked at some of the high-energy compounds utilized by cells and seen how these substances can be used to drive endergonic reactions. Before considering glucose metabolism in some detail, the chemical mechanisms by which cells take energy from reactions and transform it into high-energy compounds will be considered.

Formation of ATP at the Substrate Level

1. Dehydrogenation of 3-Phosphoglyceraldehyde

$$\underset{\text{3-Phosphoglyceraldehyde}}{HC(=O)-HC(OH)-H_2C-O-Ⓟ} + HS-ENZ \longrightarrow H(OH)C(-S-Enz)-HC(OH)-H_2C-O-Ⓟ \xrightarrow[NAD^+ \;\to\; NADH + H^+]{} O=C(-S-Enz)-HC(OH)-H_2C-O-Ⓟ$$

$$O=C(-S-Enz)-HC(OH)-H_2C-O-Ⓟ + P_i \longrightarrow \underset{\text{1,3-Diphosphoglyceric acid}}{O=C(-O-PO_3H_2)-HC(OH)-H_2C-O-Ⓟ} + HS-Enz$$

$$\underset{\text{1,3-Diphosphoglyceric acid}}{O=C(-O-PO_3H_2)-HC(OH)-H_2C-O-Ⓟ} \xrightarrow[ADP \;\to\; ATP]{} \underset{\text{3-Phosphoglyceric acid}}{COOH-HC(OH)-H_2C-O-Ⓟ}$$

The first reaction is the addition of the —SH group on the enzyme to the carbonyl of glyceraldehyde. Two electrons and two protons are then transferred from this group to NAD, forming NADH + H^+ and the 3-phosphoglyceric acid thiol ester of the enzyme. This is a high-energy compound which can exchange the enzyme for phosphate to form 1,3-diphosphoglyceric acid, another high-energy compound. The phosphate is then transferred to ADP to form ATP.

2. Dehydration

2-Phosphoglyceric acid is dehydrated to form the phosphate ester of enolpyruvic acid. The enzyme involved is enolase.

$$\underset{\text{2-Phosphoglyceric acid}}{\begin{array}{l}\text{COOH}\\ |\\ \text{HC—O—PO}_3\text{H}_2\\ |\\ \text{H}_2\text{C—OH}\end{array}} \xrightarrow{\text{enolase}} \underset{\text{2-Phosphoenolpyruvic acid}}{\begin{array}{l}\text{COOH}\\ |\\ \text{C—O—PO}_3\text{H}_2\\ \|\\ \text{CH}_2\end{array}} \xrightarrow[\text{ADP} \curvearrowright \text{ATP}]{} \underset{\text{Enolpyruvic acid}}{\begin{array}{l}\text{COOH}\\ |\\ \text{C—OH}\\ \|\\ \text{CH}_2\end{array}} \longrightarrow \underset{\text{Pyruvic acid}}{\begin{array}{l}\text{COOH}\\ |\\ \text{C=O}\\ |\\ \text{CH}_3\end{array}}$$

ATP Formation through Oxidative Phosphorylation

In the discussion of the oxidative decarboxylation of pyruvic acid (page 79) it was pointed out that NAD was reduced to NADH, and that the electrons from NADH were then transferred through a series of steps to O_2. The ΔF of the overall reaction from NADH to O_2 was greater than 50,000 calories.

A mechanism for obtaining most of this energy in the form of high-energy compounds would then be a chemical machine which could be fed with reduced NAD, and ATP would be produced as long as oxygen was available. Large amounts of reduced NAD are available from glucose and fat metabolism, and most tissues are well supplied with oxygen.

The mitochondria of most cells contain the electron transport system with the enzymes arranged in a definite spatial arrangement to facilitate their efficient interaction. If NADH is added to a mitochondrial suspension containing inorganic phosphate, ADP, and Mg^{++}, $\frac{1}{2}$ mole of O_2 will be reduced for each mole of NADH oxidized and 3 moles of inorganic phosphate will appear in ATP.

The overall reaction is

$$\text{NADH} + \text{H}^+ + 3\ \text{Pi} + 3\ \text{ADP} + \tfrac{1}{2}\,\text{O}_2 \longrightarrow \text{NAD}^+ + 3\ \text{ATP} + \text{H}_2\text{O}$$

The electron transport system is outlined below and the steps at which ATP is formed are indicated.

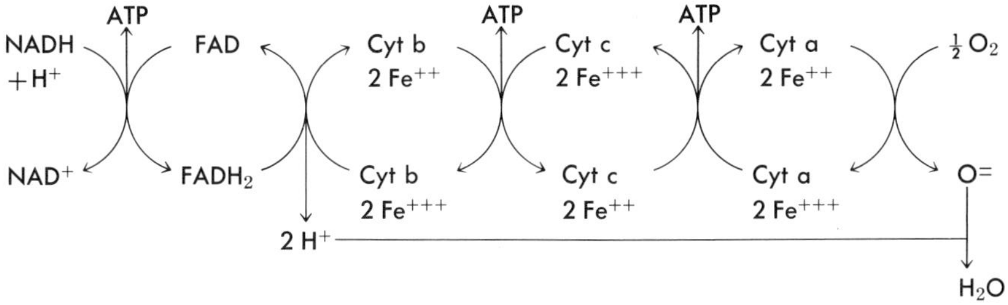

ATP ATP ATP

$NADH + H^+$ → NAD^+; FAD → $FADH_2$; Cyt b $2\ Fe^{++}$ → Cyt b $2\ Fe^{+++}$; Cyt c $2\ Fe^{+++}$ → Cyt c $2\ Fe^{++}$; Cyt a $2\ Fe^{++}$ → Cyt a $2\ Fe^{+++}$; $\frac{1}{2}O_2$ → $O^=$

$2\ H^+$ → H_2O

$E_0' = -0.32$ volt -0.12 0.12 0.22 0.29 0.82

Keeping in mind that the E_0' values are given for unit concentrations, which probably is not the case in the cell, one can still see that the overall potential difference is $0.82 - (-0.32) = 1.10$ volts, and that $\Delta F = -52{,}256$ calories.

The detailed mechanism of oxidative phosphorylation is not completely understood. Some quinones such as coenzyme Q participate in the transformation.

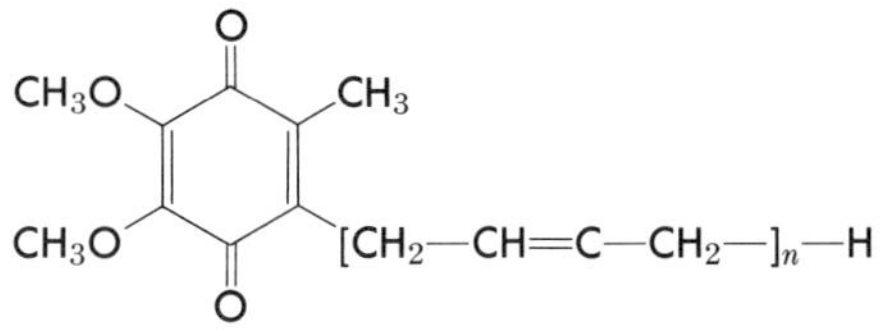

CH_3O, CH_3O, CH_3, $[CH_2—CH=C—CH_2—]_n—H$

Coenzyme Q
$n = 0$ to 10

Coenzyme Q is reduced readily to a hydroquinone. As shown below it participates as a oxido-reductant in the electron transport system. In addition it has been suggested that the coenzyme reacts with inorganic phosphate at the same time it is reduced to form a high-energy phosphate ester which can be transferred to ADP to form ATP.

So far in our discussion we have considered substrates which transfer electrons to NAD. There are a few substrates (for example, succinate, α-glycerol phosphate) which are oxidized by enzymes which transfer their electrons to a flavoprotein complex. These electrons then enter the electron transport system at a higher level, that is, at a point in the respiratory chain with a voltage higher than -0.320 volt, which is the E_0' of the NAD/NADH system. The production of ATP from those oxidations is only two molecules rather than three.

The following diagram incorporates coenzyme Q into the system and shows where succinic dehydrogenase enters the pathway. It also shows where malonate, amytal (a barbiturate), antimycin (an antibiotic), carbon monoxide, and cyanide inhibit the system. These inhibitors have been useful for determining the sequence of reactions involved in the transport system.

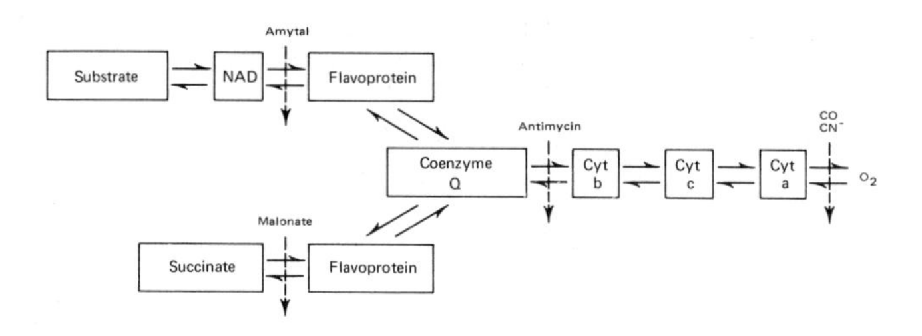

All of the enzymes needed for electron transport are found in the mitochondria of cells. The invaginated inner membrane (cristae) of the mitochondria have numerous small particles attached which have been called elementary particles (EP). The membrane and the attached particles appear to be the site of the enzymes of electron transport and of oxidative phosphorylation. Green and his coworkers at the University of Wisconsin have isolated the EP (M.W. 1,560,000) and have further separated the particle into four units, or complexes, which are isolated as relatively pure lipoproteins. The four complexes carry out the following characteristic reactions.

$$\text{NADH} + \text{H}^+ + \text{CoQ} \xrightarrow{\text{complex I}} \text{NAD}^+ + \text{CoQH}_2$$

$$\text{Succinate} + \text{CoQ} \xrightarrow{\text{complex II}} \text{Fumarate} + \text{CoQH}_2$$

$$\text{CoQH}_2 + 2\ \text{Cyt c (Fe}^{+++}) \xrightarrow{\text{complex III}} \text{CoQ} + 2\ \text{Cyt c (Fe}^{++})$$

$$4\ \text{Cyt c (Fe}^{++}) + \text{O}_2 \xrightarrow{\text{complex IV}} 4\ \text{Cyt c (Fe}^{+++}) + 2\text{H}_2\text{O}$$

Diagramatically, the functions of the four complexes are shown as

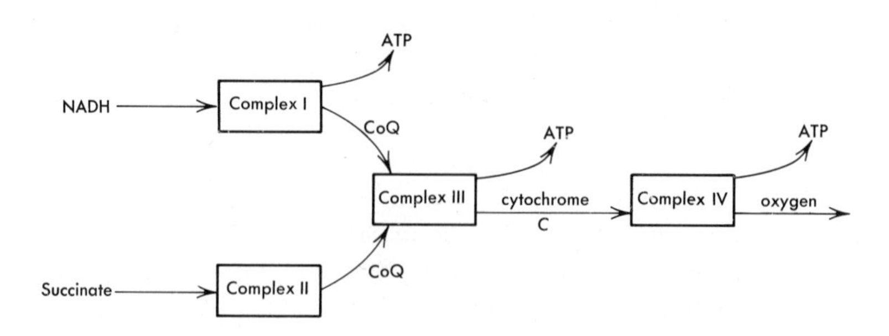

When the complexes are recombined stoichiometrically, in relatively high concentrations together with CoQ and cytochrome c, a reconstituted system is produced which is capable of electron transport.

The mechanisms and reactions by which ATP is produced during electron transport are not well understood. There are two probable mechanisms which may operate. In the first, analogous to substrate-level phosphorylation, during oxidation of a component in the electron transport chain there may be formed a high-energy, nonphosphorylated intermediate, I, and

$$I_{\text{high energy}} + P_i + ADP \longrightarrow I_{\text{low energy}} + ATP$$

The similarity to the formation of 1,3-diphosphoglyceric acid is apparent. In the second mechanism, during oxidation, a high-energy, phosphorylated intermediate (I-P) may be formed, and

$$I\text{-}P_{\text{high energy}} + ADP \longrightarrow I_{\text{low energy}} + ATP$$

It appears likely that the intermediates are probably protein and there may be several steps involved in the phosphorylation mechanism.

REFERENCES

Chargaff, E., and Davidson, J. N.: *The Nucleic Acids,* Vols. 1–3. Academic Press, New York, 1955 and 1960.

Green, D. E.: The Mitochondrion. *Sci. Am.* **210:**63, 1964.

Greenberg, D. M. (ed.): *Metabolic Pathways,* Vol. 1, 3rd ed., Chap. 1. Academic Press, New York, 1967.

Mahler, H. R. and Cordes, E. H.: *Biological Chemistry,* Chap. 4. Harper and Row, New York, 1966.

West, E. S.; Todd, W. R.; Mason, H. S.; and Van Bruggen, J. T.: *Textbook of Biochemistry,* 4th ed., Chap. 9. The Macmillan Co., New York, 1966.

White, A.; Handler, P.; and Smith, E. L.: *Principles of Biochemistry,* 4th ed., Chaps. 15 and 16. McGraw-Hill, New York, 1968.

Chapter 5

CHEMISTRY AND METABOLISM OF CARBOHYDRATES

Chemistry

THE NAME carbohydrate arose from the belief that sugars with the empirical formula $C(H_2O)$ were hydrates of carbon. With increasing knowledge, it became clear that the definition was faulty and that compounds of differing composition could also be classified as carbohydrates; for example, glucosamine, which contains nitrogen, is a carbohydrate. Broadly defined carbohydrates are polyhydroxcarbonyl compounds, their derivatives, and their condensation products. With many sugars possessing the same empirical formulas, the main differences in their physical and chemical behavior depend upon the configuration of the chemical groups about the carbon atoms in the sugar molecule, that is, many sugars are stereoisomers, or more specifically, optical isomers.

STEREOISOMERISM

The phenomenon of stereoisomerism is familiar to students of organic chemistry. Examples of four different types are

1. Structural isomers: *o*-dichlorobenzene and *p*-dichlorobenzene
2. Functional isomers: ethyl alcohol and dimethyl ether
3. Geometric isomers: fumaric acid and maleic acid
4. Optical isomers: L- and D-glyceraldehyde

The structures of the two isomers of glyceraldehyde are shown below.

```
  H   O                 H   O
   \ //                  \ //
    C                     C
    |                     |
H—C—OH              HO—C—H
    |                     |
H—C—OH               H—C—OH
    |                     |
    H                     H
```

$[\alpha]_D^{20} = +13.5°$ $\qquad$ $[\alpha]_D^{20} = -13.5°$

D(+)-Glyceraldehyde $\qquad$ L(−)-Glyceraldehyde

Glyceraldehyde, a triose (three-carbon aldose), is considered the simplest sugar since it possesses an aldehyde group and primary and secondary alcohol groupings. Of greatest importance, however, is that glyceraldehyde contains an asymmetric carbon atom (that is, a tetrahedral carbon with four different groups attached) and exists in two forms which are mirror images of each other. Carbon atom 1 is not a tetrahedral carbon and is not asymmetric; carbon 3 is tetrahedral but not asymmetric. If the two forms of glyceraldehyde are drawn as tetrahedra, with the four bonding orbitals of the asymmetric carbons directed to the corners of the tetrahedra, it is apparent that the two forms cannot be superposed, and are, therefore, different compounds.

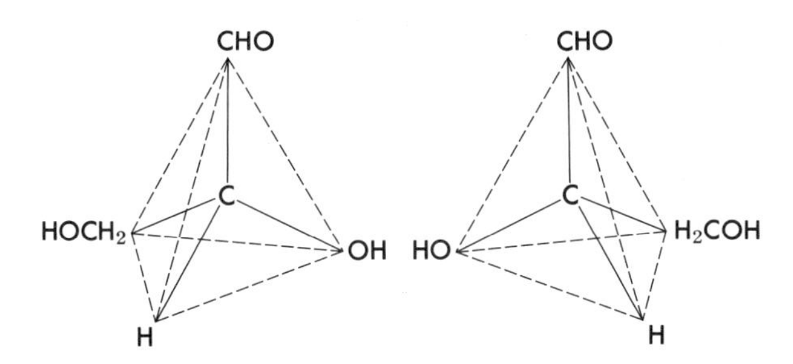

The asymmetry of the carbon atoms in glyceraldehyde, and in most other compounds with asymmetric carbons, is manifested by their optical activity. Optically active compounds have the property of rotating the plane of polarization of polarized light (light rays in which the vibrations are in one plane only) through some angle. If, for example, the polarized light ray (coming directly out of the plane of the page) is vibrating in the direction aa′ before entering the optically active material, the plane of vibration on leaving will be bb′, the plane having rotated through the angle α. Thus, passage of polarized light through separate solutions of the isomers of

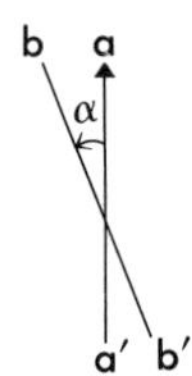

glyceraldehyde results in optical activity of equal magnitude ($+\alpha$ or $-\alpha$) but opposite directions (*dextro* to the right and *levo* to the left). Optical activity is measured in an instrument called a polarimeter.

The extent of rotation of the plane of polarized light depends upon the compound and its concentration, the wavelength of the light, the length of the light path in the solution, the solvent, and the temperature. The specific rotation, $[\alpha]$, of an optically active compound is defined as the rotation in circular degrees produced by 1 g of substance in 1 ml of solution in a 10 cm (1 decimeter) tube. The following equation summarizes these relationships:

$$[\alpha]_D^T = \frac{\alpha \times 100}{1 \times c}$$

where $[\alpha]_D^T$ is the specific rotation at temperature T, using the sodium D-line (589 mμ), where α is the observed angle of rotation in a tube 1 decimeter long, at a concentration of c grams in 100 ml of solution.

Example

A solution of glucose containing 10 mg per ml showed a rotation of $+0.525°$ in a polarimeter tube 1 decimeter long, at 20°C.

$$[\alpha]_D^{20°} = \frac{+0.525° \times 100}{1 \text{ dec} \times 1 \text{ g}}$$
$$= +52.5°$$

The spatial configurations of atoms in sugars, amino acids, and other optically active substances have been related to the glyceraldehyde isomers. Substances relating spatially to *dextro*-rotating glyceraldehyde are designated D- and those similar to the *levo*-rotating isomer are designated L-. These designations, however, do not refer to the direction of rotation, but to the spatial relationship only. The direction of rotation is designated by (+) for *dextro* and (−) for *levo*. Thus, the isomer of glyceric acid which is *levo* rotatory belongs to the D series. The same is true for lactic acid.

```
    CHO                  COOH                 COOH
     |                    |                    |
H—C—OH               H—C—OH               H—C—OH
     |                    |                    |
   CH2OH                CH2OH                 CH2
D(+)Glycer-          D(−)Glyceric         D(−)Lactic
 aldehyde               acid                 acid
```

A racemic mixture contains equal amounts of two optically active antipodes and shows no optical activity because the two rotations cancel one another. A compound with asymmetric carbon atoms may also be inactive optically because of internal compensation. The classical example of such a *meso* compound is *meso*-tartaric acid, where the top and bottom halves of the molecule are mirror images; that is, there is a plane of symmetry. The optical activity of one asymmetric carbon atom is cancelled by the equal and opposite rotation of the other carbon atom.

```
    COOH                 COOH                 COOH
     |                    |                    |
HO—C—H               H—C—OH               H—C—OH
     |                    |            -------+-------- plane of symmetry
 H—C—OH              HO—C—H               H—C—OH
     |                    |                    |
    COOH                 COOH                 COOH
D-Tartaric           L-Tartaric          meso-Tartaric
   acid                 acid                 acid
```

If there are several asymmetric carbon atoms in a compound and the end groups are different, the number of optical isomers is 2^n, where n = the number of asymmetric carbon atoms. For the structure $CH_2OH(CHOH)_4CHO$, there are 2^4 or 16 optical isomers.

SIMPLE SUGARS

The carbohydrates may be depicted and formed by the addition of one —CHOH at a time to glycerose (glyceraldehyde). Each addition produces another asymmetric carbon atom.

```
                         CHO
                          |
                      H—C—OH
                          |      ⟶ 2 D-Pentoses ⟶ 4 D-Hexoses
                      H—C—OH
                          |
    CHO                 CH2OH
     |               D-Erythrose
 H—C—OH      ⟶           +
     |                   CHO
   CH2OH                  |
D-Glycerose          HO—C—H
                          |      ⟶ 2 D-Pentoses ⟶ 4 D-Hexoses
                      H—C—OH
                          |
                        CH2OH
                      D-Threose
```

Starting with L-glycerose, a similar L-series of compounds is formed. The hydroxyl on the carbon atom derived from glycerose determines whether a compound belongs to the D or L series. Restated, the hydroxyl on the asymmetric carbon farthest removed from the carbonyl function determines whether a compound belongs to the D or L series.

In the hexoses shown below, the position of the hydroxyl on carbon atom 5 determines the series to which the compound belongs.

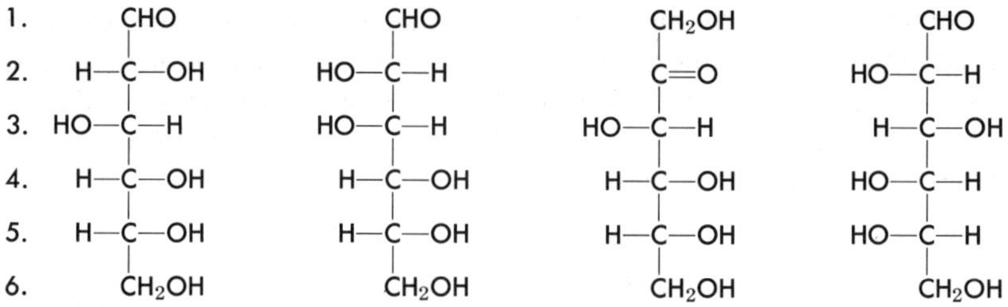

	D-Glucose	D-Mannose	D-Fructose	L-Glucose
1.	CHO	CHO	CH_2OH	CHO
2.	H—C—OH	HO—C—H	C=O	HO—C—H
3.	HO—C—H	HO—C—H	HO—C—H	H—C—OH
4.	H—C—OH	H—C—OH	H—C—OH	HO—C—H
5.	H—C—OH	H—C—OH	H—C—OH	HO—C—H
6.	CH_2OH	CH_2OH	CH_2OH	CH_2OH

These formulas show the sugars as carbonyl compounds, fructose as a ketose and the others as aldehydes.

The Structure Of Sugars

D-Glucose, as drawn above, should react as an aldehyde. The carbonyl group is oxidized to an acid or reduced to an alcohol, but other typical aldehyde reactions are not always demonstrated. For example, glucose does not react with Schiff's reagent or add sodium bisulfite. Upon acetylation, two different pentaacetates are formed. Two optically active forms of glucose are crystallized, one having $[\alpha]_D = +19°$ and the other $[\alpha]_D = +113°$. When either form is dissolved in water, the rotation of the solution changes until a value of $+52.5°$ is reached. This phenomenon, called mutarotation, is best explained by assuming intramolecular ring formation.

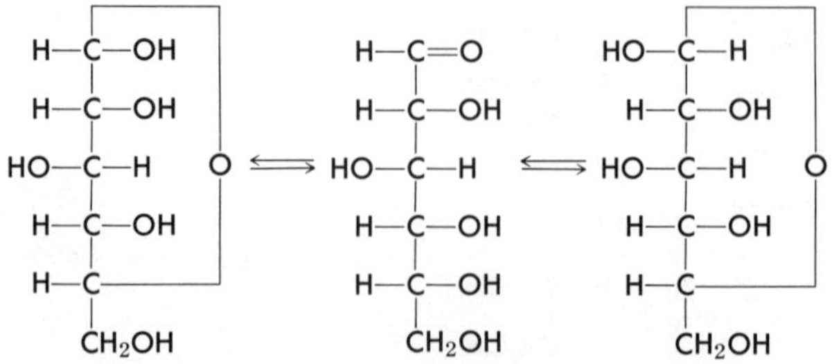

α-D-Glucopyranose +113°		Chain form of D-Glucose		β-D-Glucopyranose +19°
H—C—OH (ring O bridge C-1 to C-5)		H—C=O		HO—C—H (ring O bridge C-1 to C-5)
H—C—OH		H—C—OH		H—C—OH
HO—C—H O	⇌	HO—C—H	⇌	HO—C—H O
H—C—OH		H—C—OH		H—C—OH
H—C—		H—C—OH		H—C—
CH_2OH		CH_2OH		CH_2OH

α-D-Glucopyranose +113° Chain form of D-Glucose β-D-Glucopyranose +19°

When either the α or β form is dissolved in water, one form is converted to the other, via the aldehydo form, until an equilibrium mixture of approximately 68 per cent α and 32 per cent β form is reached with a rotation of $+52.5°$. The α and β forms are *anomers,* and carbon 1 is called the *anomeric carbon.* The open chain form of glucose is present only in small amounts ($\ll 1$ per cent).

Sugar ring structures are intramolecular acetals. With simple aldehydes, the following relationship is shown.

$$CH_3CH{=}O \xrightarrow{HOH} CH_3CH(OH)(OH) \xrightarrow[H^+]{C_2H_5OH} CH_3CH(OH)(OC_2H_5) \xrightarrow[H^+]{C_2H_5OH} CH_3CH(OC_2H_5)(OC_2H_5)$$

Acetaldehyde (hydrated form) — Hemiacetal — Acetal

Ketones form hemiketals and ketals in a similar manner. Both hemiacetals and acetals are formed in strong acid and hydrolyzed in weak acid. They are stable to alkaline hydrolysis. Hemiacetals are oxidizable, do not add sodium bisulfite, and react more slowly with the usual carbonyl reagents than aldehydes do. Acetals, on the other hand, are not oxidizable and do not react, even slowly, with carbonyl reagents.

With monosaccharides such as D-glucose, a hemiacetal may be formed between the aldehyde and the alcohol group on carbon 5, giving a stable, six-membered ring. Such sugars are called *pyranoses* from their relationship to the parent compound pyran. When the hemiacetal is formed on carbon 4, a less stable five-membered ring of a *furanose* sugar is formed. Fructose in the hemiketal form is a furanose sugar.

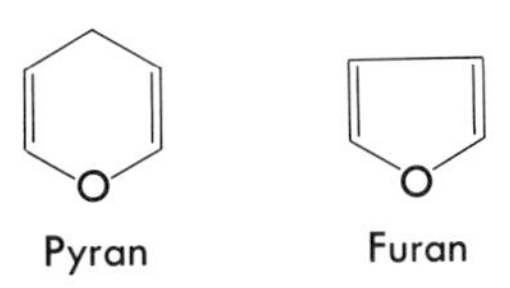

Pyran Furan

The configuration of the anomeric carbon atom of the hemiacetal form of glucose is denoted by α and β. In the D series of sugars, the α and β forms are depicted by writing the —OH group to the right or left, respectively. The configuration of the rest of the molecule is given by the prefix attached to pyranose and furanose; for example, α-D-glucopyranose or β-D-galactofuranose.

Structural Representation Of Sugars

The structures used above to show the configurations of the sugars are the Fisher projection formulas and should be visualized with the H— and —OH groups extending out of the plane of the paper and the C—C bonds forming the chain as extending back from the plane; for example

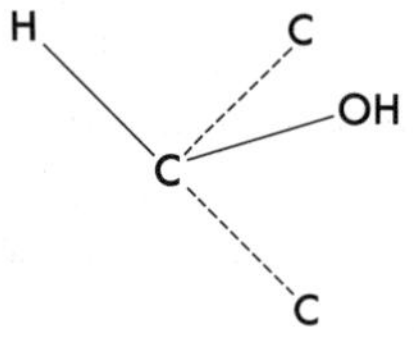

The oxygen in the hemiacetal ring should be visualized as lying below the plane of the paper.

Haworth devised a more representational system for illustrating carbohydrate structures. These structures are three-dimensional perspective structures and should be pictured with the ring projecting out of the plane of the paper and the substituent groups above or below the plane of the ring.

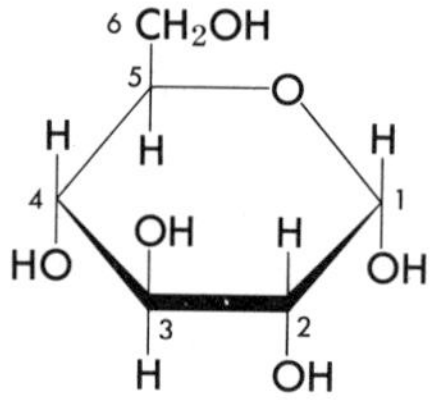

α-D-Glucopyranose

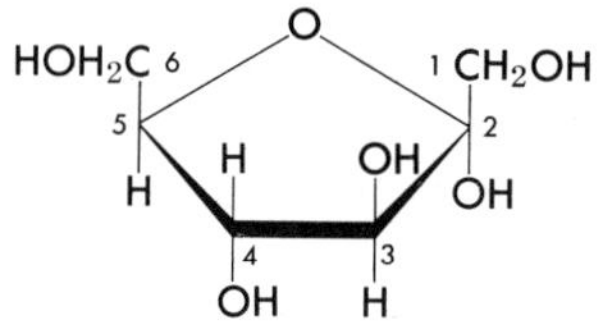

α-D-Fructofuranose

The lower edge of the ring, which is often shaded, is the part of the ring closest to the viewer. The Haworth formulas are numbered clockwise when drawn with the anomeric carbon to the right and substituent groups which appear to the right in Fisher formulas are drawn down in the Haworth structures.

Haworth structures are often depicted without the carbons and hydrogens, and with lines representing the —OH groups.

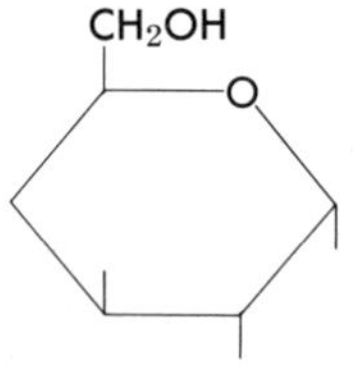

α-D-Glucopyranose

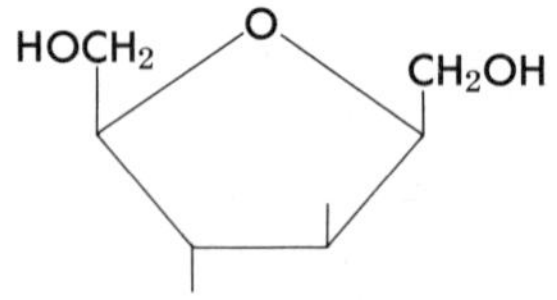

β-D-Fructofuranose

These abbreviated representations will be used at some places in this book.

The spatial relationships of the hydrogens and hydroxyls are not evident in the Haworth models and a different representation described below is often used.

The pyranoside ring may exist in eight different structures which are formed by simple rotation about the carbon-carbon or carbon-oxygen bonds. Those structures are called *conformers* and the study of their structure and reactions is called *conformational analysis*. The ring conformers of pyranoses may exist as the chair or boat forms, of which there are two chair forms and six boat forms. The existence of the boat forms among hexopyranosides is highly unusual, however, except in special derivatives of some sugars.

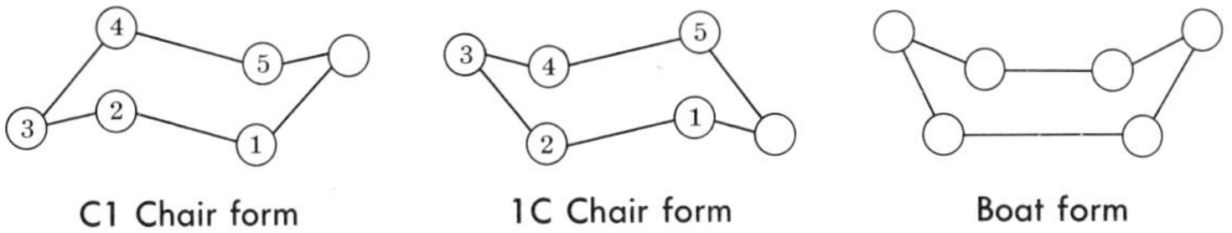

C1 Chair form   1C Chair form   Boat form

The hexopyranosides can exist in either of the two conformations of the two chair forms, the C1 and 1C forms. The hydrogens in the C1 conformation of β-D-glucopyranoside are perpendicular to the plane of the ring and are *axial groups* (solid lines); in the 1C form they are parallel to the sides of the ring and are *equatorial groups* (dotted lines). The axial groups in one conformation are equatorial in the other conformation.

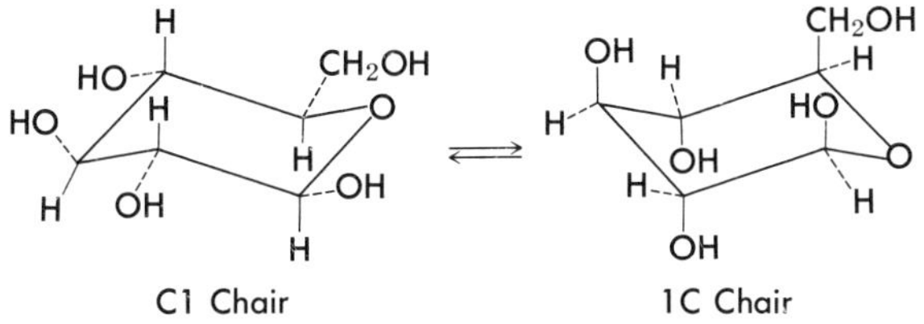

C1 Chair   1C Chair

β-D-Glucopyranose

Since the two conformers are in equilibrium, the form which is more stable will predominate. The stability of conformers depends upon the spatial arrangements and interactions between substituent groups. A discussion of the relationships between conformation and stability is, however, too long for inclusion here. Generally, the α- and β-methylglycosides of the hexoses exist in the C1 conformation.

SELECTED REACTIONS OF CARBOHYDRATES

Figure 5.1 summarizes some reactions of carbohydrates which result in carbohydrate derivatives of importance in metabolism and structural chemistry. The reactions on the left side of the diagram show glucose reacting primarily as a polyhydroxy compound. For example, reaction of sugars with acid anhydrides or acid chlorides results in ester formation. Thus, one can obtain two fully acetylated pentaacetates from glucose which are α and β isomers. Partial esters of mineral acids may be

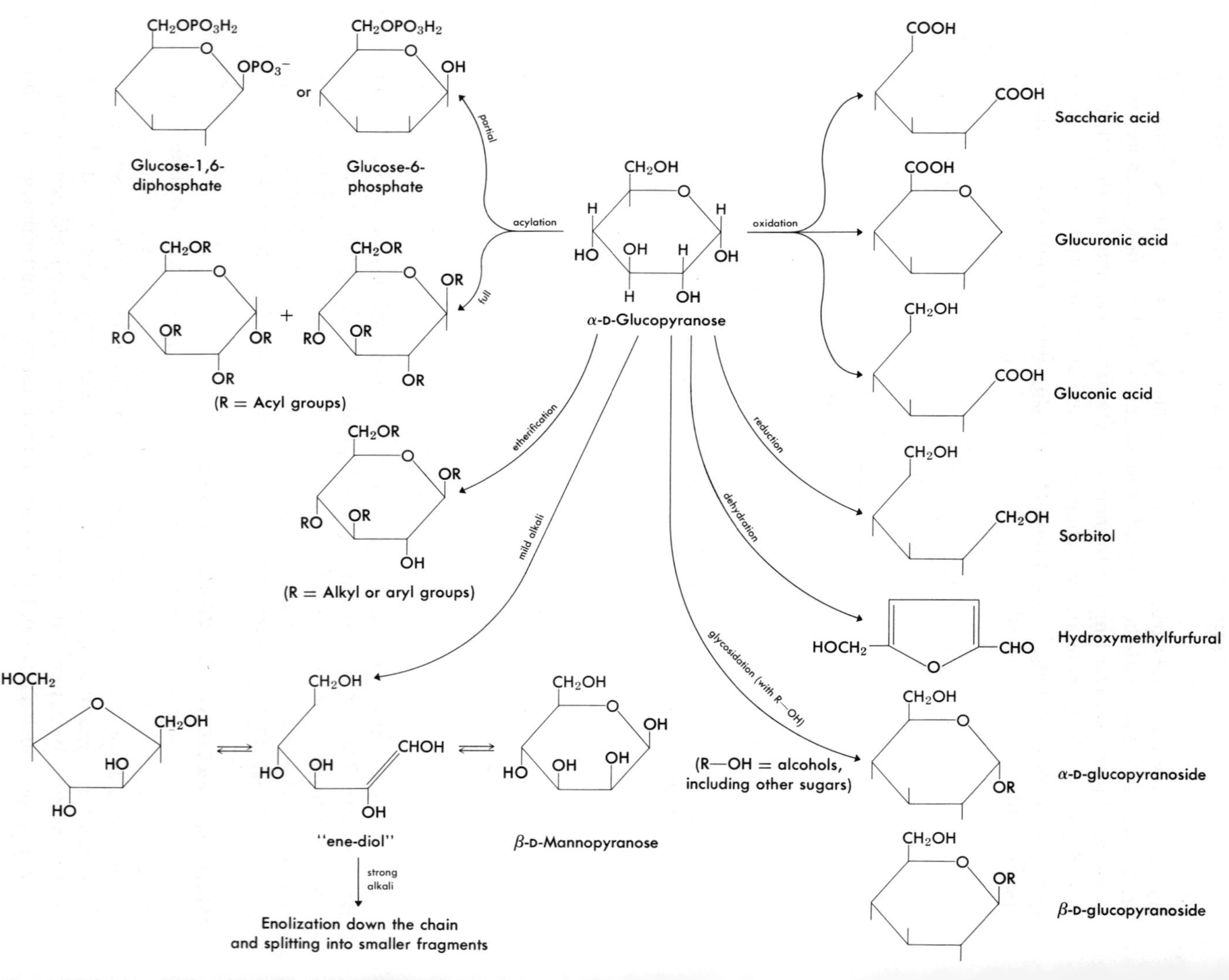
Glucose-1,6-diphosphate
or
Glucose-6-phosphate
partial
acylation
full
(R = Acyl groups)
α-D-Glucopyranose
oxidation
Saccharic acid
Glucuronic acid
Gluconic acid
etherification
(R = Alkyl or aryl groups)
mild alkali
reduction
Sorbitol
dehydration
Hydroxymethylfurfural
glycosidation (with R—OH)
(R—OH = alcohols, including other sugars)
α-D-glucopyranoside
β-D-glucopyranoside
"ene-diol"
β-D-Mannopyranose
strong alkali
Enolization down the chain and splitting into smaller fragments

produced chemically or enzymatically, and some monophosphates and monosulfates are encountered as important intermediates in carbohydrate metabolism (page 122).

Ethers of sugars such as the methoxy ethers are valuable, especially in the elucidation of the structure of polysaccharides (page 108). Other ethers, such as the fully and partially etherified benzyl ethers, are useful in carbohydrate syntheses.

Mild alkaline treatment of sugars results in the so-called Lobry de Bruyn-von Eckenstein rearrangement. This procedure is sometimes used in the synthesis of 2-keto sugars. Strong alkali breaks down sugars extensively to a variety of decomposition products of lower molecular weight.

The right side of Figure 5.1 illustrates some reactions of sugars as aldehydes and as primary alcohols. Oxidation of carbons 1 and 6, or of either alone gives different acids, as shown in the figure. Glucuronic acid occurs naturally in a variety of roles (page 111) and is also produced chemically in bulk. The oxidation of sugars by ions such as Cu^{++}, or $Fe(CN)_6^{=}$ is the basis for many analytical methods for determining sugars, such as measuring blood glucose. Note, however, that these oxidations often occur in strongly alkaline solutions so that there is extensive enolization and decomposition, and the reaction is not merely a stoichiometric oxidation of carbon 1. Historically, the oxidizability of hemiacetal groups of free sugars led to their classification as *reducing sugars,* as contrasted with naturally occurring nonreducing sugars such as sucrose, where the hemiacetals of the constituent monosaccharides are blocked.

Reduction of the anomeric carbon of glucose produces sorbitol, readily and quantitatively. Sorbitol is produced in quantity for the production of detergents and of vitamin C. The latter process is interesting because it combines chemical and enzymatic methods in a large-scale, modern manufacturing process. Sorbitol, produced by catalytic hydrogenation, is fermented by the bacterium *Acetobacter suboxydans* to the keto-sugar, sorbose, which is further oxidized to vitamin C.

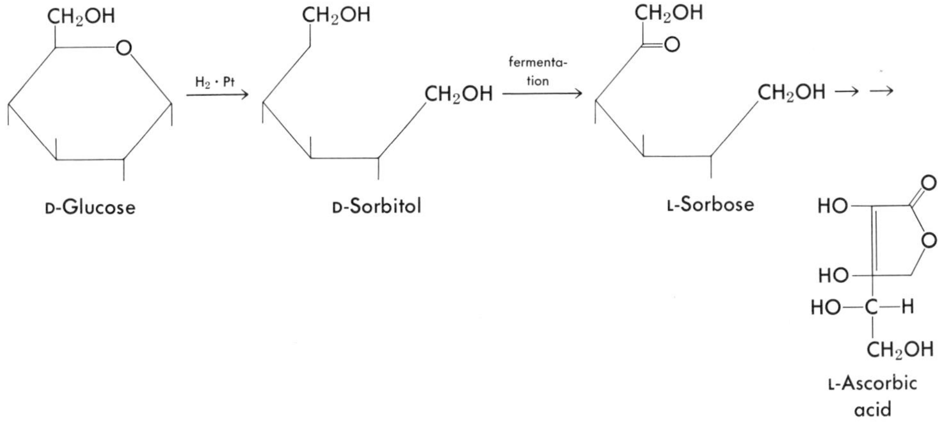

CH_2OH $\xrightarrow{H_2 \cdot Pt}$ CH_2OH ... CH_2OH $\xrightarrow{\text{fermentation}}$ CH_2OH ... $CH_2OH \rightarrow \rightarrow$

D-Glucose

D-Sorbitol

L-Sorbose

HO— ... HO— ... HO—C—H ... CH_2OH

L-Ascorbic acid

Note that oxidation of carbon 5 of D-sorbitol converts the D-alcohol to a L-keto-sugar.

Heating sugars in strong mineral acids such as H_2SO_4 dehydrates the sugars to furfural and its derivatives.

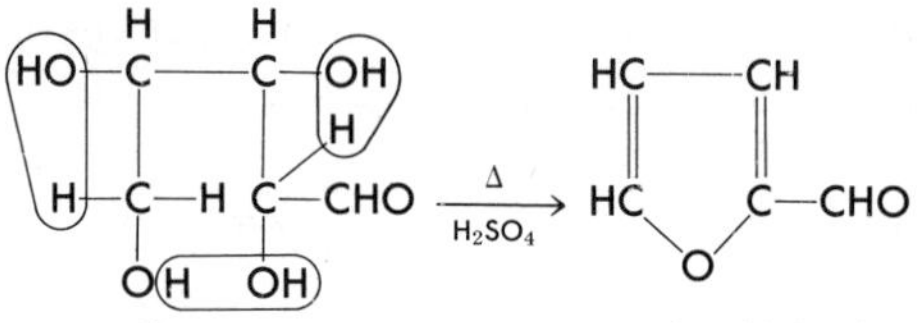

$$\text{Pentose} \xrightarrow[H_2SO_4]{\Delta} \text{Furfuraldehyde}$$

As shown diagramatically, pentoses produce furfural; hexoses produce hydroxymethylfurfural, where the HO—CH_2— group is attached to position 4 of the furfural ring. The dehydration reaction, with subsequent coupling of the furfural derivative with phenols or sulfhydryl compounds is the basis for many colorimetric methods of sugar analysis. One can often measure pentoses, hexoses, or ketoses specifically in the presence of the other sugars because of differences in the dehydration products and in their reactivities with the phenols.

Glycosidation is one of the most important reactions which sugars undergo, whether *in vitro,* or in the body. The hemiacetal and hemiketal sugars react with alcohols to form full acetals which are called glycosides. The glycosides, then, contain a sugar portion and an *aglycon,* which is the nonsugar part. Glycosidation produces α- and β-glycosides of aliphatic and aromatic alcohols. Other sugars may also function as the aglycon (Figure 5.2). For example, lactose, commonly called milk sugar, is a β-galactoside where the aglycon is a molecule of glucose and the glycosidic linkage is from carbon 1 of galactose to carbon 4 of the glucose.

Other common disaccharides listed in Figure 5.2, are maltose, a product of starch hydrolysis by pancreatic amylase or acid; sucrose, or cane sugar, which is nonreducing and nonmutarotating because the glycosidic linkage is formed between the reducing groups of both the glucose and fructose portions of the sucrose molecule.

POLYSACCHARIDES

The polysaccharides constitute a large group of substances, found widely distributed in all living organisms, which serve as reserve nutrients and skeletal materials. Most important quantitatively are the starches, glycogens, dextrins, and celluloses, all of which are polymers of glucose. Inulin, a plant polysaccharide, contains fructose units. Chitin is composed of N-acetylglucosamine.

All of the above polymers are *homopolysaccharides,* that is, composed of one type of sugar unit. Many polysaccharides are made up of more than one kind of sugar unit; hyaluronic acid and heparin are examples of such *heteropolysaccharides.*

Maltose

CH₂OH CH₂OH

4-O-α-D-Glucopyranosyl-α-D-glucopyranose

Lactose

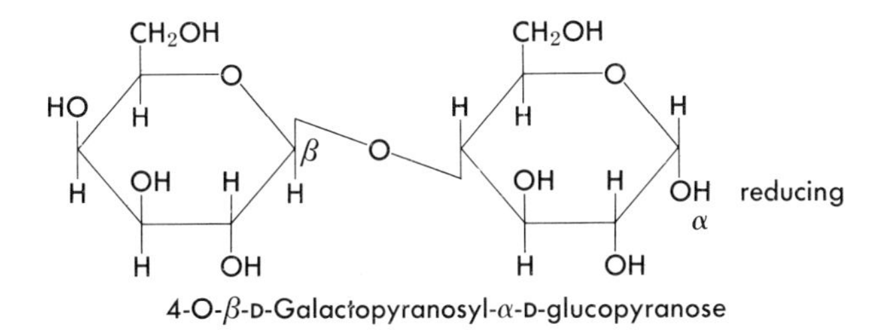

4-O-β-D-Galactopyranosyl-α-D-glucopyranose

Sucrose

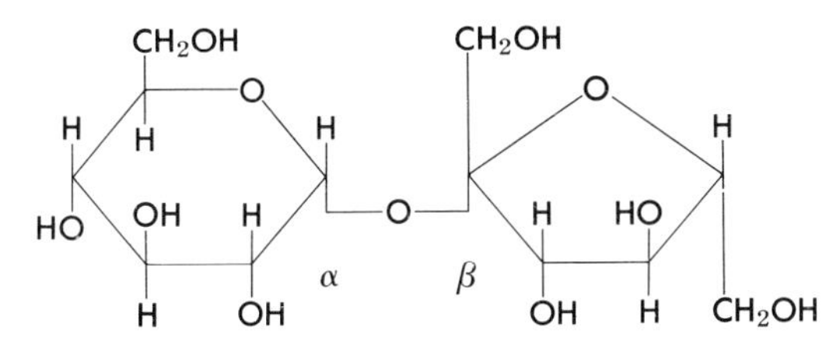

1-α-D-Glucopyranosyl-β-D-fructofuranoside

Figure 5.2. Structures of some common disaccharides.

Starches

Starches are found as reserve food materials in plants and occur as granules, the characteristic shape of which are often used to identify the source of the starch.

Complete hydrolysis of starch with strong acids yields glucose. Partial acid hydrolysis gives a complex mixture of dextrins, maltose, and glucose. Enzymatic hydrolysis with amylase produces, primarily, maltose. About 20 per cent of the starch granule is composed of amylose which is soluble in water. The remainder is insoluble amylopectin which absorbs water and swells to form starch paste.

Amylose

Amylose is a straight chain polymer composed of glucose units bound by α-1,4-linkages

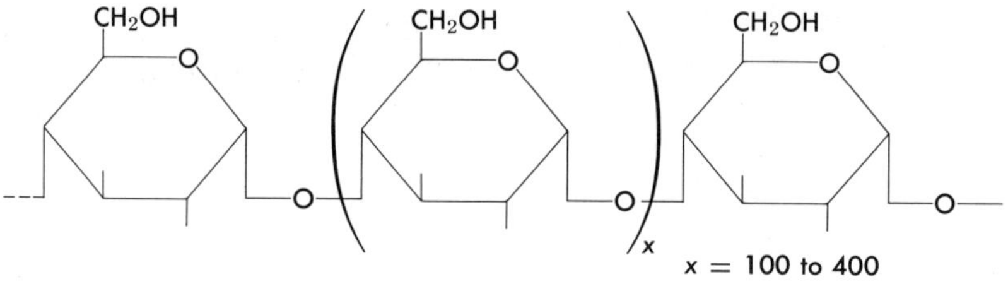

$x = 100$ to 400

The compound forms a dark blue complex with I_2 and is completely hydrolyzed to maltose by β-amylase. Methylation of the above amylose, with subsequent acid hydrolysis of the completely methylated polymer, should yield 2,3,4,6-tetra-O-methylglucopyranose and 2,3,6-tri-O-methylglucose in the molecular ratio of 1/102 to 1/402, depending upon the value of x.

Amylopectin

Amylopectin (Figure 5.3) has a molecular weight of 200,000 to 1,000,000. It gives a red-violet color with I_2, and is hydrolyzed enzymatically by β-amylase to yield about 60 per cent of glucose as maltose. A combination of β-amylase which catalyzes splitting of α-1,4-linkages, and α-amylase which splits α-1,6-linkages gives yields of maltose approaching 90 to 100 per cent. On the basis of enzymatic studies and chemical and physical studies including methylation, the structure of amylopectin is visualized as a branched polymer with α-1,4- and α-1,6-linkages.

Glycogens

Glycogens occur in animal tissues, particularly in liver and muscle. Liver glycogen is a reserve polysaccharide which is formed when excess glucose is available, and which is broken down to glucose which passes into the blood to maintain blood glucose levels.

Like amylopectin, glycogen is a branched polymer containing glucose units attached through α-1,4- and α-1,6-linkages (Figure 5.3). Glycogen is more highly branched, however, with exterior branches of six or seven units and interior chain lengths of about three units. It forms a complex with I_2, also, but the color of the complex is a reddish color, reflecting its more highly branched character. Glycogen obtained from various sources has a minimum molecular weight of about 5,000,000, and the molecular weight of glycogen in tissues may be several times this figure.

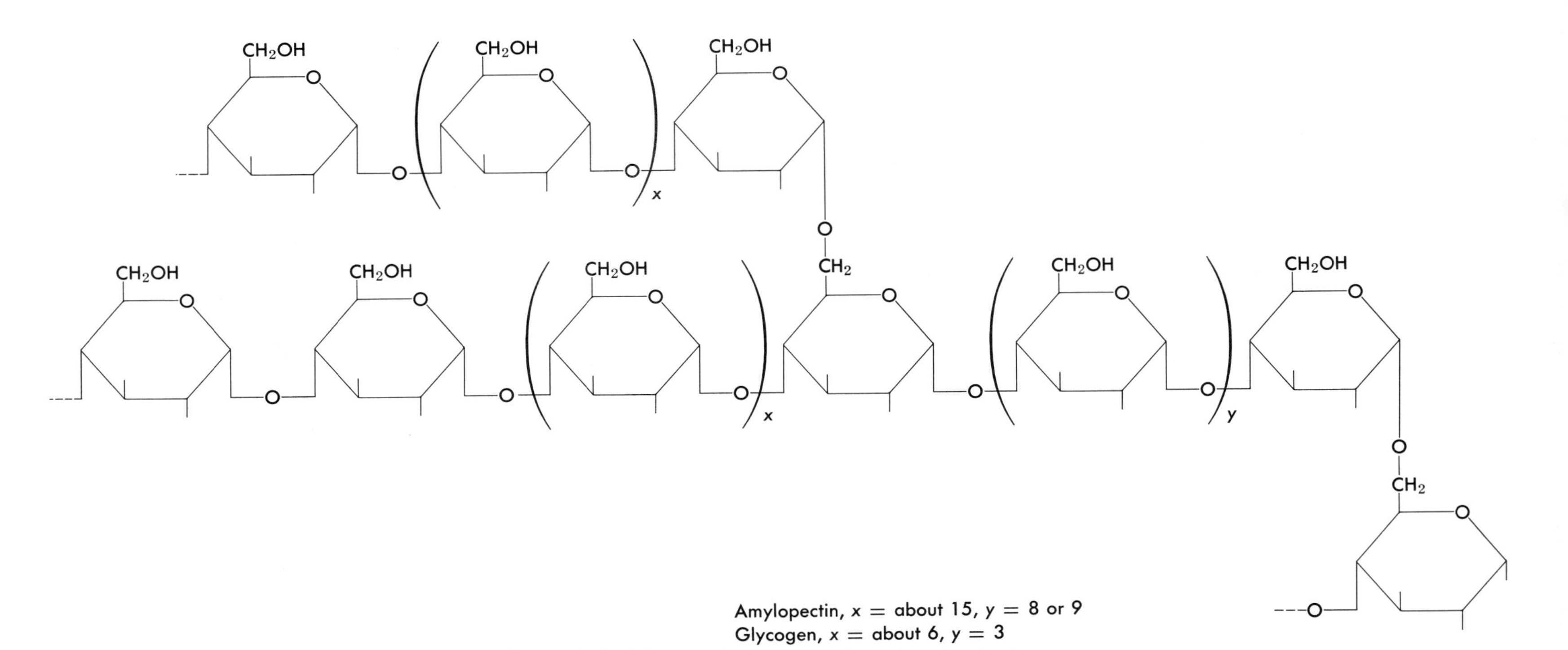

Amylopectin, x = about 15, y = 8 or 9
Glycogen, x = about 6, y = 3

Figure 5.3. Diagram of glycogen and amylopectin structures.

Cellulose

Cellulose is the chief constituent of fibrous plants and of wood. It is, therefore, the most abundant organic chemical in the world. Hydrolysis with strong acid gives D-glucose, but hydrolysis over a long period is usually required. Most animals do not possess enzymes for the hydrolysis of cellulose. Bacteria and other lower forms of life contain enzymes that hydrolyze cellulose to glucose, and herbivorous animals utilize cellulose through the activity of such organisms in the intestine.

The molecular weight of cellulose is about 500,000. Structurally it is composed of glucose units attached by β-1,4-linkages. Interestingly, the major structural difference between cellulose and amylose is that one has β-1,4-linkages, the other α-1,4-linkages, yet there are enormous differences in solubility and physical appearance and characteristics between fibrous cellulose and soft amylose powders.

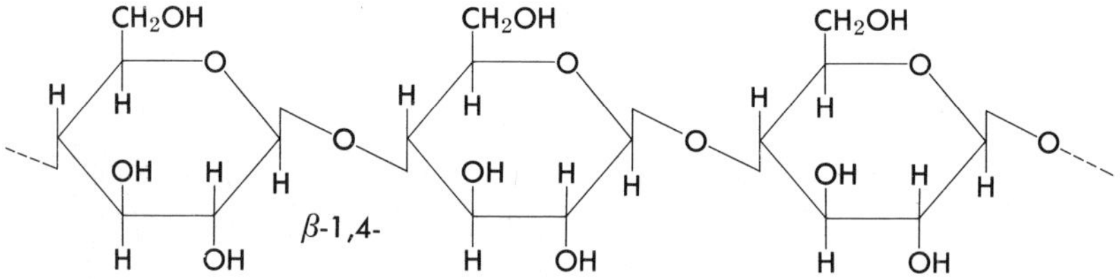

β-1,4-

MUCOPOLYSACCHARIDES

The term mucopolysaccharide is commonly used to describe heteropolysaccharides that contain residues of both a uronic acid and a hexosamine. The acidic mucopolysaccharides, hyaluronic acid, the chondroitin sulfates, and heparin, are found in the connective tissues of animals. Although complexed with proteins, the pure polysaccharides can be isolated.

The mucopolysaccharides have diverse functions in the animal body. Connective tissues have three principal components; the cells, extracellular fibers (collagen), and the ground substance. The acid mucopolysaccharides occur in the ground substance and in the form of protein complexes give connective tissues their main characteristics, for example, toughness and flexibility.

Heparin acts as a blood anticoagulant and as an antilipemic agent and may play an important part in the breakdown and synthesis of the ground substance. Connective tissues have physiological functions other than that of a supporting medium. The acid mucopolysaccharides probably have a role in calcification, control of the electrolyte composition of extracellular fluids, lubrication, and wound healing.

Hyaluronic Acid

Hyaluronic acid is found in certain streptococci, vitreous humor, and synovial fluid and is a part of the ground substance of connective tissue. It is a high molecular weight substance composed of hyalobiuronic acid units; hyalobiuronic acid, in turn, is a disaccharide formed from D-glucuronic acid and N-acetylglucosamine.

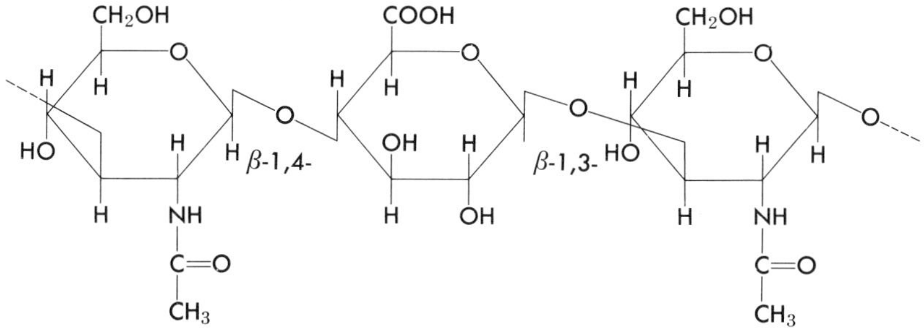

β-1,4- β-1,3-

N-Acetyl-D-glucosamine D-Glucuronic acid N-Acetyl-D-glucosamine

Hyalobiuronic acid
Repeating unit

Heparin

Heparin is a polymer of glucosamine and glucuronic acid in which some of the hydroxyl and amino groups are combined with sulfuric acid. The basic unit is best represented by the structure shown below in which all the glycosidic linkages are α-(1,4).

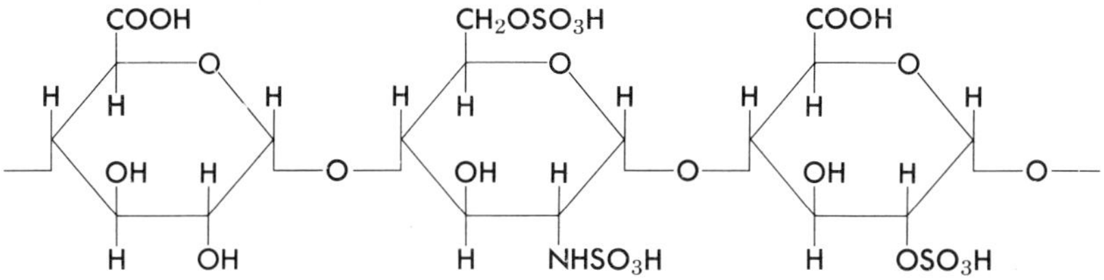

Chondroitin Sulfates

Chondroitin sulfate has a disaccharide composed of uronic acid and a D-galactosamine derivative as the repeating unit.

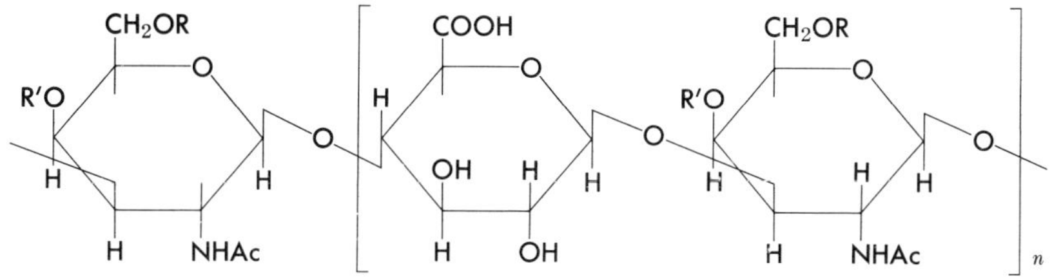

N-Acetyl-D-Galactosamine Sulfate D-Glucuronic acid N-Acetyl-D-Galactosamine Sulfate

Chondrosin

Chondroitin sulfate A: R = H, R′ = SO_3H
Chondroitin sulfate C: R = SO_3H, R′ = H

MISCELLANEOUS CARBOHYDRATES

Sialic Acids

The sialic acids are comprised of the various N-acylated and N-acyl-O-acetyl-neuraminic acids. These compounds are widely distributed in animals and occur predominantly in the bound form. The carboxyl group of sialic acid is responsible for the high negative charge upon the surfaces of erythrocytes and tumor cells. Sialic acids are found in many glycoproteins, including serum proteins and blood group specific substances, and are responsible for the high viscosity of salivary proteins as well as the biological activity of gonadotropins and other hormones. The influenza viruses and some other myxoviruses contain the enzyme neuraminidase which cleaves sialic acids from receptor sites of cells during the infectious process. Neuraminidases are also found in a number of bacterial species. Sialic acids usually occupy a nonreducing terminal position in heteropolysaccharides and are ketosidically linked to either N-acetylgalactosamine or D-galactose. The linkage is easily split by dilute acids or by neuraminidases.

The structure of the nine-carbon neuraminic acid contains a hemiketal ring extending between the keto group on carbon 2 and the hydroxyl on carbon 6.

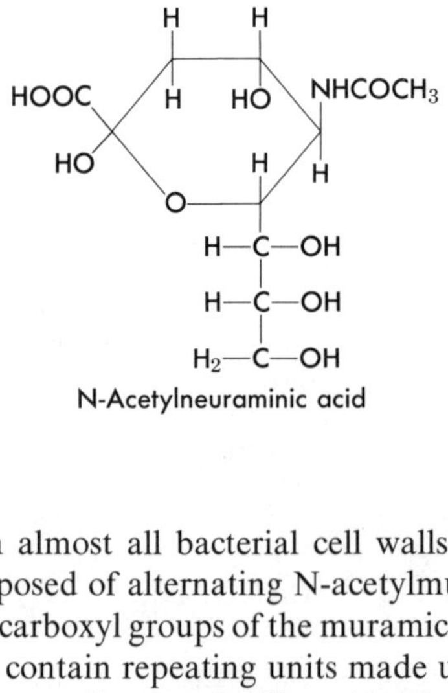

H H
HOOC H HO $NHCOCH_3$
HO H H
O
H—C—OH
H—C—OH
H_2—C—OH

N-Acetylneuraminic acid

Muramic Acid

Muramic acid occurs in almost all bacterial cell walls where it is a part of the so-called "backbone" composed of alternating N-acetylmuramic acid and N-acetyl-glucosamine residues. The carboxyl groups of the muramic acids are in amide linkage with peptide chains which contain repeating units made up usually of D-glutamine, D- and L-alanine, L-lysine, or diaminopimelic acid. The extensively cross linked polysaccharide-peptide complex is called mucopeptide, or peptidoglycan, or murein, and constitutes the insoluble material imparting rigidity and strength to the bacterial cell wall.

The specific action of penicillin upon bacteria is due to the antibiotic's inhibition of mucopeptide synthesis in bacteria. An illustration of muramic acid's structure and position in the mucopeptide is shown.

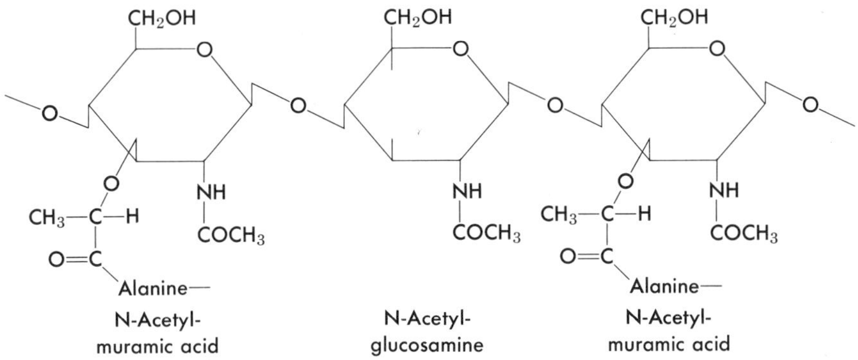

CH_2OH CH_2OH CH_2OH

O O O O

NH NH NH

CH_3—C—H $COCH_3$ $COCH_3$ CH_3—C—H $COCH_3$

O=C O=C

Alanine— Alanine—

N-Acetyl-muramic acid N-Acetyl-glucosamine N-Acetyl-muramic acid

Vitamin C (Ascorbic Acid)

Ascorbic acid occurs in high concentrations in citrus fruits, green vegetables, and tomatoes. The relatively low concentrations in potatoes represent a large portion of dietary vitamin C in the United States.

A deficiency of vitamin C in the human diet results in the disease scurvy, which was the earliest recognized deficiency disease. The disease is characterized by anemia, joint pains, and bleeding of the gums and gastrointestinal tract. A prominent underlying cause for the gross symptoms is a general weakening of the capillary walls.

Scurvy is a deficiency disease seen mostly in primates and guinea pigs. Animals such as rats, mice, and other laboratory animals synthesize ascorbic acid (page 150) and, therefore, have no dietary vitamin C requirement. Guinea pigs maintained on a vitamin C free diet for two weeks develop scurvy. Growth ceases, joints are hemorrhagic, the teeth loosen and break, and the animals lie in a "scurvy position," with hind legs sprawled.

Vitamin C is not stored in humans, unlike vitamins A and D. A simple procedure which reflects the level of vitamin C in an individual is the vitamin C saturation test. The urinary excretion of ascorbic acid is followed after administration of a test dose of vitamin C. In an individual with no, or very little reserve, the vitamin is taken into the tissues and not excreted; whereas an individual with adequate reserves will excrete a large part of the test dose.

Ascorbic acid is a lactonized enediol of L-gulonic acid. Its acidity is due to the enediol structure. The vitamin is easily oxidized to dehydroascorbic acid. Both forms

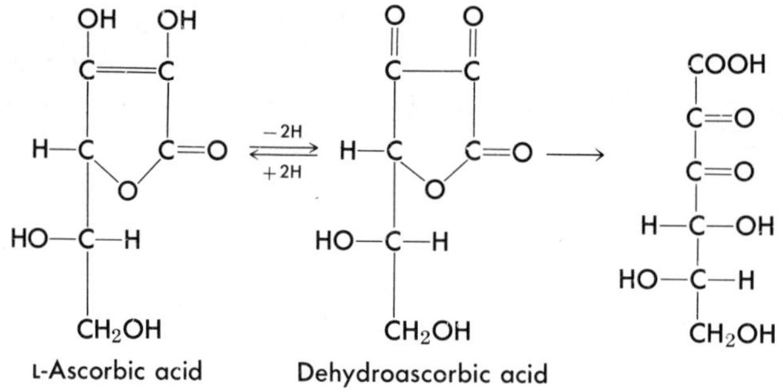

$\xrightleftharpoons[+2H]{-2H}$

L-Ascorbic acid　　　Dehydroascorbic acid

are active biologically. When the lactone ring of dehydroascorbic acid opens in mild alkaline solution, the acid is biologically inactive since the animal cannot convert it back to the lactone.

The specific biological action of ascorbic acid has not been determined. Adequate amounts of the acid are required for normal collagen synthesis. The acid is specifically required for hydroxylation of proline to hydroxyproline *in vitro*, but conclusive evidence of its requirement *in vivo* is lacking. The outstanding chemical reaction of vitamin C is its facile reversible oxidation and reduction, and suggestions about the vitamin's natural role have centered around this characteristic. Ascorbic acid has been implicated in tyrosine metabolism and iron absorption.

Inositol

Inositol is not a carbohydrate by strict definition but is a polyhydroxy compound which is synthesized and metabolized in biologic systems very much as if it were a carbohydrate. It is a growth factor for yeast and certain microorganisms. Mice on inositol-deficient diets develop alopecia. The formation of fatty livers because of large cholesterol intake can be cured or prevented by administration of inositol. It also appears that it may have some lipotropic effects under other conditions, such as very low fat diets.

Inositol is a polyhydroxycyclohexane

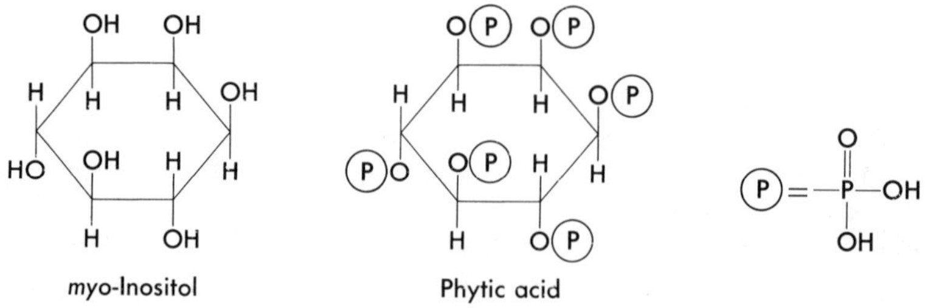

myo-Inositol　　　Phytic acid

It has the possibility of *cis-trans* isomerism. It occurs in optically inactive forms and one pair of active isomers. The calcium and magnesium salts of phytic acid (phytin) constitute the most abundant source of inositol in grains. The consumption of large amounts of phytic acid can lead to negative calcium balances because it forms insoluble calcium salts, making the calcium unavailable for absorption.

Streptomycin

A substance that is produced by a microorganism and that inhibits the growth of other microorganisms is an antibiotic. Many antibiotics contain a carbohydrate moiety as a part of their structures and streptomycin is one of them. It is a trisaccharide with the structure shown.

It should be noted that the glucosamine portion of the molecule belongs to the L series, which does not normally occur. The presence of an unnatural isomer in an antibiotic is a common phenomenon.

Streptomycin has been useful in the treatment of tuberculosis and certain infections caused by gram-negative organisms.

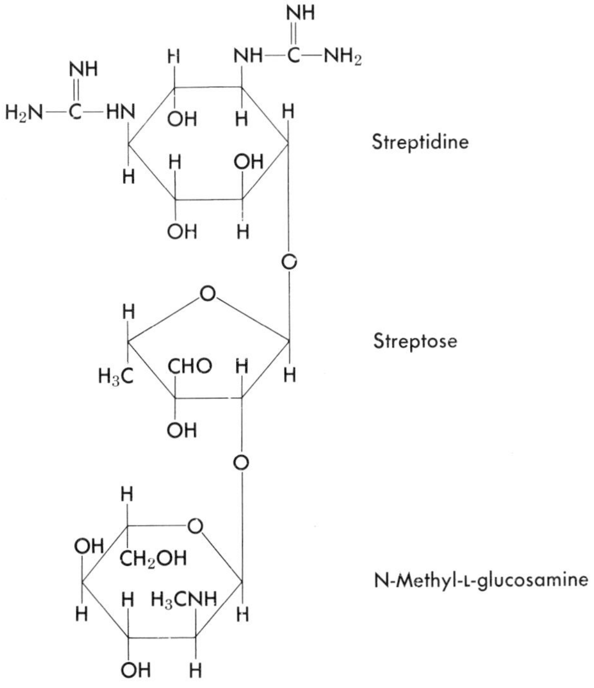

Streptidine

Streptose

N-Methyl-L-glucosamine

Streptomycin

Metabolism

The chief foodstuffs taken in the diet contain many carbohydrates, lipids, and proteins. As noted in Chapter 4, the carbohydrates and lipids are the primary source of energy for running the cellular machinery. The proteins serve as a source of amino acids for the synthesis of protein and other special compounds. If there is an excess of protein or a lack of the energy-yielding substances in the diet, the amino acids may be deaminated and oxidized. Also, several of the intermediates in carbohydrate and lipid metabolism may be used for the synthesis of needed compounds (for example, hormones).

The chart below (Figure 5.4) is an overall picture of metabolism and will serve as an outline for study. Central in the scheme is the tricarboxylic acid cycle, the Krebs cycle or the citric acid cycle, the purpose of which is to supply reduced NAD to the electron transport system. Note that from 1 mole of glucose 12 pairs of electrons (12 $\times$ 2 H) are available as fuel for this system. If all of these electrons could be transferred from NADH to oxygen, there would be a total of 36 moles of ATP available from oxidative phosphorylation for each mole of glucose. As will be shown later, there are 34 moles of ATP available from this source. Additional ATP molecules are formed at the substrate levels. It should be kept in mind that the central theme of metabolism is the production of ATP which represents useful work for the cell. There is a tendency to become lost in a maze of reactions and lose an overall point of view.

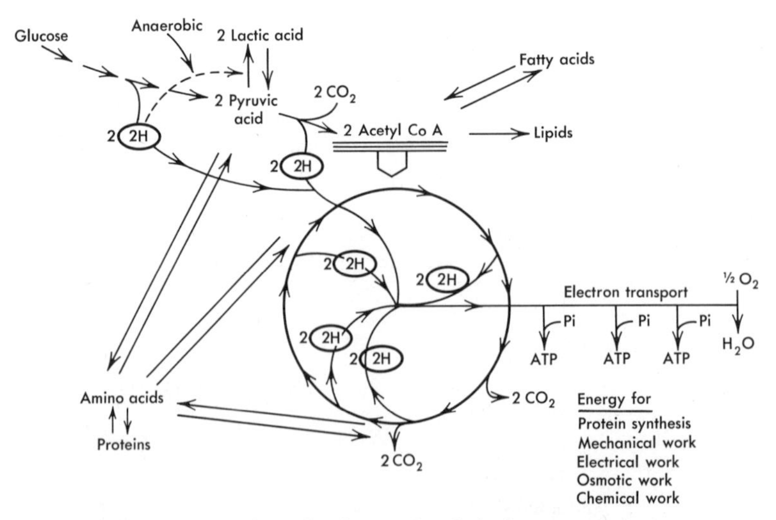

Figure 5.4. Overall scheme of carbohydrate metabolism.

The second item of special note is acetyl CoA. Just as reduced NAD is fuel for the electron transport system, acetyl CoA is fuel for the tricarboxylic acid dynamo. The products from 2 moles of acetyl CoA are 4 moles of CO_2, eight pairs of electrons (8×2 H), and water. Besides serving as the fuel for the tricarboxylic acid cycle, acetyl CoA is the key intermediate for fat and lipid synthesis. Since the reaction step from pyruvic acid to acetyl CoA is not readily reversible, acetyl CoA does not serve as a source of glucose.

There are three major points where carbohydrate and amino acid metabolism are interconnected, and it is through these reactions that the carbon skeletons of the amino acids can be oxidized or intermediates from carbohydrate metabolism can serve as carbon skeletons for certain of the amino acids.

Some aspects of the control mechanism can be seen from this scheme. If oxygen is not available, all the available NAD becomes reduced, the tricarboxylic acid cycle stops, and the pyruvate formed is converted to lactic acid. In the presence of oxygen and excess glucose, there would be an abundance of acetyl CoA and energy for the synthesis of fat.

In the absence of glucose, there would be a shift toward the utilization of fat and amino acids for the production of ATP. This is the situation in diabetes or starvation.

It is possible to organize the presentation of metabolism in several ways. The order of discussion might be: the tricarboxylic acid cycle, lipid metabolism, carbohydrate metabolism, amino acid metabolism, or any permutation of these. Most of them have been used. It is the nature of metabolism that there is no beginning or end and there is no easy way to compartmentalize the various aspects. All the reactions are running at the same time in a very carefully controlled and coordinated manner. Glucose metabolism is presented first, partially because of personal preference and partially as a result of making the production of ATP the central theme of metabolism. The amount of acetyl CoA available to our machine is dependent on the amount of glucose available. First, we will consider the mechanism by which glucose is made available and the mechanism by which the glucose level is very carefully controlled. Following this we will look at the production of ATP in the glycolytic portion of the scheme—that is, the pathway from glucose to lactic acid—and then the tricarboxylic acid cycle will be presented. At this point a more detailed look at the production of ATP will be taken. Fatty acid and lipid metabolism will be discussed, then amino acid metabolism.

Digestion

Starch and other digestible carbohydrates are converted to monosaccharides by a series of hydrolytic enzymes which include

1. Salivary amylase, an α-1,4-glucosidase which acts on starch or glycogen to give a series of oligosaccharides. Very little, if any, maltose or glucose is produced.

2. A very active pancreatic amylase which splits starch and dextrins to maltose.
3. A pancreatic maltase which catalyzes the hydrolysis of maltose to glucose.
4. Maltase, sucrase, and lactase, which split the corresponding disaccharides, are present in the intestinal mucosa.

Absorption

Only monosaccharides are absorbed through the intestinal wall and can be utilized. Disaccharides and higher polysaccharides are eliminated without change when injected into the blood stream. The monosaccharides are absorbed at different rates. If the rate of absorption of glucose is given a value of 100, the relative rates of absorption of some common sugars are

D-Galactose	110	D-Mannose	19
D-Glucose	100	D-Xylose	15
D-Fructose	43	D-Arabinose	9

Glucose, galactose, and fructose are "actively" absorbed against a concentration gradient by an unknown process. The other carbohydrates are transported, mainly, by diffusion.

Following absorption, the carbohydrates are carried by the portal circulation to the liver. Figure 5.5 gives a general pathway for glucose in the body. The monosaccharides are carried to the liver, where they are converted to glucose. The glucose may be converted to glycogen or pass into the general circulation and be carried to the tissues. Glycogen is a storage form of glucose and is important as a reservoir for maintaining a normal level of glucose in the blood. The glucose in the tissues can be utilized or converted to muscle glycogen which serves as a source of readily available energy for the muscle. The lactic acid, which results from glycogen breakdown, is carried to the liver where it is converted to liver glycogen. The series of transformations from liver glycogen, to blood glucose, to muscle glycogen, to lactic acid, and to liver glycogen is known as the Cori cycle.

Blood Glucose

The concentration of glucose is normally about 90 mg per 100 ml (90 mg per cent) and ranges from 65 to 110 mg. Figure 5.6 summarizes the reactions contributing to and withdrawing glucose from the blood. The most important process for maintaining a constant glucose level is the formation and breakdown of glycogen. Conditions in which blood glucose levels are higher than normal are called *hyperglycemias* and in extreme cases the renal threshold (about 170 mg) is reached and glucose is excreted into the urine. Conditions characterized by blood levels below normal are called *hypoglycemias*.

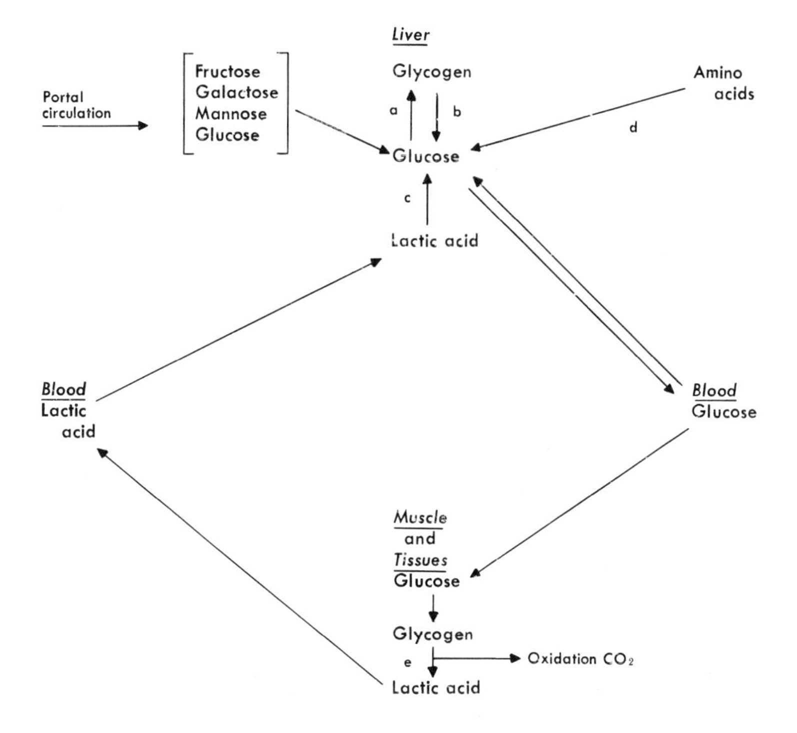

Processes involved:
- a. Glycogenesis
- b. Glycogenolysis
- c. Glucogenesis
- d. Gluconeogenesis
- e. Glycolysis

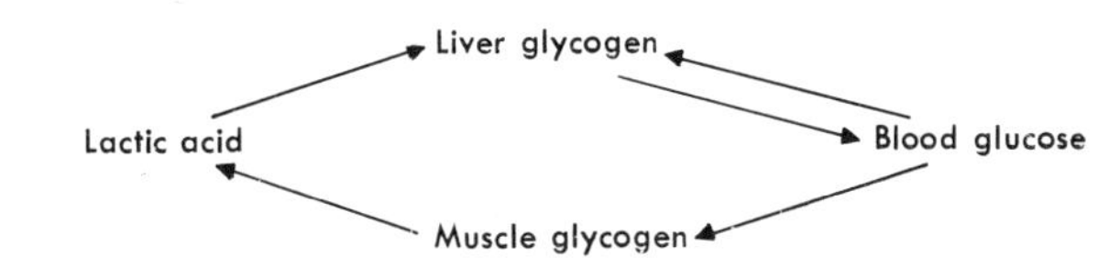

Figure 5.5. General scheme of glucose metabolism. Cori cycle.

Glycogenolysis

Liver and muscle contain enzymes which split glycogen by a process known as phosphorolysis. The enzymes are phosphorylases. The reaction is comparable to hydrolysis, in which water is added to the products. In phosphorolysis, phosphoric

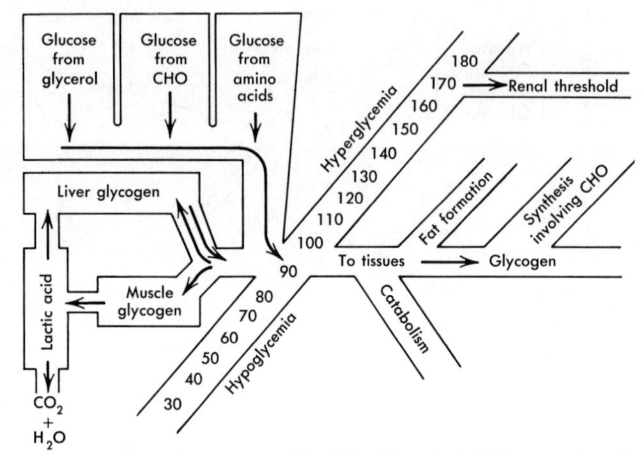

Figure 5.6. Factors controlling blood sugar. (Modified from I. S. Kleiner and J. M. Orten: *Biochemistry,* 7th ed. C. V. Mosby Co., St. Louis, 1966.)

acid is added to the products; for example

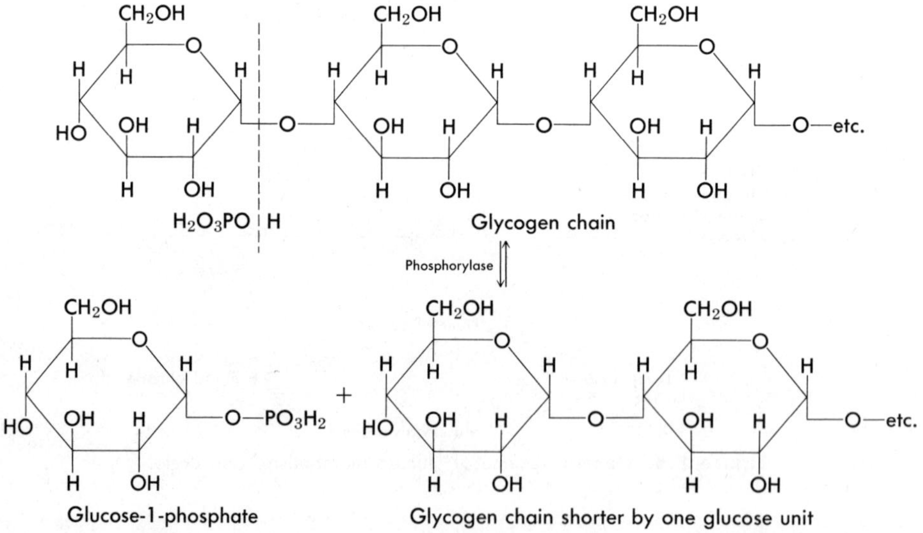

Glycogen chain + H_2O_3PO—H

Phosphorylase ⇅

Glucose-1-phosphate ($—O—PO_3H_2$) + Glycogen chain shorter by one glucose unit

The reaction is reversible, but the synthesis requires a primer—such as a small amount of glycogen or a polysaccharide chain. The enzyme is limited to forming or splitting α-1,4- linkages.

Phosphorylase exists in two forms: (1) phosphorylase a, the active form; and (2) phosphorylase b, an inactive form which can be activated. The activation of phosphorylase b by phosphokinase, ATP, and Mg^{++} is as follows.

$$\text{2-phosphorylase b} + 4\ \text{ATP} \xrightarrow[\substack{Mg^{++} \\ \text{cyclic-3}',5'\text{-AMP}}]{\text{phosphokinase}} \text{Phosphorylase a} + 4\ \text{ADP}$$

In the reaction, two molecules of phosphorylase b form a dimer. Cyclic-3′,5′-AMP is involved in the activation process, and epinephrine and glucagon activate phosphorylase by increasing the formation of this compound. Epinephrine activates both the liver and muscle phosphorylase, whereas glucagon has no effect on the muscle enzyme.

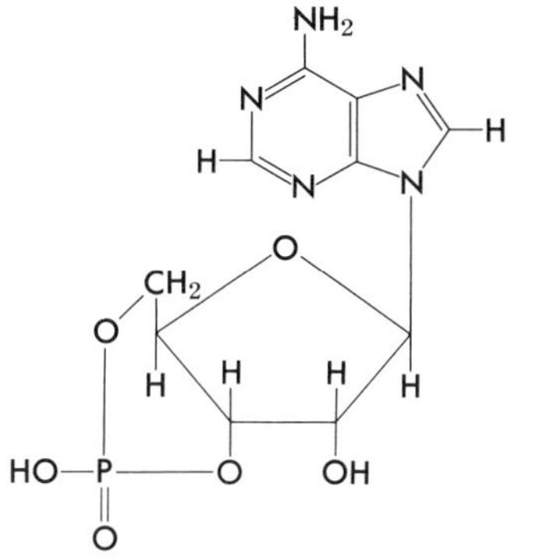

Cyclic-3′,5′-AMP

There is evidence that cyclic-3′,5′-AMP also acts by inhibiting the inactivation of phosphorylase by a phosphatase.

$$\underset{\text{active}}{\text{Phosphorylase a}} \xrightarrow[\substack{\uparrow \\ \text{inhibited by} \\ \text{cyclic-3}',5'\text{-AMP}}]{\text{phosphatase}} \underset{\text{inactive}}{\text{Phosphorylase b}}$$

Since phosphorylase splits α-1,4- bonds and glycogen is a highly branched molecule with 1,6- bonds as well as 1,4- bonds, a second enzyme is needed for extensive degradation of the molecule. The combined action of phosphorylase and amylo-1,6-glucosidase can account for the complete breakdown of glycogen. The branched chains are split to glucose-1-phosphate, and the exposed glucose units at the branch points are liberated as glucose. The synthesis of glycogen requires a branching enzyme. The enzyme is a transferring enzyme which acts on chains of glucose-1,4-glucose units and transposes a part of them to 1,6- linkages. The enzyme is an amylo-(1,4 $\longrightarrow$ 1,6)-transglucosidase.

In the liver glucose-1-phosphate is isomerized to glucose-6-phosphate, which is split by a phosphatase to form glucose and inorganic phosphate (Pi).

$$\text{Glucose-1-}\mathbf{P} \xrightarrow{\text{phosphoglucomutase}} \text{Glucose-6-}\mathbf{P} \xrightarrow{\text{phosphatase}} \text{Glucose} + \mathbf{P}_i$$

The glucose can then enter the circulation and be transported to the tissues. Very little glucose-1-**P** or glucose-6-**P** is found in the blood.

Muscle is devoid of a phosphatase so there is no way by which blood glucose can arise from muscle glycogen.

Glucose as such enters into very few biologic reactions. Before it can be utilized for glycogen synthesis or oxidized to CO_2, it is converted to phosphoric acid esters. The first step is phosphorylation (addition of phosphoric acid) by ATP with the enzyme hexokinase to form glucose-6-**P**

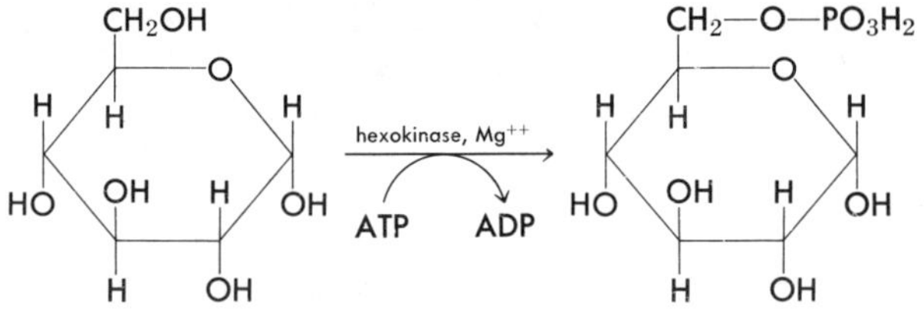

Glucose   Glucose-6-phosphate

The phosphate is then transferred to the 1 position

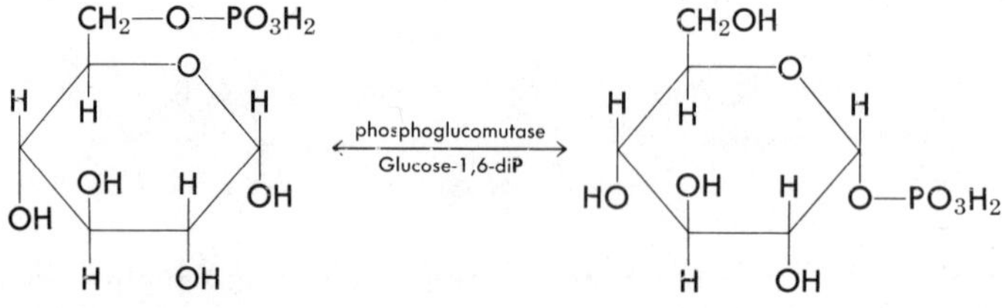

Glucose-6-phosphate   Glucose-1-phosphate

Phosphoglucomutase is an enzyme requiring magnesium and catalytic amounts of glucose-1,6-diphosphate for activity. The active enzyme contains one phosphate per enzyme molecule. In this form it can convert both glucose-1-phosphate and glucose-6-phosphate to glucose-1,6-diphosphate. The enzyme functions in the following manner.

$$\text{G-1-}\mathbf{P} + \text{Phosphoenzyme} \rightleftharpoons \text{G-1,6-di}\mathbf{P} + \text{Dephosphoenzyme}$$
$$\text{G-1,6-di}\mathbf{P} + \text{Dephosphoenzyme} \rightleftharpoons \text{G-6-}\mathbf{P} + \text{Phosphoenzyme}$$

The result is the conversion of glucose-1-phosphate to glucose-6-phosphate and the regeneration of the enzyme. Glucose-1,6-diphosphate acts catalytically by producing the phosphoenzyme by the reversal of the above reactions. The equilibrium favors the formation of glucose-6-phosphate.

Glycogenesis

There is evidence that the phosphorylase enzyme is not effective for the synthesis of glycogen *in vivo*. High activity of phosphorylase is always associated with the breakdown of glycogen rather than its synthesis. Both liver and muscle glycogen are synthesized through uridine diphosphoglucose (UDPG).

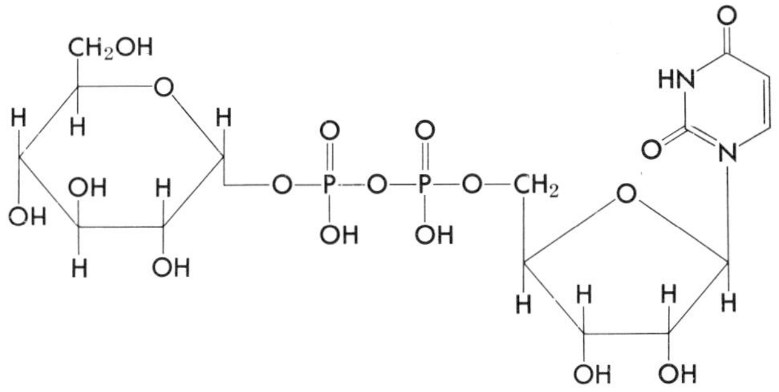

Uridine diphosphoglucose
UDPG

$$\underset{}{\text{UDPG}} + \underset{\text{Glycogen}}{(\text{Glucose})_n} \xrightarrow{\text{glycogen synthetase}} \underset{\text{Glycogen}}{(\text{Glucose})_{n+1}} + \text{UDP}$$

Glucose is added to the nonreducing end of the chain by an α-1,4- linkage. In order to obtain branched glycogen, the branching enzyme must act along with the synthetase.

The synthesis of glycogen from glucose (glycogenesis) and the breakdown of glycogen to glucose (glycogenolysis) are summarized in the following diagram.

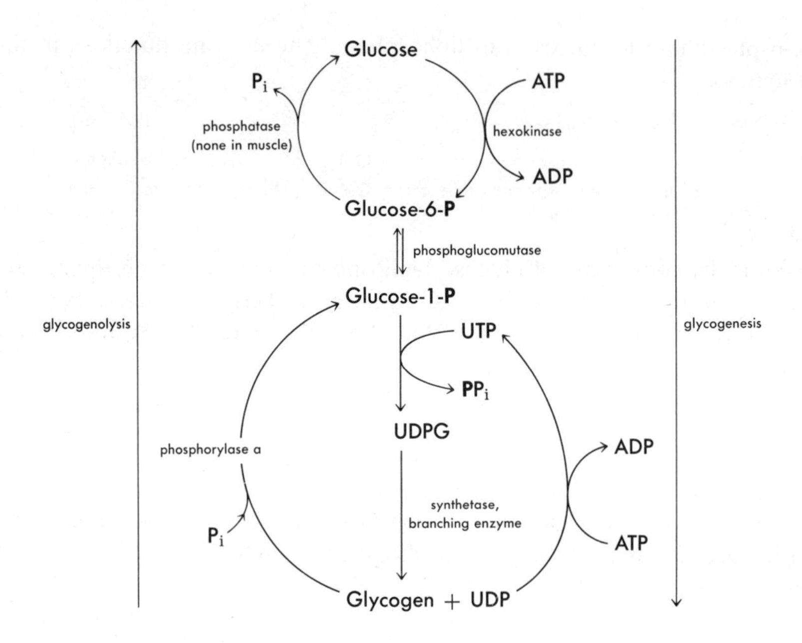

The reactions shown are characteristic of many found in biologic systems. It is possible to go from glucose to glucose-6-**P** and from glucose-6-**P** to glucose. However, this is not strictly a reversible reaction, since the mechanism for each of the reactions is different. The reaction from glucose to glucose-6-**P** has a $\Delta F_0'$ of about $+3000$ calories. By coupling it with ATP the $\Delta F_0'$ of the step becomes about -4000 calories. The hydrolysis of glucose-6-**P** has a $\Delta F_0' = -3000$ calories. The reaction of glucose-1-**P** to glycogen is reversible *in vitro* but the equilibrium is toward breakdown of glycogen. This may be due to the relatively high concentration of inorganic phosphate in cells. The formation of UDPG from uridine triphosphate (UTP) is a non-reversible reaction. The cell synthesizes glycogen through UDPG which is a nonreversible system and splits it by phosphorolysis. Again there is apparent reversibility, but the mechanisn is different in the two directions. Not shown in the diagram, but discussed above, is the conversion of phosphorylase b to phosphorylase a by one mechanism and conversion of phosphorylase b to phosphorylase a by another.

Systems of this type are important in the control of reactions. Glycogen is synthesized when the supply of glucose is available for this purpose as well as for the production of ATP which is required for at least two of the steps. The breakdown of glycogen requires very little energy, and the activity of one of the enzymes (phosphorylase) is controlled partially by hormones (epinephrine and glucagon).

At least six types of glycogen storage diseases are known. Each involves a deficiency of an enzyme required in the metabolism of glycogen. The types are

Type I, von Gierke's disease, is due to a deficiency of glucose-6-phosphatase. The glycogen has a normal structure.

Type II, Pompe's disease, is characterized by high glycogen content in the heart and muscle. There is a defect in glycogen breakdown and the enzyme involved is not known.

Type III has a deficiency of amylo-1,6-glucosidase, the debranching enzyme.

Type IV results from a lack of amylo-1,4 $\longrightarrow$ 1,6-*trans*-glucosidase. Glycogen has long outer chains and accumulates in the liver and red blood cells.

Type V is a glycogen disease of striated muscle and is characterized by a lack of phosphorylase.

Type VI is due to a lack of phosphorylase in the liver.

GLYCOLYSIS

General Aspects

The term glycolysis is usually restricted to a series of reactions involved in the biologic transformation of either glucose or glycogen to lactic acid. The series of reactions is also known as the Embden-Meyerhof pathway. Glycolysis operates in many tissues in the absence of oxygen. The following discussion will be limited to muscle and then compared to alcoholic fermentation in yeast.

Figure 5.7 reviews some of the general aspects of glycolysis (see Figures 5.4 and 5.5). One mole of glucose is converted to two moles of lactic acid by a series of ten reactions. The first four reactions involve the phosphorylation of the glucose molecule, using two molecules of ATP. Two trioses are formed, which are then converted to pyruvic acid yielding four molecules of ATP at the substrate level. The net yield of ATP is then 2 moles per mole of glucose. Thus, 14,000 calories of energy has been transferred to a form which can be utilized by the muscle for mechanical work. Since the $\Delta F_0'$ for the reaction of glucose to lactic is about $-56{,}000$ calories, the efficiency of the process is about 25 per cent.

The free energy change for the complete oxidation of glucose is 686,000 calories. Only about 8 per cent (56,000/686,000) becomes available when the product is lactic acid. The remainder of this energy is delivered when the lactic acid is oxidized in the presence of oxygen.

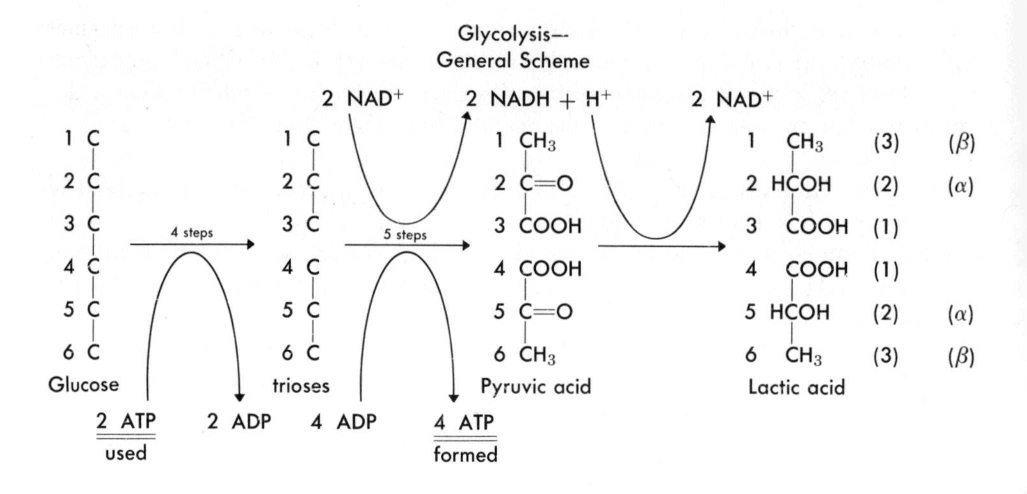

Summation: Glucose + 2 ADP $\xrightarrow{Pi}$ 2 Lactic acid + 2 ATP

2 ATP = 14,000 calories captured

$\Delta F_0'$ for Glucose $\longrightarrow$ 2 Lactic acid = −56,000 calories

Efficiency of glycolysis $\frac{14,000}{56,000} = 25\%$

Figure 5.7. Overall aspects of glycolysis.

In the steps to pyruvic acid (Figure 5.7) 2 moles of NAD are reduced. These are then used for the reduction of pyruvic acid to lactic acid.

It will be of help in organizing the reactions in this sequence and later fitting these compounds into other reactions, if it is remembered that carbon atoms 1 and 6 from glucose become the methyl carbons (3) of pyruvic and lactic acids and that carbon atoms 3 and 4 from glucose become the carboxyl groups (1) of these acids. If either the 1 carbon atom or the 6 carbon atom of glucose was labeled with radioactive C^{14}, the methyl carbon (3) of pyruvic acid would be labeled. These relationships are shown in Figure 5.7.

Formation of Triosephosphates

A more detailed scheme for glycolysis is shown in Figure 5.8. The first three reactions—a phosphorylation, an isomerization, and a phosphorylation—are reactions that change the glucose molecule to a structure that is at the necessary energy level and conformation for splitting to yield two triosephosphates. The details of

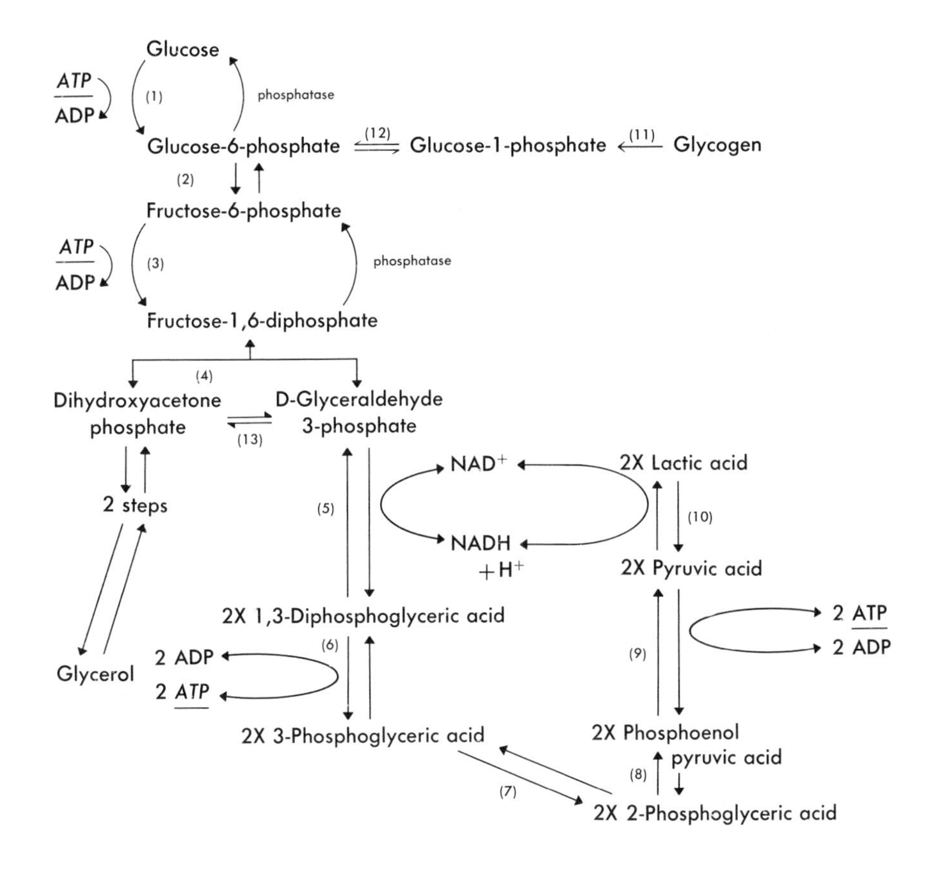

1. Glucokinase
2. Phosphohexoseisomerase
3. Phosphohexokinase
4. Aldolase
5. Glyceraldehyde-3-P-dehydrogenase
6. Phosphokinase
7. Phosphoglyceromutase
8. Enolase
9. Phosphokinase
10. Lactic dehydrogenase
11. Phosphorylase
12. Phosphoglucomutase
13. Isomerase

Figure 5.8. Detailed scheme of glycolysis.

these reactions are as follows

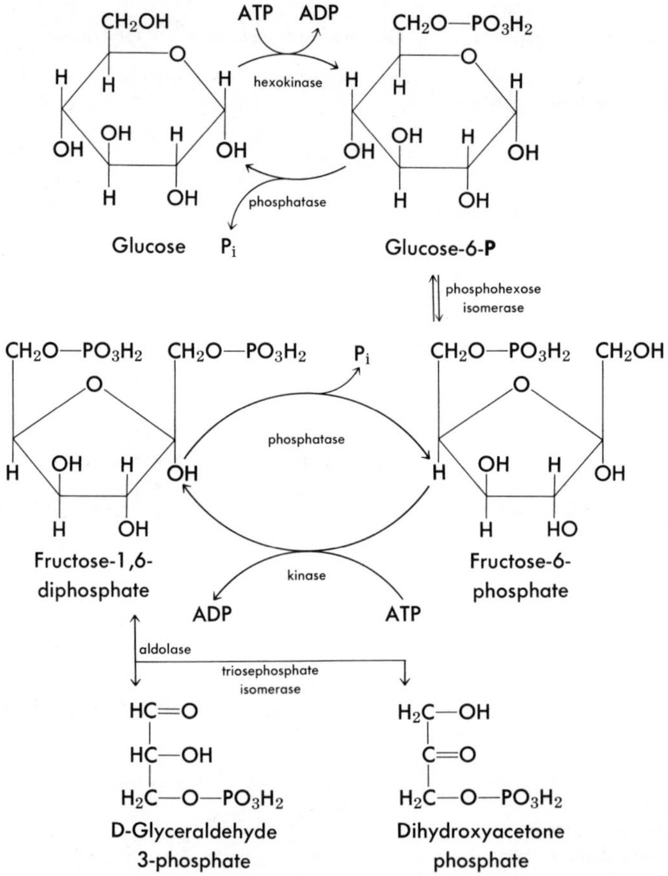
CH2OH
ATP ADP
CH2O—PO3H2
hexokinase
phosphatase
Glucose Pi
Glucose-6-P
phosphohexose isomerase
CH2O—PO3H2 CH2O—PO3H2
Pi
CH2O—PO3H2 CH2OH
phosphatase
kinase
Fructose-1,6-diphosphate
Fructose-6-phosphate
ADP
ATP
aldolase
triosephosphate isomerase
HC=O
HC—OH
H2C—O—PO3H2
D-Glyceraldehyde 3-phosphate
H2C—OH
C=O
H2C—O—PO3H2
Dihydroxyacetone phosphate

The reaction from glucose to glucose-6-phosphate has been discussed. The isomerization of glucose-6-phosphate to fructose-6-phosphate is readily reversible and at equilibrium the ratio of the glucose-6-phosphate to fructose-6-phosphate is 70:30. The phosphorylation of fructose-6-phosphate to fructose-1,6-diphosphate is similar to the phosphorylation of glucose since by coupling it with ATP the reaction is exergonic and not reversible. There is a separate enzyme, a phosphatase, that catalyzes the reverse reaction.

The cleavage of fructose-1,6-diphosphate is catalyzed by the enzyme aldolase. The reaction is reversible and the equilibrium favors the condensation of the two trioses.

The enzyme is stereospecific, forming only fructose of the four possible isomers. The enzyme requires dihydroxyacetone phosphate but other aldehydes may be substitutes for 3-phosphoglyceraldehyde. Formaldehyde may be used to form D-erythrulose-1-phosphate.

The equilibrium between the triosephosphates is catalyzed by a very active enzyme. Dihydroxyacetone phosphate is converted to 3-phosphoglyceraldehyde as rapidly as the latter is used. Only 3-phosphoglyceraldehyde is used in subsequent reactions in the glycolytic pathway. Thus, glucose can supply 2 moles of 3-phosphoglyceraldehyde for conversion to lactic acid.

Metabolism of 3-Phosphoglyceraldehyde

A summation of the reactions required to convert 3-phosphoglyceraldehyde to pyruvic acid follows

$$\underset{\text{3-Phosphoglyceraldehyde}}{\begin{array}{c}H\\|\\C{=}O\\|\\HC{-}OH\\|\\H_2C{-}O{-}PO_3H_2\end{array}} \xrightleftharpoons[P_i]{NAD \;\to\; NADH + H^+} \underset{\text{1,3-Diphosphoglyceric acid}}{\begin{array}{c}O\\\|\\C{-}O{-}PO_3H_2\\|\\HC{-}OH\\|\\H_2C{-}O{-}PO_3H_2\end{array}} \xrightleftharpoons{ADP \;\to\; ATP} \underset{\text{3-phosphoglyceric acid}}{\begin{array}{c}O\\\|\\C{-}OH\\|\\HC{-}OH\\|\\H_2C{-}O{-}PO_3H_2\end{array}}$$

$$\Updownarrow$$

$$\underset{\text{Pyruvic acid}}{\begin{array}{c}COOH\\|\\C{=}O\\|\\CH_3\end{array}} \xleftrightarrow{ATP \;\leftarrow\; ADP} \underset{\text{Phosphoenol pyruvic acid}}{\begin{array}{c}O\\\|\\C{-}OH\\|\\C{-}O{-}PO_3H_2\\\|\\CH_2\end{array}} \longleftrightarrow \underset{\text{2-Phosphoglyceric acid}}{\begin{array}{c}O\\\|\\C{-}OH\\|\\HC{-}O{-}PO_3H_2\\|\\H_2C{-}OH\end{array}}$$

Phosphoglyceraldehyde dehydrogenase contains sulfhydryl groups which function to form an intermediate compound with the aldehyde group. The mechanism has been explained on page 91.

One mole of ATP is formed at the substrate level by transfer of phosphate from 1,3-diphosphoglyceric acid to ADP, leaving 3-phosphoglyceraldehyde as a product. Since 1 mole of glucose can furnish 2 moles of 3-phosphoglyceraldehyde, 2 moles of ATP would be obtained from glucose at this step.

The enzyme phosphoglyceromutase catalyzes the conversion of 3-phosphoglyceric acid to 2-phosphoglyceric acid. The reaction is readily reversible and requires 2,3-diphosphoglyceric acid for activity. The mechanism is probably similar to that for phosphoglucomutase (page 122), in which glucose-1,6-diphosphate is required for the conversion of glucose-1-phosphate to glucose-6-phosphate.

The next reaction is a dehydration reaction catalyzed by the enzyme enolase and is freely reversible. Phosphoenolpyruvic acid is an ester of an enol and is a high-energy compound. The transfer of the phosphate to ADP to form ATP is catalyzed by a kinase. Although the latter reaction is reversible, there are other methods for forming phosphoenolpyruvic acid. Phosphoenolpyruvic acid enters into a number of synthetic reactions other than the formation of pyruvic acid. It can be considered a phosphorylated form of pyruvic acid at an energy level needed for further activity and comparable to the phosphorylated forms of glucose.

Reactions of Pyruvic Acid

The enzyme lactic dehydrogenase catalyzes the reduction of pyruvic acid to lactic acid in muscle. The production of NADH by the dehydrogenation of 3-phosphoglyceraldehyde is equivalent to the need for NADH for the reduction of pyruvic acid. From 1 mole of glucose, 2 moles of NADH would be produced, and 2 moles of NADH would be used to convert 2 moles of pyruvic acid to lactic acid (Figure 5.8).

The pathway for alcoholic fermentation in yeast is identical to glycolysis in muscle up to the point of formation of pyruvic acid. Yeast contains a carboxylase which catalyzes the following reaction

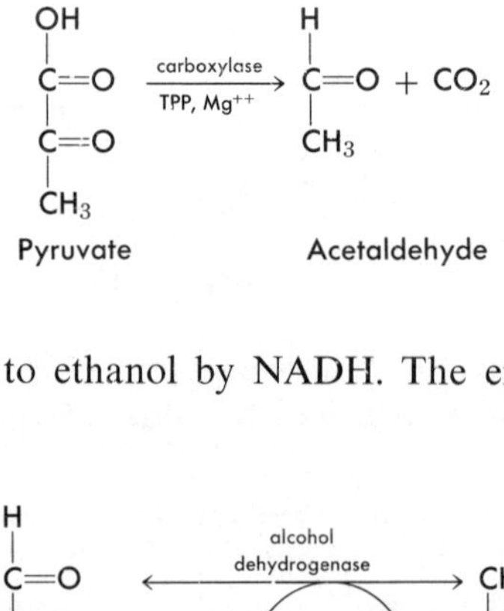

$$\underset{\text{Pyruvate}}{HO{-}C(=O){-}C(=O){-}CH_3} \xrightarrow[\text{TPP, } Mg^{++}]{\text{carboxylase}} \underset{\text{Acetaldehyde}}{H{-}C(=O){-}CH_3} + CO_2$$

Acetaldehyde is reduced to ethanol by NADH. The enzyme is alcohol dehydrogenase.

$$\underset{\text{Acetaldehyde}}{H{-}C(=O){-}CH_3} \xrightleftharpoons[NADH + H^+ \;\rightarrow\; NAD^+]{\text{alcohol dehydrogenase}} \underset{\text{Ethanol}}{CH_2OH{-}CH_3}$$

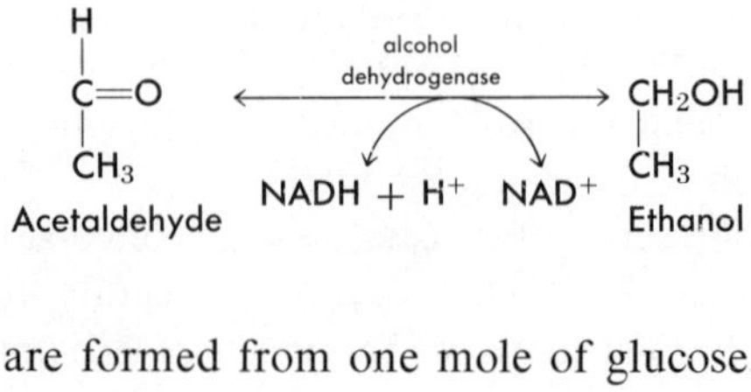

Two moles of alcohol are formed from one mole of glucose.

Muscle Contraction

If a muscle fiber is stimulated under anaerobic conditions, it will contract and lactic acid will be formed until the muscle is fatigued. Glycogen will disappear. The fatigued muscle in the presence of oxygen will recover its ability to contract, lactic acid will disappear, and glycogen will be formed. A small portion of the lactic acid is used to supply energy for glycogen synthesis. Muscle also contains a high-energy compound, creatine phosphate, which is degraded during contraction and restored during rest.

It has been shown that ATP is the direct source of energy for muscular work and that the function of creatine phosphate and glycolysis is to supply ATP. The concentration of ATP in muscle is low and cannot meet the demand for muscular activity for more than a fraction of a second. The concentration of creatine phosphate is several times that of ATP and serves as a ready source of energy. As shown in the equation

$$HN{=}C(NH{-}PO_3H_2)(N(CH_3){-}CH_2{-}COOH) \underset{ADP \quad ATP}{\longleftrightarrow} NH{=}C(NH_2)(N(CH_3){-}CH_2COOH)$$

Creatine phosphate — Creatine

creatine phosphate reacts reversibly with ADP to form ATP. When ATP is in demand, the reaction proceeds to the right. During restoration, when ATP is being formed, the reaction proceeds to the left. The amount of creatine phosphate is not sufficient to support muscular activity for sustained periods of time. The energy obtained from glycolysis sustains activity for much longer periods of time.

Muscular activity is thus dependent upon ATP, which can be sustained (1) by creatine phosphate, (2) by glycolysis, and (3) through restoration of creatine phosphate and glycogen. This type of backup for the production of a needed compound is characteristic of biologic systems. Such a scheme is important in giving resiliency to the system which would be impossible with the supply dependent on a single pathway. It has been observed previously that the supply of pyruvic acid can be supported from more than one reaction.

In the *in vitro* experiments referred to, the lactic acid is converted to glycogen during periods of relaxation. *In vivo,* the lactic acid diffuses out of the muscle and is transported to the liver where it is used for glycogen synthesis. Glucose from the circulation is used by the muscle for the synthesis of glycogen and the energy needed for restoration.

Formation of Glycerol

A two-step synthesis of glycerol from dihydroxyacetone phosphate was indicated in Figure 5.8. The reactions follow

$$\begin{array}{c} CH_2OH \\ | \\ C{=}O \\ | \\ CH_2{-}O{-}PO_3H_2 \end{array} \xrightleftharpoons[\text{NADH} + \text{H}^+ \;\to\; \text{NAD}^+]{\text{glycerophosphate dehydrogenase}} \begin{array}{c} CH_2OH \\ | \\ HO{-}C{-}H \\ | \\ CH_2{-}O{-}PO_3H_2 \end{array} \xrightleftharpoons[\text{phosphatase} \;\to\; P_i]{\text{ADP} \;\leftarrow\; \text{kinase} \;\leftarrow\; \text{ATP}} \begin{array}{c} CH_2OH \\ | \\ CHOH \\ | \\ CH_2OH \end{array}$$

Dihydroxyacetone phosphate; NADH + H^+; NAD^+; L-α-Glycerophosphate; Glycerol

Very little glycerol is formed during alcohol fermentation because the alcohol dehydrogenase is more active than glycerophosphate dehydrogenase and all the NADH is used for the reduction of acetaldehyde. If bisulfite is added to the fermentation, the acetaldehyde is tied up as a bisulfite addition compound and NADH becomes available for the reduction of dihydroxyacetone phosphate. Germany used this process for the production of glycerol in World War I.

Metabolism of Other Hexoses

In Figure 5.5 it was indicated that mannose, fructose, and galactose are converted to glucose for metabolism. Up to this point only glucose has been considered. The interrelationships between these carbohydrates and glucose metabolism are shown in the following outline.

Galactose ⟶ Galactose-1-P ⇌ UDP-Galactose

UDP-Galactose ⇅ UDPG

Glucose-1-P ⟶ UDPG ⟶ [Glycogen, Sucrose, Glucuronic acid, etc.]

Glucose-1-P ⇅ Glucose-6-P

Glucose ⇌ Glucose-6-P ⇌ Fructose-6-P ⇌ Mannose-6-P

Fructose ⟶ Fructose-6-P

Mannose ⟶ Mannose-6-P

Fructose-1-P ⇌ Fructose

Fructose

Fructose may be phosphorylated directly by a fructokinase to fructose-6-phosphate which is an intermediate in the glycolytic pathway. Muscle and liver contain a second fructokinase which converts fructose to fructose-1-phosphate. Fructose-1-phosphate cannot be isomerized to fructose-6-phosphate or phosphorylated to fructose-1,6-diphosphate. It is metabolized as follows

$$\text{Fructose-1-P} \longrightarrow \text{Dihydroxyacetone phosphate} + \text{Glyceraldehyde}$$
$$\text{Glyceraldehyde} + \text{ATP} \longrightarrow \text{Glyceraldehyde-3-phosphate} + \text{ADP}$$

Dihydroxyacetone phosphate and glyceraldehyde-3-phosphate then enter reactions previously discussed.

Mannose

Mannose is phosphorylated to mannose-6-phosphate which isomerizes into fructose-6-phosphate.

Galactose

Galactose is phosphorylated to galactose-1-phosphate and converted into glucose-1-phosphate as follows

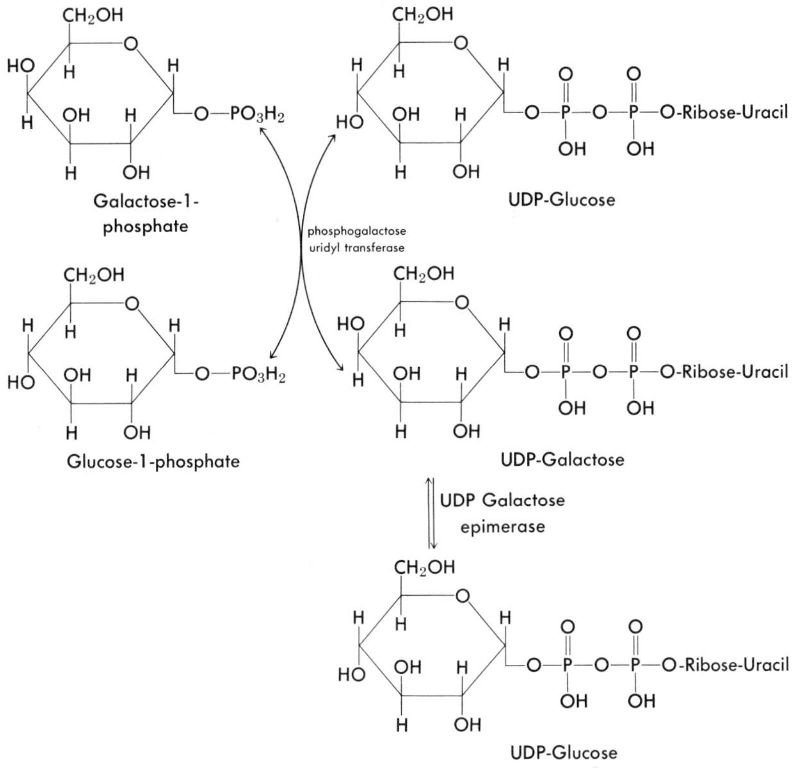

Galactose-1-phosphate + UDP-Glucose ⇌ (phosphogalactose uridyl transferase) Glucose-1-phosphate + UDP-Galactose

UDP-Galactose ⇌ (UDP Galactose epimerase) UDP-Glucose

Glucose-1-phosphate then enters the pathways indicated previously.

Children who have congenital galactosemia have little or no uridyl transferase. As a result, galactose-1-phosphate accumulates, causing damage to liver, brain, lens, and other tissues. The epimerization reaction requires NAD and involves formation of the 4-keto sugar as an intermediate. A number of similar epimerizations are known, for example

UDP-D-glucosamine ⇌ UDP-D-galactosamine;
UDP-D-xylose ⇌ UDP-L-arabinose
and UDP-D-glucuronate ⇌ UDP-D-galacturonate.

Nucleotide Diphosphate Sugars

Since 1955 it has become known that a large number of carbohydrate interconversions and syntheses proceed by way of the nucleotide diphosphate sugars. Their structures are similar to the formula for uridine diphosphate glucose, (page 123). The general reaction by which these derivatives are formed is illustrated by the formation of uridine diphosphate glucose. In the presence of the appropriate *pyrophosphorylase*

$$\text{Glucose-1-}\mathbf{P} + \text{UTP} \longrightarrow \text{UDP gluclose} + \text{PP}_i$$

(UTP:glucose-1-**P** uridyl transferase) and the requisite nucleotide triphosphate (uridine triphosphate), the nucleotide diphosphate sugar is formed and undergoes a variety of reactions.

Two examples of nucleotide diphosphate sugar reactions have been mentioned previously: the epimerization of galactose to glucose; and the glucosyl donor role of UDPG in the synthesis of the homopolysaccharide glycogen. Examples of other reactions of this class of sugar derivatives are shown below.

Oxidation

$$\text{UDP Glucose} \xrightarrow{2\text{NAD}^+} \text{UDP Glucuronic acid} + 2\text{NADH}$$

This oxidation is catalyzed by the enzyme UDP glucose dehydrogenase, found in liver as well as in plants and bacteria. The reaction appears to be a one-stage reaction, without appearance of an intermediate 6-aldehydo compound.

6-Deoxyhexose Formation

The reactions leading to L-fucose illustrate 6-deoxyhexose formation.

CH_2OH — O-PPG → (H_2O) → CH_3, O= — O-PPG →

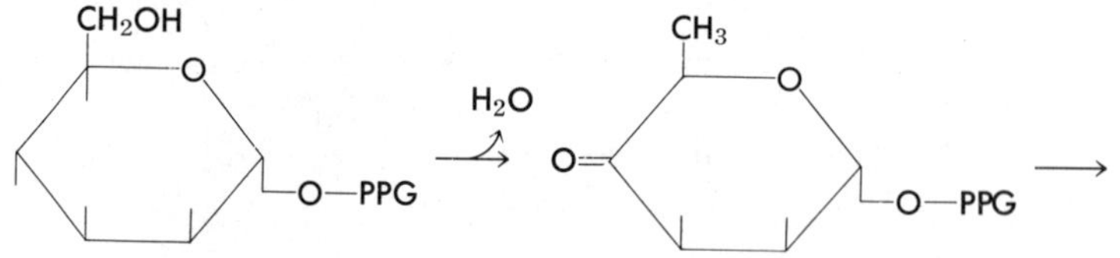

Guanosine diphosphate mannose

GDP-4-Keto-6-deoxy-D-mannose

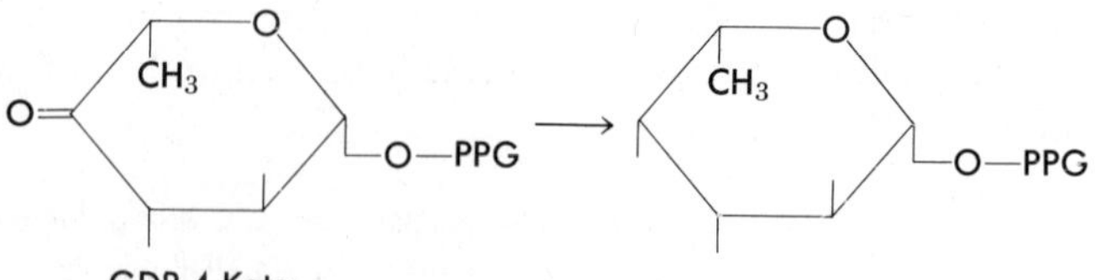

O= — CH_3 — O-PPG → CH_3 — O-PPG

GDP-4-Keto-6-deoxy-L-glucose

GDP-L-fucose

Synthesis Of Glycosidic Bonds

The most important role of the nucleotide diphosphate sugars is as glycosyl donors in synthetic reactions. Some examples of disaccharide synthesis are

$$\text{Fructose-6-P} \xrightarrow{\text{UDP glucose}} \text{Sucrose-6-P} \longrightarrow \text{Sucrose}$$

$$\text{Glucose-1-P} \xrightarrow{\text{UDP galactose}} \text{Lactose-1-P} \longrightarrow \text{Lactose}$$

The usual routes of polysaccharide synthesis are illustrated by glycogen synthesis (page 123). Heteropolysaccharides appear to be synthesized from different glycosyl donors acting in turn, although it is also possible that oligosaccharides are synthesized first and then coupled together. Hyaluronic acid synthesis requires the specific enzymes UDP N-acetylglucosamine and UDP glucuronic acid. Usually, a primer is required, at least under *in vitro* conditions.

Amino Sugars

Carbohydrates containing an amino function are found in a variety of glycoproteins, mucopolysaccharides, and in bacterial cell walls. Glucosamine is formed from D-fructose-6-phosphate. The amino group is transferred from glutamine

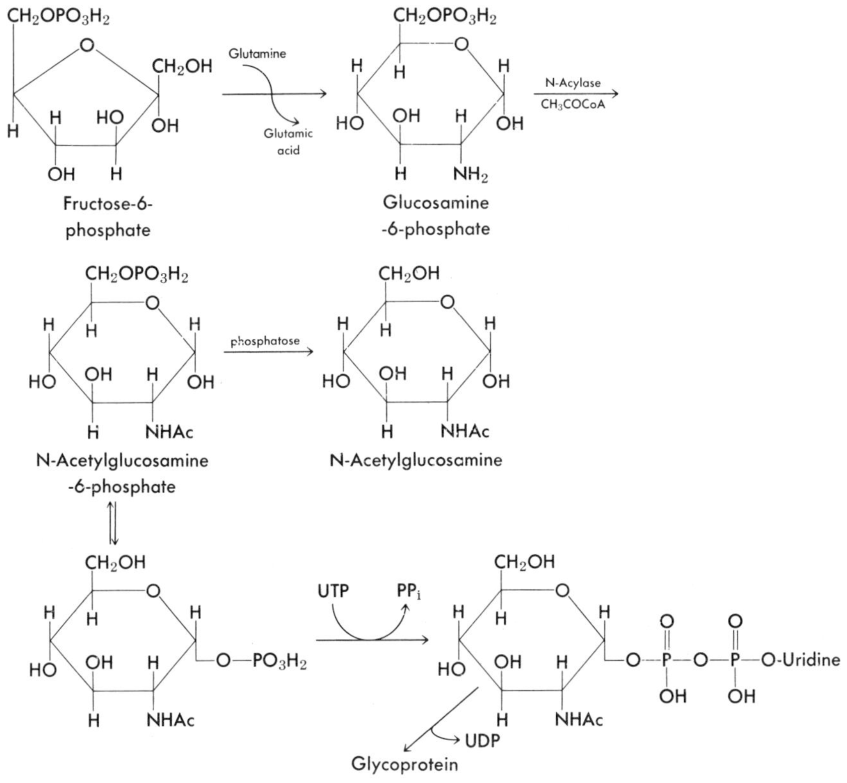

$CH_2OPO_3H_2$ O CH_2OH H H HO OH OH H

Fructose-6-phosphate

Glutamine → Glutamic acid

$CH_2OPO_3H_2$ O H H H HO OH H OH H NH_2

Glucosamine-6-phosphate

N-Acylase, CH_3COCoA →

$CH_2OPO_3H_2$ O H H H HO OH H OH H NHAc

N-Acetylglucosamine-6-phosphate

phosphatase →

CH_2OH O H H H HO OH H OH H NHAc

N-Acetylglucosamine

⇅

CH_2OH O H H H HO OH H O—PO_3H_2 H NHAc

UTP → PP_i

CH_2OH O H H H HO OH H —O—P(=O)(OH)—O—P(=O)(OH)—O-Uridine H NHAc

→ UDP

Glycoprotein

N-Acetylneuraminic is synthesized from N-acetylmannosamine-6-phosphate and phosphoenolpyruvate. The cytidine monophosphate (CMP) derivative is utilized from the introduction of N-acetylneuraminic acid into polysaccharides or proteins.

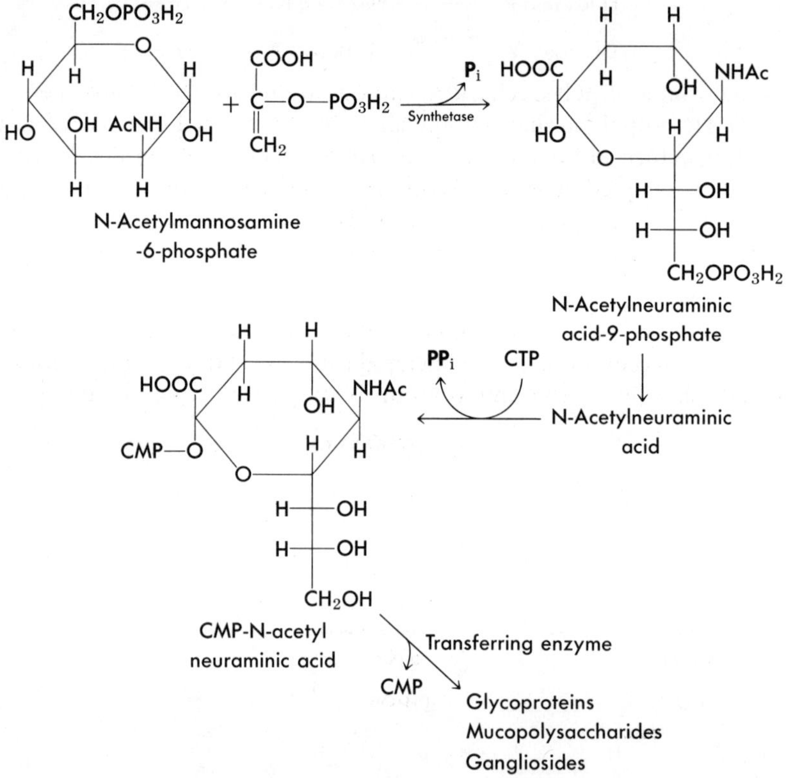

N-Acetylmannosamine-6-phosphate + $CH_2=C(COOH)-O-PO_3H_2$ —(Synthetase; P_i)→ N-Acetylneuraminic acid-9-phosphate

N-Acetylneuraminic acid-9-phosphate → N-Acetylneuraminic acid

N-Acetylneuraminic acid —(CTP → PP_i)→ CMP-N-acetyl neuraminic acid

CMP-N-acetyl neuraminic acid —(Transferring enzyme; CMP)→ Glycoproteins, Mucopolysaccharides, Gangliosides

It is interesting that the active compound for transferring the glycosyl group is a monophosphate. All other known transfer reactions involve derivatives of nucleoside diphosphates. Uridine diphosphoglucose and uridine diphospho-N-acetylglucosamine are typical examples.

TRICARBOXYLIC ACID CYCLE (KREBS CYCLE: CITRIC ACID CYCLE)

At the beginning of this section a general scheme of metabolism was presented. The portion of the scheme from glucose to pyruvic acid has been considered in some detail. The sources of glucose, control of glucose concentration, phosphorylation of glucose, conversion to trioses, and the transferring of energy from these 3-carbon compounds to ATP have been studied. It was shown that in the absence of oxygen, pyruvate was reduced to lactate.

In aerobic metabolism, the electrons are transferred through the electron transport systems, maintaining NAD in the oxidized form. Under these conditions pyruvate is not reduced to lactate but is further oxidized to CO_2 and H_2O through the tricarboxylic acid cycle. It is this pathway that provides most of the energy for endergonic processes. The fuel is acetyl CoA and is provided by carbohydrates and fats.

Figure 5.9 is a modification of Figure 5.4 showing the tricarboxylic acid cycle in more detail. Again the emphasis is on the production of ATP. Twelve pairs of electrons are transported from one molecule of glucose through the electron transport system. The steps by which these can give rise to 34 molecules of ATP will be considered in the next few paragraphs. In addition to these 34 molecules, six additional molecules are available, four from glycolysis and two from reactions at the substrate level in the tricarboxylic acid cycle. Two molecules of ATP are used in the phosphorylation of glucose in the glycolytic pathway giving a net of 38 molecules of ATP from one molecule of glucose. This represents 266,000 calories (38×7000) of the 686,000 calories available from glucose or an efficiency of about 40 per cent.

There are other features of this cyclic system which should be noted. The 12 pairs of electrons find their way to oxygen, giving 12 molecules of water, and six molecules of CO_2 are produced. These products then account for the complete oxidation of glucose. Figure 5.9 shows the formation of two molecules of pyruvate, two molecules of acetate, and two molecules of trioses from one molecule of glucose, but shows only one molecule of citrate and other members of the cycle. Oxaloacetate can react with one molecule of acetyl CoA, go around the cycle, become regenerated, and pick up the second molecule of acetyl CoA. In this way one molecule of oxaloacetate can handle many molecules of acetyl CoA. Thus, the nine compounds in the cycle along with the enzymes constitute a chemical machine for the metabolism of acetyl CoA. After considering the functioning of the catalytic machinery, we will consider the mechanisms for maintenance of the machinery.

Most of the information about the tricarboxylic acid cycle has come from a study of tissue homogenates. If pyruvate is added to a brei of broken cells, it will be oxidized to CO_2 and water. The addition of any one of the members of the tricarboxylic acid cycle will act catalytically; for example, several equivalents of pyruvate will be oxidized for each equivalent of oxaloacetate added. If the tissue homogenate is fractionated by centrifugation, the tricarboxylic acid cycle enzymes will be found in the mitochondria.

The first five compounds (Figure 5.9) in the cycle are shown with two of the carbons enclosed. These are the carbons which originate from the acetyl CoA. The CO_2 is formed from the carboxyl groups that originated from oxaloacetate. The

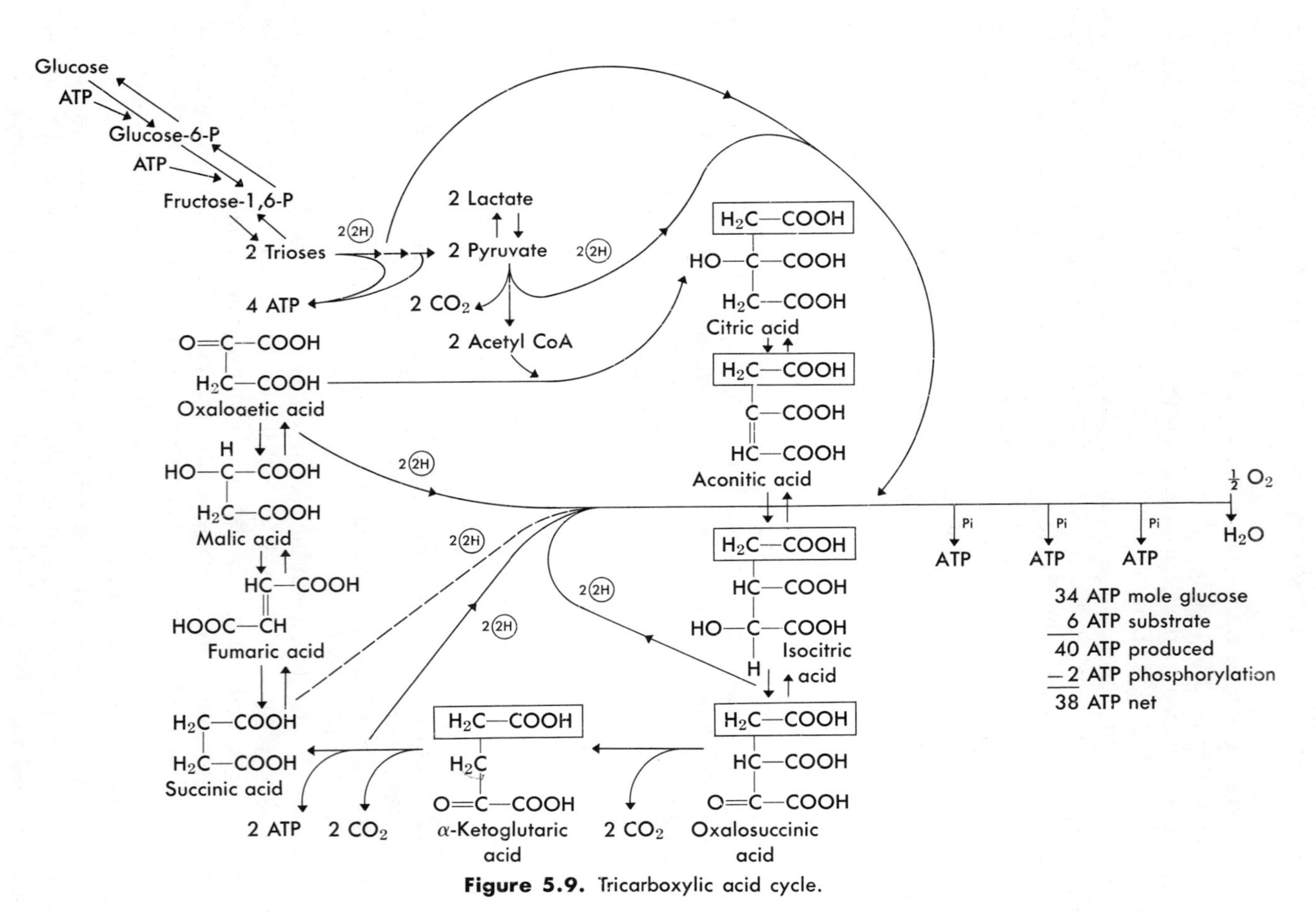

Figure 5.9. Tricarboxylic acid cycle.

carbons from acetyl CoA do not appear as CO_2 until the second turn of the cycle. Succinic acid and fumaric acid exhibit symmetrical behavior as expected, in contrast to citric acid which appears to be an asymmetric compound. This problem is presented on page 140.

Individual Reactions of the Tricarboxylic Acid Cycle

Formation of Acetyl CoA

The details of the following reaction were presented in pages 72–79 and should be reviewed.

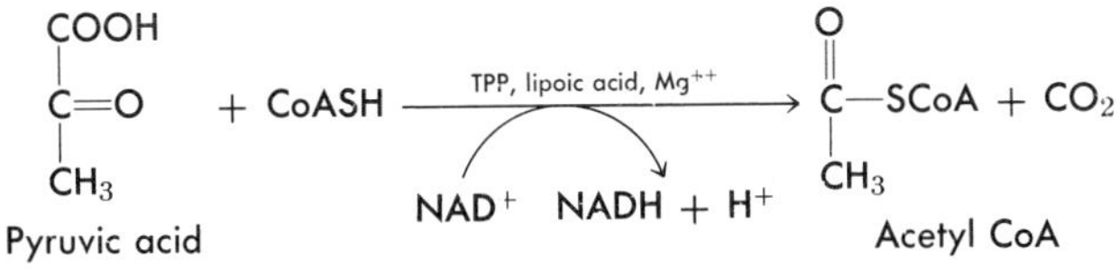

$$\underset{\text{Pyruvic acid}}{CH_3-CO-COOH} + CoASH \xrightarrow[NAD^+ \;\rightarrow\; NADH + H^+]{\text{TPP, lipoic acid, } Mg^{++}} \underset{\text{Acetyl CoA}}{CH_3-\overset{O}{\overset{\|}{C}}-SCoA} + CO_2$$

The major sources of acetyl CoA are pyruvic acid and fatty acids. There is a very small amount of acetic acid in the tissues. This can be activated by reacting with ATP in the following manner

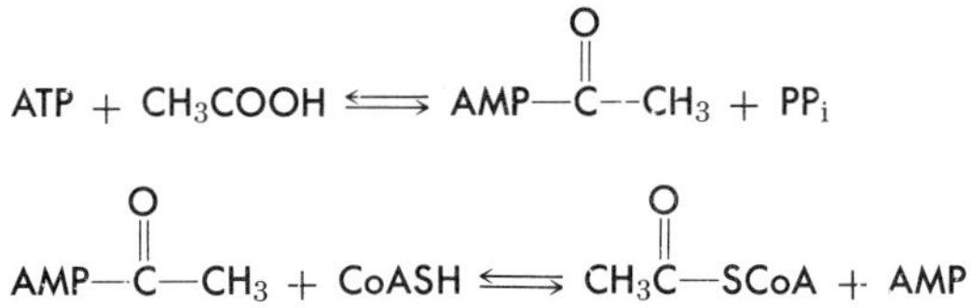

$$ATP + CH_3COOH \rightleftharpoons AMP-\overset{O}{\overset{\|}{C}}-CH_3 + PP_i$$

$$AMP-\overset{O}{\overset{\|}{C}}-CH_3 + CoASH \rightleftharpoons CH_3\overset{O}{\overset{\|}{C}}-SCoA + AMP$$

Formation of Citric Acid

Acetyl CoA is a high-energy compound which can react as a biologic acetylating agent, or can enter into condensation reactions. The enzyme catalyzing the reaction of acetyl CoA with oxaloacetate is the condensing enzyme. The methyl group is activated in the reaction

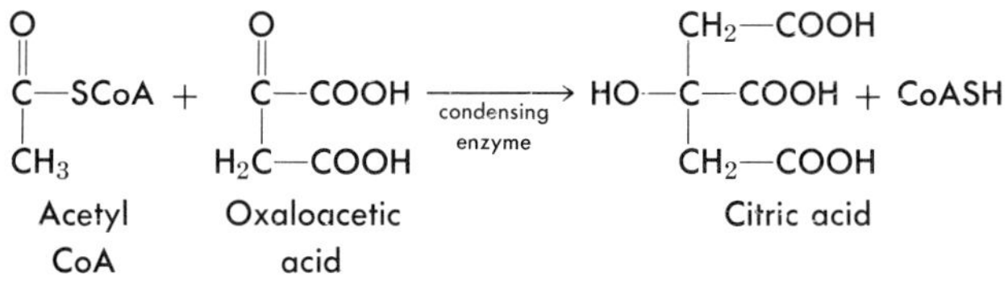

$$\underset{\text{Acetyl CoA}}{CH_3-\overset{O}{\overset{\|}{C}}-SCoA} + \underset{\text{Oxaloacetic acid}}{H_2C(COOH)-\overset{O}{\overset{\|}{C}}-COOH} \xrightarrow[\text{enzyme}]{\text{condensing}} \underset{\text{Citric acid}}{HO-C(COOH)(CH_2-COOH)_2} + CoASH$$

Although the equilibrium for this reaction lies far to the right, it can be reversed and citric acid can be a source of acetyl CoA.

Citric Acid, cis-*Aconitic Acid, and Isocitric Acid*

The following sequence of reactions

$$\begin{array}{c} H_2C-\overset{*}{C}OOH \\ | \\ HO-C-COOH \\ | \\ H_2C-COOH \end{array} \underset{-H_2O}{\overset{\text{aconitase}\atop +H_2O}{\rightleftharpoons}} \begin{array}{c} H_2C-\overset{*}{C}OOH \\ | \\ C-COOH \\ \| \\ HC-COOH \end{array} \underset{+H_2O}{\overset{\text{aconitase}\atop -H_2O}{\rightleftharpoons}} \begin{array}{c} H_2C-\overset{*}{C}OOH \\ | \\ HC-COOH \\ | \\ HO-C-COOH \\ H \end{array}$$

Citric acid — *cis*-Aconitic acid — Isocitric acid

is catalyzed by one enzyme, aconitase. At equilibrium the respective amounts of the three acids are 90 per cent, 4 per cent, and 6 per cent. The reaction proceeds to the right in respiring tissues because isocitric acid is removed in the next reaction of the cycle.

Aconitase is an Fe^{++} containing enzyme which, according to a recent proposal, may form a carbonium ion with citric acid, *cis*-aconitic acid, or isocitric acid.

If the carboxyl group of acetate is labeled with carbon C^{14}, the label will appear as shown (*). Ogston explained this phenomenon by suggesting that citric acid became unsymmetric when attached to the enzyme surface. If the enzyme has three specific points of attachment, one for each, the $—CH_2COOH$, the —OH, and the —COOH of citric acid, then the two $—CH_2COOH$ groups are no longer equivalent to each other and the enzyme can act unsymmetrically. This situation is illustrated in Figure 5.10.

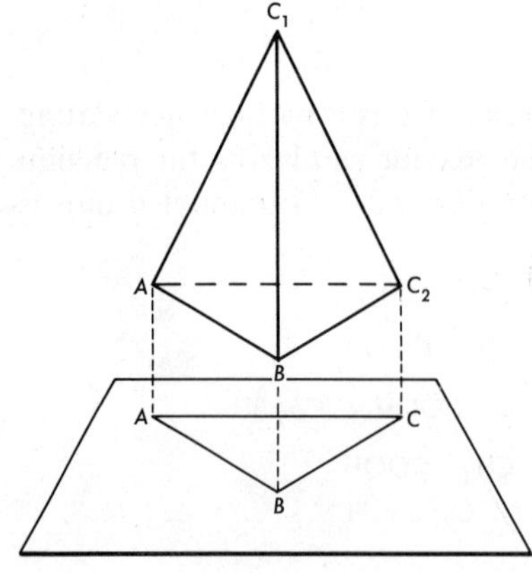

Figure 5.10. Three-point attachment of enzyme.

In the diagram citric acid is represented by a tetrahedron in which A, B, C_1 and C_2 are —OH, —COOH, —CH_2COOH, and —CH_2COOH, respectively. The only face of the tetrahedron which will fit on the fixed sites on the enzyme surface is ABC_2. If a face containing C_1 is fixed at point C on the enzyme, it is impossible to fit A and B to their respective points on the enzyme.

When fluoroacetic acid (FCH_2COOH) is introduced into the system in the place of acetate, it condenses with oxaloacetate to form fluorocitric acid which competitively inhibits the conversion of citric acid to *cis*-aconitic acid. If one of the hydrogens on the methylene carbon of oxaloacetate is replaced with fluorine, the condensation with acetyl CoA takes place to form a fluorocitric acid which behaves differently. It does not inhibit aconitase but inhibits isocitric dehydrogenase.

Formation of α-Ketoglutaric Acid

Apparently the following two-step reaction is catalyzed by one enzyme.

$$
\begin{array}{c}
CH_2\text{—}COOH \\
| \\
HC\text{—}COOH \\
| \\
HO\text{—}C\text{—}COOH \\
| \\
H \\
\text{Isocitric acid}
\end{array}
\xrightleftharpoons[\text{NADP}^+ \;\to\; \text{NADPH} + H^+]{\text{isocitric dehydrogenase}}
\begin{array}{c}
CH_2\text{—}COOH \\
| \\
HC\text{—}COOH \\
| \\
O{=}C\text{—}COOH \\
\text{Oxalosuccinic acid}
\end{array}
\xrightleftharpoons[\to\; CO_2]{\text{carboxylase}}
\begin{array}{c}
CH_2\text{—}COOH \\
| \\
CH_2 \\
| \\
O{=}C\text{—}COOH \\
\alpha\text{-Ketoglutaric acid}
\end{array}
$$

Again the reaction is reversible, but the equilibrium is far to the right.

It is at this point in the cycle that for each mole of acetate, 1 mole of reduced NADPH and 1 mole CO_2 are produced. Three moles of ATP are produced per mole of NADPH, the same as for NADH. In some tissues NAD is used in place of NADP for the above reaction.

Oxidative Decarboxylation of α-Ketoglutarate

The formation of succinyl CoA from α-ketoglutarate is analogous to the oxidative decarboxylation of pyruvic acid. This reaction also requires TPP, Mg^{++}, lipoic acid and CoASH:

$$
\begin{array}{c}
\boxed{CH_2\text{—}COOH} \\
| \\
CH_2 \\
| \\
O{=}C\text{—}COOH \\
\alpha\text{-Ketoglutaric acid}
\end{array}
+ \text{CoASH}
\xrightarrow[\text{NAD}^+ \;\to\; \text{NADH} + H^+]{\text{TPP, } Mg^{++}\text{, lipoic acid}}
\begin{array}{c}
\boxed{CH_2\text{—}COOH} \\
| \qquad O \\
| \qquad \| \\
CH_2C\text{—}SCoA + CO_2 \\
\text{Succinyl CoA}
\end{array}
$$

The reaction is not reversible, although other reactions in the cycle can be reversed. Therefore, the entire cycle becomes irreversible.

This is the step in the cycle where it returns to a four-carbon compound. The second CO_2 is liberated in this reaction and another pair of electrons is transferred to NAD. Two carbons were introduced as acetyl CoA and two carbons have been liberated. However, neither of the liberated carbons came from acetate.

Succinyl CoA is a high-energy compound which enters into reactions other than its cyclic function; for example, it is an intermediate in the biosynthesis of the porphyrin molecule. The energy can be transferred to ATP by the following reactions.

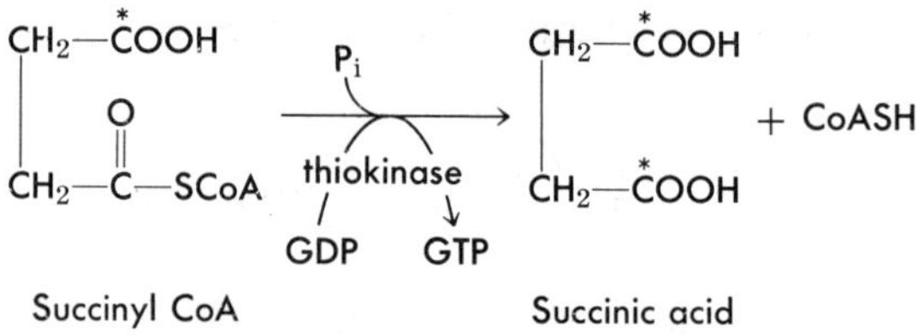

$$\underset{\text{Succinyl CoA}}{\begin{array}{l} CH_2-\overset{*}{C}OOH \\ | \\ CH_2-\overset{O}{\overset{\|}{C}}-SCoA \end{array}} \xrightarrow[\text{GDP} \searrow \text{GTP}]{P_i \text{ thiokinase}} \underset{\text{Succinic acid}}{\begin{array}{l} CH_2-\overset{*}{C}OOH \\ | \\ CH_2-\overset{*}{C}OOH \end{array}} + CoASH$$

$$GTP + ADP \rightleftharpoons ATP + GDP$$

Succinic acid, being symmetrical, behaves as a symmetrical compound in its enzymatic reactions. Thus, the carbons become randomized. If the carboxyl of acetyl CoA is labeled, the label is found in both carboxyl groups of succinic acid. The specific activity of each carbon is one half that of the acetate carboxyl. Only one carbon is labeled in a single molecule. The situation is as if the labeled carboxyl group at one end of the molecule were diluted by an equal number of unlabeled molecules from the other end.

Succinic Dehydrogenase

Succinic dehydrogenase is a flavin-containing protein which accepts electrons in the form of hydride from succinic acid to form fumaric acid.

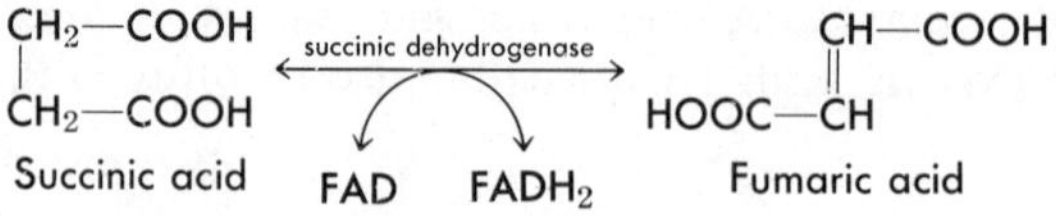

$$\underset{\text{Succinic acid}}{\begin{array}{l} CH_2-COOH \\ | \\ CH_2-COOH \end{array}} \underset{\text{FAD} \searrow \text{FADH}_2}{\overset{\text{succinic dehydrogenase}}{\rightleftharpoons}} \underset{\text{Fumaric acid}}{\begin{array}{r} CH-COOH \\ \| \\ HOOC-CH \end{array}}$$

Malonic acid is a competitive inhibitor for succinic dehydrogenase (page 62) and was an important tool in the study of the tricarboxylic acid cycle. If malonic acid is added to a homogenate of a tissue which is oxidizing pyruvic acid to CO_2 and H_2O, pyruvic acid will soon cease to be utilized and succinic acid will accumulate. If oxaloacetate is then added, pyruvic acid will be used and succinic acid appear.

However, 1 mole of oxaloacetate will be required for each mole of pyruvate oxidized. Oxaloacetate thus no longer behaves in a catalytic manner. What would be the effect of adding α-ketoglutaric acid or malic acid to the inhibited system?

This is the first time we have encountered the formation of a double bond by the removal of (2 H). This type of reaction usually requires a FAD-containing enzyme rather than NAD. Reduced FAD then transfers its electrons to the electron transport system. Since the electrons are entering the system at a more positive potential, only 2 moles of ATP are formed for each pair of electrons. This accounts for the overall production of 34 ATP moles from 12 pairs of electrons (12×2 H) instead of 36 as would be the case if all the electrons were transferred through NAD (see Figure 5.9).

Formation of Malic Acid

The next reaction is the addition of water to fumaric acid; the enzyme is fumarase.

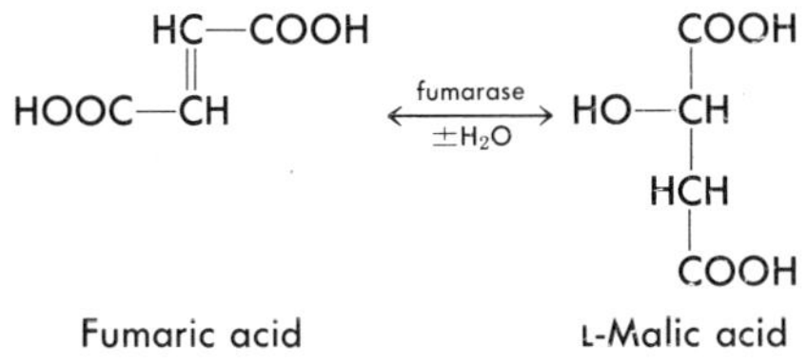

$$\underset{\text{Fumaric acid}}{\begin{matrix} HC{-}COOH \\ \| \\ HOOC{-}CH \end{matrix}} \quad \underset{\pm H_2O}{\overset{\text{fumarase}}{\longleftrightarrow}} \quad \underset{\text{L-Malic acid}}{\begin{matrix} COOH \\ | \\ HO{-}CH \\ | \\ HCH \\ | \\ COOH \end{matrix}}$$

Formation of Oxaloacetic Acid

The last step in the cycle is the dehydrogenation of malic acid to form oxaloacetic acid; the enzyme is malic dehydrogenase.

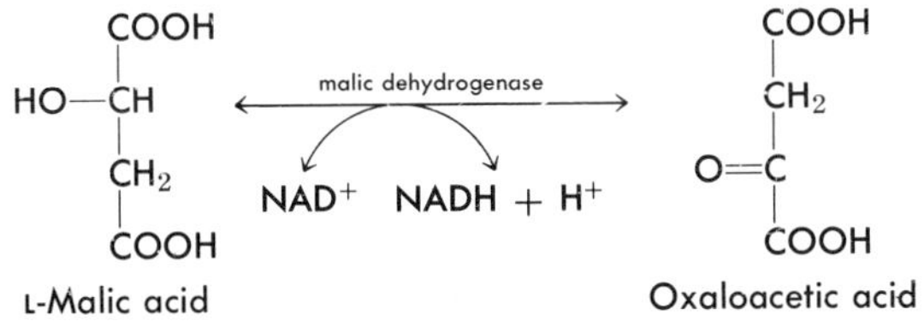

$$\underset{\text{L-Malic acid}}{\begin{matrix} COOH \\ | \\ HO{-}CH \\ | \\ CH_2 \\ | \\ COOH \end{matrix}} \quad \underset{NAD^+ \curvearrowright NADH + H^+}{\overset{\text{malic dehydrogenase}}{\longleftrightarrow}} \quad \underset{\text{Oxaloacetic acid}}{\begin{matrix} COOH \\ | \\ CH_2 \\ | \\ O{=}C \\ | \\ COOH \end{matrix}}$$

The fourth and last pair of electrons from the cycle is transferred to NAD in this step.

Oxaloacetate can now react with acetyl CoA and repeat the cycle. The carbon atoms from the first acetate will appear as CO_2 during subsequent turns of the cycle. The student should follow the carboxyl carbon and the α-carbon of an acetate through two or more turns of the cycle and note the amount of each carbon liberated at each turn.

Table 5.1
Yield of ATP

Reaction	Coenzyme	Yield ATP/mole glucose
Electron Transport		
3-Phosphoglyceraldehyde ⟶ 3-phosphoglycerate	NAD	6
Pyruvate ⟶ acetyl CoA	NAD	6
Isocitrate ⟶ α-ketoglutarate	NADP	6
α-Ketoglutarate ⟶ succinyl CoA	NAD	6
Succinate ⟶ fumarate	FAD	4
Malate ⟶ oxaloacetate	NAD	6
Substrate Level		
3-Phosphoglyceraldehyde ⟶ 3-phosphoglycerate		2
Phosphoenolpyruvate ⟶ pyruvate		2
Succinyl CoA ⟶ succinate		2
	Total produced	40
	Used for phosphorylation of glucose	−2
	Net	38

Energy From Oxidation of Glucose

Table 5.1 summarizes the yield of ATP obtained from the oxidation of glucose.

Maintenance of the Tricarboxylic Acid Cycle

It has been pointed out that certain of the compounds in the pathway of glucose oxidation could be used for purposes other than glucose utilization. Succinyl CoA is an intermediate in porphyrin synthesis, α-ketoglutarate can be removed from the cycle by the transamination reaction (page 220), oxaloacetate can be converted to aspartic acid, pyruvic acid can be converted to alanine, and so forth. Although most of these reactions are reversible and can contribute to the cycle, such reactions probably constitute a removal of compounds needed for its operation. In the same way that addition of a small amount of one of the compounds of the cycle can lead to the oxidation of large amounts of glucose, the removal of small amounts of one of the compounds will decrease the rate of oxidation of glucose by a large amount.

An effective means of maintaining the cycle is by CO_2 fixation to pyruvic acid or phosphoenolpyruvic acid. The malic enzyme has been considered for this purpose. However, reduced NADP is required, and very little NADPH is available in the mitochondria. This enzyme is more likely involved in extramitochondrial lipogenesis by converting via malate to pyruvate and generating NADPH.

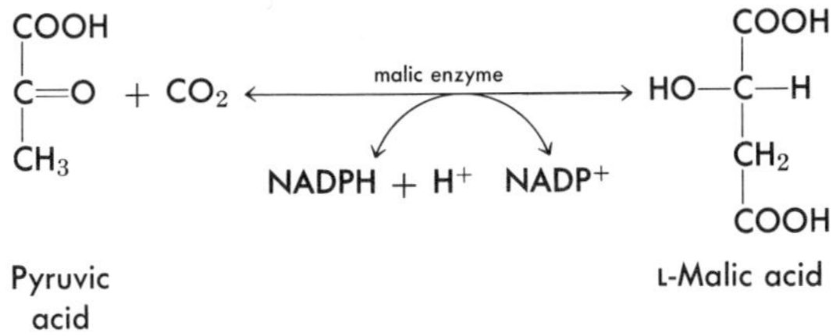

$$\underset{\text{Pyruvic acid}}{\begin{matrix}COOH\\|\\C{=}O\\|\\CH_3\end{matrix}} + CO_2 \xleftrightarrow[\text{NADPH + H}^+ \;\; \text{NADP}^+]{\text{malic enzyme}} \underset{\text{L-Malic acid}}{\begin{matrix}COOH\\|\\HO{-}C{-}H\\|\\CH_2\\|\\COOH\end{matrix}}$$

A second enzyme of importance is phosphoenolpyruvate carboxykinase which proceeds by the following reaction and utilizes either inosine triphosphate (ITP) or guanosine triphosphate (GTP). Again this reaction is important for the production of phosphoenolpyruvate and is the key reaction in the reversal of gluconeogenesis (page 119).

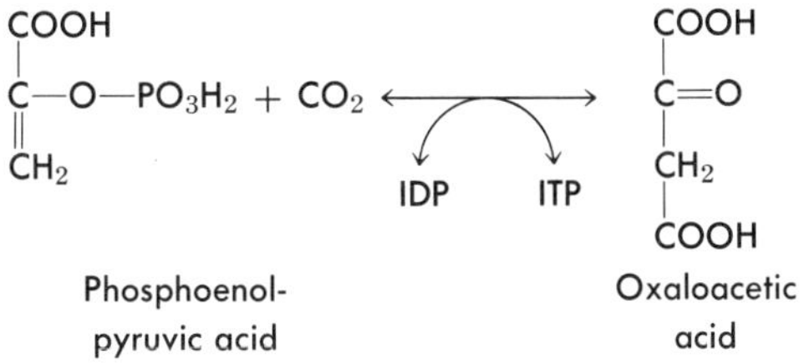

$$\underset{\text{Phosphoenol-pyruvic acid}}{\begin{matrix}COOH\\|\\C{-}O{-}PO_3H_2\\||\\CH_2\end{matrix}} + CO_2 \xleftrightarrow[\text{IDP} \;\; \text{ITP}]{} \underset{\text{Oxaloacetic acid}}{\begin{matrix}COOH\\|\\C{=}O\\|\\CH_2\\|\\COOH\end{matrix}}$$

Pyruvate carboxylase is located predominately in the mitochondria and is probably responsible for the synthesis of oxalacetate by the following reaction

$$\begin{matrix}COOH\\|\\C{=}O\\|\\CH_3\end{matrix} + CO_2 + ATP \rightleftharpoons \begin{matrix}COOH\\|\\C{=}O\\|\\CH_2\\|\\COOH\end{matrix} + ADP + P_i$$

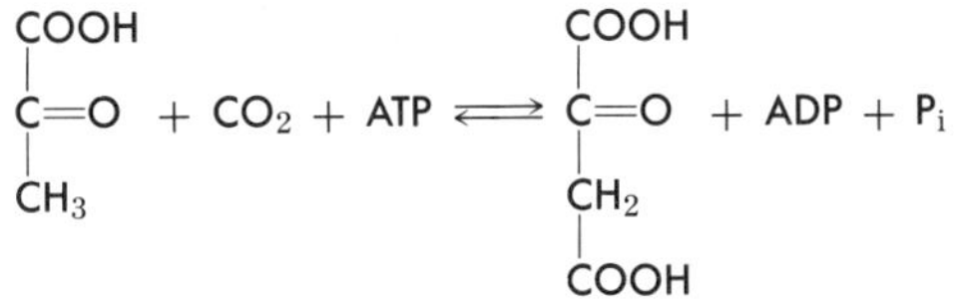

OTHER PATHWAYS FOR CARBOHYDRATE METABOLISM

The so-called "hexose monophosphate shunt," or "phosphogluconate oxidative pathway," is an important system for the metabolism of glucose. Whereas glycolysis and the citric acid cycle operate to produce NADH for electron transport, this series of reactions produces important chemical intermediates and reduced NADP for biologic reductions.

The pertinent aspects of the system are shown in Figure 5.11.

From 1 mole of glucose-6-phosphate, 2 moles of reduced NADP and 1 of CO_2 are produced in the steps leading to ribulose-5-phosphate. These are the only de-

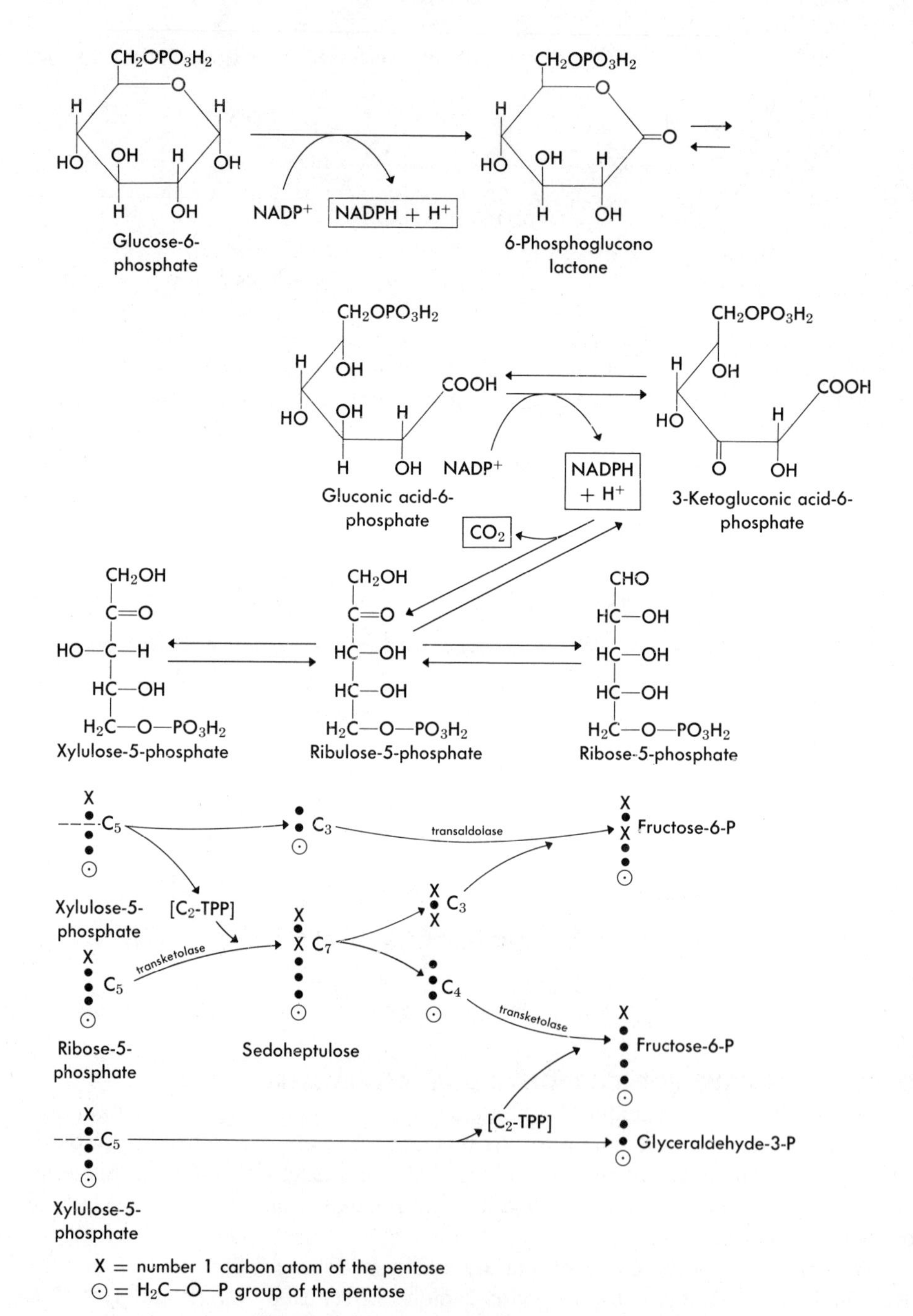

Figure 5.11. Hexose monophosphate shunt.

hydrogenation and decarboxylation reactions in the entire system. There is not a progressive degradation of glucose to CO_2 and H_2O. Ribulose-5-phosphate, the product of the decarboxylation of gluconic acid-6-phosphate, is isomerized to ribose-5-phosphate an important intermediate in the synthesis of nucleotides and nucleic acids.

The lower portion of the figure shows the fate of pentose phosphates. Starting with three pentose phosphates, a series of transformations takes place furnishing 2 moles of fructose-6-phosphate and 1 mole of glyceraldehyde-3-phosphate. There are 15 carbons in the three pentoses and 15 in the two hexoses and one triose. These compounds then enter into reactions previously discussed.

The reaction may be summarized as follows

$$3 \text{ glucose-6-P} + 6 \text{ NADP}^+ \rightleftharpoons 2 \text{ fructose-6-P} + 1 \text{ glyceraldehyde-3-P} + 6 \text{ NADPH} + 6 \text{ H}^+$$

The conversion of the pentoses to hexoses requires three transfer reactions. In two of the reactions the 1 and 2 carbon atoms of a pentose are transferred to an aldose. The enzymes are transketolases and require thiamine pyrophosphate as a coenzyme. The reactions are

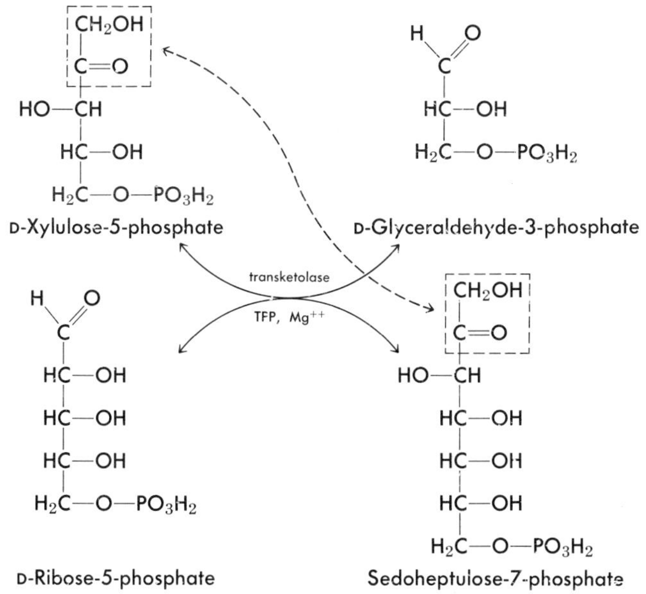

```
CH2OH                                H   O
 |                                    \ //
 C=O                                   C
 |                                     |
HO—CH                                 HC—OH
 |                                     |
 HC—OH                                H2C—O—PO3H2
 |
 H2C—O—PO3H2
D-Xylulose-5-phosphate               D-Glyceraldehyde-3-phosphate

                 transketolase
                 TFP, Mg++

 H   O                                CH2OH
  \ //                                 |
   C                                   C=O
   |                                   |
  HC—OH                              HO—CH
   |                                   |
  HC—OH                               HC—OH
   |                                   |
  HC—OH                               HC—OH
   |                                   |
  H2C—O—PO3H2                         HC—OH
                                       |
                                      H2C—O—PO3H2
D-Ribose-5-phosphate                 Sedoheptulose-7-phosphate
```

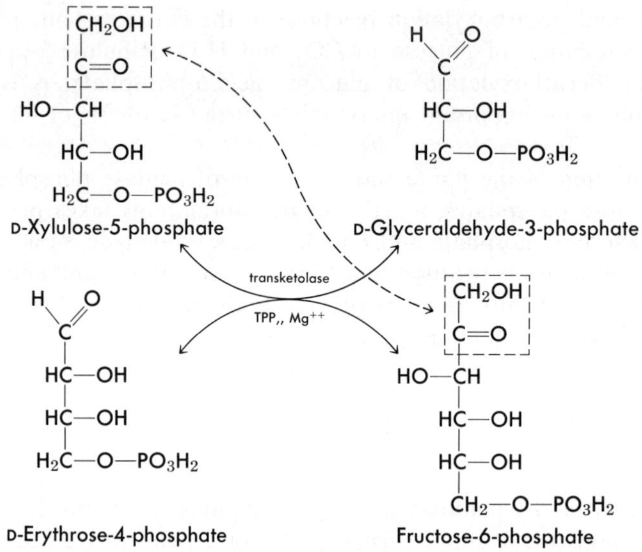

CH_2OH
$C{=}O$
HO—CH
HC—OH
$H_2C—O—PO_3H_2$
D-Xylulose-5-phosphate

H O
C
HC—OH
$H_2C—O—PO_3H_2$
D-Glyceraldehyde-3-phosphate

transketolase
TPP,, Mg^{++}

H O
C
HC—OH
HC—OH
$H_2C—O—PO_3H_2$
D-Erythrose-4-phosphate

CH_2OH
$C{=}O$
HO—CH
HC—OH
HC—OH
$CH_2—O—PO_3H_2$
Fructose-6-phosphate

The two carbon atoms are first transferred to TPP to form an active glycolaldehyde (see page 72).

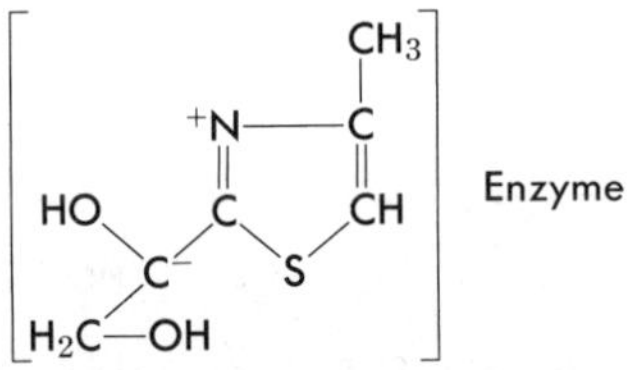

CH_3
^{+}N—C
HO C CH
C^{-} S
$H_2C—OH$
Enzyme

The third transfer, catalyzed by a transaldolase, shifts a dihydroxyacetone residue to an aldose.

The transketolases and the transaldolase are specific for optical configurations shown at carbon atoms 3 and 4. Ribulose-5-phosphate does not serve as a substrate for transketolase.

Significance of the Hexose Monophosphate Shunt

The evaluation of the role of the shunt pathway in the metabolism of glucose in tissues has been studied with labeled compounds. If glucose is labeled with C^{14} in the 1 position, the label is not immediately available as CO_2 in the tricarboxylic acid cycle pathway, whereas it is all liberated as CO_2 early in the shunt pathway. It can be seen from Figure 5.11 that glucose labeled with C^{14} in the 2 position would give pentoses labeled in the 1 position and that two thirds of the label will appear

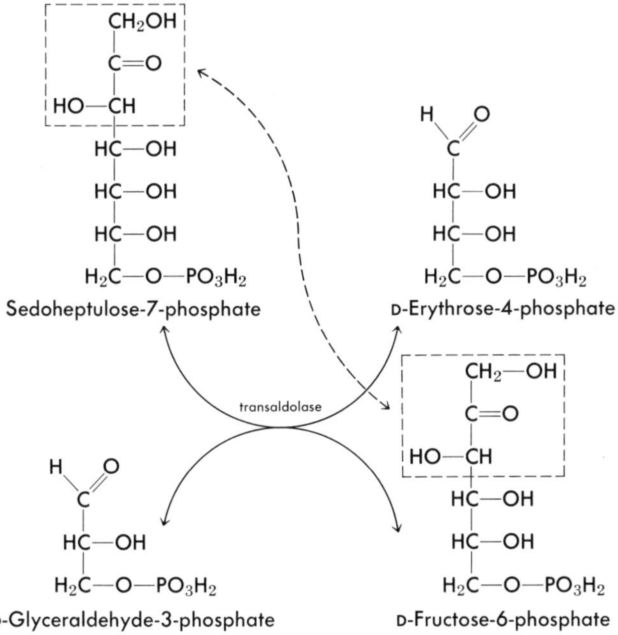

Sedoheptulose-7-phosphate

D-Erythrose-4-phosphate

transaldolase

D-Glyceraldehyde-3-phosphate

D-Fructose-6-phosphate

in the 1 position and one third in the 3 position of fructose-6-phosphate.

Studies based on these labeling patterns have shown that the shunt pathway is a minor pathway in muscle, accounting for no more than 10 or 15 per cent of the oxidation of glucose. About one half of the glucose metabolized by the kidney may enter the shunt mechanism. The shunt pathway may be the major course for the metabolism of glucose in the liver.

The separate roles of NAD and NADP deserve comment. If NADP could enter the electron transport system and 3 moles of ATP were produced for each mole, the phosphogluconate pathway could furnish 36 moles of ATP for each mole of glucose-6-phosphate oxidized. Although this is possible, the evidence indicates that reduced NADP is very slowly oxidized. In normal respiring cells NAD is largely in the oxidized form and NADP in the reduced form. Reduced NADP can then provide reduced coenzymes for essential reduction reactions in cells which are actively respiring.

The principal sources of NADPH are the isocitric dehydrogenase reaction in the Krebs cycle and the phosphogluconate pathway. The reduced NADP is used for fatty acid synthesis (page 194), steroid synthesis (page 205), CO_2 fixation (page 145), maintenance of glutathione in the reduced form, and many other reactions.

The Glucuronate Pathway

This pathway for the catabolism of glucose appears important as a source of certain sugars, and of ascorbic acid as well. Figure 5.12 illustrates the essential reactions involved in these transformations. The enzymes for this pathway are found in animal tissues and in the tissues of higher plants. Among the primates a block occurs in the transformation of L-gulonolactone to 2-keto-L-gulonate, and the primate cannot synthesize ascorbic acid. For this reason ascorbic acid is an essential vitamin for primates, and for guinea pigs as well, although other animals such as rats synthesize their own ascorbic acid.

Another block is known to occur at the L-xylulose to xylitol reaction. Among humans an hereditary condition known as *idiopathic pentosuria* is characterized by the excretion of relatively large amounts of L-xylulose in the urine. This condition is a benign one and occurs because of a hereditary inability to synthesize the enzyme which reduces L-xylulose to xylitol.

The two oxidations in this pathway require NAD, whereas the two reduction reactions require NADPH. The interconversion of the D- to L-xylulose proceeds by way of xylitol and the operation of an oxidation and a reduction.

This catabolic pathway is not of great significance in energy production, except as it feeds D-xylulose-5-phosphate into the hexose monophosphate shunt. It does, however, supply ascorbate in animals which possess the entire complement of enzymes, and produces other metabolically significant sugars such as D-glucuronate and D-xylose.

Photosynthesis

Many of the reactions used by the plant for the conversion of CO_2 to glucose are similar to those in the "shunt pathway." An authoritative review is available, and we recommend its use (see References, page 157). The pathway of CO_2 fixation in photosynthesis is illustrated in the Appendix, Figure A-12.

Cellular Regulation of Carbohydrate Metabolism

In the chapter on enzymes (page 51), it was pointed out that small molecules could behave as effectors in modulating the activity of enzymes. Cells producing abundant ATP will be low in AMP, and cells deficient in ATP will contain large quantities of AMP and ADP. A number of recent observations suggest that stimulation by AMP and inhibition by ATP are important in regulatory mechanisms in carbohydrate metabolism. Phosphofructokinase is stimulated by AMP and cyclic AMP and inhibited by ATP. AMP and ADP are positive effectors for isocitric dehydrogenase. Isocitric dehydrogenase and citrate synthetase are strongly inhibited by ATP. Atkinson has integrated these observations along with others into the regulatory scheme shown in Figure 5.13. This diagram is not complete but serves to show the role of adenylic acids in regulation.

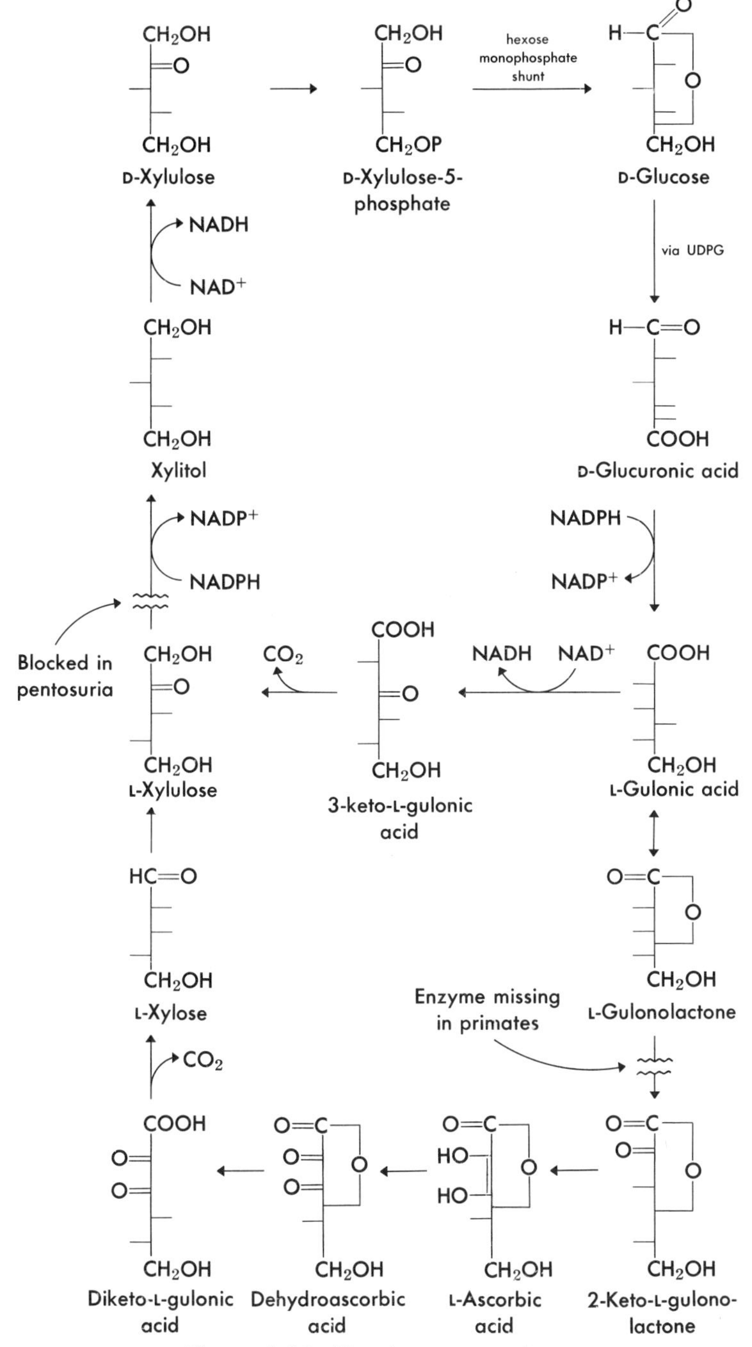

Figure 5.12. The glucuronate pathway.

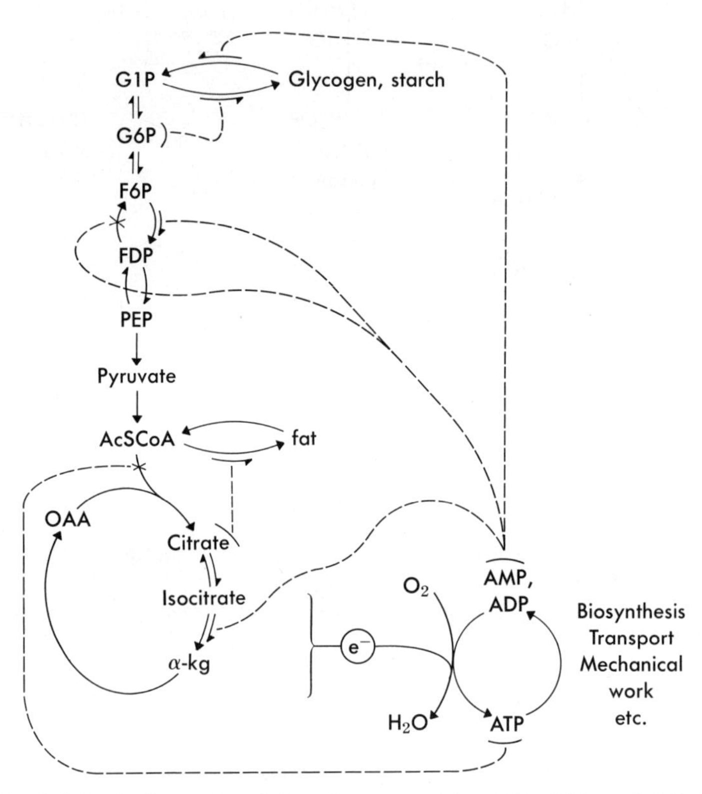

Figure 5.13. Schematic illustration of the role proposed for AMP, ADP, and ATP, in regulation of energy metabolism. Broken lines connect effector compounds (indicated by heavy arcs) to the enzymes which they modulate. Positive effector action is denoted by a heavy arrow, and negative effector action by a heavy cross. Abbreviations are: G1P, glucose-1-phosphate; G6P, glucose-6-phosphate; F6P, fructose-6-phosphate; FDP, fructose-1,6-diphosphate; PEP, phosphoenolpyruvate; AcSCoA, acetyl coenzyme A; α-kg, α-ketoglutarate; OAA, oxaloacetate. Supply of electrons (from oxidative reactions in the glycolytic and Krebs cycle pathways) to the electron transport phosphorylation system is indicated by the symbol e^-. (From D. E. Atkinson: *Science*, **150:**851, 1965.)

Let us consider first the effect of a high concentration of ATP on the utilization of acetyl CoA. The affinity of citrate synthetase for acetyl CoA will be diminished and iscitric dehydrogenase activity will be depressed. The concentration of citrate will be relatively high, and it is known that citrate stimulates fatty acid synthesis. The net result is an increased utilization of acetyl CoA for fatty acid biosynthesis.

The conversion of fructose-6-phosphate to fructose diphosphate is modulated by the positive effect of AMP and the negative effect of ATP. The metabolically reverse reaction is catalyzed by fructose diphosphate phosphatase. This enzyme is inhibited by AMP and stimulated by ATP. High levels of ATP would then lead to high levels of glucose-6-phosphate which has been shown to stimulate the action of the enzyme for transfering glycosyl groups to glycogen. Under these conditions the enzymes are modulated in the direction of glycogen synthesis. A high level of AMP would stimulate the enzymes in the direction of carbohydrate breakdown and ATP formation.

In addition to metabolic control by effectors, regulation is also dependent upon compartmentalization. As noted previously the citric acid cycle enzymes are found in the mitochondria, whereas glycolytic enzymes are found in the cytoplasm. One of the effects of such compartmentalization is illustrated with respect to gluconeogenesis.

Glucogenesis occurs mainly in the kidney or liver, where lactate or pyruvate arising from exercising muscle are converted to glucose and returned to muscle for utilization. Gluconeogenesis from three- and four-carbon compounds, such as the acids of the citric acid cycle, is also important in normal animals and becomes rapid during diabetes. The enzyme phosphoenol carboxykinase (PEP carboxykinase, page 145) is generally considered to be the key enzyme in gluconeogenesis from those compounds, although pyruvate carboxylase (page 145) also operates in some tissues. It should be noted, also, that reversal of PEP carboxykinase is important in replenishing the four-carbon acids of the cycle which are depleted by use in synthetic reactions.

The complications in gluconeogenesis which may arise by compartmentalization are seen in comparing liver cells of various animals. In rabbits and chickens PEP carboxykinase is found in mitochondria, whereas it is in the cytoplasm in rat and mouse. Therefore, in the first two animals, phosphoenolpyruvate is formed in the mitochondria from four-carbon precursurs, and then released to the cytoplasm where glucogenesis occurs. In the rat and mouse, the process described by Lardy and coworkers (Figure 5.14) operates.

Within the mitochondria, oxalacetate is transformed to malate or aspartate or both; the products are transported to the cytoplasm where they are reconverted to oxalacetate, then phosphoenolpyruvate and, ultimately, glucose. The malic enzyme may aid somewhat, but is primarily a source of NADPH for synthetic reactions. Pyruvate carboxylase does not operate to any extent in the process of gluconeogenesis.

When we consider the role of effectors, compartmentalization and hormones in the regulation of cellular metabolism, it is apparent that control of metabolic reactions is complex and precise. The above discussion does not include all of our present knowledge and new information is being obtained at a rapid pace. The discussion does not consider the effect of metabolites on enzyme levels. This will be dealt with in connection with protein synthesis.

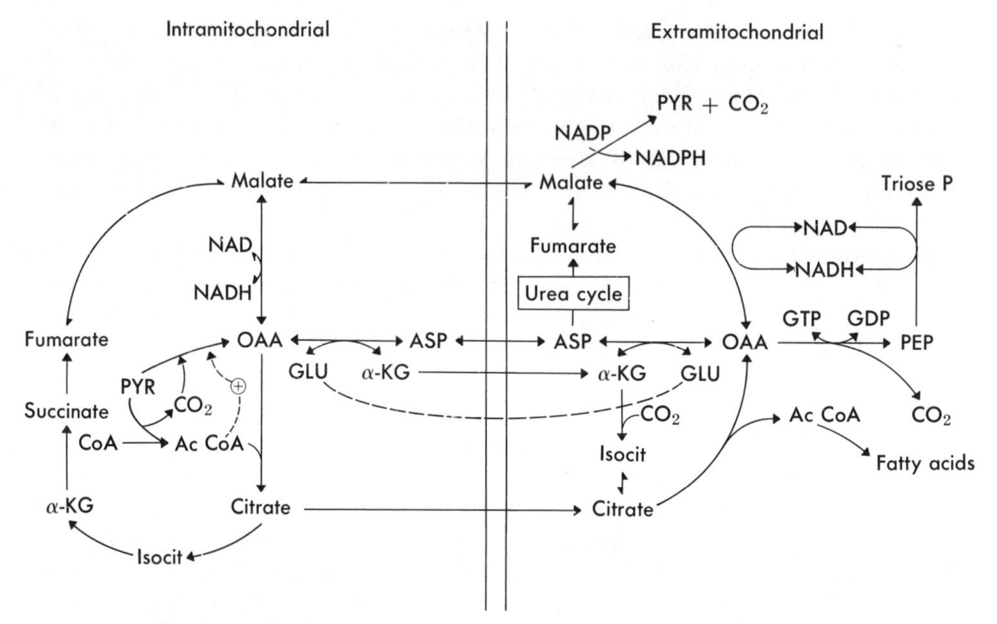

Figure 5.14. (From H. A. Lardy et al.: *Biochemistry,* **53:**1413, 1965.)

HORMONAL CONTROL OF CARBOHYDRATE METABOLISM

Insulin

It had been known from late in the nineteenth century that removal of the pancreas from a dog would result in diabetes. In 1922, Banting and Best prepared extracts from the pancreas which would relieve the diabetic condition. Insulin has been prepared in crystalline form from such extracts. The proof of its structure is one of the classic advances in biochemistry (pages 30–33).

The removal of the pancreas from a dog produces the following changes: severe hyperglycemia and glucosuria; an increase in the breakdown of tissue proteins; ketone bodies (page 188) appear in both the blood and urine resulting in severe acidosis; an increase in volume of water excreted in the urine resulting in dehydration; and the excretion of ingested glucose in the urine. Injection of insulin will correct all these changes. The amount of glycogen in the liver of a diabetic animal is normal.

Insulin acts on muscle and adipose tissue by increasing the rate of transport of glucose across membranes into cells. Most of the changes seen in the diabetic animal can be explained on the basis of the depletion of intermediate compounds normally available from glucose metabolism; for example, NADPH, required for fatty acid synthesis, is supplied by carbohydrate metabolism. There will be a deficiency of tricarboxylic acid cycle intermediates not only for metabolizing acetyl CoA but for furnishing intermediates for other important pathways. It is apparent that a disturbance at one point in carbohydrate metabolism can lead to major changes in many metabolic pathways. This intertwining of the biochemical reactions makes it difficult to pinpoint the mechanism of action of insulin.

Although there is general agreement that the primary effect of insulin is on the transport of glucose across membranes into cells of the muscle, it is possible that there may be a direct effect on reactions that control the concentration of glucose-6-phosphate. The effects of insulin on liver tissue seem to be much more complex than on extrahepatic tissue. Liver contains two kinases, hexokinase and glucokinase, for phosphorylating glucose. Hexokinase activity is not decreased in the diabetic animal, whereas glucokinase activity is low. Apparently the primary effect of insulin on liver cells is not on transport.

Severe diabetes can be produced in animals by feeding alloxan which destroys the β-cells of the islets of Langerhans. Diabetic animals of this type have been useful for the study of carbohydrate metabolism.

Except for the hyperglycemia and glucosuria, severe starvation will result in most of the changes seen in the diabetic animal. The diabetic animal is a starving animal bathed in a sea of glucose.

Certain sulfonylureas stimulate the β cells to increase insulin production. In order for these compounds to be effective, functional tissue must be present. These compounds are used for the oral treatment of diabetes, particularly in mild cases among older patients. Tolbutamide is an example of this class of substances.

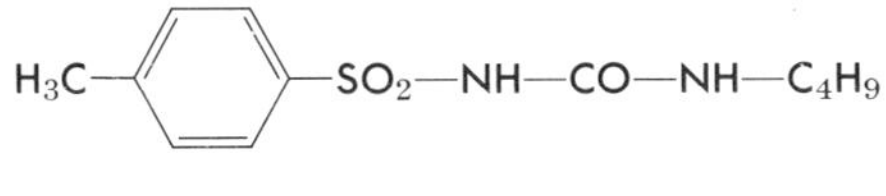

$$H_3C-C_6H_4-SO_2-NH-CO-NH-C_4H_9$$

Glucagon

Glucagon is a polypeptide secreted by the α cells of the pancreas. It increases the breakdown of liver glycogen to glucose through an increase of cyclic-3′,5′-AMP (page 121). It has no effect on muscle glycogen and apparently has no effect on glucose utilization.

Epinephrine

Epinephrine is a hormone secreted by the adrenal medulla. The injection of epinephrine brings about a rapid conversion of liver glycogen and muscle glycogen to lactic acid. There is a rapid rise in blood glucose and blood lactic acid. It acts by increasing the amount of cyclic-3′,5′-AMP which increases active phosphorylase. Epinephrine acts to make glucose quickly available to the cells under conditions of stress, whereas glucagon gives the normal release of glucose from liver glycogen.

Cortical Hormones

Certain steroids secreted by the adrenal cortex are concerned with carbohydrate metabolism. Cortisone (page 172) and cortisol (cortisone with an —OH in the 11 position in place of the carbonyl oxygen) are the most active glucosteroids. These compounds act to increase glucose production in the liver by increasing gluconeogenesis from amino acids. The mechanism of action is unknown.

Thyroid Hormones

The administration of thyroxin and other thyroid hormones increases the rate of absorption of glucose from the intestine. Patients with hyperthyroidism will show high blood glucose levels following a meal; otherwise the utilization of glucose appears to be normal. The increase in the severity of diabetes when thyroid hormone is given to a diabetic animal is probably due to the stimulation of the metabolic rate leading to increased lipid and protein catabolism.

Anterior Pituitary Hormones

A number of hormones secreted by the anterior lobe of the pituitary gland affect carbohydrate metabolism. ACTH (page 20) acts indirectly by controlling the production of the cortical hormones. The thyrotropic hormone controls the production of thyroxin by the thyroid gland. The growth hormone has a direct effect on carbohydrate metabolism. When administered in large doses for a prolonged time, the hormone has a strong diabetogenic effect and causes atrophy of the β cells in the islet. The growth hormone causes increased output of glucose by the liver and increased uptake of glucose by the tissues. The latter effect may be due to an increase in insulin activity. The growth hormone has been studied in many *in vivo* and *in vitro* systems. The effects on carbohydrate metabolism are very complex.

Glucose Tolerance Curves

It is apparent that the ability of a patient to metabolize glucose is dependent on many factors. The determination of tolerance to a test dose of glucose has been used to obtain information about impaired carbohydrate metabolism (Figure 5.15). A dose of glucose (100 to 150 g) is given in the postabsorptive state and the amount of glucose in the blood is determined in one half, one, two, and three hours. The blood sugar levels are plotted against time. Curves for normal, diabetic and hyperthyroid patients are shown.

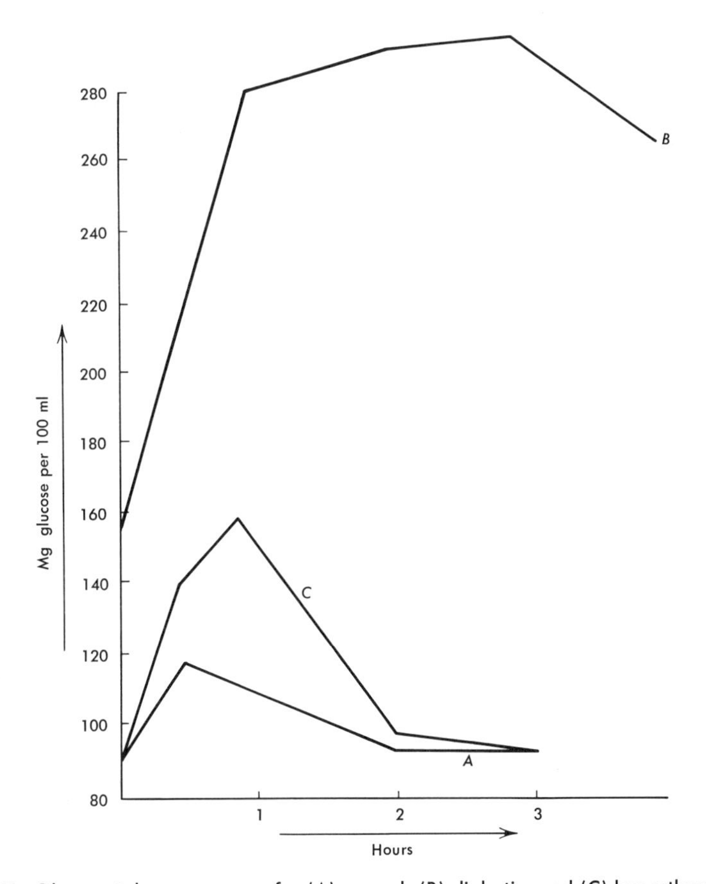

Figure 5.15. Glucose tolerance curves for (A) normal, (B) diabetic, and (C) hyperthyroid patients.

REFERENCES

Bassham, J. A.: The Path of Carbon in Photosynthesis. *Sci. Am.* **206:**88, 1962.

Greenberg, D. M. (ed.): *Metabolic Pathways,* Vol. 1, 3rd ed., Chaps. 2, 3, 4, 5 and 6. Academic Press, New York, 1967.

Mahler, H. R., and Cordes, E. H.: *Biological Chemistry,* Chaps. 10 and 11. Harper and Row, New York, 1966.

West, E. S.; Todd, W. R.; Mason, H. S.; and Van Bruggen, J. T.: *Textbook of Biochemistry,* 4th ed., Chap. 24. The Macmillan Co., New York, 1966.

White, A., Handler, P. and Smith, E. L.: *Principles of Biochemistry.* 4th ed., Chaps. 18 and 19. McGraw-Hill, New York, 1968.

Chapter 6

CHEMISTRY AND METABOLISM OF LIPIDS

Chemistry

THE LIPIDS are composed of a wide variety of substances that are (1) insoluble in water and soluble in fat solvents such as chloroform, ether, and benzene and (2) utilized by living organisms. Most of these substances are actually or potentially fatty acid esters. In this chapter the following substances will be discussed.

- I. Fatty Acids
  - A. General properties
  - B. Chemical properties
- II. Neutral fats
  - A. Physical properties
  - B. Chemical properties
- III. Waxes
- IV. Phosphoglycerides
  - A. Phosphatidyl cholines
  - B. Phosphatidyl ethanolamines and phosphatidyl serines
  - C. Phosphatidyl inositols
- V. Sphingolipids
  - A. Sphingomyelins

   B. Cerebrosides
   C. Gangliosides
VI. Steroids
   A. Cholesterol and related compounds
   B. Bile acids
   C. Hormones
VII. Fat-soluble vitamins
   A. Vitamin A
   B. Vitamin D
   C. Vitamin E
   D. Vitamin K

FATTY ACIDS

The fatty acids are usually among the products obtained from the hydrolysis of lipids. Most of them are straight-chain monocarboxylic acids having from two to more than 20 carbon atoms in the chain. With few exceptions they have an even number of carbon atoms, and some are unsaturated. A few fatty acids contain a hydroxy, keto, methyl, or other substituent. Of the saturated fatty acids listed in Table 6.1, myristic, palmitic, and stearic acids are most abundant and are found in most fats.

Table 6.1
Common Saturated Fatty Acids

Name	No. of Carbon Atoms	Formula	Occurrence
Acetic	2	CH_3COOH	Vinegar
Butyric	4	C_3H_7COOH	Butter
Caproic	6	$C_5H_{11}COOH$	Butter, palm oil, coconut oil
Caprylic	8	$C_7H_{15}COOH$	Butter, palm oil, coconut oil
Capric	10	$C_9H_{19}COOH$	Butter, palm oil, coconut oil
Lauric	12	$C_{11}H_{23}COOH$	Laurel oil, coconut oil
Myristic	14	$C_{13}H_{27}COOH$	Coconut oil, animal fats
Palmitic	16	$C_{15}H_{31}COOH$	Plant and animal fats
Stearic	18	$C_{17}H_{35}COOH$	Plant and animal fats
Arachidic	20	$C_{19}H_{39}COOH$	Peanut oil

The monounsaturated fatty acids are related to the saturated fatty acids; they differ in having a double bond between the ninth and tenth carbon atoms. The most common is oleic acid, an 18-carbon compound. Certain fatty acids having two or more double bonds are required in the diet, and these acids are known as essential fatty acids. Examples of important unsaturated fatty acids are

Name	No. of Carbon Atoms	No. of Double Bonds	Position of Double Bonds†
Palmitoleic	16	1	Δ9
Oleic	18	1	Δ9
Linoleic*	18	2	Δ9,12
Linolenic*	18	3	Δ9,12,15
Arachidonic*	20	4	Δ5,8,11,14

* Essential fatty acids.
† Δ9 indicates a double bond between the ninth and tenth carbon atoms, Δ12 between the twelfth and thirteenth carbon atoms, and so on.

The unsaturated fatty acids exhibit *cis-trans* isomerism about the double bond; for example, oleic acid is a *cis* compound.

$$\begin{array}{r} CH_3(CH_2)_7-CH \\ \| \\ HOOC(CH_2)_7-CH \end{array} \qquad \begin{array}{l} CH_3(CH_2)_7-CH \\ \qquad\qquad\quad \| \\ \qquad\qquad HC-(CH_2)_7COOH \end{array}$$

Cis form, oleic acid — *Trans* form, elaidic acid

The melting point of the saturated acids increases with increase in chain length from −8°C for butyric acid (C_4) to 70°C for stearic acid (C_{18}). The melting point decreases with increasing unsaturation—for example, stearic acid (C_{18}), 70°C; oleic acid (C_{18}; Δ9), 14°C; linoleic (C_{18}; Δ9,12), −5°C; and linolenic (C_{18}; Δ9,12,15), −111°C. The saturated fatty acids give characteristic physical properties to animal fats, whereas the unsaturated fatty acids are responsible for the physical properties of vegetable fats.

Soaps

The sodium and potassium salts of the longer-chain fatty acids are soluble and are known as *soaps*. They react with calcium ions and magnesium ions to form insoluble compounds. Soaps are good emulsifying agents.

Reduction

To make some detergents, the carboxyl group of the fatty acid is reduced catelytically to an alcohol group, which is then esterified with sulfuric acid. The salts of these sulfuric acid esters are excellent detergents which can be used in hard water or solutions which are acidic enough to convert the soaps to the free fatty acids.

$$RCOOH \xrightarrow[cat]{H_2} RCH_2OH \xrightarrow{H_2SO_4} RCH_2OSO_3H \xrightarrow{NaOH} RCH_2OSO_3Na$$

Hydrogenation

Unsaturated fatty acids may be converted to saturated acids by catalytic hydrogenation.

$$\underset{\text{Oleic acid}}{CH_3(CH_2)_7CH{=}CH(CH_2)_7COOH} \xrightarrow[catalyst]{H_2} \underset{\text{Stearic acid}}{CH_3(CH_2)_7CH_2CH_2(CH_2)_7COOH}$$

Halogenation

Bromine, chlorine, iodine, or iodine chloride may be added to the double bonds of the unsaturated acids.

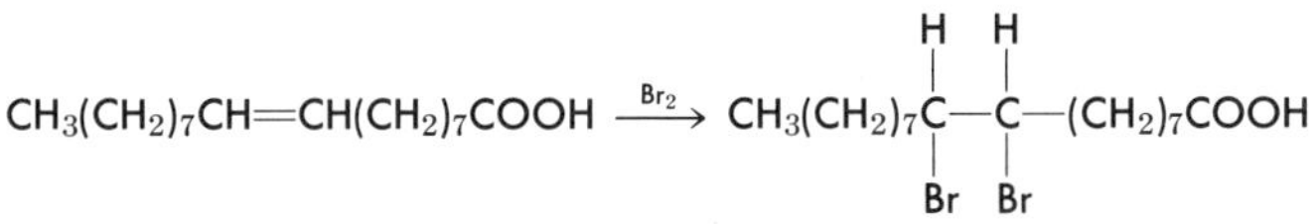

$$\underset{\text{Oleic acid}}{CH_3(CH_2)_7CH{=}CH(CH_2)_7COOH} \xrightarrow{Br_2} \underset{\text{Dibromostearic acid}}{CH_3(CH_2)_7\overset{H}{\underset{Br}{C}}{-}\overset{H}{\underset{Br}{C}}{-}(CH_2)_7COOH}$$

The quantitative uptake of iodine by a fatty acid may be used as a measure of the number of double bonds present.

NEUTRAL FATS

The neutral fats are esters of glycerol. The triglycerides, the most abundant of the neutral fats, have a structure in which all three hydroxyl groups are esterfied with fatty acids.

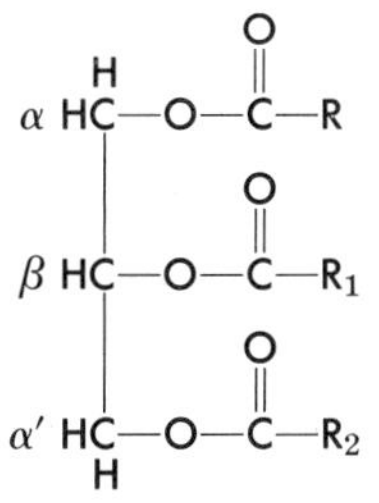

$$\begin{array}{l} \alpha\ H_2C{-}O{-}\overset{O}{\overset{\|}{C}}{-}R \\ \quad\ \ | \\ \beta\ HC{-}O{-}\overset{O}{\overset{\|}{C}}{-}R_1 \\ \quad\ \ | \\ \alpha'\ H_2C{-}O{-}\overset{O}{\overset{\|}{C}}{-}R_2 \end{array}$$

Triglyceride

When R, R_1, and R_2 are the same, the compound is a simple triglyceride; for example, tristearin, triolein, and so on. In natural fats the acids differ, giving mixed glycerides; for example, if R = oleic acid, R_1 = stearic acid, and R_2 = palmitic acid, then the glyceride would by α-oleyl, β-stearyl, α'-palmitin.

The melting point of a fat depends on its composition. Glycerides of the saturated fatty acids melt at a higher temperature than unsaturated ones. The vegetable fats, olive oil, cottonseed oil, linseed oil, and peanut oil contain more unsaturated fatty acids than animal fats such as lard or tallow. The fat is characteristic of the species of animal or plant from which is is obtained.

Chemical Properties

Saponification

The ester groups are readily broken by alkali in a process called saponification. This is the commercial method for the production of glycerol and soap.

$$\begin{array}{l} CH_2-O-\overset{\displaystyle O}{\overset{\|}{C}}-R \\ | \\ CH-O-\overset{\displaystyle O}{\overset{\|}{C}}-R \\ | \\ CH_2-O-\overset{\displaystyle O}{\overset{\|}{C}}-R \\ \text{Triglyceride} \end{array} \xrightarrow{3\,KOH} \begin{array}{l} CH_2OH \\ | \\ CHOH \\ | \\ CH_2OH \\ \text{Glycerol} \end{array} + \begin{array}{c} 3\,R\overset{\displaystyle O}{\overset{\|}{C}}-O^- \\ \text{A soap} \end{array} + 3\,K^+$$

The hydrolysis may also be catalyzed by lipases, which are widely distributed in both plants and animals.

The saponification reaction can be performed quantitatively and used to determine the average molecular weight of a neutral fat. The saponification number is the weight in milligrams of potassium hydroxide (KOH) required to saponify 1 g of fat. From this it follows that,

$$\text{Average mol. wt.} = \frac{3 \times 56 \times 1000}{\text{saponif. number}}$$

Glycerol is a water-soluble, fat-insoluble, colorless, sweet, odorless liquid. It resembles the carbohydrates in its chemical properties and is metabolized by the same biologic mechanism as glucose. A characteristic reaction of glycerol is the production of acrolein when heated with a dehydrating agent.

$$\begin{array}{l} CH_2OH \\ | \\ CHOH \\ | \\ CH_2OH \\ \text{Glycerol} \end{array} \xrightarrow[KHSO_4]{} \begin{array}{l} CHO \\ | \\ CH \\ \| \\ CH_2 \\ \text{Acrolein} \\ \text{(acrid, irritating odor)} \end{array} + 2\,H_2O$$

Halogenation

The double bonds in a neutral fat may be halogenated in the same manner as in the fatty acid. The iodine number is a measure of the number of double bonds in a fat and is defined as the grams of iodine which will add to the double bonds in 100 g fat.

Hydrogenation

The unsaturated glyceride may be hydrogenated to saturated glycerides by means of hydrogen with a nickel catalyst. In the commercial production of oleomargarine and shortenings, oils such as cottonseed oil are partially hydrogenated to a fat of the correct softness.

Rancidity

The unpleasant taste and odor which develop in fats are caused by (1) the presence of free acids which are produced by slow hydrolysis of the fat, and (2) oxidation at the double bonds which first forms peroxides. The peroxides decompose to form aldehydes, which give objectionable odors and flavors. Rancidity is accelerated by light and prevented by natural antioxidants such as vitamin E. The production of lipases by bacterial contaminants may be an additional factor in the production of fatty acids.

WAXES

Waxes such as beeswax, carnauba wax, sperm oil, and lanolin are widespread in nature. They are esters of fatty acids and high molecular weight monohydroxy alcohols. The alcohols range in chain length from 14 to 34 carbon atoms.

PHOSPHOGLYCERIDES

The phosphoglycerides, sometimes called phospholipids, are an abundant form of lipid. They are esters formed by the union of different alcohols with phosphatidic acid, which is itself an ester of glycerol in which the α' and β positions are esterified with fatty acids and the α position is esterified with phosphoric acid.

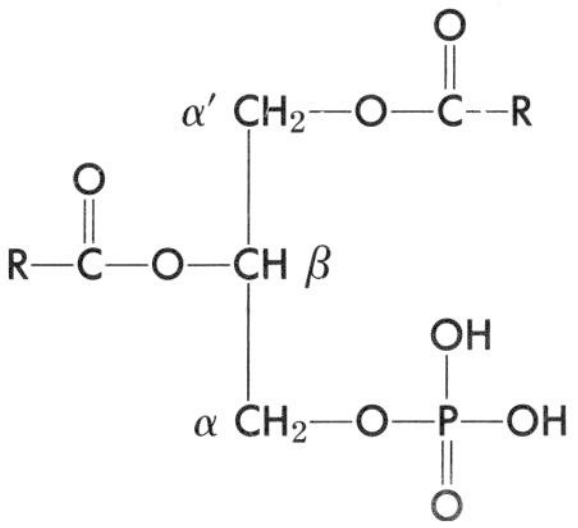

```
                    O                                O
                    ‖                                ‖
          α' CH2—O—C—R                   α' CH2—O—C—R
     O         |                                |      O
     ‖         |                                |      ‖
R—C—O—CH β                              β HC—O—C—R
               |        OH                      |    OH
               |        |                       |    |
           α CH2—O—P—OH                     α CH2—O—P—OH
                        ‖                            ‖
                        O                            O
```

L-α-Phosphatidic acid

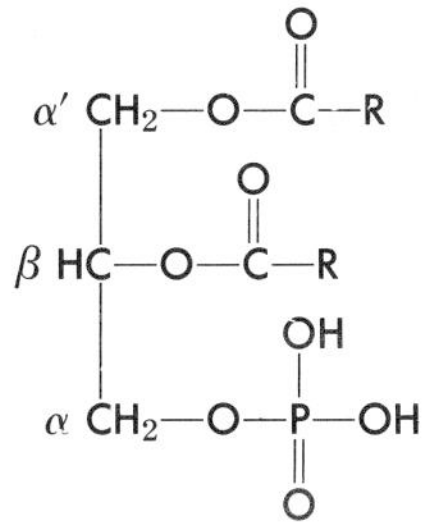

D-α-Phosphatidic acid

The β carbon atom in phosphatidic acid is asymmetric, and the natural form has the same configuration as that of L-glyceraldehyde.

Phosphatidyl Chlolines (lecithins)

Formerly *lecithins* referred to the alcohol-soluble, acetone-insoluble tissue extracts which were predominantly phosphatidyl cholines. Further work showed that there were other phosphoglycerides in the lecithin fraction so that now the term is sometimes used to mean the phosphatidyl cholines, but it is often used in its older meaning.

$$
\begin{array}{l}
CH_2{-}O{-}C({=}O){-}R \\
\vert \\
R{-}C({=}O){-}O{-}CH \\
\vert \\
CH_2{-}O{-}P(O^-)({=}O){-}O{-}CH_2CH_2\overset{+}{N}(CH_3)_3
\end{array}
$$

L-α-Phosphatidyl choline
α-Lecithin (zwitterion form)

The phosphatidyl cholines are zwitterions. The fatty acids in the phosphatidyl cholines are less varied than in the neutral fats and are usually palmitic, stearic, oleic, linoleic, linolenic, and arachidonic acids.

The lecithins are soluble in all the usual fat solvents except acetone. They are white waxy substances which form emulsions. Soybean lecithin is used in large quantities as an emulsifying agent in the food industry.

The phosphatidyl cholines are the most abundant phosphoglyceride in animal tissues; they occur in every cell and make up about 50 per cent of the total phosphoglyceride.

Mild acid hydrolysis separates choline and phosphatidic acid, and more stringent hydrolysis with boiling alkali or acid results in free fatty acids and glycerophosphate. Very strong hydrolysis splits glycerophosphate to glycerol and phosphoric acid.

Choline is a quaternary ammonium base which is about as alkaline as sodium hydroxide. The acetyl derivative is important in the transmission of nerve impulses.

$$
CH_3C({=}O){-}O{-}CH_2CH_2\overset{+}{N}(CH_3)_3 \quad OH^-
$$

Acetylcholine

Phosphatidyl Ethanolamines and Phosphatidyl Serines (cephalins)

Cephalin is an old term referring to the acetone- and alcohol-insoluble phosphoglycerides which accompanied lecithins in tissue extracts. The phosphatidyl ethanolamines and phosphatidyl serines were the major components in the cephalin fraction and sometimes they are referred to collectively as cephalins.

Phosphatidyl ethanolamine and phosphatidyl serine are found in all tissues and cells but are particularly abundant in brain and other nerve tissue. These compounds show hydrolytic properties very much like the lecithins.

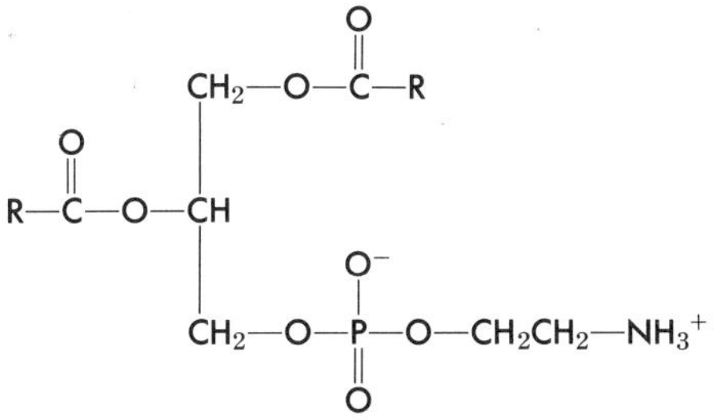

Phosphatidyl ethanolamine

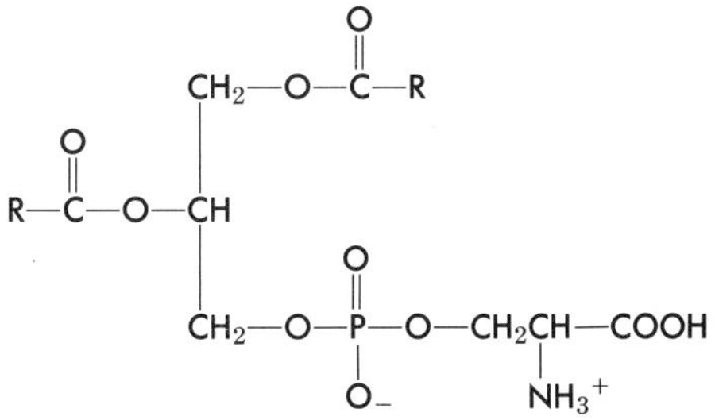

Phosphatidyl serine

Phosphatidyl Inositols (phosphoinositides)

Phosphatidyl inositols occur in all cells and tissues and have the structure

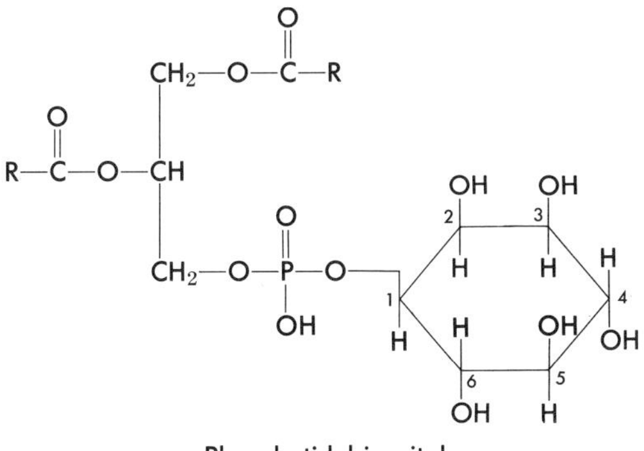

Phosphatidyl inositol

Diphosphoinositides, containing a second phosphate esterified to position 4, and triphosphoinositides esterified at positions 3 and 4 are found in significant amounts in brain tissue.

The lecithins, cephalins, and sphingomyelins are the principal phospholipids found in tissues. However, a number of chemically related compounds are found in lesser amounts. One such group that has received attention is the plasmologens. Hydrolysis of these compounds yields 1 mole each of a long-chain aliphatic aldehyde, fatty acid, glycerol phosphate, and a nitrogenous base. The plasmologen containing ethanolamine has the structure

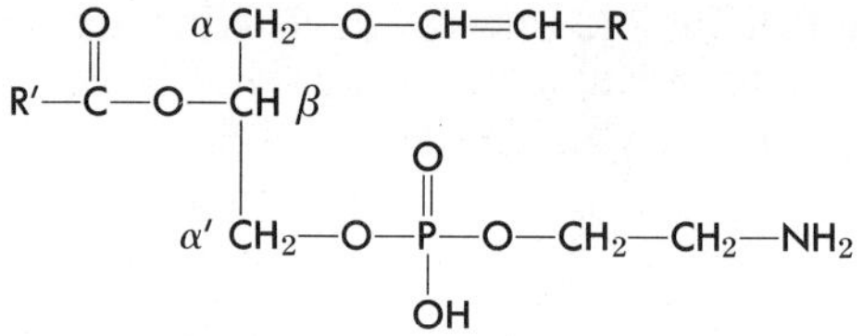

```
      O    α CH2—O—CH=CH—R
      ‖      |
R′—C—O—CH β
             |        O
             |        ‖
          α′ CH2—O—P—O—CH2—CH2—NH2
                      |
                      OH
```

Sphingolipids

The common structure in sphingolipids is an amide, called a ceramide, formed by a fatty acid, R, and the compound sphingosine (D-*erythro*-1,3-dihydroxy-2-amino-4-*trans*-octadecene).

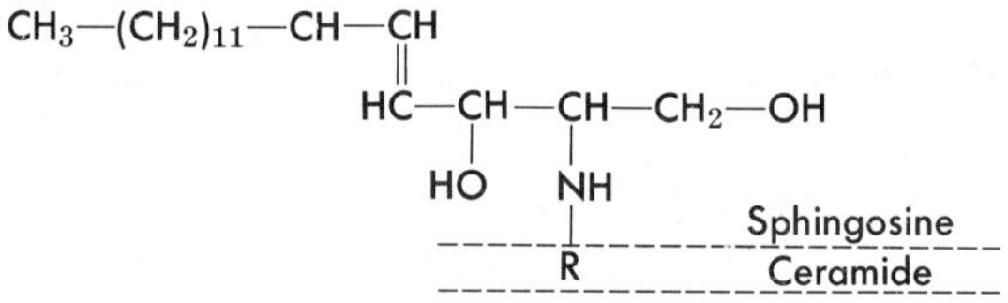

```
CH3—(CH2)11—CH—CH
                ‖
               HC—CH—CH—CH2—OH
                   |   |
                  HO   NH
                       |        Sphingosine
                       R        Ceramide
```

The ceramids are differentiated by the nature of the acyl residue. Sphingolipids are composed of a ceramide with a substituent attached to carbon 1 through the hydroxyl group. Some lipids contain dihydrosphingosine which is sphingosine reduced at the double bond.

Sphingomyelins

These lipids have the structure

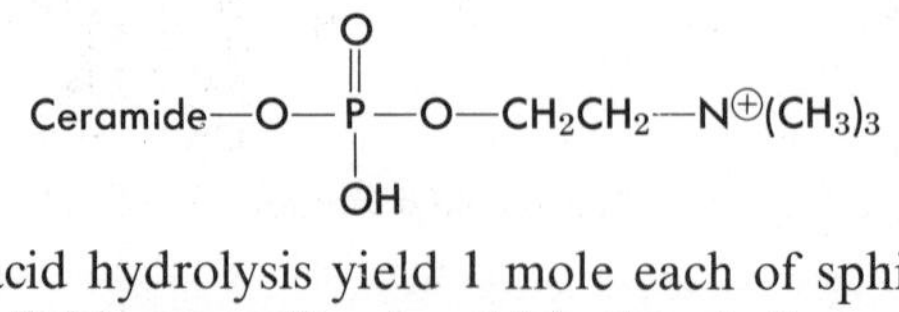

```
              O
              ‖
Ceramide—O—P—O—CH2CH2—N⊕(CH3)3
              |
              OH
```

and upon complete acid hydrolysis yield 1 mole each of sphingosine, a fatty acid, H_3PO_4, and choline. Sphingomyelins in which the choline molecule has been replaced with ethanolamine are known.

The sphingomyelins are found in all tissues, but are especially abundant in brain and nervous tissue. They accumulate in large amounts in the brain, liver, and spleen of persons with Niemann–Pick disease. The metabolism of other phospholipids appears to be normal in this condition.

Cerebrosides (galactosyl ceramides)

The cerebrosides are hydrolyzed to yield 1 mole each of sphingosine, galactose, and a fatty acid. Their structure is

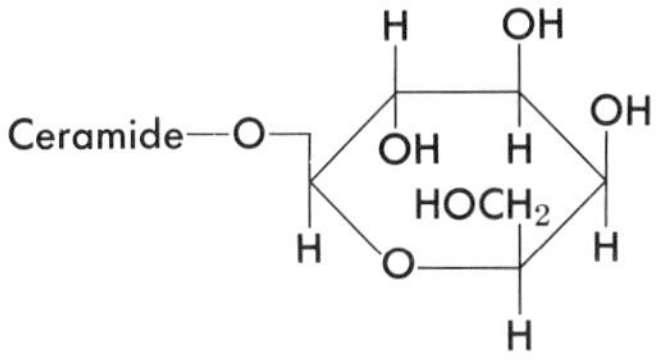

Ceramide—O— (galactose ring: H, OH; OH, H; OH; $HOCH_2$; H; O; H; H)

The cerebrosides are found in brain tissue, particularly in the white matter and are characterized by long-chain fatty acid components, namely lignoceric acid $CH_3(CH_2)_{22}COOH$, cerelonic acid $CH_3(CH_2)_{21}CHOHCOOH$, and nervonic acid $CH_3(CH_2)_7CH{=}CH(CH_2)_{13}COOH$. Glucose may replace galactose in pathological conditions, for example, Gaucher's disease. The brain also contains cerebroside sulfates, sometimes called *sulfatides,* in which C_3 of the galactose is esterified to a sulfate group.

Gangliosides

The gangliosides are complex sphingolipids composed of sphingosine, a fatty acid, and one or more carbohydrates. They are characterized by the presence of at least one molecule of N-acetylneuraminic acid. A ganglioside isolated from ox brain has the structure

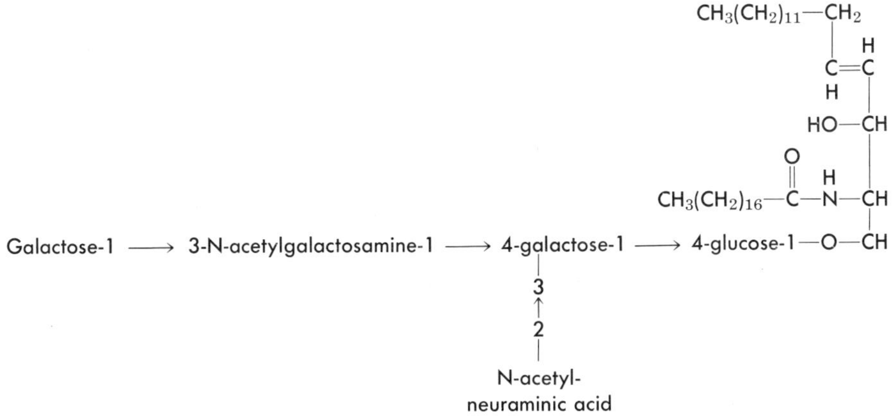

$CH_3(CH_2)_{11}—CH_2$

H—C=C—H

HO—CH

$CH_3(CH_2)_{16}—C(=O)—N(H)—CH$

Galactose-1 ⟶ 3-N-acetylgalactosamine-1 ⟶ 4-galactose-1 ⟶ 4-glucose-1—O—CH

(4-galactose-1) 3 ← 2 N-acetylneuraminic acid

Steroids

When tissues are extracted with solvents for lipids and the extract is saponified, it is found that a portion of the material is not saponified. This portion, called the *nonsaponifiable fraction*, contains the steroids. The main constituent from animal tissues is cholesterol. Cholesterol is particularly abundant in brain, nerve tissue, and glandular tissue. Cholesterol may be the chief component of gallstones. Normal whole blood contains about 200 mg per 100 ml, a portion of which is bound to the plasma proteins.

The structure of cholesterol is as follows

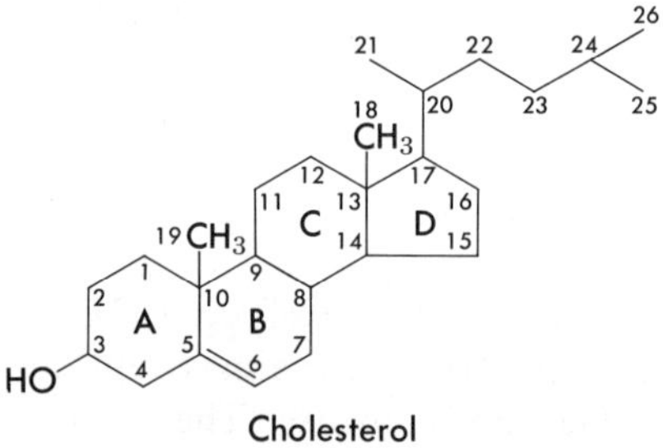

Cholesterol

The rings are designated by letters, and the carbon atoms are numbered as shown.

Owing to the presence of eight asymmetric carbon atoms many stereoisomers are possible; for example the α and β forms of cholesterol are

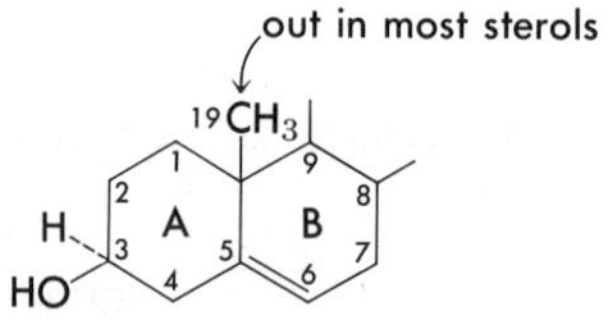

out in most sterols

β form
Hydroxyl *cis* to methyl

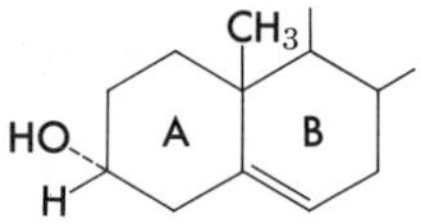

α form
Hydroxyl *trans* to methyl

The β forms with free hydroxyls are precipitated by digitonin.

When the double bond in cholesterol is reduced, two isomers are formed.

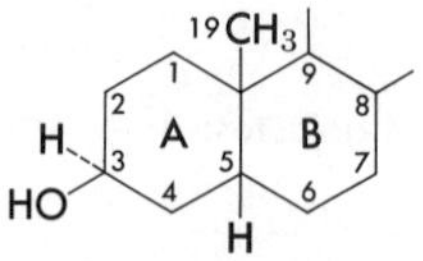

Normal form
A/B *cis*
Coprostanol structure
3-β-hydroxycoprostane
H at 5 *cis* to methyl

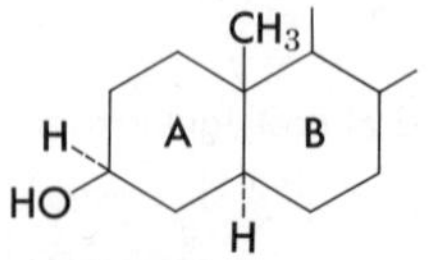

Allo-form
A/B *trans*
Cholestanol structure
3-β-hydroxycholestanol
H at 5 *trans* to methyl

Almost all the natural steroids belong to these two series, the cholestanol type perdominating.

By convention, the CH_3 group on C_{10} is designated as projecting above the plane of the ring (out of the page); this group and all other groups with the same orientation are β oriented and are attached to the ring with solid lines. Groups with the orientation opposite to the β are α oriented (projecting below the ring) and attached by dotted lines.

The most probable conformations of the two sterols are shown, using the dotted and solid lines as explained below.

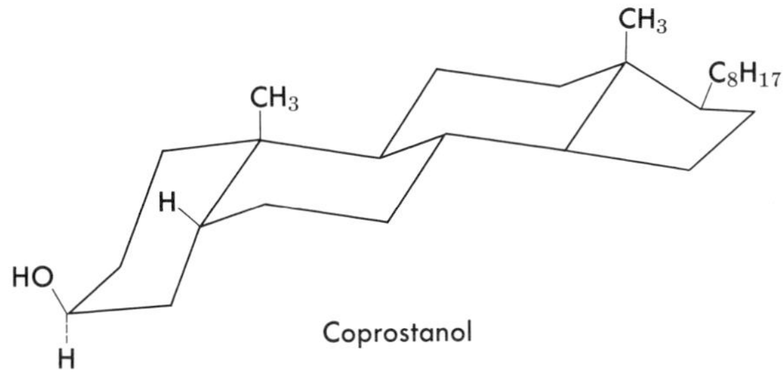

Coprostanol

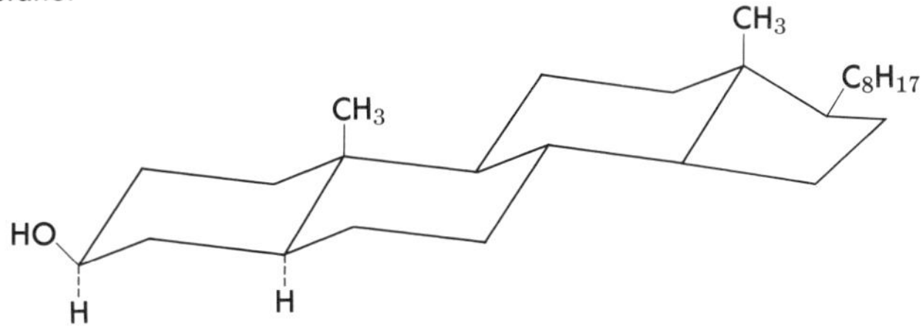

Cholestanol

Cholesterol reacts with acetic anhydride and sulfuric acid in chloroform solution to give a characteristic green color. This is the basis of the Liebermann–Burchard reaction for the quantitative determination of cholesterol. Cholesterol is the precursor of the bile acids, the steroidal sex hormones, the steroidal hormones of the adrenal cortex, and vitamin D.

BILE ACIDS

Bile is formed continuously in the liver and secreted into the gallbladder which empties intermittently into the small intestine. The chief constituents of bile are the bile pigments, bile salts, and cholesterol. Other materials present include salts, urea, and proteins. The bile pigments are formed from hemoglobin and will be considered under the metabolism of hemoglobin.

The bile salts are steroid in nature and eight or ten in number. Salts of glycocholic acid and taurocholic acid are present in the largest amounts.

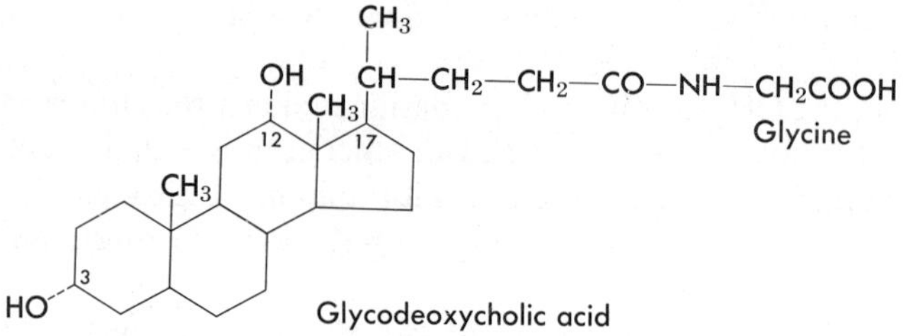

Glycine

Glycodeoxycholic acid

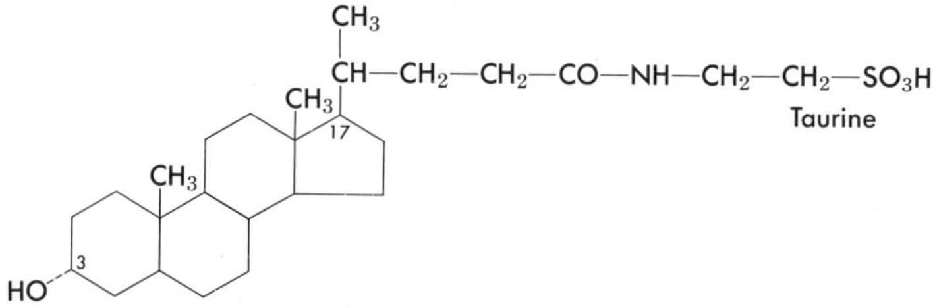

Taurine

Taurolithocholic acid

The bile salts are strong emulsifying agents which help disperse fatty materials into very small particles which may be absorbed or more rapidly broken down by enzymes. A patient whose gallbladder has been removed may exhibit poor breakdown and absorption of intestinal fats and even develop deficiency symptoms of fat-soluble vitamins such as vitamin K (page 177). Bile salts are reabsorbed and returned by portal circulation to the liver. The reabsorbed salts stimulate the liver to secrete bile. Bile salts also stimulate intestinal motility.

STEROIDAL HORMONES

Sex Hormones

Both the female and the male sex hormones are steroids. The male hormone, testosterone, is elaborated by the testes and has been isolated in crystalline form from extracts of this organ. Several other closely related compounds have androgenic

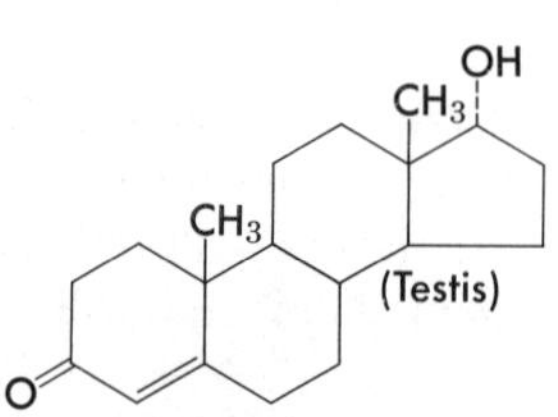

(Testis)

Testosterone

(Urine)

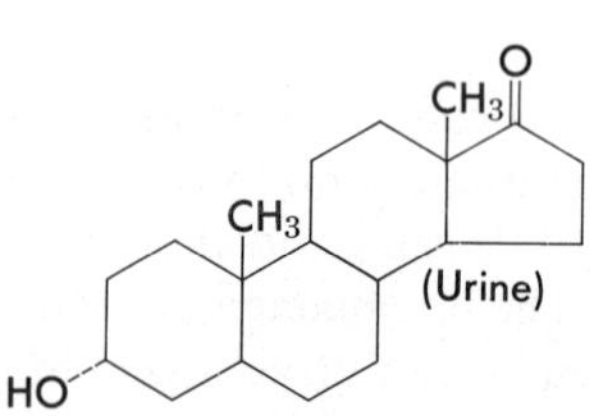

Androsterone

activity. These compounds are usually found in the urine and are probably metabolic products of testosterone. The most important of these is androsterone. Androgens are formed in small amounts by the ovaries, placenta, and adrenals.

There are two types of female sex hormones: estrogens and progesterone. At least three estrogens have been identified, estradiol, estrone, and estriol. Ring A in these compounds is aromatic, and the hydroxyl in position 3 is acidic, having the characteristics of a phenol. Progesterone, the progestational hormone, has a side chain at position 17. The chief metabolic product of progesterone found in the urine is pregnanediol.

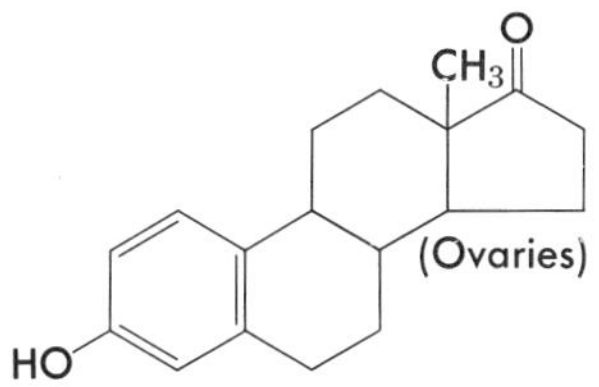

(Ovaries)
Estrone

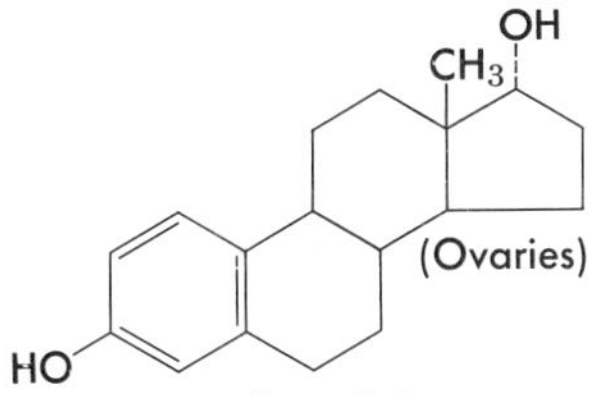

(Ovaries)
α-Estradiol

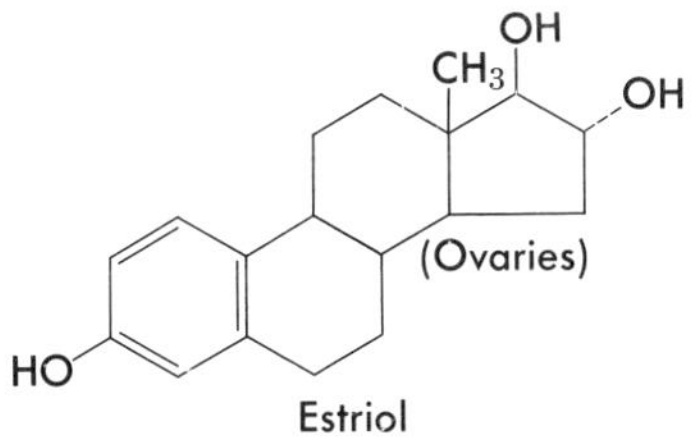

(Ovaries)
Estriol

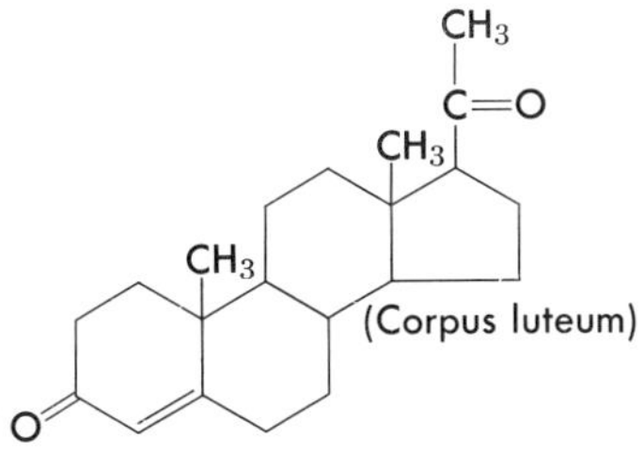

(Corpus luteum)
Progesterone

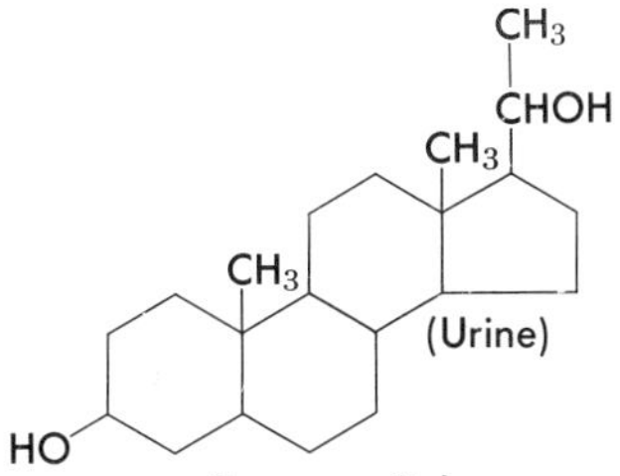

(Urine)
Pregnanediol

The estrogenic hormones are secreted by the ovaries as a result of stimulation by three hormones (the gonadotropic hormones) secreted by the pituitary gland (hypophysis). The gonadotropic hormones are polypeptides (pages 19–20).

Adrenocortical Hormones

If the adrenal cortex is removed, the animal dies. A partial loss of function of the gland results in retention of K^+, increased excretion of Na^+ and Cl^-, muscle weakness, decreased liver glycogen, lowered resistance to insulin, and greater sensitivity to cold and stress. Addison's disease is a result of adrenocortical insufficiency and responds to treatment with extracts from the adrenal cortex.

Although many steroids have been isolated from the adrenal cortex, seven of them account for most of the activity of the gland. Each of these compounds exhibits more than one type of activity. The substances are usually divided into two classes, one having a pronounced effect on carbohydrate metabolism and the other on mineral metabolism. The first class is represented by cortisone and cortisol; the second by aldosterone.

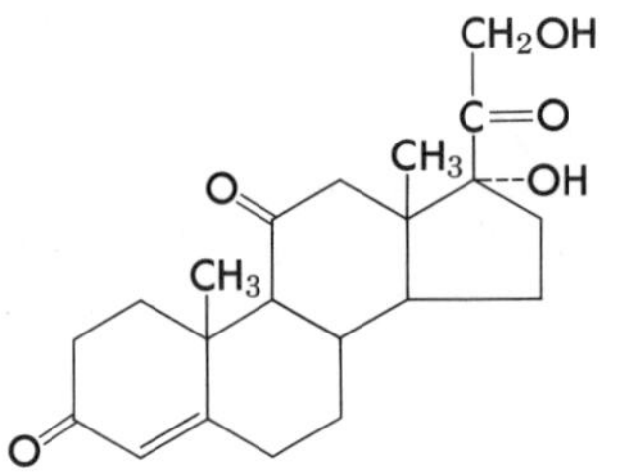

Cortisone

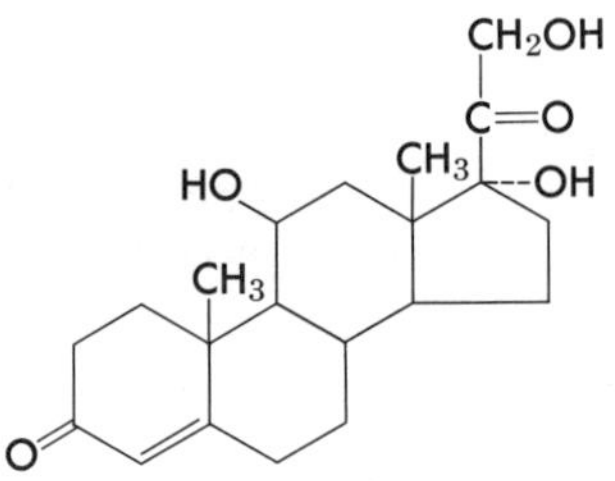

Cortisol

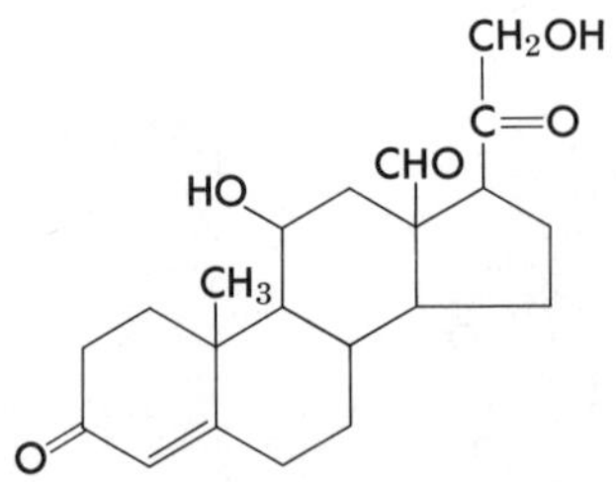

Aldosterone

Like hormone production in the gonads, the production of hormones by the adrenal cortex is under control of the pituitary gland. Adrenocorticotropic hormone (ACTH) secreted by the pituitary stimulates the adrenal cortex to produce steroid hormones which in turn exert an inhibitory effect on the pituitary gland.

ACTH, which stimulates the production of cortisone, as well as cortisone itself is useful for the treatment of rheumatoid arthritis, acute rheumatic fever, acute asthma, Addison's disease, and a number of other ailments. The administration of cortisone can produce many side effects, some of which are characteristic of Cushing's syndrome, a disease in which there is excessive production of adrenal steroids.

FAT-SOLUBLE VITAMINS

Our present knowledge of the vitamins came as a result of the need to find the cause of certain diseases such as scurvy and beriberi and as a result of animal experimentation. Before the role of substances other than protein, fat, and carbohydrate in the diet was recognized, the Japanese found they could prevent beriberi in their navy by adding meat, vegetables, and milk to a diet which had been mostly polished rice; the English Navy used lemons and limes to prevent scurvy.

At the turn of the century Eijkman showed that an extract of rice polishings could cure beriberi in chickens that had been fed polished rice. Funk isolated a crystalline substance from rice polishings and showed that it was effective in very small quantities for curing beriberi in pigeons. It was also found to be useful for preventing beriberi when included in the diet. This substance was called a "vitamin" and is now known as thiamin or vitamin B_1. It was soon recognized that rickets, scurvy, and pellagra were diseases caused by a lack of a vitamin in the diet.

By 1910 nutrition was on a firm experimental basis, and by 1940 most of the vitamins had been isolated and something about the nutritional significance of each was known.

The vitamins are usually divided into two groups: the water-soluble vitamins and the fat-soluble vitamins. Most of the water-soluble vitamins (except vitamin C, page 113), have a known function in enzymatic reactions and will be considered as coenzymes (pages 70–81). The fat-soluble vitamins A, D, E, and K will be considered as lipids.

Vitamin A

Vitamin A is found in cod liver oil, fish livers, butter, and eggs. Carotenes, which can be converted to vitamin A by the intestinal mucosa and liver, are found in carrots, yellow vegetables, and many green vegetables. The efficiency of the conversion of carotene to vitamin A is usually less than 50 per cent of that theoretically available. This inefficiency may be due to poor absorption or lack of conversion.

Chemistry of Vitamin A

Vitamin A is soluble in fat solvents and insoluble in water. It is a polyunsaturated compound composed of isoprene units. Other compounds possessing vitamin A activity are known; they differ in the number of double bonds in the ring or configuration about one of the double bonds and are less active. It should be noted that vitamin A has an all *trans* structure.

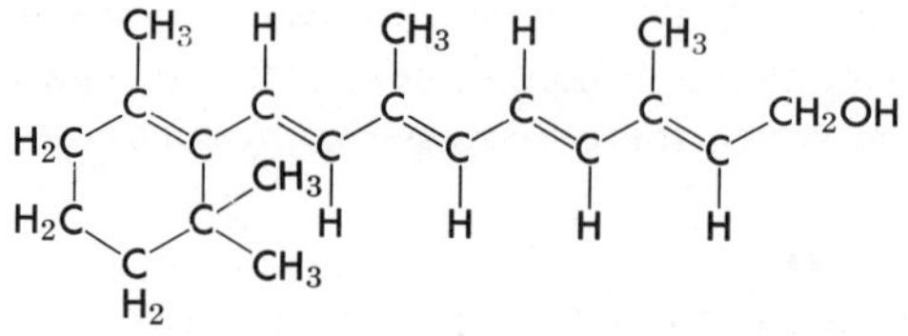

All *trans* vitamin A

The carotenes are polyunsaturated hydrocarbons. The structure of β-carotene, the most active of the group, is shown below. It has an all *trans* structure.

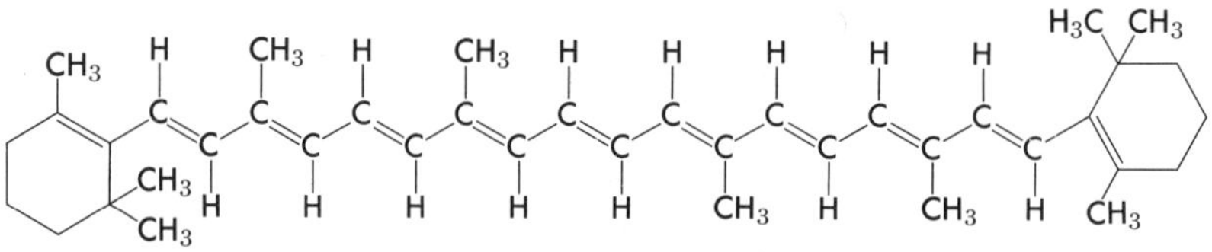

β-Carotene

Function of Vitamin A

Animals maintained on a diet deficient in vitamin A will gradually exhaust the vitamin A stored in the liver, fail to grow, and exhibit epithelial changes in many organs. The function of vitamin A in epithelial cells is not known.

The function of vitamin A in nightblindness is well known. Wald has shown that the rhodopsin in the rods is a complex composed of retinene and a protein. Retinene is an oxidized form of vitamin A in which the alcoholic functional group has been converted to an aldehyde. When rhodopsin is exposed to light, retinene is isomerized at the eleven position from a *trans* configuration to a *cis* configuration and is then split to opsin (protein) and *trans* retinene. In order to maintain normal vision, rhodopsin must be regenerated. The time required to become adapted to dim light when passing from a room with bright light is the time needed to restore rhodopsin to normal amounts. The regeneration process is outlined in the rhodopsin cycle (Figure 6.1). There is a loss of vitamin A during this cycle which has to be replenished by the vitamin in the blood. The rate of rhodopsin synthesis depends on the concentration of vitamin A in the blood.

Vitamin D

Rickets is caused by a lack of vitamin D in the diet or a lack of production of vitamin D in the skin. Irradiation by ultraviolet light is necessary for the conversion of a steroid in the skin to vitamin D. Two forms of vitamin D are important: (1) vitamin D_2 (calciferol), and (2) vitamin D_3.

Ergosterol → (ultraviolet light) → Vitamin D_2 (Calciferol)

7-Dehydrocholesterol → (ultraviolet light) → Vitamin D_3

Vitamin D_3 is formed by irradiation of 7-dehydrocholesterol from animals. Other forms are also known. The conversion of ergosterol to vitamin D_2 by ultraviolet light involves isomerization at the methyl group on carbon atom 10 followed by

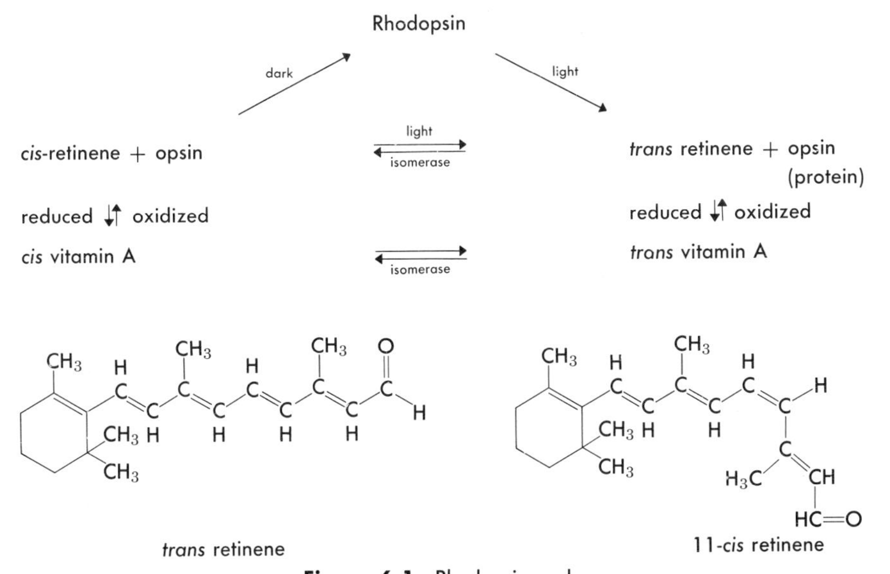

Figure 6.1. Rhodopsin cycle.

the opening of ring B between carbons 9 and 10 and establishing the conjugated double bond system shown. Note that these formulations of vitamins D_2 and D_3 do not indicate the configurations about the conjugated double bond system. The structures of the physiologically active forms more accurately represent the structures of the vitamins.

Most naturally occurring foods contain little vitamin D. Cod liver oil is a good source of the vitamin and is often used in infant feeding. Practically all milk in the United States is fortified with either vitamin D_2 or vitamin D_3.

Function of Vitamin D

It is known that vitamin D is required for the proper deposition of calcium phosphate in bone, but there is no evidence that it participates in the calcification process itself. Instead, vitamin D appears to aid in supersaturating body fluids with calcium phosphate by (1) inducing a calcium transport system in the intestine and (2) mobilizing old or deep bone. Other effects of vitamin D which may add to the blood levels of calcium and phosphate are increased reabsorption of phosphate by the kidney tubule and increased phosphatase activity.

Recent work indicates that vitamins D_2 and D_3 are converted to different forms after absorption and these forms may be the physiologically active forms of the vitamin. The conversion occurs in the liver and results in hydroxylation at the position 25 in the side chain.

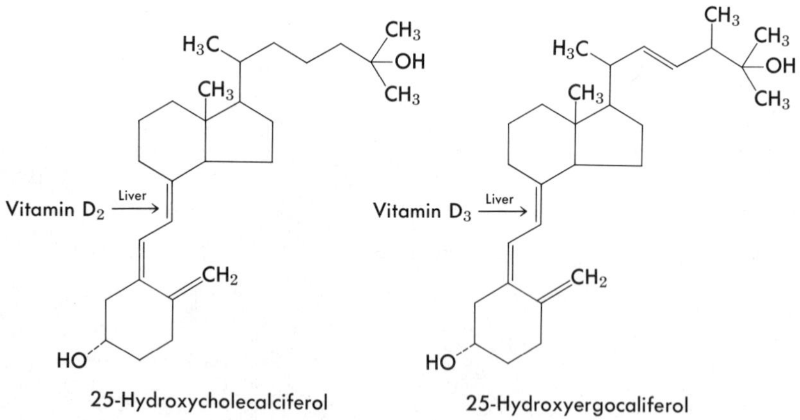

Vitamin D_2 $\xrightarrow{\text{Liver}}$ 25-Hydroxycholecalciferol

Vitamin D_3 $\xrightarrow{\text{Liver}}$ 25-Hydroxyergocaliferol

Active Forms of Vitamin D

The active form is then transported to other tissues such as bone and to the intestine where it induces a calcium transport system which involves at least one protein as the active carrier of calcium across the intestinal wall. The parathyroid hormone, which is also concerned with calcium and phosphorous metabolism, appears to augment bone metabolism and, possibly, intestinal transport of vitamin D.

Vitamin E

Diets composed of starch, casein, lard, salts, cod liver oil, and yeast support normal growth in rats, but the animals will not reproduce. Ovulation and fertilization take place, but the fetus dies and is reabsorbed. Wheat germ oil has been found to contain a substance which in small amounts permits normal reproduction. The substance is known as vitamin E.

Several compounds exhibit vitamin E activity. They are all "tocopherols," differing in the position of the methyl groups in the benzene ring.

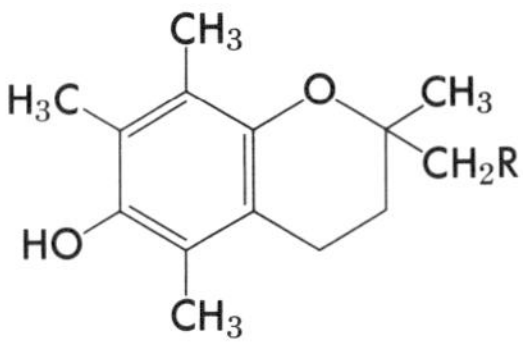

α-Tocopherol

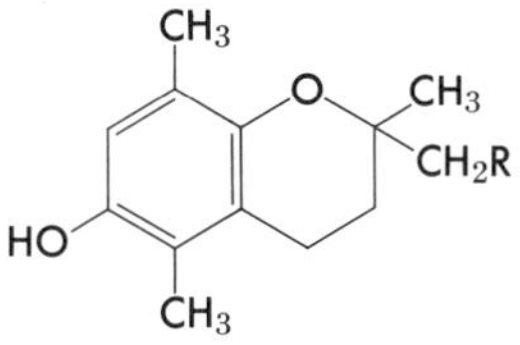

β-Tocopherol

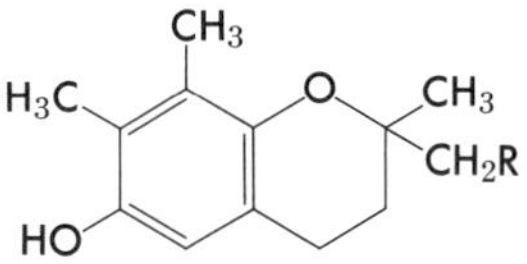

γ-Tocopherol

where $R = -CH_2CH_2CH(CH_3)CH_2CH_2CH_2CH(CH_3)CH_2CH_2CH_2CH(CH_3)CH_3$

The β- and γ-tocopherols are about one fourth as active as α-tocopherol.

Experimental animals on vitamin E deficient diets may develop muscular dystrophy, creatinuria, and anemia in addition to the changes in the reproductive organs. Red blood cells from deficient animals are readily hemolyzed. It appears that the activity of the tocopherols is a result of their antioxidant properties. They are very effective in preventing the oxidation of vitamin A and the unsaturated fatty acids.

Vitamin K

Dam (1929) demonstrated that chicks that had been fed diets containing the fat-soluble vitamins A, D, and E developed a hemorrhagic condition characterized by a prolonged clotting time and internal bleeding. He found that the syndrome could be prevented or cured by feeding a fat-soluble substance obtained from liver and green plants. The substance was named vitamin K (Koagulations-vitamin).

Vitamin K_1 is 2-methyl-3-phytyl-1,4-naphthoquinone.

$-CH_2-CH=C(CH_3)-(CH_2)_3-[CH(CH_3)-(CH_2)_3]_2-CH(CH_3)-CH_3$

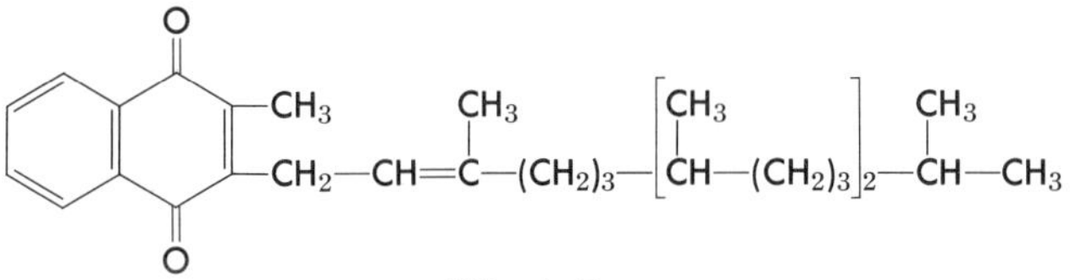

Vitamin K_1
2-methyl-3-phytyl-1,4-naphthoquinone

2-Methyl-1,4-naphthoquinone, 2-methyl-1,4-naphthohydroquinone, and a number of derivatives have vitamin K activity. Water-soluble esters of 2-methyl-1,4-naphthohydroquinone are useful for parental administration.

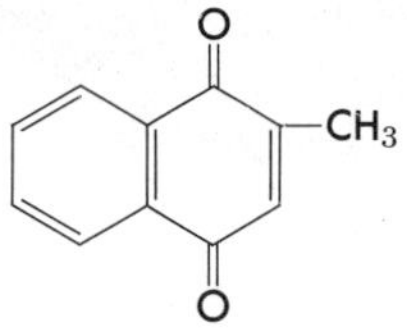

2-Methyl-1,4-naphthoquinone

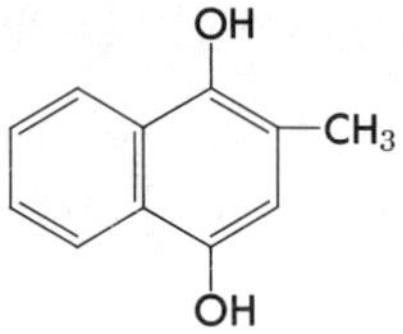

2-Methyl-1,4-naphthohydroquinone

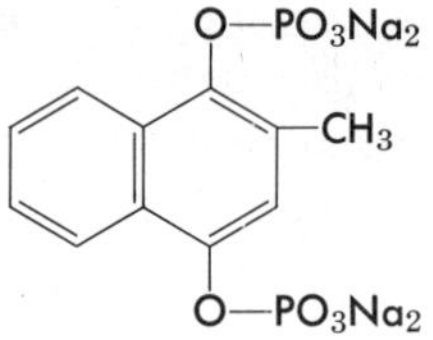

Tetra sodium salt of 2-methyl-1,4-naphthohydroquinone-1,4-diphosphate

The blood from animals deficient in vitamin K has a prolonged clotting time because of a lack of prothrombin. In the clotting process prothrombin is converted to thrombin, which is necessary for changing fibrinogen to fibrin. The role of vitamin K in prothrombin formation by the liver is unknown. There is evidence that the vitamin functions in electron transport systems.

The anticoagulants Dicumarol and Warfarin are used clinically to decrease susceptibility to blood clot formation. Continued administration of such anticoagulants to animals results in the same symptoms as vitamin K deficiency, and a prothrombin deficiency also occurs. The similarities in structure of vitamin K and the 4-hydroxycoumarin anticoagulants is apparent and is probably the basis for the anticoagulant action.

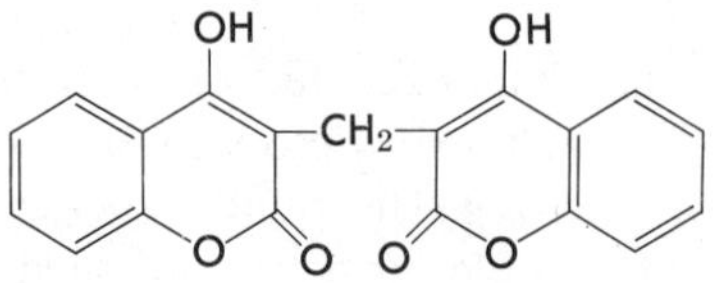

Dicumarol

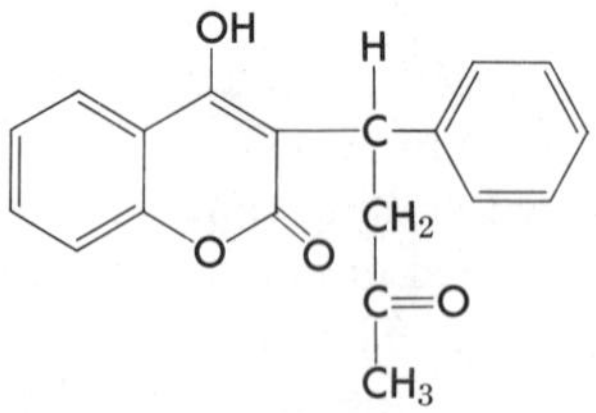

Warfarin

Metabolism

The metabolism of lipids involves the metabolism of a heterogeneous and complex array of compounds. As noted earlier, the principal classes of lipids are the *neutral fats* (triglycerides), *phospholipids, sphingolipids, fatty acids,* and *steroids.* Although our knowledge of the metabolism and function of certain of these compounds has increased dramatically in recent years, there remains much that is unknown or not understood.

An important function of fats is that they serve as the chief storage form of caloric energy in the animal body. Approximately 9 kilocalories (kcal) can be obtained from the oxidation of 1 g of fat versus 4 kcal for carbohydrates and proteins. In addition the storage of fat, unlike that of protein and carbohydrate, is accompanied by little water. Thus fat is an efficient and concentrated storage form of potential caloric energy. It has been estimated that fat accounts for over 85 per cent of the store of available energy in the average man.

In addition to insulating the organism against heat loss and mechanical trauma, lipids have an important role as structural elements of cell membranes and membranes of intracellular organelles such as nuclei, microsomes, and mitochondria. Lipid derivatives are associated with the electron transport system in mitochondria, acting to stabilize proteins of the respiratory chain and to enhance their interactions with factors required for oxidative phosphorylation.

The nutritional aspects of lipids are also important but not completely understood. Unsaturated fatty acids containing a double bond within the terminal seven carbon atoms are required in the diet (page 160) of animals for normal growth and function. Otherwise lipids can be synthesized by the organism.

Digestion

When fat enters the upper portion of the small intestine, the hormone cholecystokinin is secreted, enters the circulation, and stimulates the gallbladder to discharge bile into the intestine. The bile acids (page 169) are excellent detergents and emulsifying agents and serve to disperse the water-insoluble lipids into very small droplets. The bile thus functions to bring the lipids into a physical state suitable for the action of lipolytic enzymes.

The flow of chyme into the duodenum stimulates the production of the hormone secretin which acts on the pancreas to promote secretion of pancreatic juice. Pancreatic juice contains lipases which catalyze the hydrolysis of fatty acid esters at both the α and β positions of triglycerides. Since the equilibrium for the hydrolytic reaction is such that hydrolysis is about 50 per cent complete, some synthesis of fatty acid esters is obtained. The products of lipase action are fatty acids, soaps, glycerol, and a mixture of mono-, di-, and triglycerides.

Absorption

Fatty acids with less than 10 to 12 carbon atoms are absorbed through the intestinal wall and transported to the liver via the portal blood. Longer-chain fatty acids and glycerides of varying degrees of esterification are converted to triglycerides and phospholipids in the intestinal mucosa and pass via the lymphatic system to the blood.

Phospholipids undergoing hydrolysis by phosphatidases (page 181) are resynthesized in the intestinal mucosa during absorption. However, a portion of the phospholipids can be absorbed into the lymph without change.

Cholesterol is readily absorbed in the small intestine and passes into the lymph where most is found in the esterified form.

Blood Lipids

There is a characteristic hyperlipemia following an average meal. This lasts for four or five hours with a peak at about two hours. A portion of the fat appears in microscopic particles called "chylomicrons." These particles range in size from 35 $m\mu$ to 1 μ in diameter and are stabilized by small amounts of protein, phospholipid, and cholesterol. These particles give the plasma a turbid appearance. As the turbid plasma passes through the various organs, it is cleared by an enzyme known as the "clearing factor" or "lipoprotein lipase." Lipoprotein lipase is present in large amounts in most tissues and is stimulated or released by heparin. A portion of the fat in the plasma is in lipoprotein which has about 50 per cent triglyceride, 20 per cent phospholipid, and 7 per cent protein. Another portion of fat appears as free fatty acid bound to albumin. This fat is rapidly metabolized and may be fat that is being transported for rapid use as a source of energy.

The lipid levels in plasma at any time will represent the overall state of lipid metabolism and will be a balance between contribution from the diet and stores, and utilization. Table 6.2 gives the approximate levels of the major lipids in the blood of a normal adult under fasting conditions.

Most of the plasma lipids are transported as complex lipoproteins which contain triglycerides, cholesterol, and phospholipids bound to α- and β-globulins. These are mixtures of high molecular weight complexes of variable composition. Apparently cholesterol and phospholipid can readily enter and leave lipoproteins. Lipid that enters the circulation following absorption goes to the liver and is converted to lipoprotein complexes for transport to tissues. The liver is the key organ in maintaining lipoprotein concentration at a relatively constant level. Conditions that interfere with phospholipid synthesis and consequently with lipoprotein synthesis cause deposition of fat in the liver. The mobilization of fat from the tissues may involve lipoprotein formation or formation of free acids attached to albumin. The mechanism is poorly understood.

At one time it was believed that a portion of ingested fat was utilized and the remainder was deposited in adipose tissue. By means of labeling experiments Schoenheimer showed that fat was in a dynamic state of flux, that is, fat was being added to and removed from depots at a rapid rate. This is in accord with the well-known observation that depot fat can be modified by dietary fat. Although the fat for each species is characteristic of that species, the inclusion of a low melting fat in the diet will lower the melting point of the depot fat. The same investigators showed that the animal can shorten and lengthen fatty acid chains and can introduce a single double bond or saturate unsaturated fats. These are the primary means of altering the composition of fats. Although the liver possesses all the enzymes for fat metabolism and is the organ most involved, adipose tissue is metabolically active

Table 6.2
Lipid Levels

	mg/100 ml of Plasma
Total lipid	450–725
Triglycerides	30–140
Phospholipids	140–225
Total cholesterol	140–280
Esterified cholesterol	70–78% total cholesterol

and may be quantitatively more important than the liver in certain reactions, for example, conversion of glucose to fat.

DEGRADATION OF LIPIDS

It was noted above that the action of lipases results in the hydrolysis of neutral fats to fatty acids and glycerol. The glycerol produced from neutral fats or phospholipids can be converted to carbohydrate via glycerol phosphate (page 132). Before presenting the oxidation of fatty acids, we will consider the breakdown of phospholipids and sphingolipids. The pathways for cholesterol metabolism will be discussed in conjunction with its biosynthesis.

Degradation of Phospholipids

Relatively little is known about the enzymatic breakdown of these complex lipids. There are several enzymes known to catalyze the hydrolysis of one or more of the ester linkages in lecithins, cephalins, and phosphatidyl serines. These enzymes are termed phosphatidases or phospholipases. Four of these enzymes, phosphatidases A, B, and C, and lysophosphatidase, appear to be of major importance in animal tissues. Although the exact specificities of these enzymes are neither completely known nor agreed upon, a current concept of their action is shown in Figure 6.2. It is possible that phosphatidase B is a combination of phospholipase A and a lysolecithinase.

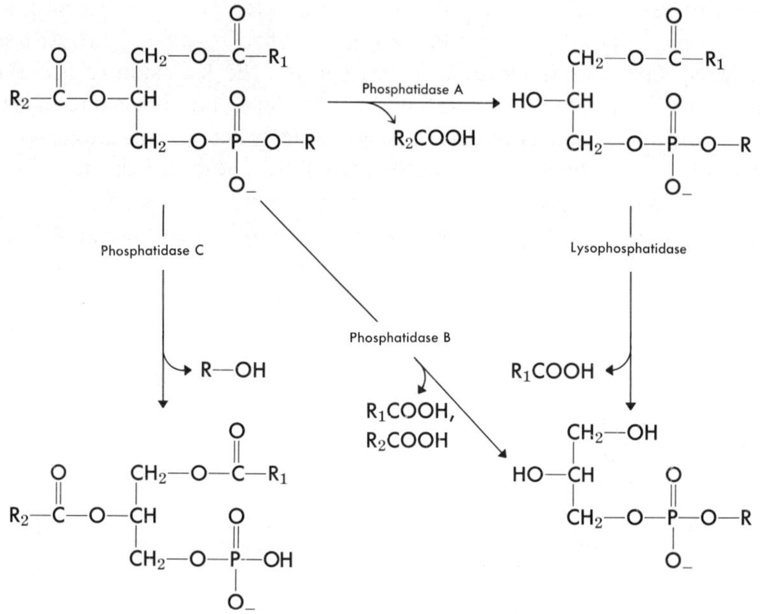

Phosphatidase A

R_2COOH

Phosphatidase C

Phosphatidase B

Lysophosphatidase

R—OH

R_1COOH, R_2COOH

R_1COOH

R_1, R_2 = Fatty acids
R = Choline, ethanolamine, serine

Figure 6.2. Action of phosphatidases on phospholipids.

Degradation of Sphingolipids

Little is known about the degradative pathways of this heterogeneous group of complex lipids (page 166). A sphingomyelinase has been found in liver, brain, kidney, and other tissues which cleaves sphingomyelin to ceramide and phosphorylcholine. Individuals with Niemann-Pick disease have a generalized diminution of this enzyme. Since synthesis of sphingomyelin is normal in these patients, the cause of the accumulation of excessive amounts of sphingomyelin is due presumably to the lack of sphingomyelinase.

Cerebrosides are cleaved normally to ceramide and free hexose by glycocerebrosidases. In Gaucher's disease, in which glucocerebrosides accumulate, there is a marked reduction in the spleen glucocerebrosidase. Since the synthesis of glucocerebrosides is normal in these individuals, the glucocerebroside accumulation in the spleen and elsewhere is thought to be due to the lack of glucocerebrosidase. The source of the glucocerebroside in Gaucher's disease may be the monosialoganglioside which has been converted by successive enzymic action to the glucocerebroside. This compound accumulates because of a deficiency of the enzyme catalyzing the penultimate step.

The sialogangliosides are degraded presumably by sequential hydrolysis of monosaccharide units by glycosidases after the removal of the terminal sialic acid residue by neuraminidase. In Tay-Sachs disease a ganglioside, differing from the normal monosialoganglioside in that the terminal galactose residue is missing, accumulates in various tissues. Although not proven, the defect may be due to the lack of a N-acetylgalactosaminidase to remove the N-acetylgalactosamine which becomes terminal upon removal of galactose from the asialoganglioside.

Oxidation of Even-Numbered Fatty Acids

β Oxidation

The major portion of the oxidative degradation of fatty acids occurs in mitochondria by a stepwise process known as β oxidation. By the sequential action of five different enzymes and the systematic repetition of the action of the last four of these enzymes, fatty acids are broken down to acetyl-SCoA.

Knoop (1905) first proposed the theory of β oxidation based on feeding experiments with ω-phenyl fatty acids. When benzoic and phenylacetic acids are fed to dogs, they are excreted as glycine conjugates in the urine, benzoic acid as hippuric acid and phenylacetic acid as phenylaceturic acid.

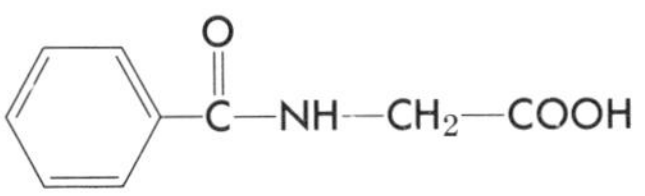

$$C_6H_5-\overset{O}{\overset{\|}{C}}-NH-CH_2-COOH \qquad C_6H_5-CH_2-\overset{O}{\overset{\|}{C}}-NH-CH_2-COOH$$

Hippuric acid 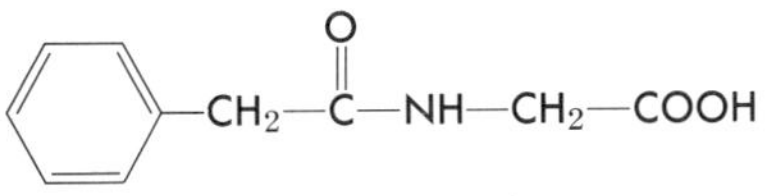 — Phenylaceturic acid

Hippuric acid was excreted in the urine when ω-phenyl fatty acids containing an odd number of carbon atoms were fed. Phenylaceturic acid was the urinary product with the labeled even-numbered fatty acids. Knoop explained these results on the basis that fatty acids were oxidized at the β position with the successive removal of a two-carbon unit.

The concept of β oxidation has been confirmed and the enzymatic pathway is now known in considerable detail. Figure 6.3 shows the pathway for the breakdown of an even-numbered fatty acid to acetyl-SCoA; hexanoic acid is used as the example.

Following the activation of hexanoic acid by conversion to hexanoyl-SCoA (reaction 1 of Figure 6.3), the fatty acid acyl-SCoA derivative is oxidized to the α,β-unsaturated CoA-derivative (reaction 2), hydrated (reaction 3), oxidized to the β-keto derivative (reaction 4), and cleaved to acetyl-SCoA and butyryl-SCoA (reaction 5). Reactions 2 through 5 are repeated on butyryl-SCoA to yield two molecules of acetyl-SCoA. Thus, two turns of the cycle have converted hexanoic acid to three molecules of acetyl-SCoA. In the case of fatty acids of long chain length the same

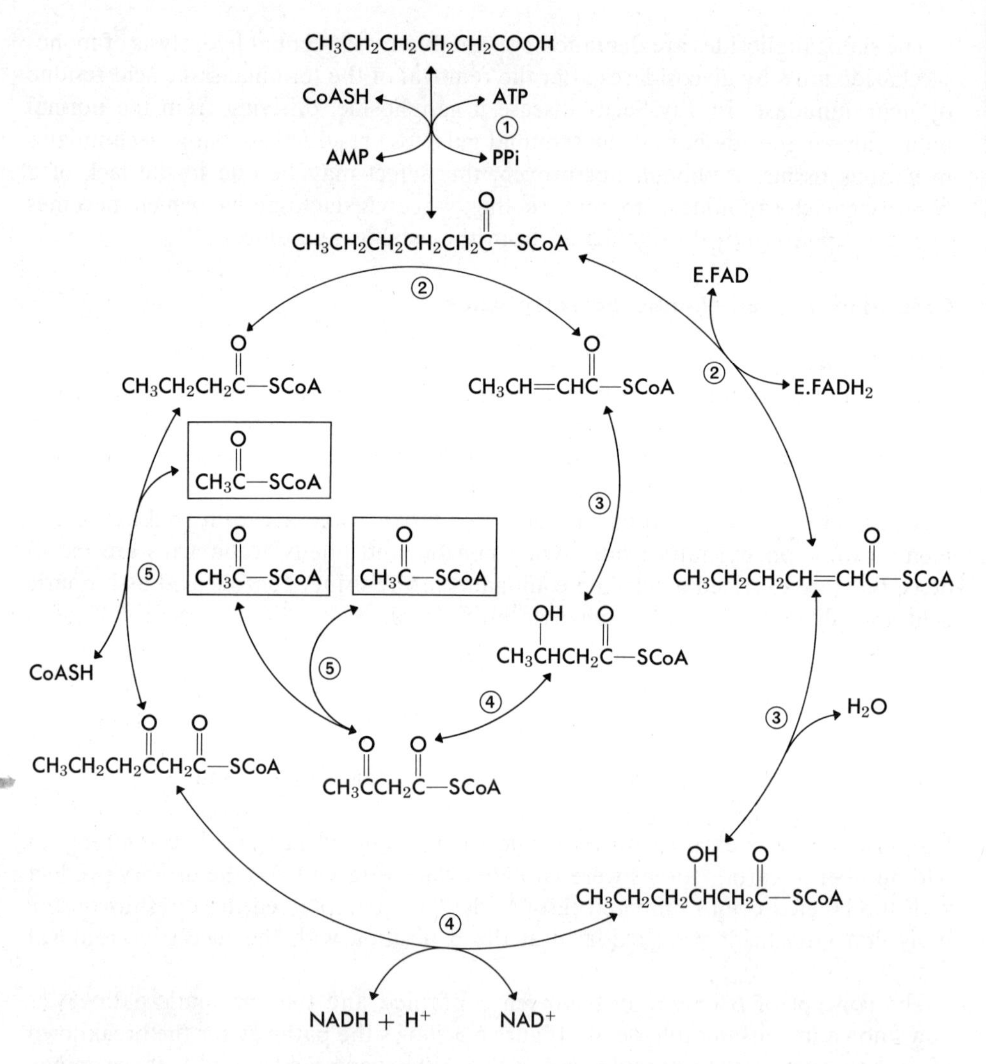

Sum:

$$CH_3(CH_2)_4COOH + ATP + 3\ CoASH + 2\ NAD^+ + 2\ FAD + 2\ H_2O$$

$$= 3\ CH_3\overset{\overset{\displaystyle O}{\|}}{C}{-}SCoA + AMP + PPi + 2\ FADH_2 + 2\ NADH + 2\ H^+$$

Figure 6.3. Breakdown of hexanoic acid to acetyl-SCoA.

series of reactions is repeated until the entire chain is broken down to acetyl-SCoA. Palmitic acid which has a chain of 16 carbon atoms forms eight molecules of acetyl-SCoA by *seven* turns of the cycle. The individual enzymes and the reactions they catalyze are discussed below.

1. *Thiokinases* catalyze a two-step reaction in which enzyme-bound fatty acyl-adenylates are formed as intermediates in the conversion of the fatty acid to the fatty acyl-SCoA derivative.

$$\text{Enzyme} + RCH_2CH_2COOH + ATP \underset{}{\overset{Mg^{++}}{\rightleftharpoons}} [RCH_2CH_2\overset{\overset{O}{\parallel}}{C}\text{—}AMP]\text{—}E + PP_i$$

$$[RCH_2CH_2\overset{\overset{O}{\parallel}}{C}\text{—}AMP]\text{—}E + HSCoA \rightleftharpoons RCH_2CH_2\overset{\overset{O}{\parallel}}{C}\text{—}SCoA + AMP + E$$

$$RCH_2CH_2COOH + ATP + HSCoA \overset{Mg^{++}}{\rightleftharpoons} RCH_2CH_2\overset{\overset{O}{\parallel}}{C}\text{—}SCoA + AMP + PP_i$$

In the process ATP is split to AMP and pyrophosphate. There are several thiokinases known which differ in their substrate specificity; in particular in the length of the carbon chain of the substrates.

Fatty acids may also be activated by *thiophorases* which catalyze the interconversion of free acids and acyl-SCoA derivatives:

$$R_1COOH + R_2\overset{\overset{O}{\parallel}}{C}\text{—}SCoA \rightleftharpoons R_1\overset{\overset{O}{\parallel}}{C}\text{—}SCoA + R_2COOH$$

The specificities of the enzymes from animal tissue are restricted for the most part to conversions between succinyl-SCoA and acetoacetate and other β-keto acids, and between acetyl-SCoA and C_3 to C_5 aliphatic acids.

2. *Fatty acyl-SCoA dehydrogenases* are mammalian enzymes differing in their substrate specificity that catalyze the oxidation of the fatty acyl-SCoAs to the α,β-unsaturated fatty acyl-SCoAs. FAD is required as the coenzyme.

$$RCH_2CH_2\overset{\overset{O}{\parallel}}{C}\text{—}SCoA + E\text{—}FAD \rightleftharpoons R\underset{H}{C}\text{=}\overset{H}{C}\text{—}\overset{\overset{O}{\parallel}}{C}\text{—}SCoA + E\text{—}FADH_2$$

The similarity of the reaction to the conversion of succinate to fumarate should be noted. Two molecules of ATP can be formed for each pair of electrons transported from a molecule of $FADH_2$ through the electron transport system (page 93).

3. *Enoyl hydrases* are found in mammalian liver. Enoyl hydrase (crotonase) catalyzes the conversion of *trans* derivatives to the corresponding L-β-hydroxyacyl-SCoAs.

$$\mathrm{RCH{=}CH{-}\overset{O}{\overset{\|}{C}}{-}SCoA + H_2O \rightleftharpoons R\underset{H}{\overset{OH}{C}}{-}CH_2{-}\overset{O}{\overset{\|}{C}}{-}SCoA}$$

4. *β-Hydroxyacyl-SCoA dehydrogenases* catalyze the oxidation of the L-β-hydroxyacyl-SCoA derivatives to the corresponding β-keto derivative.

$$\mathrm{R\overset{OH}{C}H{-}CH_2{-}\overset{O}{\overset{\|}{C}}{-}SCoA + NAD^+ \rightleftharpoons R{-}\overset{O}{\overset{\|}{C}}{-}CH_2{-}\overset{O}{\overset{\|}{C}}{-}SCoA + NADH + H^+}$$

Reoxidation of the reduced NAD through the electron transport system can furnish three molecules of ATP (page 92).

5. *β-Ketothiolases* catalyze the CoA-dependent cleavage of β-ketoacyl-SCoA.

$$\mathrm{R{-}\overset{O}{\overset{\|}{C}}{-}CH_2{-}\overset{O}{\overset{\|}{C}}{-}SCoA + HS{-}CoA \rightleftharpoons R{-}\overset{O}{\overset{\|}{C}}{-}SCoA + CH_3\overset{O}{\overset{\|}{C}}{-}SCoA}$$

The thermodynamic equilibrium strongly favors cleavage. The fatty acyl-SCoA produced in this reaction can be metabolized further via reactions 2 through 5 until the entire chain has been converted to acetyl-SCoA. The acetyl-SCoA can be oxidized to CO_2 and H_2O via the tricarboxylic acid cycle or used in reactions requiring acetyl-SCoA.

Certain aspects of the overall scheme deserve comment.

1. All the substrates are acyl-SCoA derivatives.

2. All the enzymes are localized in the mitochondrion along with the enzymes of the tricarboxylic acid cycle and the systems for electron transport and oxidative phosphorylation. Such an arrangement should provide for the efficient capture of energy released by the oxidation of the acetyl-SCoA formed in the breakdown of the fatty acids.

3. The fatty acid has to be activated only *once* regardless of the length of the carbon chain. Since the pyrophosphate formed in the activation step must be hydrolyzed to orthophosphate, the formation of each high-energy molecule of activated fatty acid requires the use of *two* high-energy phosphate bonds.

Energetics of β Oxidation

The energy derived from the complete oxidation of fatty acids can be readily calculated. The *net* yield of high-energy bonds from the complete oxidation of hexanoic acid may be estimated as follows

	High-Energy Bonds
1. Hexanoic Acid ⟶ Hexanoyl-SCoA	−2
2. Hexanoyl-SCoA ⟶ 3 Acetyl-SCoA	
a. 2 $FADH_2$ ⟶ 2 FAD (2 × 2 = 4)	+4
b. 2 NADH + H^+ ⟶ 2 NAD^+ (2 × 3 = 6)	+6
3. 3 Acetyl-SCoA ⟶ 6 CO_2 + 3 H_2O + 3 CoASH (3 × 12 = 36)	
	+36
	Net = 44

This yield of 44 high-energy bonds from the oxidation of hexanoic acid should be compared with the 38 high-energy bonds obtained from the oxidation of glucose, also a six-carbon compound.

Oxidation of Odd-Numbered and Branched-Chain Fatty Acids

In general, odd-numbered fatty acids are oxidized by the conventional pathways given above for the even-numbered fatty acids. It is at the terminal step in which the cleavage of β-ketovaleryl-SCoA yields one molecule of acetyl-SCoA and one molecule of propionyl-SCoA that the differences arises.

$$CH_3CH_2\overset{O}{\overset{\|}{C}}CH_2\overset{O}{\overset{\|}{C}}\text{—SCoA} + \text{HS—CoA} \longrightarrow CH_3CH_2\overset{O}{\overset{\|}{C}}\text{—SCoA} + CH_3\overset{O}{\overset{\|}{C}}\text{—SCoA}$$

Propionyl-SCoA can be converted to succinyl-SCoA in mammalian tissues via the following series of reactions.

$$\underset{\text{Propionyl CoA}}{CH_3CH_2\overset{O}{\overset{\|}{C}}\text{—SCoA}} + HCO_3^- + ATP \xrightleftharpoons[\text{biotin}]{\text{propionyl-CoA carboxylase}} \underset{\text{S-Methylmalonyl CoA}}{CH_3\text{—}\overset{HOOC}{\underset{H}{C}}\text{—}\overset{O}{\overset{\|}{C}}\text{—SCoA}} + ADP + P_i$$

$$\text{S-Methylmalonyl-SCoA} \xleftrightarrow{\text{racemase}} \text{R-Methylmalonyl-SCoA}$$

$$\text{R-Methylmalonyl-SCoA} \xleftrightarrow[B_{12}\text{ coenzyme}]{\text{mutase}} \underset{\text{Succinyl-SCoA}}{\text{HOOC—}CH_2\text{—}CH_2\text{—}\overset{O}{\overset{\|}{C}}\text{—SCoA}}$$

In the first reaction, biotin-bound HCO_3^- (page 192) is added to propionyl-SCoA by the propionyl-CoA carboxylase to yield one isomer of methylmalonyl-SCoA. Methylmalonyl-CoA racemase converts the *S* (sinister) isomer to the *R* (rectus) isomer of methylmalonyl-SCoA which serves as the substrate for the methylmalonyl-CoA mutase. Dimethylbenzimidazolyl cobamide (page 240) is the coenzyme for this reaction. The succinyl-SCoA may be metabolized via the tricarboxylic acid cycle, used for porphyrin synthesis (page 237) or for activation of short-chain fatty acids in the thiophorase reaction.

Branched-chain fatty acids such as isobutyrate, α-methylbuyrate, and isovalerate are oxidized by pathways identical to those for the degradation of the branched-chain amino acids (page 248) and will be discussed at that point. Two other pathways of propionyl-CoA metabolism occur in animal and plant tissues. In the lactate pathway, proprionyl-CoA is dehydrogenated to acrylyl-CoA, α-hydrated to L-lactoyl-CoA and then hydrolyzed and converted to pyruvate and to acetyl-CoA.

$$CH_3{-}CH_2{-}\overset{\overset{\displaystyle O}{\|}}{C}{-}SCoA$$

$$\downarrow$$

$$CH_2{=}CH{-}\overset{\overset{\displaystyle O}{\|}}{C}{-}SCoA \xrightarrow{\beta\ \text{hydration}} \overset{\overset{\displaystyle HO}{|}}{CH_2}{-}CH_2{-}\overset{\overset{\displaystyle O}{\|}}{C}{-}SCoA$$

$$\downarrow \alpha\text{-hydration} \qquad\qquad\qquad\qquad \downarrow$$

$$CH_3{-}\overset{\overset{\displaystyle OH}{|}}{CH}{-}\overset{\overset{\displaystyle O}{\|}}{C}{-}SCoA \qquad\qquad H{-}\overset{\overset{\displaystyle O}{\|}}{C}{-}CH_2{-}\overset{\overset{\displaystyle O}{\|}}{C}{-}OH$$

$$\downarrow$$

$$\text{Lactate} \qquad\qquad (-CO_2)$$

$$\downarrow$$

$$\text{Pyruvate} \longrightarrow \text{Acetyl CoA} \xleftarrow{-CO_2} HO{-}\overset{\overset{\displaystyle O}{\|}}{C}{-}CH_2{-}\overset{\overset{\displaystyle O}{\|}}{C}{-}SCoA$$

In the third pathway, β hydration of the acrylyl-CoA occurs and the resulting β-hydroxypropionyl-CoA is oxidized and hydrolyzed to malonyl semialdehyde. Malonyl semialdehyde may be oxidatively decarboxylated directly to acetyl-CoA or it may be oxidized to malonyl-CoA and decarboxylated to form acetyl-CoA.

Formation and Metabolism of Ketone Bodies

The ketone bodies are acetoacetic acid, β-hydroxybutyric acid, and acetone. In the normal individual ketone bodies are produced in the liver and transported to extrahepatic tissues where they are utilized. This is a normal pathway for the utilization of fat for energy. Normally, peripheral tissues utilize ketone bodies at about the same rate as they are produced by the liver. The blood level is very low (less than 1 mg per 100 ml) and less than 0.1 g is excreted in the urine per day.

Ketone bodies accumulate in the blood under conditions of impaired carbohydrate metabolism, for example, diabetes and starvation. In starvation the carbohydrate stores are soon depleted. The fat stores are then mobilized and proteins are degraded to furnish fats and carbohydrates. Some of the amino acids are ketogenic (page 233)—that is, when they are metabolized they furnish acetyl-SCoA. The result of the mobilization of fat and amino acids is the production of ketone bodies at a rate that exceeds utilization by the tissues (ketosis).

In diabetes, blood glucose is elevated, but the glucose is not available to the cells. The result is mobilization of fat stores from the tissues. The production of ketone bodies is greater than utilization, leading to an accumulation of ketone bodies in the blood (ketosis). In the severe diabetic the level may reach 80 mg per 100 ml. Ketone bodies are excreted in the urine under these conditions (ketonuria) and may reach a level of a 100 g or more per day. The administration of insulin can restore ketone bodies to a normal concentration in the diabetic individual.

The overproduction of ketone bodies probably results in part from an impairment of utilization due to a lack of four-carbon intermediates of the tricarboxylic acid cycle. Glucose is needed for the formation of oxaloacetate, the acceptor for the carbon unit, acetyl-SCoA. In addition there is a decrease in lipogenesis. When glucose is not being utilized or is not available, there is a decreased production of NADPH which is required for fatty acid synthesis (page 194). Also, acetyl-CoA carboxylase (page 192) which catalyzes the first step in fatty acid synthesis is inhibited by long-chain fatty acids. This enzyme is activated by citrate and other intermediates of the tricarboxylic acid cycle. This may explain the decrease in lipogenesis in starvation and in diabetes and the restoration of lipogenesis when carbohydrate is provided or made available to the organism.

Each of the ketone bodies is derived from the metabolism of acetoacetyl-SCoA. The primary pathway for the metabolism of acetoacetyl-SCoA in liver is conversion to the cholesterol precursor β-hydroxy-β-methylglutaryl-SCoA (page 204).

$$CH_3{-}\overset{O}{\overset{\|}{C}}{-}CH_2{-}\overset{O}{\overset{\|}{C}}{-}SCoA + CH_3\overset{O}{\overset{\|}{C}}{-}SCoA \longrightarrow HOOC{-}CH_2{-}\underset{OH}{\overset{CH_3}{C}}{-}CH_2{-}\overset{O}{\overset{\|}{C}}{-}SCoA + CoASH$$

The β-hydroxy-β-methyglutaryl-SCoA is cleaved to acetyl-SCoA and free aceto-acetate by a liver enzyme distinct from the one involved in its synthesis.

$$HOOC{-}CH_2{-}\underset{OH}{\overset{CH_3}{C}}{-}CH_2{-}\overset{O}{\overset{\|}{C}}{-}SCoA \longrightarrow HOOC{-}CH_2{-}\overset{O}{\overset{\|}{C}}{-}CH_3 + CH_3\overset{O}{\overset{\|}{C}}{-}SCoA$$

The combined action of these two enzymes is the major route for the formation in liver of free acetoacetate. A liver deacylase exists

$$CH_3{-}\overset{\overset{\large O}{\|}}{C}{-}CH_2{-}\overset{\overset{\large O}{\|}}{C}{-}SCoA \xrightarrow{H_2O} CH_3{-}\overset{\overset{\large O}{\|}}{C}{-}CH_2{-}COOH + CoASH$$

but this reaction appears to be of minor significance. Since liver is deficient in the activating enzyme system, free acetoacetate is not efficiently reconverted to aceto-acetyl-SCoA, and the major portion of the free acetoacetate is transported to the muscles and other tissues where it is reconverted to the CoA derivative by thio-phorase reactions.

$$CH_3{-}\overset{\overset{\large O}{\|}}{C}{-}CH_2{-}COOH + HOOC{-}CH_2{-}CH_2{-}\overset{\overset{\large O}{\|}}{C}{-}SCoA \rightleftharpoons CH_3{-}\overset{\overset{\large O}{\|}}{C}{-}CH_2{-}\overset{\overset{\large O}{\|}}{C}{-}SCoA + HOOC{-}CH_2{-}CH_2{-}COOH$$

A soluble liver enzyme catalyzes the reduction of free acetoacetate to β-hydroxy-butyrate.

$$CH_3{-}\overset{\overset{\large O}{\|}}{C}{-}CH_2{-}COOH + NADH + H^+ \rightleftharpoons CH_3{-}\underset{\underset{\large H}{|}}{\overset{\overset{\large OH}{|}}{C}}{-}CH_2{-}COOH + NAD^+$$

The reaction is stereospecific; only D-(−)-hydroxybutyrate is formed.

The enzyme acetoacetic carboxylyase catalyzes the conversion of acetoacetate to acetone and CO_2 via the intermediate formation of a Schiff base between substrate and enzyme.

$$CH_3{-}\overset{\overset{\large O}{\|}}{C}{-}CH_2{-}COOH + H_2N{-}E \rightleftharpoons CH_3{-}\underset{\underset{\large H\diagup N \diagdown E}{\|}}{C}{-}CH_2{-}C\overset{\diagup O^-}{\underset{\diagdown O}{}} \xrightarrow{\;CO_2\;}$$

$$CH_3{-}\underset{\underset{\large H\diagup N \diagdown E}{|}}{C}{=}CH_2 \rightleftharpoons CH_3{-}\underset{\underset{\large N \diagdown E}{\|}}{C}{-}CH_3 \overset{H_2O}{\rightleftharpoons} CH_3{-}\overset{\overset{\large O}{\|}}{C}{-}CH_3 + H_2N{-}E$$

BIOSYNTHESIS OF LIPIDS

Biosynthesis of Saturated Fatty Acids

Since each step of the enzymatic pathway for fatty acid degradation is reversible (page 184), it was assumed at one time that fatty acid biosynthesis from acetyl-SCoA could be achieved by simple reversal of the oxidative pathway. Subsequent investigations showed that, although there is a synthesis of a portion of the fatty acids in mitochondria by a pathway similar to the reverse of fatty acid oxidation, the major pathway for the *de novo* synthesis of fatty acids is located in the soluble portion of the cell. The soluble system differs from the oxidation system in having an absolute requirement for bicarbonate and using NADP rather than NAD. Although the fatty acid molecule is built from acetyl-SCoA, acetyl-SCoA contributes directly only the two carbon atoms at the methyl end of the carbon chain, whereas the remaining two-carbon units are derived from malonyl-SCoA. Since malonyl-SCoA is formed by the carboxylation of acetyl-SCoA, the entire carbon chain of the fatty acid is derived ultimately from acetyl-SCoA. The following structure shows the origin of the carbon atoms of palmitic acid.

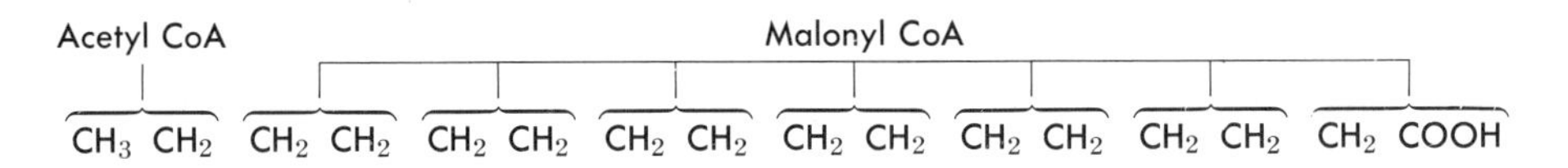

Since the major cellular source of acetyl-SCoA is the intramitochondrial oxidation of pyruvate and only about half of the NADPH required for fatty acid synthesis appears to be derived from reactions of the hexosemonophosphate shunt (page 145), the source of acetyl-SCoA and NADPH for the extramitochondrial system presented a problem in view of their slow diffusion across the mitochondrial membrane. We will discuss pathways for the transport of these substances later in this chapter under control mechanisms (page 197).

Fatty Acid Synthetase of Cytoplasm

In several organisms fatty acid synthesis has been shown to be catalyzed by a cytoplasmic enzyme complex which contains at least seven different proteins. These proteins are tightly bound together in yeast in which the purified synthetase behaves as a single particulate component with a molecular weight of 2,3000,000. The pigeon liver system has been separated into two components, whereas the *E. coli* synthetase has been resolved into several components. This includes an acyl carrier protein (ACP) that binds acyl intermediates during the formation of long-chain fatty acids, an acetyl transacylase, a malonyl transacylase, a condensing enzyme, a β-ketoacyl-ACP-reductase, an enoyl-ACP-hydrase, and a crotonyl-ACP-reductase. ACP has a molecular weight of about 9700 and contains a single sulfhydryl group, that of the prosthetic group 4′-phosphopantetheine which has the structure

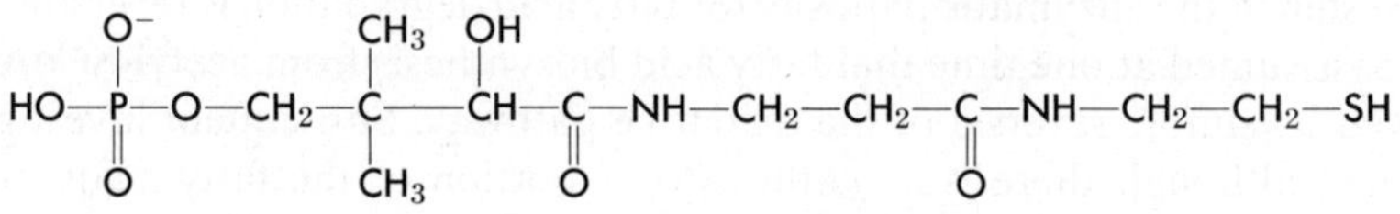

$$HO-\overset{O^-}{\underset{O}{\overset{|}{\underset{\|}{P}}}}-O-CH_2-\overset{CH_3}{\underset{CH_3}{\overset{|}{\underset{|}{C}}}}-\overset{OH}{\overset{|}{CH}}-\underset{O}{\underset{\|}{C}}-NH-CH_2-CH_2-\underset{O}{\underset{\|}{C}}-NH-CH_2-CH_2-SH$$

During fatty acid synthesis all acyl intermediates are bound in thiolester linkage to the sulfhydryl group of ACP. Although the role of ACP has not been demonstrated in the yeast and pigeon liver systems, the available evidence indicates that its involvement is likely. Since the *E. coli* system has been most thoroughly studied, we will consider this system as being a representative one for fatty acid synthesis. Although the detailed mechanisms of all the reactions involved in fatty acid synthesis are not known, the sequence of events may be presented as shown in Figure 6.4.

A discussion of the individual reactions follows.

1. FORMATION OF MALONYL-SCoA. The biotin-dependent enzyme acetyl-CoA carboxylase catalyzes the ATP-dependent synthesis of malonyl-SCoA from acetyl-SCoA and HCO_3^-. The reaction occurs in two steps involving the intermediate formation of enzyme-bound carboxybiotin.

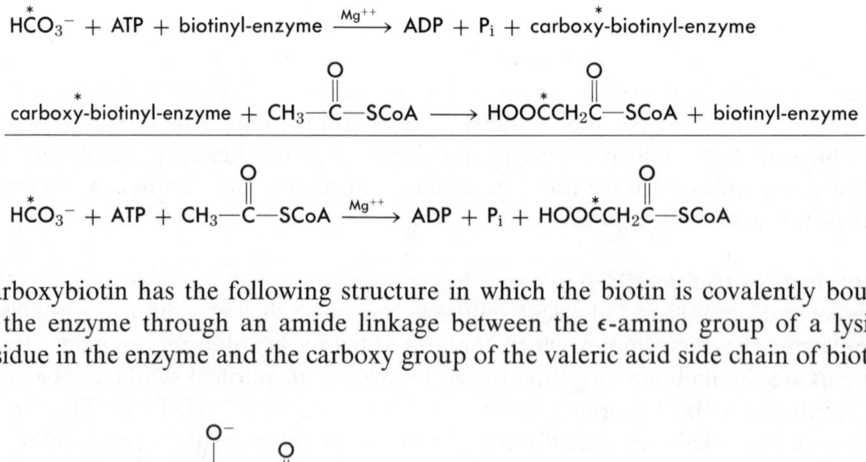

$$H\overset{*}{C}O_3^- + ATP + \text{biotinyl-enzyme} \xrightarrow{Mg^{++}} ADP + P_i + \text{carbox}\overset{*}{y}\text{-biotinyl-enzyme}$$

$$\text{carbox}\overset{*}{y}\text{-biotinyl-enzyme} + CH_3-\overset{O}{\overset{\|}{C}}-SCoA \longrightarrow HOO\overset{*}{C}CH_2\overset{O}{\overset{\|}{C}}-SCoA + \text{biotinyl-enzyme}$$

$$H\overset{*}{C}O_3^- + ATP + CH_3-\overset{O}{\overset{\|}{C}}-SCoA \xrightarrow{Mg^{++}} ADP + P_i + HOO\overset{*}{C}CH_2\overset{O}{\overset{\|}{C}}-SCoA$$

Carboxybiotin has the following structure in which the biotin is covalently bound to the enzyme through an amide linkage between the ε-amino group of a lysine residue in the enzyme and the carboxy group of the valeric acid side chain of biotin.

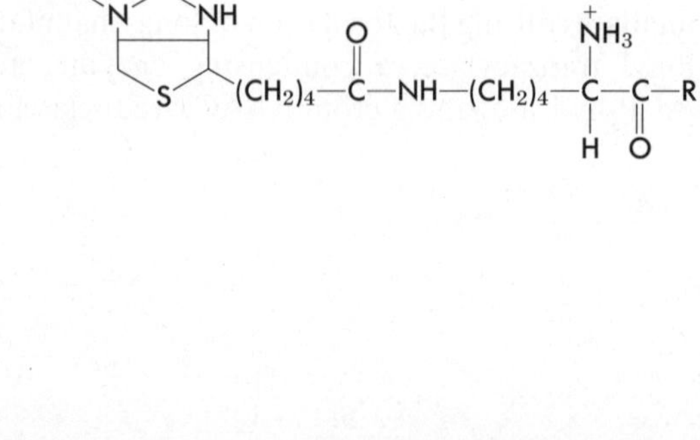

$$O=\overset{O^-}{\overset{|}{C}}-N\overset{\overset{O}{\|}}{\frown}NH \quad \text{S}-(CH_2)_4-\overset{O}{\overset{\|}{C}}-NH-(CH_2)_4-\overset{\overset{+}{N}H_3}{\underset{H}{\overset{|}{\underset{|}{C}}}}-\underset{O}{\underset{\|}{C}}-R$$

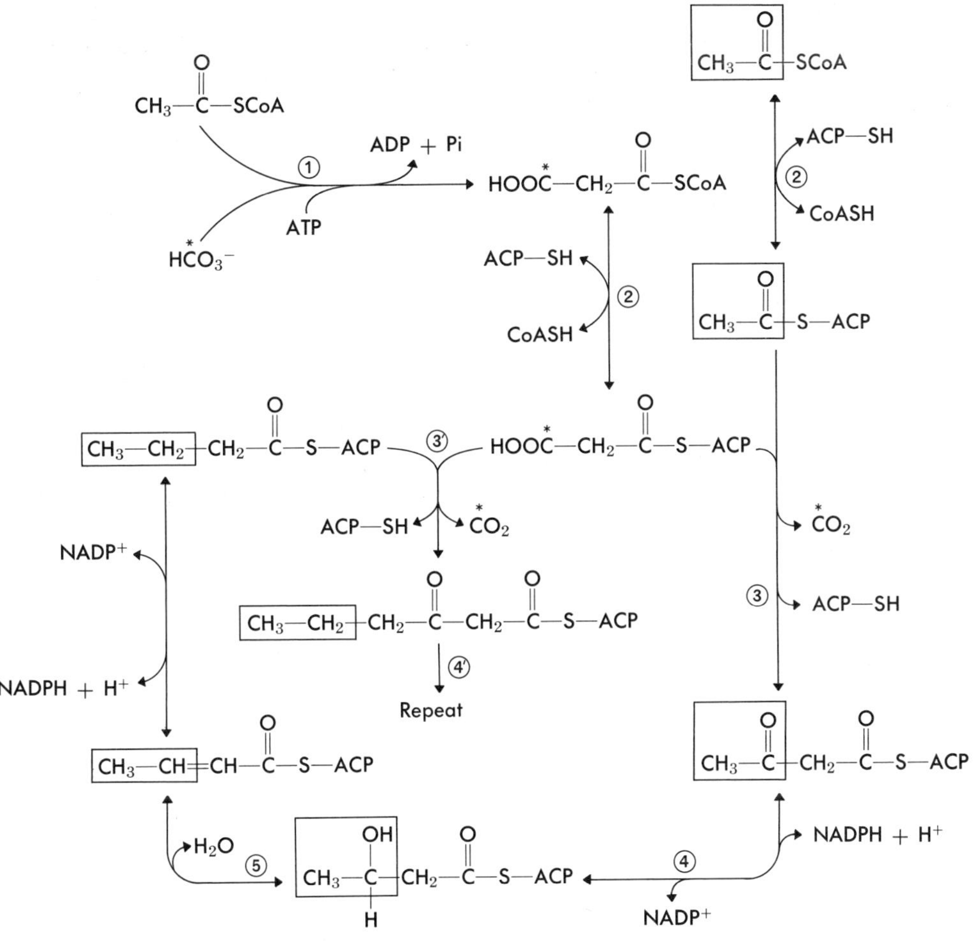

Figure 6.4. Biosynthesis of fatty acids.

2. TRANSFER OF ACETYL-SCoA AND MALONYL-SCoA TO ACYL CARRIER PROTEIN (ACP-SH). Two distinct transacylases catalyze the transfer of acetyl and malonyl groups to ACP-SH.

$$CH_3—\overset{O}{\overset{\|}{C}}—SCoA + HS—ACP \underset{\text{transacylase}}{\overset{\text{acetyl}}{\rightleftharpoons}} CH_3—\overset{O}{\overset{\|}{C}}—S—ACP + CoASH$$

$$HOO\overset{*}{C}CH_2\overset{O}{\overset{\|}{C}}—SCoA + HS—ACP \underset{\text{transacylase}}{\overset{\text{malonyl}}{\rightleftharpoons}} HOO\overset{*}{C}CH_2\overset{O}{\overset{\|}{C}}—S—ACP + CoASH$$

3. CONDENSATION REACTION. The use of malonyl-SCoA as the condensing unit with an acyl-SCoA provides a very favorable thermodynamic advantage in that the decarboxylation "drives" the reaction in the direction of synthesis. The enzyme is β-ketoacyl-ACP synthetase.

$$CH_3\overset{O}{\overset{\|}{C}}-S-ACP + HOO\overset{*}{C}CH_2\overset{O}{\overset{\|}{C}}-S-ACP \longrightarrow \underset{\text{Acetoacetyl-S-ACP}}{CH_3\overset{O}{\overset{\|}{C}}CH_2\overset{O}{\overset{\|}{C}}-S-ACP} + ACPSH + \overset{*}{C}O_2$$

It should be noted that the same CO_2 utilized in the acetyl-SCoA carboxylase reaction is liberated in this step. Carbon dioxide is therefore not incorporated into fatty acids and is given a catalytic function. It is important to note also that the acetate carbon atoms are localized in the methyl and penultimate carbon atoms of the acetoacetyl-S-ACP and will remain there as all other carbon atoms are subsequently provided by malonyl-SCoA.

4. REDUCTION OF ACETOACETYL-S-ACP. The reduction of acetoacetyl-S-ACP and other β-ketoacyl-S-ACP derivatives is catalyzed by a NADPH-dependent β-ketoacyl-ACP reductase.

$$\underset{\text{Acetoacetyl-S-ACP}}{CH_3\overset{O}{\overset{\|}{C}}CH_2\overset{O}{\overset{\|}{C}}-S-ACP} + NADPH + H^+ \rightleftharpoons \underset{\text{D-(}-\text{)-}\beta\text{-Hydroxybutyryl-S-ACP}}{CH_3\overset{OH}{\overset{|}{C}}HCH_2-\overset{O}{\overset{\|}{C}}-S-ACP} + NADP^+$$

5. FORMATION OF α,β-UNSATURATED ACYL-S-ACP. An enoyl hydrase specific for ACP thiolesters catalyzes the reversible dehydration of D-(−)-β-hydroxybutyryl-S-ACP to yield crotonyl-S-ACP. The enzyme is β-hydroxyacyl-ACP dehydrase.

$$CH_3\overset{OH}{\overset{|}{C}}H-CH_2\overset{O}{\overset{\|}{C}}-S-ACP \rightleftharpoons \underset{\text{Crotonyl-S-ACP}}{CH_3CH=CH\overset{O}{\overset{\|}{C}}-S-ACP} + H_2O$$

6. REDUCTION OF α,β-UNSATURATED ACYL-ACP. This reaction is catalyzed by a crotonyl-ACP reductase.

$$CH_3CH=CH\overset{O}{\overset{\|}{C}}-S-ACP + NADPH + H^+ \rightleftharpoons \underset{\text{Butyryl-S-ACP}}{CH_3CH_2CH_2\overset{O}{\overset{\|}{C}}-S-ACP} + NADP^+$$

Repetition of the above steps can elongate the chain, the free acids being liberated from ACP by specific deacylases.

The stoichiometry for the synthesis of palmitic acid from acetyl-SCoA and malonyl-SCoA is

$$CH_3\overset{\overset{O}{\|}}{C}{-}SCoA + 7\ HOOCCH_2\overset{\overset{O}{\|}}{C}{-}SCoA + 14\ NADPH + 14\ H^+ \longrightarrow$$
$$CH_3(CH_2)_{14}COOH + 14\ NADP^+ + 8\ CoASH + 6\ H_2O + 7\ CO_2$$

From acetyl-SCoA the equation is

$$8\ CH_3\overset{\overset{O}{\|}}{C}{-}SCoA + 7\ \overset{*}{C}O_2 + 7\ ATP + 14\ NADPH + 14\ H^+ \longrightarrow$$
$$CH_3(CH_2)_{14}COOH + 7\ \overset{*}{C}O_2 + 8\ CoASH + 14\ NADP^+ + 6\ H_2O + 7\ ADP + 7\ P_i$$

Some of the differences in the substrates and reactions of the mitochondrial oxidation system and the extramitochondrial synthetase are shown in Table 6.3.

Table 6.3
Comparison of Pathways for β-Oxidation and Synthesis of Fatty Acids

Reaction or Component	β-Oxidation	Synthetase
Sulfhydryl component	CoASH	Acyl carrier protein (ACP)
C—C Bond formation	Acetyl-SCoA	Malonyl-SCoA
Reduction of β-ketoacyl derivative	L(+)-hydroxyacyl-SCoA, NAD	D(−)-hydroxyacyl-SACP, NADP
Formation of α,β-unsaturated derivative	L(+)-hydroxyacyl-SCoA	D(+)-hydroxyacyl-SACP
Reduction of α,β-unsaturated derivative	FAD	NADP

The degradative and synthetic pathways for the fatty acids show characteristics common to catabolic and anabolic routes of metabolism in general. It is usual to find steps located at the end of a degradative pathway and the beginning of a synthetic pathway which are thermodynamically virtually irreversible. In this case there are the reactions catalyzed by β-ketothiolase which forms acetyl-SCoA and by acetyl-SCoA carboxylase which utilizes acetyl-SCoA by conversion of malonyl-SCoA. The latter enzyme which initiates the biosynthetic pathway is liable to various kinds of control such as feedback inhibition. Neither the substrates nor the enzymes involved in the anabolic pathway are identical to those of the catabolic pathway. The enzymes are in fact localized in different cellular compartments.

Alternate Pathways of Fatty Acid Biosynthesis

There are in addition to the extramitochondrial enzyme complexes for *de novo* fatty acid synthesis, mechanisms for the elongation of long-chain acyl-SCoA derivatives in particulate fractions. Mitochondria contain an avidin-insensitive elongation system which adds acetyl-SCoA to saturated or unsaturated long-chain acyl-SCoA derivatives in the presence of NADH and NADPH. Malonyl-SCoA, HCO_3^-, and biotin are not required. The products are C_{18} to C_{24} fatty acids. The details of the mechanism are not known, but may involve some of the enzymes of the β-oxidation pathway.

Microsomes contain a system which elongates both saturated and unsaturated long-chain acyl-SCoA derivatives by the malonyl-SCoA pathway.

Biosynthesis of Odd-Numbered and Branched-Chain Fatty Acids

Odd-numbered and branched-chain fatty acids are believed to arise by a modification of the initial condensation reaction of the straight-chain pathway. Propionyl-SCoA (instead of acetyl-SCoA) is thought to condense with malonyl-SCoA to give rise to the odd-numbered acid, for example

$$C_3 + C_3 \xrightarrow{-CO_2} C_5;\; C_5 + C_3 \xrightarrow{-CO_2} C_7 \text{ etc.}$$

The branched-chain acids are thought to arise from the condensation of malonyl-SCoA with a branched, short-chain fatty acyl-SCoA as follows

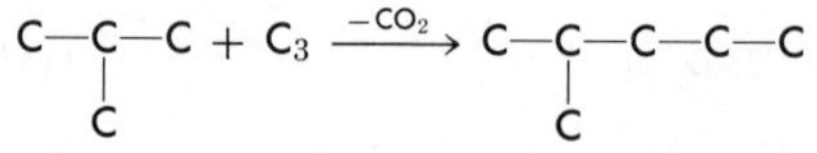

```
C—C—C + C3 --(-CO2)--> C—C—C—C—C
   |                      |
   C                      C
```

Biosynthesis of Unsaturated Fatty Acids

It has been known for some time that animal tissues may form certain unsaturated fatty acids from saturated precursors. For example, a microsomal system from liver can form oleic acid and palmitoleic acid from stearic and palmitic acids. The system requires the fatty acyl-SCoA derivatives, NADPH, and molecular oxygen. Although animal tissues contain a considerable variety of polyunsaturated fatty acids, only one series can be formed *de novo* by the organism. These are fatty acids made from oleic acid by alternate desaturation and chain elongation. All the double bonds lie between the carboxyl group and the seventh carbon atom from the terminal methyl group. An example is shown in the following reactions.

$$CH_3(CH_2)_7CH{=}CH(CH_2)_7COOH \xrightarrow{-2\,H} CH_3(CH_2)_7CH{=}CHCH_2CH{=}CH(CH_2)_4COOH \xrightarrow{+C_2}$$

Oleic acid

$$CH_3(CH_2)_7CH{=}CHCH_2CH{=}CH(CH_2)_6COOH \xrightarrow{-2\,H}$$

$$CH_3(CH_2)_7CH{=}CHCH_2CH{=}CHCH_2CH{=}CH(CH_2)_3COOH$$

5,8,11-Eicosatrienoic acid

Other polyunsaturated fatty acids such as linoleic or linolenic acids cannot be formed *de novo* and must be provided in the diet. These fatty acids contain one or more double bonds within the terminal seven carbon atoms. Linoleic acid (C_{18},Δ9,12) can be converted by alternate desaturation and elongation reactions to arachidonic acid (C_{20},Δ5,8,11,14), and linolenic acid (C_{18},Δ9,12,15) to docosahexenoic acid (C_{22},Δ4,7,10,13,16,19).

Elongation is presumably achieved by the mitochondrial acetyl-SCoA elongation system and the desaturation by the microsomal system which forms oleic and palmitoleic acids.

Control of Fatty Acid Metabolism

We noted earlier (page 191) that the principal source of acetyl-SCoA was from intramitochondrial oxidations. Since the permeability of the mitochondrial membrane to acetyl-SCoA is slight, the source of extramitochondrial acetyl-SCoA for fatty acid synthesis was in question. A similar problem exists for cytoplasmic NADPH. Four pathways have been suggested to account for the transport of intramitochondrial acetyl-SCoA into the cytoplasm: (1) transport of citrate followed by its cleavage to oxaloacetate and acetyl-SCoA by the citrate cleaving enzyme; (2) diffusion of free acetate followed by its conversion to acetyl-SCoA via the acetate thiokinase reaction; (3) transport as acetyl-carnitine; and (4) direct transport of acetyl-SCoA. The presently available data suggest that the citrate cleavage mechanism is the major source of cytoplasmic acetyl-SCoA. The citrate cleaving enzyme of the cytoplasm catalyzes the following reaction.

$$\text{Citrate} + \text{ATP} + \text{HS-CoA} \rightleftharpoons \text{Acetyl-SCoA} + \text{Oxaloacetate} + \text{ADP} + P_i$$

Citrate also activates acetyl-SCoA carboxylase, the enzyme catalyzing the formation of malonyl-SCoA, the initial step in fatty acid biosynthesis. This enzyme is in turn inhibited by the products of the synthetic sequence, long-chain fatty acyl-SCoAs.

Oxaloacetate, the other product of the citrate cleavage enzyme, is thought to give rise to cytoplasmic NADPH via the following reactions.

$$\text{Oxaloacetate} + \text{NADH} + \text{H}^+ \xrightarrow[\text{malic dehydrogenase}]{\text{cytoplasmic}} \text{Malate} + \text{NAD}^+$$

$$\text{Malate} + \text{NADP}^+ \xrightarrow[\text{malic enzyme}]{\text{cytoplasmic}} \text{Pyruvate} + \text{CO}_2 + \text{NADPH} + \text{H}^+$$

It is the malic enzyme that serves to generate NADPH for fat synthesis. The level of this enzyme has been shown to vary directly with the amount of citrate cleavage enzyme under various physiological conditions. The relationships of the citrate cleavage enzyme, malic dehydrogenase, and malic enzyme in fatty acid biosynthesis are shown in Figure 6.5.

Carnitine (β-hydroxy-γ-trimethylaminobutyrate) is known to stimulate fatty acid oxidation in mitochondria. Fatty acyl esters of carnitine are intermediates in this process and probably act as a carrier of acyl groups between the mitochondrial and

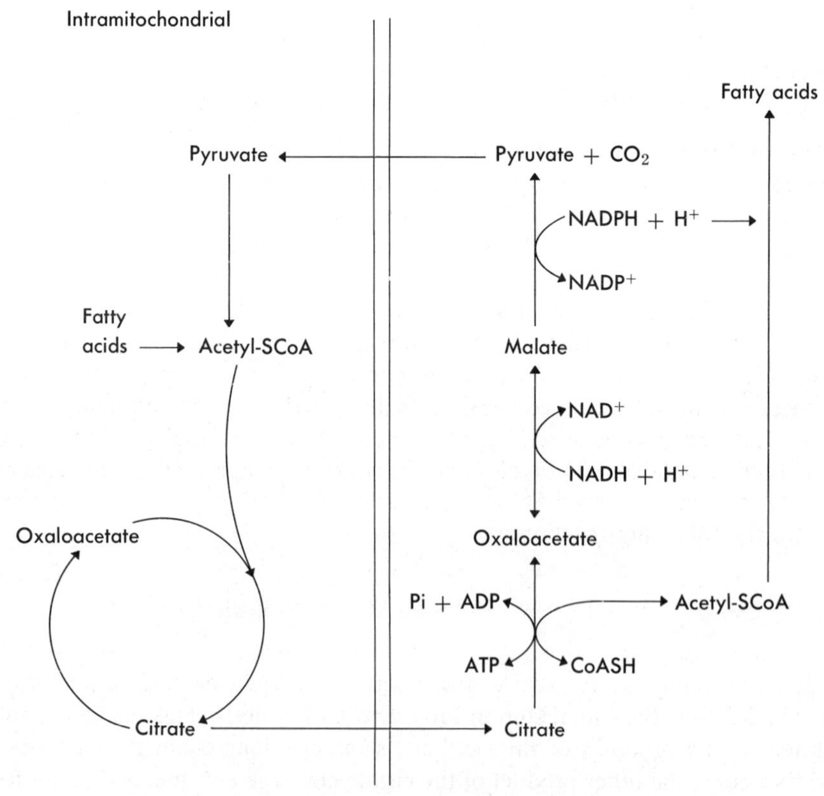

Figure 6.5. Origin of carbon and NADPH for fatty acid synthesis.

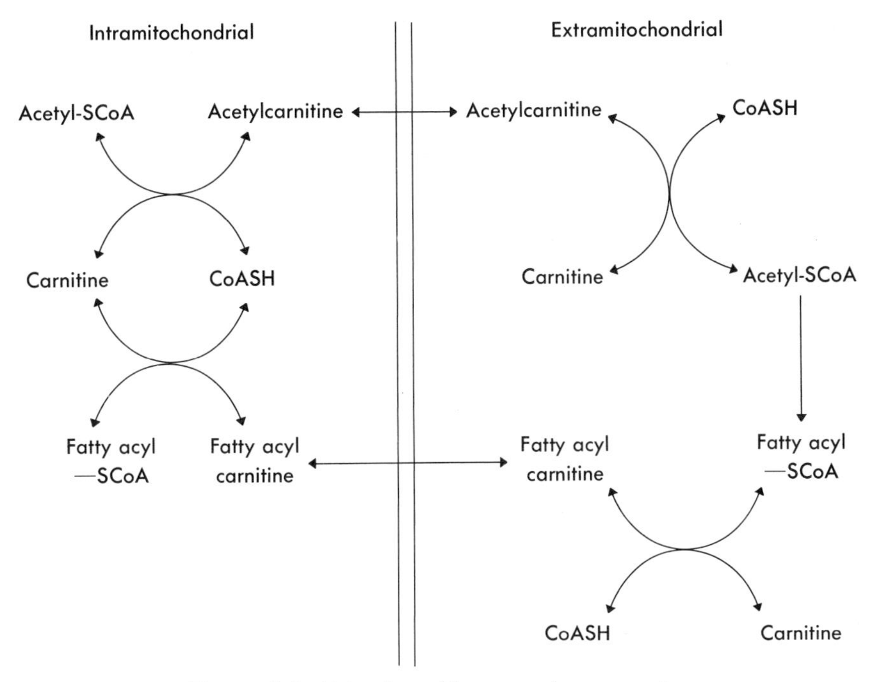

Figure 6.6. Role of carnitine as acyl group carrier.

extramitochondrial compartments. Acylated carnitines readily pass the mitochondrial membrane, whereas CoA derivatives do not. Carnitine-O-acetyltransferase is found both in mitochondria and cytoplasm and catalyzes the reaction

$$\underset{\text{Carnitine}}{(CH_3)_3\overset{+}{N}CH_2\underset{\displaystyle OH}{\underset{|}{C}}HCH_2COO^-} + \text{Acetyl-SCoA} \rightleftharpoons \underset{\text{Acetylcarnitine}}{(CH_3)_3\overset{+}{N}CH_2\underset{\displaystyle O=C-CH_3}{\underset{|}{\underset{O}{\underset{|}{C}}}}HCH_2COO^-} + \text{CoASH}$$

A number of carnitine-O-acyltransferases are known which catalyze the same type of reversible reaction. The major role of carnitine is probably the transport of acyl derivatives from the cytoplasm to the mitochondria for the purpose of oxidation, although a portion of the extramitochondrial acetyl-SCoA may be derived from the mitochondrial pool through carnitine-mediated transport. Figure 6.6 shows the proposed functions of carnitine as a group carrier.

Biosynthesis of Triglycerides

The biosynthetic pathway for the formation of triglycerides is shown in Figure 6.7. The first step in the synthesis is the formation of L-α-glycerophosphate by either the glycerokinase reaction (reaction 1 of Figure 6.7) or the NADH-dependent reduction of dihydroxyacetone phosphate (reaction 2). Liver and kidney employ the former reaction, whereas intestinal mucosa and adipose tissue use the latter reaction. The glycerophosphate is acylated with 2 moles of fatty acyl-SCoA to yield a phosphatidic acid (reaction 3). A phosphatase hydrolyzes (reaction 4) the phosphatidic acid to the 1,2-diglyceride, which is acylated (reaction 5) with a third mole of fatty acyl-SCoA to yield the triglyceride.

Biosynthesis of Phospholipids

The biosynthetic pathways for the phospholipids have been summarized in Figure 6.7. Phosphatidyl ethanolamines and phosphatidyl cholines (lecithins) are both formed in reactions with the D-1,2-diglyceride and the corresponding cytidine diphosphate derivative (Figure 6.7, reactions 6 and 10). Cytidine diphosphate choline (CDP-choline) and cytidine diphosphate ethanolamine (CDP-ethanolamine) are formed by the reactions shown in which R = H = ethanolamine and R = CH_3 = choline.

$HO{-}CH_2{-}CH_2{-}\overset{+}{N}(R)_3\ OH^-$ (Ethanolamine or choline) + ATP →(kinases)→ $HO{-}P(=O)(O^-){-}O{-}CH_2{-}CH_2{-}\overset{+}{N}(R)_3$ (Phosphoethanolamine or phosphorylcholine) + ADP

Phosphoethanolamine or phosphorylcholine + Cytidine triphosphate →(transferases)→ CDP-ethanolamine or CDP-choline + PP_i

CDP-ethanolamine or CDP-choline: cytidine (cytosine ring with NH_2; ribose with H, OH, OH, H) $-CH_2{-}O{-}P(=O)(OH){-}O{-}P(=O)(O^-){-}O{-}CH_2{-}CH_2{-}\overset{+}{N}(R)_3$

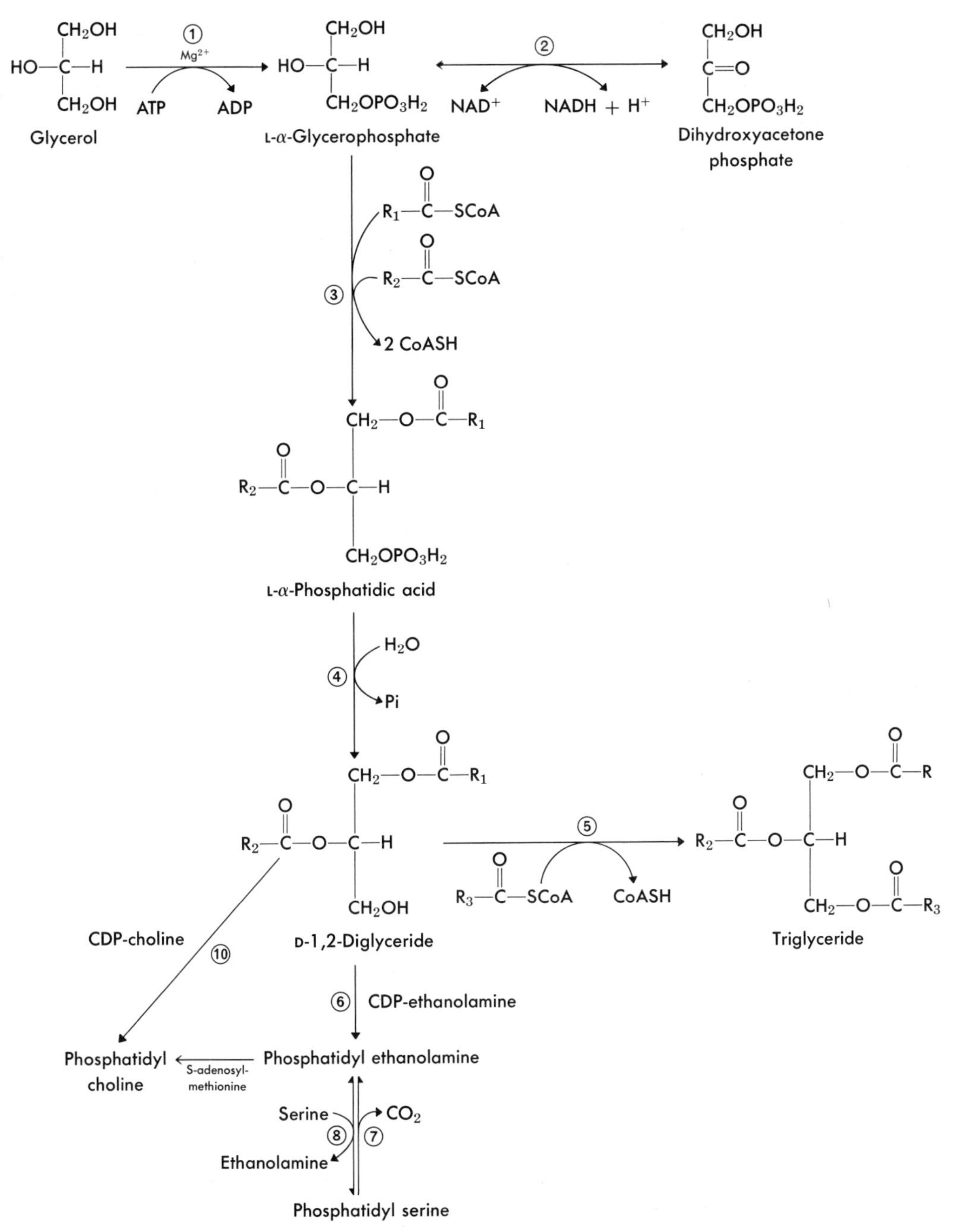

Figure 6.7. Biosynthesis of triglycerides and relationships to phospholipids.

The CDP-ethanolamine or CDP-choline then reacts with the D-1,2-diglyceride to yield phosphatidyl ethanolamine or phosphatidyl choline.

$$\text{CDP-ethanolamine or CDP-choline} + \text{D-1,2-Diglyceride} \xrightarrow{\text{glyceride transferases}} \text{CMP} + R_2{-}\overset{O}{\overset{\|}{C}}{-}O{-}CH(CH_2{-}O{-}\overset{O}{\overset{\|}{C}}{-}R_1)(CH_2{-}O{-}\overset{O}{\overset{\|}{P}}(O^-){-}CH_2{-}CH_2{-}\overset{+}{N}(R)_3)$$

Phosphatidyl ethanolamine
or
phosphatidyl choline

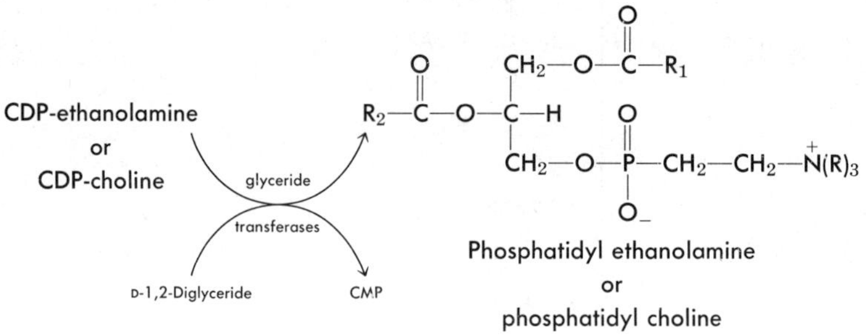

Phosphatidyl ethanolamine can be formed from phosphatidyl serine by decarboxylation (reaction 7 of Figure 6.7) and may be converted to phosphatidyl choline (reaction 9) by sequential transmethylation reactions involving S-adenosylmethionine (page 228). Phosphatidyl serine is formed by an exchange reaction (reaction 8) between phosphatidyl ethanolamine and serine (page 242).

Biosynthesis of Sphingolipids

Sphingomyelins are thought to be synthesized in rat brain via the following sequence of reactions

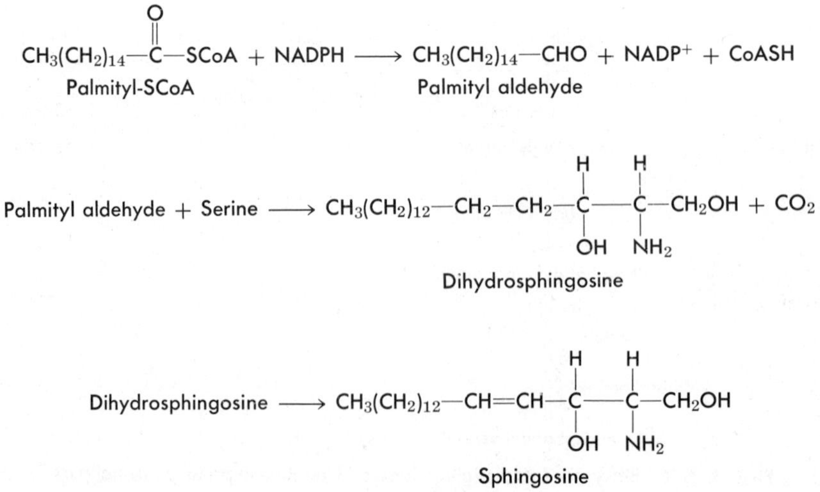

$$\underset{\text{Palmityl-SCoA}}{CH_3(CH_2)_{14}{-}\overset{O}{\overset{\|}{C}}{-}SCoA} + NADPH \longrightarrow \underset{\text{Palmityl aldehyde}}{CH_3(CH_2)_{14}{-}CHO} + NADP^+ + CoASH$$

$$\text{Palmityl aldehyde} + \text{Serine} \longrightarrow \underset{\text{Dihydrosphingosine}}{CH_3(CH_2)_{12}{-}CH_2{-}CH_2{-}\underset{OH}{\overset{H}{C}}{-}\underset{NH_2}{\overset{H}{C}}{-}CH_2OH} + CO_2$$

$$\text{Dihydrosphingosine} \longrightarrow \underset{\text{Sphingosine}}{CH_3(CH_2)_{12}{-}CH{=}CH{-}\underset{OH}{\overset{H}{C}}{-}\underset{NH_2}{\overset{H}{C}}{-}CH_2OH}$$

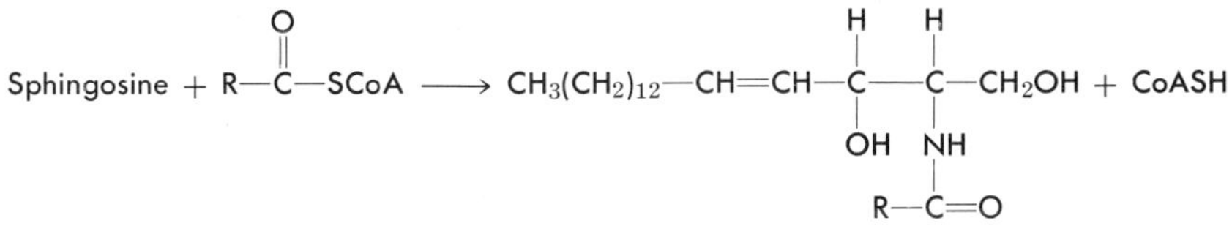

$$\text{Sphingosine} + R{-}\overset{\overset{\displaystyle O}{\|}}{C}{-}SCoA \longrightarrow CH_3(CH_2)_{12}{-}CH{=}CH{-}\underset{\underset{\displaystyle OH}{|}}{\overset{\overset{\displaystyle H}{|}}{C}}{-}\underset{\underset{\underset{\underset{\displaystyle R{-}C{=}O}{|}}{NH}}{|}}{\overset{\overset{\displaystyle H}{|}}{C}}{-}CH_2OH + CoASH$$

N-Acylsphingosine
(ceramide)

$$\text{N-Acylsphingosine} + \text{CDP-choline} \longrightarrow$$

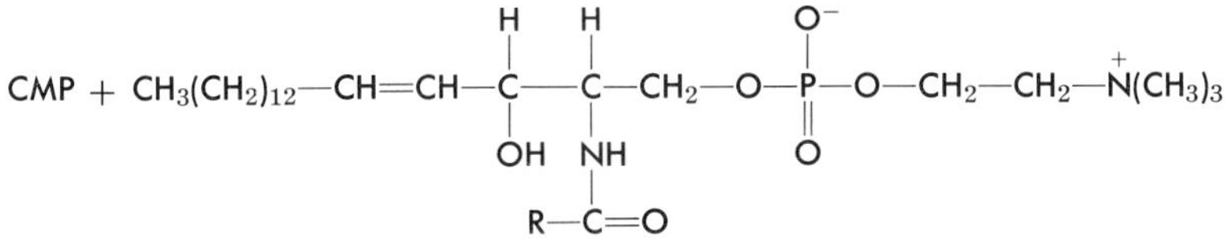

$$CMP + CH_3(CH_2)_{12}{-}CH{=}CH{-}\underset{\underset{\displaystyle OH}{|}}{\overset{\overset{\displaystyle H}{|}}{C}}{-}\underset{\underset{\underset{\underset{\displaystyle R{-}C{=}O}{|}}{NH}}{|}}{\overset{\overset{\displaystyle H}{|}}{C}}{-}CH_2{-}O{-}\underset{\underset{\displaystyle O}{\|}}{\overset{\overset{\displaystyle O^-}{|}}{P}}{-}O{-}CH_2{-}CH_2{-}\overset{+}{N}(CH_3)_3$$

Sphingomyelin

Cerebrosides (ceramide monohexosides, page 167) are formed from sphingosine in two steps. In the first reaction of sugar moiety from uridine diphosphate sugars (page 134) is added to sphingosine. The 1-O-glycosylsphingosine (psychosine) is acylated in the second reaction to form the cerebroside.

$$\text{Sphingosine} + \text{UDP-galactose} \longrightarrow \text{UDP} + \text{1-o-galactosylsphingosine}$$

$$\text{1-o-galactosylsphingosine} + R{-}\overset{\overset{\displaystyle O}{\|}}{C}{-}SCoA \longrightarrow$$

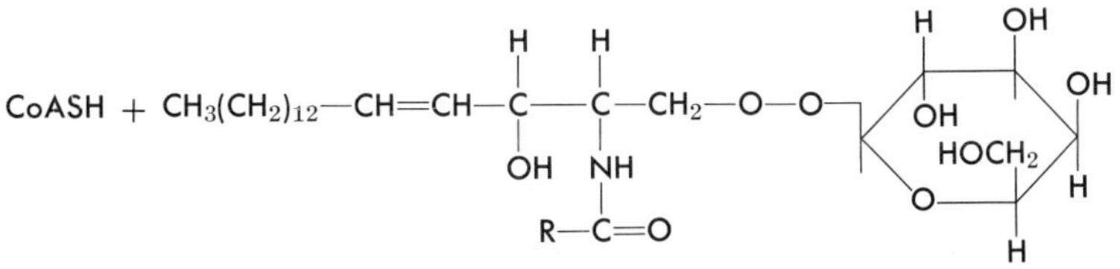

$$CoASH + CH_3(CH_2)_{12}{-}CH{=}CH{-}\underset{\underset{\displaystyle OH}{|}}{\overset{\overset{\displaystyle H}{|}}{C}}{-}\underset{\underset{\underset{\underset{\displaystyle R{-}C{=}O}{|}}{NH}}{|}}{\overset{\overset{\displaystyle H}{|}}{C}}{-}CH_2{-}O{-}O{-}[\text{galactose ring: H, OH, OH, OH, HOCH}_2\text{, H, O, H}]$$

Galactocerebroside

Polyhexoside derivatives of ceramide are also found in nature, but their biosynthetic pathway is unknown.

Relatively little is known about the synthesis of gangliosides. The monosialoganglioside is believed to be synthesized by the following sequential series of reactions.

Glucose-ceramide $\xrightarrow{\text{UDP-galactose}}$ Galactose-glucose-ceramide

$\xrightarrow{\text{CMP-N-acetylneuraminic acid}}$ Galactose-glucose-ceramide
|
N-acetylneuraminic acid

$\xrightarrow{\text{UDP-N-acetylgalactosamine}}$ Acetylgalactosamine-galactose-glucose-ceramide
|
N-acetylneuraminic acid

$\xrightarrow{\text{UDP-galactose}}$ Galactose-acetylgalactosamine-galactose-glucose-ceramide
|
N-acetylneuraminic acid

Biosynthesis and Metabolism of Cholesterol

Cholesterol is synthesized from acetyl-SCoA with the methyl group (M) providing 15 carbon atoms and the carboxyl group (C) 12 of the 27 carbon atoms. The numbering and origin of the cholesterol carbon atoms are shown in the following diagram.

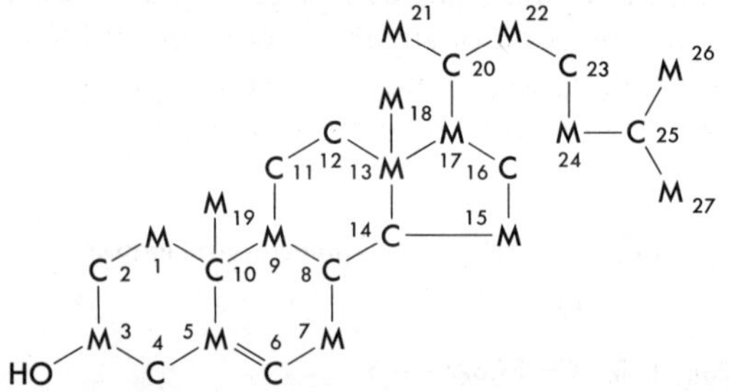

The enzymatic steps involved in the conversion of acetyl-SCoA to isopentenyl pyrophosphate and dimethallyl pyrophosphate are shown in Figure 6.8.

Two related pathways for the biosynthesis of mevalonic acid are known. The pathway represented by reactions 10, 11, and 3 of Figure 6.8 has acetoacetyl-SCoA and β-hydroxy-β-methylglutaryl-SCoA (HMG-SCoA) as intermediates. Alternately, malonyl-SCoA and acetoacetyl-SCoA react to produce enzyme-bound acetoacetate (reaction 1) which reacts with another acetyl-SCoA to give enzyme-bound HMG (reaction 2) which is reduced to mevalonic acid (reaction 3). Reactions 9 and 10 serve to interconvert the intermediates of the two pathways.

Mevalonic acid is phosphorylated with ATP in three successive reactions catalyzed by mevalonic kinase (reaction 4), phosphomevalonic kinase (reaction 5), and pyrophosphoryl mevalonic kinase (reaction 6). This last reaction may involve the forma-

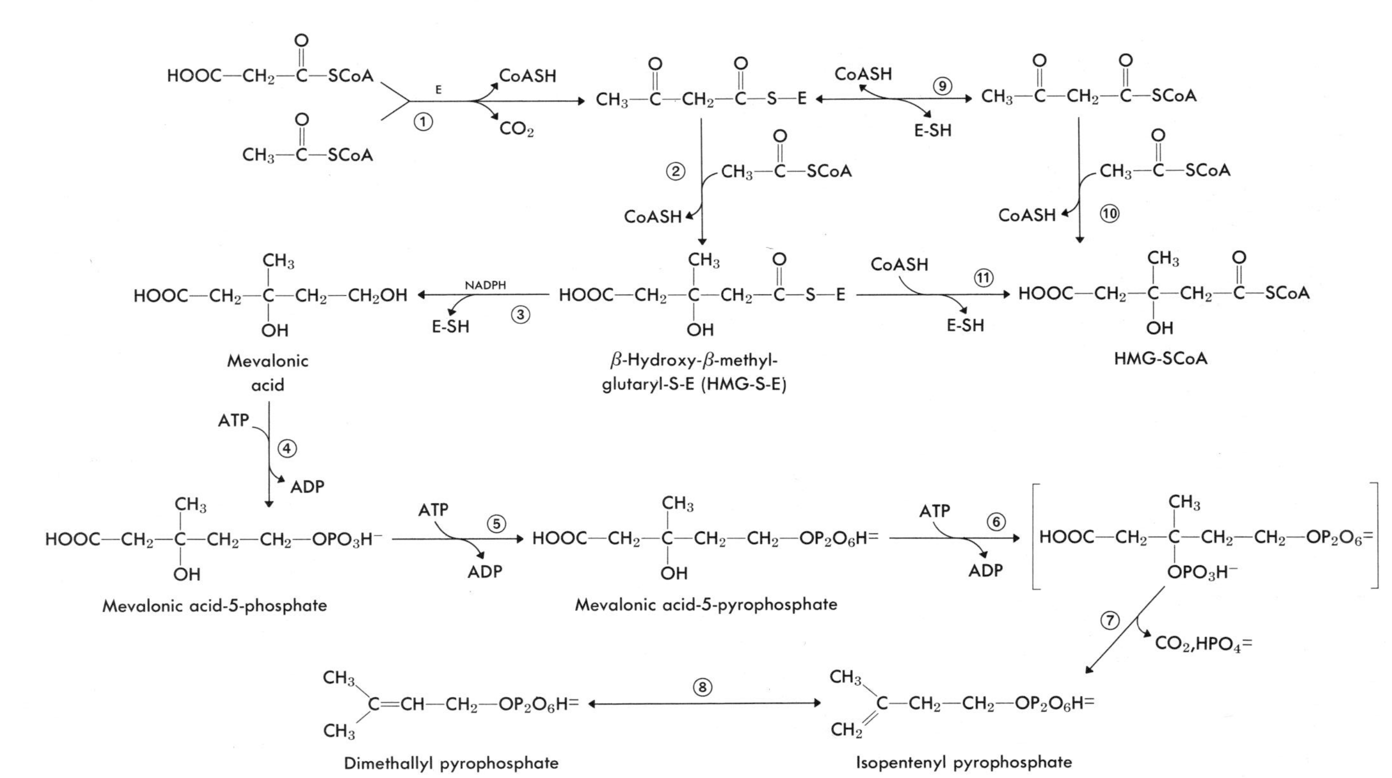

Figure 6.8. Biosynthetic pathways from acetyl-SCoA to isopentenyl pyrophosphate and dimethallyl pyrophosphate.

tion of 3-phosphomevalonic acid-5-pyrophosphate as an intermediate which undergoes simultaneous decarboxylation and loss of phosphate to give isopentenyl pyrophosphate (reaction 7). Isopentenyl pyrophosphate is converted reversibly by isopentenyl pyrophosphate isomerase to dimethallyl pyrophosphate (reaction 8).

Isopentenyl pyrophosphate and dimethallyl pyrophosphate are converted to squalene by successive condensation reactions (Figure 6.9). In the first reaction, isopentenyl pyrophosphate and dimethallyl pyrophosphate are condensed to form *trans*-geranyl pyrophosphate (reaction 1 of Figure 6.9). Geranyl pyrophosphate is then condensed with isopentenyl pyrophosphate to yield *trans-trans*-farnesyl pyrophosphate (reaction 2). Two molecules of farnesyl pyrophosphate are coupled in a NADPH-dependent reaction to form squalene, a C_{30} symmetrical triterpene (reaction 3). The cyclization of squalene to lanosterol (reaction 4) is a complex reaction requiring NADPH and molecular oxygen. In this reaction the oxygen atom of the 3-β-hydroxy group is derived from the molecular oxygen. There are two 1:2 methyl group shifts from position 14 to 13 and from 8 to 13 and various H migrations. An intermediate of the type shown may be involved. The conversion of lanosterol to cholesterol is not completely understood. This reaction involves the oxidative removal of three methyl groups, saturation of the side chain and the double bond at the 8,9-position, and introduction of the double bond in the 5,6-position (reaction 5). Several pathways for this conversion appear to exist each involving a number of related intermediates.

The rate of cholesterol synthesis is influenced by its availability in the body. Cholesterol appears to inhibit by a negative feedback mechanism the NADP reduction of hydroxymethylglutarate to mevalonate. When the cholesterol level is reduced by fasting, the rate of cholesterol synthesis is increased.

The role of cholesterol in atherosclerosis has received much attention. Although it is possible to produce experimental atherosclerosis in rabbits by feeding high cholesterol diets and the condition in man is often associated with high blood cholesterol, other factors of equal importance are probably involved. Cholesterol in the blood is present as free cholesterol and esters of fatty acids. The ratio of the two is nearly constant. The state of the cholesterol as such as well as the physical state of the lipoprotein of which it is a part are important in atherosclerosis.

Since cholesterol is synthesized by tissues, it is difficult to control the blood level except by very strict diet. There is some evidence that dietary saturated fats tend to increase cholesterol levels. Populations that have a low intake of animal fats have a low incidence of atherosclerosis.

The major metabolic fate of cholesterol is its oxidation in the liver to the various bile acids (page 169). Cholesterol can be converted by various tissues to a number of biologically active compounds. Examples are the adrenocortical steroid hormones (page 172), the estrogens (page 170), and the androgens (page 170).

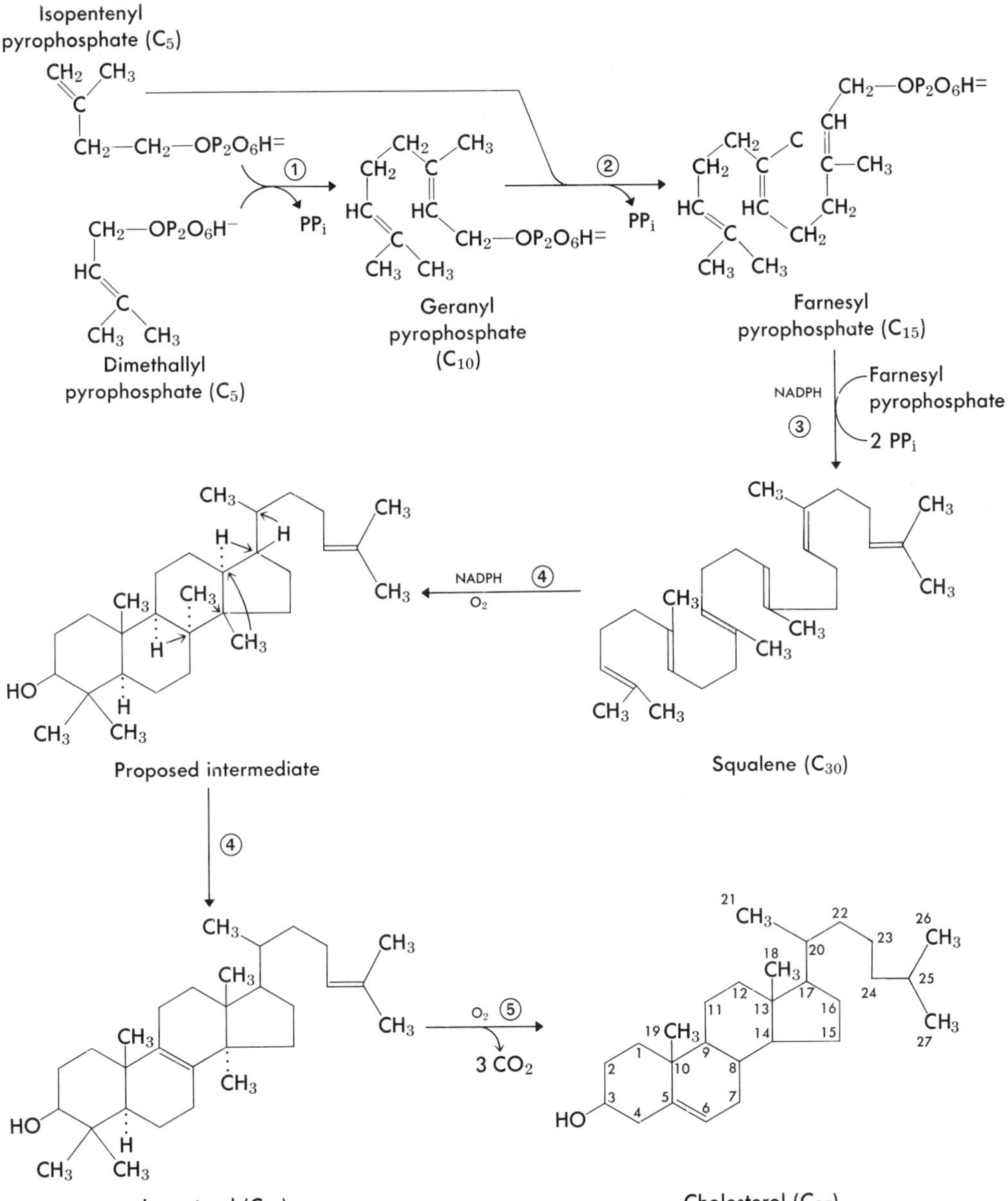

Isopentenyl pyrophosphate (C_5)

Dimethallyl pyrophosphate (C_5)

Geranyl pyrophosphate (C_{10})

Farnesyl pyrophosphate (C_{15})

Squalene (C_{30})

Proposed intermediate

Lanosterol (C_{30})

Cholesterol (C_{27})

Figure 6.9. Biosynthetic pathway from dimethallyl pyrophosphate and isopentenyl pyrophosphate to cholesterol.

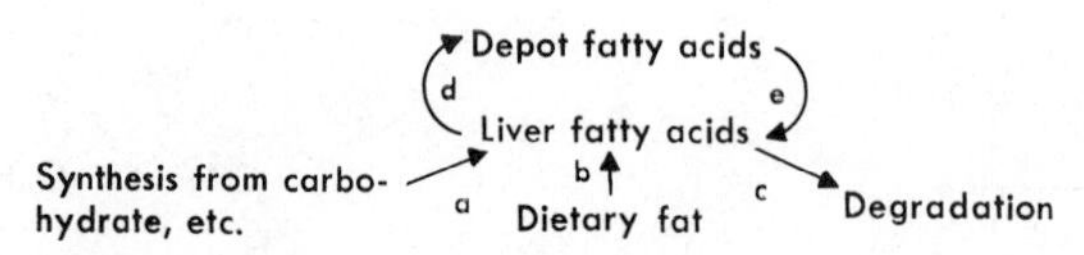

Figure 6.10. Regulation of liver fatty acids. (Modified from I. S. Kleiner and J. M. Orten: *Biochemistry*, 6th ed. C. V. Mosby Co., St. Louis, 1962.)

Fatty Livers

The accumulation of fat in livers may be due to one of a number of causes. Figure 6.10 serves as a basis for explanation of these conditions. A common reason for the accumulation of fat in the liver is a partial block in reaction d. In order for the fats to be transported from the liver, a large portion has to be converted to phospholipids. The synthesis of phospholipids requires choline or the constituents for making choline, for example, methionine for supplying the methyl groups. Fatty livers which develop under these conditions are called *physiologic fatty livers*. Fatty livers of this type can be prevented or cured with choline, methionine, betaine, casein, and raw pancreas. These substances are lipotropic agents.

Livers damaged by phosphorus or chloroform poisoning accumulate fat which comes from the depots (reaction c). Although choline is available for synthesis of phospholipids, the liver is unable to carry out the synthesis. This condition does not respond to lipotropic agents.

Fatty livers can be caused by feeding excess lipids particularly cholesterol (reaction b). The accumulation of neutral fats will respond to choline and the accumulation of cholesterol will respond to inositol. Vitamins B, riboflavin, and cystine increase the rate of fat synthesis (reaction a), and certain types of fatty livers are not produced in thiamine-deficient animals. In conditions where carbohydrate metabolism is low, as in diabetes or starvation, the rate of mobilization of fat from the depots is greater than the rate of utilization.

REFERENCES

Annual Review of Biochemistry. Annual Reviews, Inc., Palo Alto, California, 1932–1970.

Dawson, R. M. C., and Rhodes, D. N. (eds): *Metabolism and Physiological Significance of Lipids*. John Wiley and Sons, New York, 1964.

Deuel, H. J., Jr.: *The Lipids, Their Chemistry and Biochemistry*, 3 vols. Interscience Publishers, Inc., New York, 1951, 1955, 1957.

Greenberg, D. M. (ed): *Metabolic Pathways*, Vol. 2, 3rd ed., Chaps. 8, 9, 10, 11, 12 and 13. Academic Press, New York, 1968.

Mahler, H. R., and Cordes, E. H.: *Biological Chemistry*, Chaps. 12 and 15. Harper and Row, New York, 1966.

Masoro, E. J.: *Physiological Chemistry of Lipids in Mammals*, W. B. Saunders Co., Philadelphia, 1968.

West, E. S.; Todd, W. R.; Mason, H. S.; and Van Bruggen, J. T.: *Textbook of Biochemistry*, 4th ed., Chap. 23. The Macmillan Co., New York, 1966.

White, A.; Handler, P.; and Smith, E. L.: *Principles of Biochemistry*, 4th ed., Chaps. 21 and 22. McGraw-Hill, New York, 1968.

Chapter 7

AMINO ACID AND PROTEIN METABOLISM

IN THE MAMMALIAN organism the principal dietary source of nitrogen is ingested protein. Dietary protein is digested to amino acids in the gastrointestinal tract from whence they are absorbed into the portal circulation and transported to the liver. A portion of these amino acids is removed by the liver, the remainder passing into the systemic circulation for transport to extrahepatic tissues. Amino acids in excess of hepatic needs for protein synthesis and other specific functions are deaminated. The amino group is converted to urea and the resultant keto acids are converted to products common to the metabolism of carbohydrate and fat. The liver is the major, although not exclusive, organ for catabolism (degradation) and anabolism (synthesis) of amino acids. It is also the site of synthesis of plasma albumin, the α- and β-glubulins, fibrinogen, and prothrombin.

Amino acids synthesized in the liver and amino acids derived from the catabolism of proteins in the liver are added to the blood and carried to other tissues for use. Similar anabolic and catabolic processes occur in all tissues, although at widely differing rates. Blood amino acids are derived from the diet via intestinal absorption, from the breakdown of cellular proteins, and from *de novo* cellular synthesis. The blood levels depend upon the balance between the processes adding and those removing amino acids. Some of these overall aspects of protein and amino acid metabolism are shown schematically in Figure 7.1.

In a series of classic experiments Schoenheimer and his colleagues established

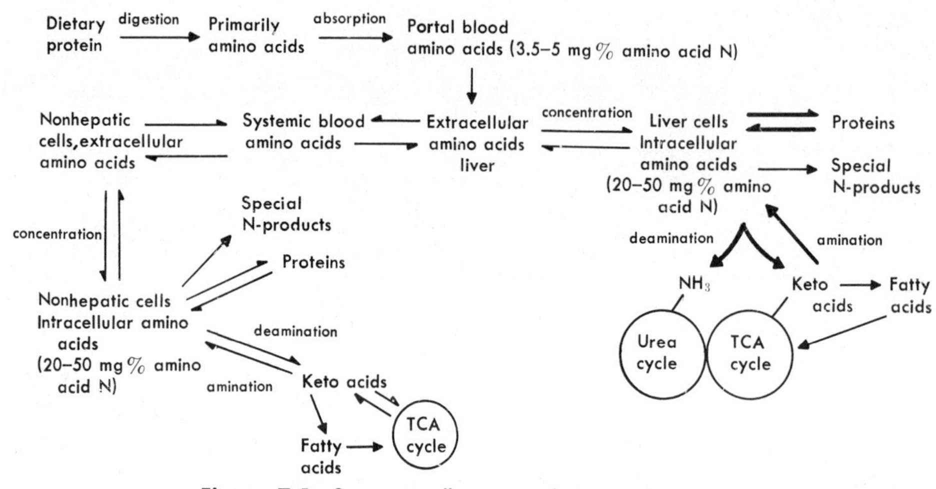

Figure 7.1. Some overall aspects of protein metabolism.

that cellular proteins and their constituent amino acids were in a constant flux, with degradation and synthesis continuously replacing the existing molecules. The apparent constancy of the adult nongrowing animal is not due to the metabolic inertness of cellular materials, but is rather the result of dynamic processes in which the sum of synthesis and degradation is zero. Schoenheimer administered N^{15}-amino acids to nongrowing rats and found that not only had the original amino acid been incorporated intact into the proteins of various tissues but that the amino group had been transferred to other amino acids which were in turn incorporated into protein. Other than the administered amino acid, glutamic acid followed by aspartic acid had the highest isotope concentrations, observations now readily explained by the active transamination (page 220) undergone by the dicarboxylic acids and the central role of glutamic dehydrogenase (page 216) in the assimilation of ammonia. Lysine (page 250) and threonine (page 244) which do not partake in transamination reactions did not contain significant amounts of isotope.

Schoenheimer concluded that amino acids derived from dietary sources merged with those derived from protein breakdown to form a *metabolic pool* of components indistinguishable as to origin which may be employed for either synthetic or degradative processes. The rates of uptake of N^{15}-amino acids acids by various tissues varied considerably; intestinal mucosa, liver, kidney, and spleen incorporated amino acids rapidly in contrast to muscle and skin which incorporated the labeled amino acids at much slower rates. The turnover rate is a measure of this dynamic state and is defined as the quantity of a substance synthesized or degraded per unit of time. A considerable portion of an organism's energy supply may be involved in protein turnover and in the turnover of other bodily components. The protein turnover rate of the human has been estimated at 1.2 g of protein per kilogram of body weight per day and may involve some 5 per cent of the total basal energy production.

It is known that individual proteins vary greatly in turnover rate; proteins of the blood, liver, and other internal organs have half-life times in the range of 2.5 to 10 days, whereas total muscle protein has a half-life of 180 days, and collagen has a half-life perhaps as long as 1000 days. Although the turnover rate of muscle protein is only 5 to 10 per cent that of liver proteins, the fact that muscle protein represents the bulk of the protein of the animal means that the amount of muscle protein synthesized per unit time is minimally equal to that formed by liver.

There are at least three possible mechanisms which operate to account for the phenomenon of protein turnover. First, cells die, their components are catabolized, and new cells with new components arise in the place of dead cells. Second, individual protein molecules are degraded and replaced by synthesis without cell death and replacement. Finally, protein which is secreted from the cellular sites of synthesis is replaced by new protein, for example, hormones, digestive enzymes, serum proteins, and so forth.

A dietary intake of protein is required to provide amino acids for the biosynthesis of various nitrogenous compounds, to replace tissue proteins which are being catabolized, and to replace nitrogen excreted by the animal body in the form of end-products such as urea. There is in addition the requirement for the presence in the diet of certain specific amino acids. These amino acids are called "indispensable" or "essential" amino acids and are defined as amino acids that cannot be synthesized by the animal from available food molecules at a rate rapid enough to meet the demands of normal growth or function. Table 7.1 shows the dietary indispensable and the dispensable amino acids for the growth of young rats.

Man has similar requirements to those of the rat with the exception that in man arginine and histidine are not required to maintain nitrogen balance. In general all higher animals studied require isoleucine, leucine, lysine, methionine, phenylalanine, threonine, tryptophan, and valine, and, with the exception of man, histidine. Certain species require arginine, and under certain circumstances, cystine and tyrosine. The human infant requires the eight amino acids needed by the adult but the quantitative requirements are higher. An animal is said to be in nitrogen balance when the nitrogen excreted equals the nitrogen ingested in a given time period. Animals are in positive nitrogen balance during growth and while recovering from debilitating diseases, and in negative nitrogen balance under conditions of starvation, during wasting illnesses, and when fed a diet lacking or deficient in an indispensable amino acid.

The indispensable amino acids are best provided by proteins of animal origin such

Table 7.1
Classification of L-Amino Acids for Growth Effects in Rats

Indispensable	Dispensable
Arginine	Alanine
Histidine	Aspartic acid
Isoleucine	Citrulline
Leucine	Cystine*
Lysine	Glutamic acid
Methionine	Glycine
Phenylalanine	Hydroxyproline
Threonine	Proline
Tryptophan	Serine
Valine	Tyrosine†

* In the presence of adequate amounts of methionine.
† In the presence of adequate amounts of phenylalanine.

as milk, meat, cheese, eggs, fish, and so forth. Plant proteins, with some exceptions, are often deficient in lysine, methionine, and tryptophan and are less digestible than animal proteins. The recommended daily intake of protein is about 1 to 1.5 g of protein per kilogram of body weight. This figure is based on the assumption that the protein is derived from diverse foods which include eggs, milk, and meat.

Although D-amino acids have not been found in the tissues of higher animals, animals can utilize to some extent the D isomers of certain amino acids. The rat has the ability to utilize partially the D isomers of histidine, leucine, methionine, phenylalanine, tryptophan, tyrosine, and valine. Man can apparently utilize D-methionine and D-phenylalanine. The ability to utilize D-amino acids depends on the rate of conversion of the L isomers. The inversion takes place by oxidative deamination to the corresponding α-keto acid followed by specific reamination to the L isomer. With the exceptions of lysine and threonine, the α-keto and α-hydroxy analogs of the indispensable amino acids may be substituted for the corresponding amino acids. We will discuss in this chapter the reactions that allow the utilization of the α-keto and α-hydroxy analogs and of the D isomers.

Digestion

Table 7.2 lists the enzymes present in the gastrointestinal tract which are concerned with the digestion of protein. The successive action of these proteolytic enzymes converts dietary protein to free amino acids which are absorbed chiefly in the small intestine. The specificities of some of these enzymes have been discussed in Chapter 3.

Table 7.2
Proteolytic Enzymes of Digestive Tract

Enzyme	Source
Pepsin	Gastric juice
Trypsin	Pancreatic juice
Chymotrypsin	Pancreatic juice
Carboxypeptidase	Pancreatic juice
Aminopeptidase	Intestinal mucosa
Tripeptidase	Intestinal mucosa
Dipeptidase	Intestinal mucosa

Intestinal Absorption

The absorption of amino acids from the small intestine appears to be an active, energy-requiring process, since it is inhibited by anaerobiosis, cyanide, and dinitrophenol. The L isomers are absorbed more rapidly than the D isomers, and, in general, the dicarboxylic and diamino acids are absorbed more slowly than the neutral amino acids. There is a competition for absorption among the amino acids actively transported which may be a factor in nutritional amino acid deficiencies. It has recently been shown that ingested protein stimulates the digestive system to release several times its mass of endogenous protein into the lumen of the small intestine. The subsequent hydrolysis of this protein mixture yields an amino acid pool with relatively constant molar ratios despite the nature of the meal ingested. This may serve to minimize fluctuations in the assortment of amino acids available for absorption and promote the efficiency of absorption.

Amino Acids of Blood and Tissues

The level of amino acids in the blood depends upon the balance between the processes adding amino acids (intestinal absorption, breakdown of cellular proteins, and *de novo* synthesis) and the processes removing amino acids (synthesis of proteins and other nitrogenous products, the rates of oxidation, and loss in urine). The plasma concentration of α-amino nitrogen in the fasting individual is usually in the range of 3.5 to 5 mg per 100 ml and will increase 2 to 5 mg per 100 ml shortly after ingesting a meal, returning to the fasting level in four to six hours.

The tissues contain 5 to 10 times the concentration of free amino acids as plasma. As with absorption from the intestine, the entry of amino acids into other tissue cells is an active, energy-requiring process.

Small amounts of free amino acids regularly appear in the urine. Approximately 1 g of free amino acids and 2 g of bound amino acids (peptides, conjugates such as hippuric acid, etc.) are excreted in the urine per day. This amounts to the loss of some 500 mg of nitrogen per day.

METABOLISM OF THE AMINO GROUP

In this section we consider reactions of the amino group which are common to most of the amino acids.

Deamination

Although the deamination of amino acids to the corresponding α-keto acids was observed early in this century, it was not until 1935 that Krebs reported the oxidative deamination of both D- and L-amino acids by homogenates of mammalian liver and kidney and a study of the enzymes involved was begun. Several D- and L-amino acid oxidases have been studied and shown to be flavoproteins, containing either flavin adenine dinucleotide (FAD) or flavin mononucleotide (FMN). They catalyze the general reaction

$$R-\underset{\underset{^{+}NH_3}{|}}{CH}-COO^- + O_2 + H_2O \longrightarrow R-\overset{\overset{O}{\|}}{C}-COO^- + NH_4^+ + H_2O_2$$

In the presence of catalase, the peroxide formed is converted to H_2O and $\frac{1}{2}O_2$ so that the overall reaction may be formulated as

$$R-\underset{\underset{^{+}NH_3}{|}}{CH}-COO^- + \tfrac{1}{2}O_2 \longrightarrow R-\overset{\overset{O}{\|}}{C}-COO^- + NH_4^+$$

D-*Amino Acid Oxidases*

With the exception of mouse liver, highly active D-amino acid oxidases have been found in the liver and kidney of all mammals studied. Crystalline preparations have been obtained which contain 1 mole of FAD per subunit of molecular weight 45,000. The enzyme oxidizes a broad spectrum of D-amino acids with some species differences in substrate specificity and the relative rates of oxidation of different amino acids. Purified D-amino acid oxidase preparations also oxidize glycine to glyoxylate and ammonia.

$$H_3\overset{+}{N}-CH_2-COO^- + O_2 + H_2O \longrightarrow H-\overset{\overset{O}{\|}}{C}-COO^- + NH_4^+ + H_2O_2$$

Although a specific glycine oxidase was reported to catalyze the above reaction, it now seems that glycine oxidase and D-amino acid oxidase are identical.

Other than its function as a glycine oxidase the physiologic role of D-amino acid oxidase in mammalian organisms is not known. Since D-amino acids have a limited distribution in nature and no evidence has been adduced to show production of D-amino acids by mammalian transamination reactions, the necessity for an extremely active D-amino acid oxidase in mammalian organisms is not clear. That the D-amino acid oxidase is active *in vivo* is supported by the fact that the D-isomer of certain dietary indispensable L-amino acids will support growth or maintain nitrogen balance in various animals. Transamination reactions (to be discussed below) can convert the α-keto acid to the appropriate L-amino acid. The α-hydroxy acids after oxidation to the keto acid are presumably metabolized in the same manner.

L-Amino Acid Oxidases

No single L-amino acid oxidase has yet been isolated from liver or kidney which has an activity that can account for the activity of crude homogenates on L-amino acids. This activity is probably explained by the coupling of transaminase systems with glutamic dehydrogenase (see below).

A purified L-amino acid oxidase has been isolated from rat kidney that oxidizes a large number of L-α-amino acids and L-α-hydroxy acids. The prosthetic group is flavin mononucleotide (FMN). Since the enzyme has very low activity with amino acids, it is doubtful that it has a significant role in the deamination of L-amino acids.

The L-amino acid oxidases of snake venoms have been well studied, and a crystalline enzyme has been obtained from rattlesnake venom which has a molecular weight of about 130,000 and contains 2 moles of FAD per mole of protein.

The usual mechanism suggested for the amino acid oxidase reactions is

$$\underset{\substack{|\\ NH_2}}{R-CH_2-COOH} + FAD \rightleftharpoons \underset{\text{Imino acid}}{\left[\underset{\substack{\|\\ NH}}{R-C-COOH}\right]} + FADH_2$$

$$\left[\underset{\substack{\|\\ NH}}{R-C-COOH}\right] + H_2O \rightleftharpoons \overset{\substack{O\\ \|}}{R-C-COOH} + NH_3$$

$$FADH_2 + O_2 \longrightarrow FAD + H_2O_2$$

While the imino acid is probably an intermediate, the actual mechanism is a complicated one in which the two FAD prosthetic groups cooperate in the dehydrogenation of the amino acid so that each of the flavins accepts one hydrogen atom to yield a half-reduced enzyme. After dissociation of the amino acid, the half-reduced enzyme may either be completely reduced by reaction with a second molecule of amino acid or reoxidized by molecular oxygen.

L-*Glutamic Acid Dehydrogenase*

This enzyme catalyzes the pyridine nucleotide-linked reversible deamination of L-glutamic acid. The enzyme is widely distributed, having been found in animals, plants, and microorganisms. It is localized in mitochondria and has been found in most mammalian tissues, liver and kidney having the highest activities. Either NAD or NADP is used as the coenzyme. The reaction is usually formulated as follows

$$\underset{\text{L-Glutamic acid}}{HOOC-CH_2-CH_2-CH(NH_2)-COOH} \xrightleftharpoons[NADH + H^+ \text{ or } NADPH + H^+]{NAD^+ \text{ or } NADP^+} \left[HOOC-CH_2-CH_2-C(=NH)-COOH\right] \xrightleftharpoons[NH_3]{H_2O} \underset{\alpha\text{-Ketoglutaric acid}}{HOOC-CH_2-CH_2-C(=O)-COOH}$$

The fact that this reaction is reversible is of considerable importance, since it provides a means for the net incorporation of ammonia into amino acids and a reversible link between carbohydrate and amino acid metabolism. By coupling with transamination reactions (page 219), the glutamic dehydrogenase can by operating in one direction bring about the general oxidative deamination of diverse L-amino acids; operating in the reverse direction the linked system can achieve the net incorporation and transfer of ammonia to amino acids.

$$R-CO-COOH + HOOC-(CH_2)_2-CH(^*NH_2)-COOH \underset{\text{transaminases}}{\rightleftharpoons} R-CH(^*NH_2)-COOH + HOOC-(CH_2)_2-CO-COOH$$

$$HOOC-(CH_2)_2-CH(^*NH_2)-COOH + H_2O + NAD(P)^+ \underset{\text{glutamic dehydrogenase}}{\rightleftharpoons} HOOC-(CH_2)_2-CO-COOH + {}^*NH_3 + NAD(P)H + H^+$$

Although the linked mechanism for the general deamination of L-amino acids has not been experimentally proven, the extent and rate of transamination reactions with

α-ketoglutarate and the rate of glutamic dehydrogenase reaction appear sufficient to account for the observed rates of deamination of L-amino acids in crude homogenates of liver and kidney.

Contrary to earlier views, glutamic dehydrogenase has some activity with other α-keto acids and with certain monocarboxylic L-amino acids, especially alanine. This is related to the structure of the enzyme. The bovine liver enzyme has a molecular weight of about 2×10^6 and is composed of five enzymatically active monomeric subunits. There is a dependence of the substrate specificity on the conformation of the enzyme which is believed to exist in two monomeric forms, *x* and *y*. The equilibrium between these monomeric forms is influenced by specific regulator molecules.

$$\text{Polymer} \overset{\text{I}}{\rightleftharpoons} \text{Monomer } x \overset{\text{II}}{\rightleftharpoons} \text{Monomer } y$$

Equilibrium I is dependent on the protein concentration, dilution producing dissociation into subunits. Equilibrium II is influenced by specific regulator molecules such as ADP, GTP, NAD, NADH, and steroids. Monomer *x* is primarily active for the glutamate reaction, and monomer *y* is more active for the alanine dehydrogenase reaction and the oxidation of other monocarboxylic L-amino acids. Activation of monocarboxylic amino acid dehydrogenase activity by a shift in equilibrium II toward monomer *y* (GTP) is accompanied by simultaneous loss of glutamic dehydrogenase activity, and stimulation of glutamic dehydrogenase activity with ADP reciprocally inhibits the monocarboxylic amino acid dehydrogenase activity. It appears that the substrate specificity of the enzyme is determined by the position of equilibrium II. Glutamic dehydrogenase appears to have a number of properties in common with typical regulatory enzymes (page 69) and considering its key position in metabolism, it could play a significant role in the partition of metabolites among alternate metabolic sequences. It has been suggested that nucleotides control the cofactor specificity of glutamic dehydrogenase and the physiologic direction of the reaction, NADPH being used for synthesis of glutamate and the NAD system for its degradation. There is evidence to show that glutamate formation in mitochondria does in fact require NADPH.

Other Reactions Producing Ammonia from Amino Acids

There are a number of enzyme-catalyzed reactions in animal tissues which produce ammonia from various amino acids by nonoxidative reactions; a partial list is given in Table 7.3

Table 7.3
Nonoxidative Deamination Reactions of Amino Acids

Enzyme	Substrate	Products
Serine dehydratase	L-Serine	Pyruvate + NH_3
Threonine dehydratase	L-Threonine	α-Ketobutyrate + NH_3
Cysteine desulfhydrase	L-Cysteine	Pyruvate + NH_3 + H_2S
Homocysteine desulfhydrase	L-Homocysteine	α-Ketobutyrate + NH_3 + H_2S
Glutaminase I	L-Glutamine	L-Glutamate + NH_3
Glutaminase II	L-Glutamine + RCOCOOH	α-Ketoglutarate + NH_3 + $RCH(NH_2)COOH$
Asparaginase I	L-Asparagine	L-Aspartate + NH_3
Asparaginase II	L-Asparagine + RCOCOOH	Oxaloacetate + NH_3 + $RCH(NH_2)COOH$
Histidase	L-Histidine	Urocanic acid + NH_3

Animal tissues are known to contain various mono-, di-, and polyamines. The simple amines presumably arise from the decarboxylation of amino acids (cysteic acid, cysteine sulfinic acid, glutamic acid, histidine, tyrosine, dihydroxyphenylalanine, and hydroxytryptophan) by the specific L-amino acid decarboxylases. The amine oxidases are widely distributed and are usually considered under two classes, monoamine and diamine oxidases. They catalyze the following overall reaction.

$$RCH_2NH_2 + O_2 + H_2O \longrightarrow RCHO + NH_3 + H_2O_2$$

There are also various enzymes that produce ammonia from nucleotides, nucleosides, purines and pyrimidines, and hexosamines.

Reamination

There are a number of systems available to the animal for the assimilation of ammonia. In addition to the glutamic acid dehydrogenase reaction (page 216), there are the reactions catalyzed by carbamyl phosphate synthetase, glutamine synthetase, and asparagine synthetase.

Carbamyl Phosphate Synthetase

This enzyme catalyzes the ATP-dependent formation of carbamyl phosphate.

$$NH_4^+ + HCO_3^- + 2\,ATP \longrightarrow H_2N{-}\overset{\overset{\displaystyle O}{\|}}{C}{-}OPO_3H^- + 2\,ADP + P_i$$

Carbamyl phosphate is the initial reactant in the formation of both urea and pyrimidines. This reaction and related reactions are discussed on page 000.

Glutamine Synthetase

This enzyme catalyzes the synthesis of glutamine according to the reaction

$$\underset{\text{Glutamic acid}}{HOOC{-}(CH_2)_2{-}\underset{NH_2}{\underset{|}{C}H}{-}COOH} \underset{NH_3 \quad P_i}{\overset{ATP \quad ADP}{\underset{Mg^{++}}{\rightleftharpoons}}} \underset{\text{Glutamine}}{H_2N{-}\overset{O}{\overset{\|}{C}}{-}(CH_2)_2{-}\underset{NH_2}{\underset{|}{C}H}{-}COOH}$$

The mechanism of the reaction appears to involve the formation of enzyme-bound γ-glutamyl phosphate as an intermediate. The amide nitrogen of glutamine forms a reservoir of ammonia nitrogen which can be used for the synthesis of hexosamines (page 135), histidine (page 375), and purines (page 292), and as the major source of ammonia in urine for the conservation of Na^+ (page 368). Glutamine is synthesized and found in all tissues of the body; plasma glutamine, which is synthesized largely by the liver, accounts for some 20 to 25 per cent of the total plasma amino acid nitrogen.

Asparagine Synthetase

Asparagine appears to be synthetized in animal tissues by a pathway that is analogous to the reaction catalyzed by the glutamine synthetase. Other than as a constituent of proteins, asparagine does not have an assigned metabolic function in animal tissues.

Transamination

Transamination is a reaction in which an amino group is transferred from one molecule to another without the intermediate formation of ammonia. Enzymatic transamination was first reported in 1937 by Braunstein and Kritzmann who observed the transfer of amino groups from α-amino acids to α-keto acids in pigeon breast muscle. Transamination reactions are extensive and are now known to involve mono- and dicarboxylic α-amino and α-keto acids, β-, γ-, and δ-amino acids, amines, and amino acid-ω-amides.

Amino Acid Transaminases

These are a ubiquitous group of enzymes, requiring pyridoxal phosphate as a cofactor, which catalyze the reversible transfer of an amino group to a keto acid or aldehyde. The process may be represented by the following general reaction.

$$R_1-\underset{H}{\overset{NH_2}{C}}-COOH \;+\; R_2-\overset{O}{\overset{\|}{C}}-COOH \;\rightleftharpoons\; R_1-\overset{O}{\overset{\|}{C}}-COOH \;+\; R_2-\underset{H}{\overset{NH_2}{C}}-COOH$$

Although complete details of the mechanism are not known, the available data support the initial formation of a Schiff base between the substrates and the enzyme-bound pyridoxal or pyridoxamine phosphate, followed by tautomerization and hydrolysis to give the products and regeneration of the enzyme. This is shown schematically in Figure 7.2. The four substrates [$R_1CH(NH_2)COOH$, $R_1CO—COOH$, $R_2CO—COOH$, and $R_2CH(NH_2)—COOH$] are mutually competitive and occupy in turn a single site on the enzyme.

The most active and widely distributed transaminases are those which involve L-glutamate and α-ketoglutarate, the glutamate-aspartate [glutamate-oxaloacetate (GOT)] and the glutamate-alanine [glutamate-pyruvate (GPT)] reactions being the most active enzymes.

Earlier evidence had indicated that amino acid transamination reactions involved only L-α-amino acids and α-keto acids and that one of the substrates had to be a dicarboxylic acid. More recent evidence shows that transamination reactions can take place between keto acids and monocarboxylic amino acids and the ω-amino acids and aldehydes are also active in transamination.

Amino Acid Amide Transaminases

Glutaminase II (page 218) catalyzes the following irreversible reaction in which transamination is followed by hydrolysis of the keto amide.

$$\underset{\text{L-Glutamine}}{HOOC-\underset{{}^*NH_2}{\overset{H}{C}}-(CH_2)_2-\overset{O}{\overset{\|}{C}}-NH_2} + R-\overset{O}{\overset{\|}{C}}-COOH \longrightarrow \underset{\alpha\text{-Ketoglutaramic acid}}{HOOC-\underset{O}{\underset{\|}{C}}-(CH_2)_2-\overset{O}{\overset{\|}{C}}-NH_2} + R-\underset{H}{\overset{{}^*NH_2}{C}}-COOH$$

$$\alpha\text{-Ketoglutaramic acid} + H_2O \longrightarrow \underset{\alpha\text{-Ketoglutaric acid}}{HOOC-\underset{O}{\underset{\|}{C}}-(CH_2)_2-COOH} + NH_3$$

An analogous asparagine-α-keto acid transamination reaction is catalyzed by as-

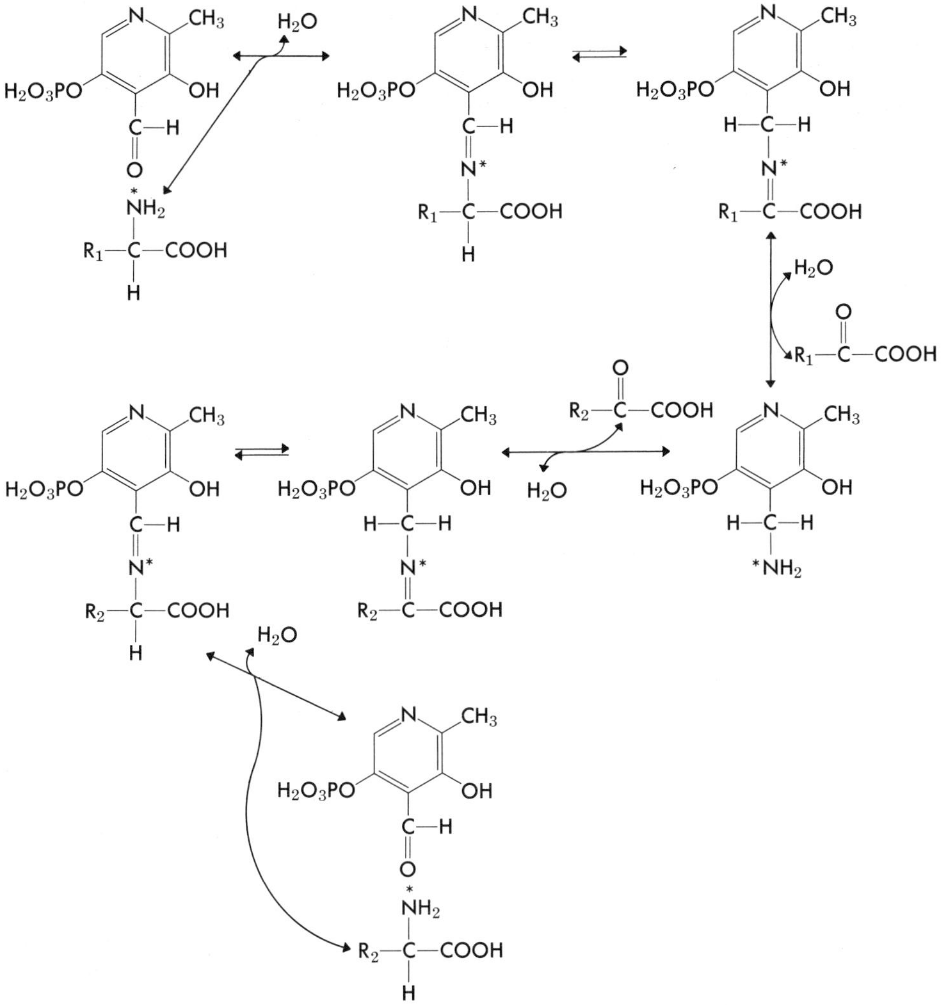

Figure 7.2. Mechanism of enzymatic transamination.

paraginase II, the intermediate being in this case α-ketosuccinamic acid. Both enzymes are relatively specific for the amino donor but can utilize as acceptors a large number of keto acids.

The metabolic roles of the transaminases are of considerable significance. First, they provide a reversible link between carbohydrate and amino acid metabolism. Second, the coupling of transamination reactions with the oxidative deamination of glutamic acid via the glutamic dehydrogenase reaction provides a plausible mechanism for the oxidative deamination of amino acids. Third, transamination is involved directly in the biosynthesis of a number of amino acids. Finally, degradation of many of the amino acids involves the initial removal of the amino group by transamination. Some of these reactions are summarized in Figure 7.3 in which the key role of glutamate and α-ketoglutarate in the assimilation of ammonia and in the redistribution of amino group is apparent.

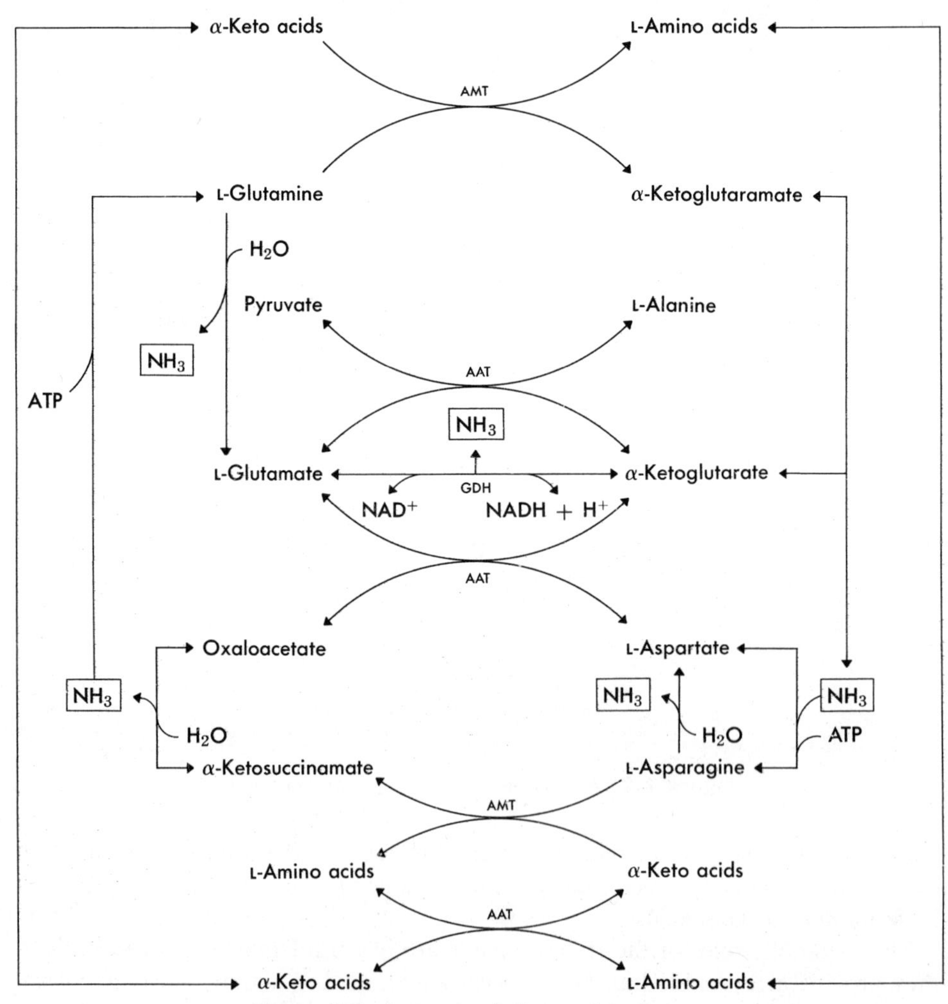

Figure 7.3. Summary of nitrogen interchange.

End Products of Nitrogen Metabolism

Urea is the predominant nitrogen end product excreted by man and other mammals, although some ammonia, creatinine, uric acid, and small amounts of other nitrogeneous substances are excreted also. Certain amphibia and the elasmobranchs (cartilagenous fish) also excrete urea. Reptiles and birds excrete uric acid and most aquatic forms (invertebrates and teleost fish) excrete ammonia. The tadpole excretes mainly ammonia. During metamorphosis of the tadpole into the frog, the various enzymes of the urea cycle appear in the liver and the frog synthesizes and excretes urea. The major form in which nitrogen is excreted appears to be determined largely by the availability of water to the organism. Thus, aquatic organisms can readily dispose of ammonia which is quite toxic, whereas terrestrial species do not have an unlimited supply of water and must detoxify the ammonia by conversion to other excretion products such as urea and uric acid.

Urea Biosynthesis

The relatively large amount of ammonia produced by deamination of amino acids and other compounds is detoxified rapidly in the liver by conversion to urea. Some 20 to 35 g of urea are excreted in the urine each day by a normal adult on a balanced diet.

The initial step in urea biosynthesis is the formation of carbamyl phosphate, which is also the initial compound in pyrimidine synthesis (page 297). This virtually irreversible reaction is catalyzed by carbamyl phosphate synthetase and requires N-acetylglutamate or other N-acylglutamate derivatives as a cofactor.

$$NH_4^+ + HCO_3^- + 2\,ATP \xrightarrow[\text{N-acetylglutamate}]{Mg^{++}} \underset{\text{Carbamyl phosphate}}{H_2N{-}\overset{\overset{\displaystyle O}{\|}}{C}{-}OPO_3H_2} + 2\,ADP + P_i \qquad (7.1)$$

N-Acetylglutamate appears to act as an allosteric cofactor as it has a marked effect both on the physical properties of the enzyme and its activity.

Ornithine transcarbamylase catalyzes the formation of L-citrulline from carbamyl phosphate and L-ornithine.

$$\underset{\text{L-Ornithine}}{\begin{array}{l} CH_2{-}NH_2 \\ | \\ (CH_2)_2 \\ | \\ H{-}C{-}NH_2 \\ | \\ COOH \end{array}} + H_2N{-}\overset{\overset{\displaystyle O}{\|}}{C}{-}OPO_3H_2 \longrightarrow \underset{\text{L-Citrulline}}{\begin{array}{l} CH_2{-}NH{-}\overset{\overset{\displaystyle O}{\|}}{C}{-}NH_2 \\ | \\ (CH_2)_2 \\ | \\ H{-}C{-}NH_2 \\ | \\ COOH \end{array}} + H_3PO_4 \qquad (7.2)$$

The conversion of L-citrulline to L-arginine involves two enzymatic steps. In the first reaction catalyzed by arginosuccinate synthetase, L-citrulline and L-aspartic acid are condensed to yield L-arginosuccinic acid.

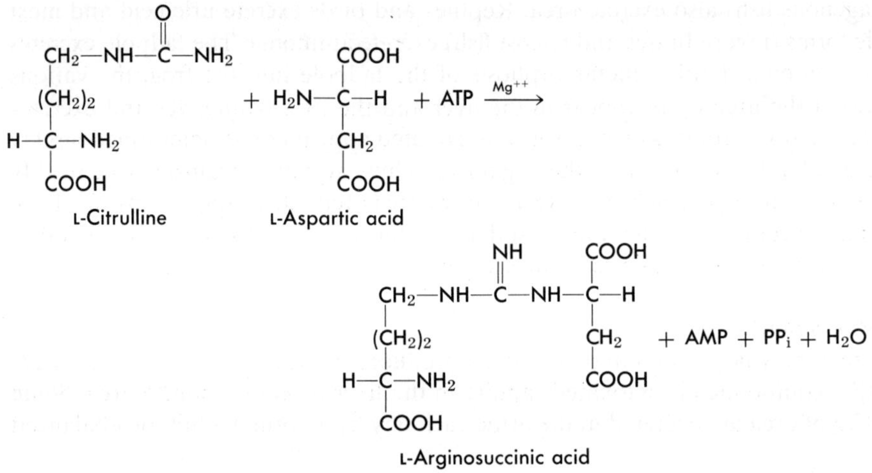

$$
\begin{array}{c}
\text{O} \\
\| \\
\text{CH}_2\text{—NH—C—NH}_2 \\
| \\
(\text{CH}_2)_2 \\
| \\
\text{H—C—NH}_2 \\
| \\
\text{COOH} \\
\text{L-Citrulline}
\end{array}
+
\begin{array}{c}
\text{COOH} \\
| \\
\text{H}_2\text{N—C—H} \\
| \\
\text{CH}_2 \\
| \\
\text{COOH} \\
\text{L-Aspartic acid}
\end{array}
+ \text{ATP} \xrightarrow{\text{Mg}^{++}}
$$

$$
\begin{array}{c}
\text{NH} \quad\quad \text{COOH} \\
\| \quad\quad\quad | \\
\text{CH}_2\text{—NH—C—NH—C—H} \\
| \quad\quad\quad\quad\quad\quad | \\
(\text{CH}_2)_2 \quad\quad\quad \text{CH}_2 \\
| \quad\quad\quad\quad\quad\quad | \\
\text{H—C—NH}_2 \quad\quad \text{COOH} \\
| \\
\text{COOH} \\
\text{L-Arginosuccinic acid}
\end{array}
+ \text{AMP} + \text{PP}_i + \text{H}_2\text{O} \qquad (7.3)
$$

L-Arginosuccinic acid is cleaved to L-arginine and fumarate by arginosuccinase.

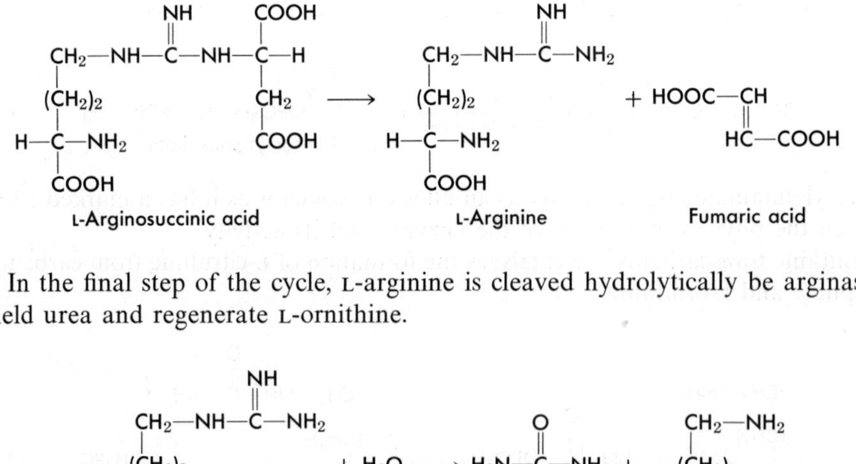

$$
\begin{array}{c}
\text{NH} \quad\quad \text{COOH} \\
\| \quad\quad\quad | \\
\text{CH}_2\text{—NH—C—NH—C—H} \\
| \quad\quad\quad\quad\quad\quad | \\
(\text{CH}_2)_2 \quad\quad\quad \text{CH}_2 \\
| \quad\quad\quad\quad\quad\quad | \\
\text{H—C—NH}_2 \quad\quad \text{COOH} \\
| \\
\text{COOH} \\
\text{L-Arginosuccinic acid}
\end{array}
\longrightarrow
\begin{array}{c}
\text{NH} \\
\| \\
\text{CH}_2\text{—NH—C—NH}_2 \\
| \\
(\text{CH}_2)_2 \\
| \\
\text{H—C—NH}_2 \\
| \\
\text{COOH} \\
\text{L-Arginine}
\end{array}
+
\begin{array}{c}
\text{HOOC—CH} \\
\| \\
\text{HC—COOH} \\
\text{Fumaric acid}
\end{array}
\qquad (7.4)
$$

In the final step of the cycle, L-arginine is cleaved hydrolytically be arginase to yield urea and regenerate L-ornithine.

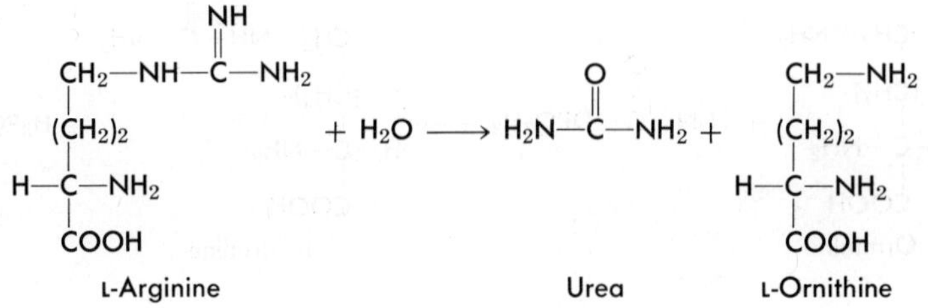

$$
\begin{array}{c}
\text{NH} \\
\| \\
\text{CH}_2\text{—NH—C—NH}_2 \\
| \\
(\text{CH}_2)_2 \\
| \\
\text{H—C—NH}_2 \\
| \\
\text{COOH} \\
\text{L-Arginine}
\end{array}
+ \text{H}_2\text{O} \longrightarrow
\begin{array}{c}
\text{O} \\
\| \\
\text{H}_2\text{N—C—NH}_2 \\
\text{Urea}
\end{array}
+
\begin{array}{c}
\text{CH}_2\text{—NH}_2 \\
| \\
(\text{CH}_2)_2 \\
| \\
\text{H—C—NH}_2 \\
| \\
\text{COOH} \\
\text{L-Ornithine}
\end{array}
\qquad (7.5)
$$

The overall reaction for the enzymic synthesis of urea (sum of reactions 1 to 5) may be written

$$NH_4^+ + HCO_3^- + 3\ ATP + \text{L-Aspartate} \longrightarrow \text{Urea 6 Fumarate} + AMP + PP_i + 2\ ADP + 2\ P_i$$

The $\Delta F_0'$ of the overall enzymic reaction has been estimated at -13 kcal, based on a value of -7.7 kcal for the $\Delta F_0'$ of the hydrolysis of ATP at pH 7.0.

Relation of Urea Cycle to Other Metabolic Reactions

The intimate association of the urea cycle with other metabolic reactions may be seen from an examination of Figure 7.4, which summarizes urea formation and some related metabolism. The glutamic dehydrogenase reaction (8), the glutamic-aspartic transaminase reaction (7), and the reactions of the tricarboxylic acid cycle (6) are of particular importance for maintaining urea synthesis. Glutamate can provide by these reactions both the amino group and carbon chain of asparate as well as the necessary supply of ATP. In the absence of direct sources of ATP and aspartate, urea formation will be dependent on the tricarboxylic acid cycle and transamination reactions. Substances that limit the formation of ATP (respiratory inhibitors), or prevent the formation of aspartate by blocking the recycling of oxaloacetate in the tricarboxylic acid cycle (malonate inhibition), or situations which deplete the supply of aspartate (the shifting of the transaminase equilibrium away from aspartate) will

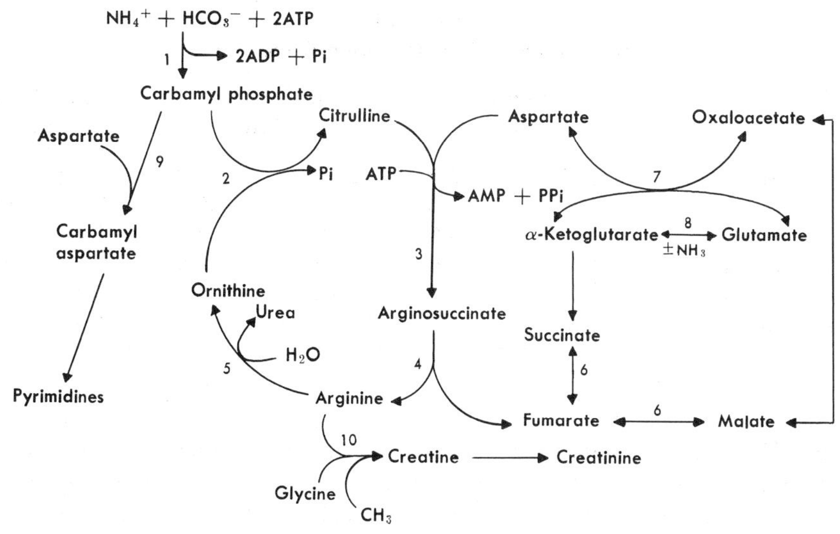

Figure 7.4. Urea cycle and related metabolism.

all limit the formation of urea. The fumarate produced by the cleavage of arginosuccinate (4) may be converted to aspartate via the tricarboxylic acid cycle and transamination. Thus, substances that are common substrates to both the urea and tricarboxylic acid cycles or are linked by appropriate enzymes may act catalytically in the formation of urea. It is worthy of comment that carbamyl phosphate via reaction sequence (9), is a precursor of the pyrimidines and, as a common precursor for two metabolic pathways, it provides a likely site for regulatory action. It is of interest that the activity as aspartate transcarbamylase (9) increases rapidly in regenerating liver and attains a level twice that of normal liver in 36 hours after partial hepatectomy. At the same time there is a decrease in ornithine transcarbamylase activity (2). This is apparently the result of greater needs for RNA and DNA for cell division. The enzyme associated with the more fundamental functions, growth and proliferation, takes precedence.

It has been established that the liver is the major site for the synthesis of urea. Although kidney contains, in amounts comparable to liver, the three enzymes necessary for conversion of citrulline to urea, it has neither significant amounts of arginase nor appreciable ability to convert ornithine to citrulline.

Other End Products of Nitrogen Metabolism

In addition to urea, there are smaller amounts of other nitrogenous products excreted in the urine. Table 7.4 shows the partition of urinary nitrogen in a human subject on normal, high-, and low-protein diets.

Table 7.4
Twenty-Four-Hour Nitrogen Excretion

	Normal Diet		High-Protein Diet		Low-Protein Diet	
Total nitrogen	13.2 g	100%	23.2 g	100%	4.2 g	100%
Urea nitrogen	11.4	86.3	20.4	88.0	2.9	69.0
Ammonia nitrogen	0.5	3.8	0.8	3.4	0.2	4.7
Creatinine nitrogen	0.6	4.6	0.6	2.6	0.6	14.3
Uric acid nitrogen	0.2	1.5	0.3	1.3	0.1	2.4
Undetermined nitrogen	0.5	3.8	1.1	4.7	0.4	9.6

Both the total quantity of urea excreted and the proportion of total urinary nitrogen excreted as urea are proportional to the extent of protein metabolism. Urea usually accounts for 80 to 90 per cent of the urinary nitrogen, but may drop to considerably lower values on a low-protein intake. Diseases of the liver which impair the ability to synthesize urea, and renal diseases in which excretion of urea is restricted, will decrease the amount of urea excreted in the urine.

Ammonia

The larger portion ($\frac{2}{3}$) of the urinary ammonia is formed in the kidney by the enzymic hydrolysis of blood glutamine, the remainder being produced by the oxidative deamination of blood amino acids. Since both ammonia and urea are derived from amino acids, an increase in one results in a decrease of the other for a given level of nitrogen excretion. The amount of ammonia in the urine is determined generally by the requirements for control of electrolyte and acid-base balance of the body. Increased ammonia excretion with increased levels of dietary proteins is due to the production of acids by the oxidation of the sulfur and phosphorus contained in the protein.

Uric Acid

Uric acid is the principal end product of purine metabolism in man and other primates. The metabolic origin of the carbon and nitrogen atoms of uric acid is discussed in Chapter 8.

The urinary excretion of uric acid (0.2 to 0.4 g of uric acid nitrogen per day) depends on the quantity of purine being metabolized. Uric acid is derived from the dietary intake of purine-containing compounds and from the metabolism of tissue nucleic acids and nucleotides.

Creatine and Creatinine

The first step in creatine biosynthesis is a transamidination reaction between arginine and glycine catalyzed by a kidney transamidinase.

$$\underset{\text{L-Arginine}}{HN{=}C(NH_2){-}NH{-}(CH_2)_3{-}CH(NH_2){-}COOH} + \underset{\text{Glycine}}{NH_2{-}CH_2{-}COOH} \longleftrightarrow \underset{\text{L-Ornithine}}{NH_2{-}(CH_2)_3{-}CH(NH_2){-}COOH} + \underset{\text{Guanidoacetic acid}}{HN{=}C(NH_2){-}NH{-}CH_2{-}COOH}$$

In a second reaction, catalyzed by a liver transmethylase, guanidoacetic acid is N-methylated to form creatine.

$$\underset{\text{Guanidoacetic acid}}{NH{=}C(NH_2){-}NH{-}CH_2{-}COOH} + \underset{\text{S-Adenosylmethionine}}{CH_3{-}\overset{+}{S}(CH_2\text{-adenosine}){-}(CH_2)_2{-}CH(NH_2){-}COO^-} \longrightarrow \underset{\text{Creatine}}{NH{=}C(NH_2){-}N(CH_3){-}CH_2{-}COOH} + \underset{\text{S-Adenosylhomocysteine}}{S(CH_2\text{-adenosine}){-}(CH_2)_2{-}CH(NH_2){-}COOH}$$

S-Adenosylmethionine, the active transmethylating form of methionine, is formed in the liver from methionine (page 270).

$$\underset{\text{L-Methionine}}{CH_3{-}S{-}(CH_2)_2{-}CH(NH_2){-}COOH} + ATP \xrightarrow[\text{enzyme}]{\text{activating}} \underset{\text{S-Adenosylmethionine}}{CH_3{-}\overset{+}{S}(CH_2\text{-adenosine}){-}(CH_2)_2{-}CH(NH_2){-}COO^-} + PP_i + P_i$$

Neither the methyl group nor the nitrogen of creatine is usable for other metabolic purposes. Creatine is therefore an end product of the metabolism of arginine, glycine, and methionine. Most of the creatine, which occurs in all body tissues, is found as the high-energy compound creatine phosphate. Only a small amount of creatine

(60 to 150 mg per day) is excreted normally in the urine. Prior to excretion, the creatine phosphate is converted to creatinine.

$$\underset{\text{Creatine phosphate}}{\begin{array}{l} NH{=}C{-}NH{-}PO_3H_2 \\ \quad\;\; | \\ CH_3{-}N{-}CH_2{-}COOH \end{array}} \xrightarrow[\text{reaction}]{\text{spontaneous}} \underset{\text{Creatinine}}{\begin{array}{l} HN{=}C{-}NH \\ \quad\;\; | \qquad\quad \diagdown \\ \quad\;\; | \qquad\quad\;\; C{=}O \\ \quad\;\; | \qquad\quad \diagup \\ CH_3{-}N{-}CH_2 \end{array}} + P_i$$

The normal adult excretes 1.0 to 1.8 g creatinine (0.37–0.67 g creatinine N) per day, and this is not significantly altered by dietary variations (see Table 7.4).

ONE-CARBON METABOLISM

It has been known for some time that formate and formaldehyde were incorporated into a number of compounds (serine, purines, methyl group of methionine, and so on), and that these compounds yielded "active" one-carbon groups related to formate and formaldehyde which could participate in several biochemical reactions. The vitamin folic acid was shown subsequently to function as a carrier of one-carbon units at the oxidation states of both formaldehyde and formate.

The structures of folic acid and its more important biologically functional forms are shown in Figure 7.5. We have considered here five biologically active one-carbon adducts of tetrahydrofolic acid (FAH_4), three at the oxidation level of formate (N^5-formyl FAH_4, N^5,N^{10}-methenyl FAH_4, N^5-formimino FAH_4), one at the oxidation level of formaldehyde (N^5,N^{10}-methylene FAH_4), and a recently discovered compound at the oxidation level of methanol (N^5-methyl FAH_4). Although this list does not include all known biologically active forms of folic acid one-carbon carriers, it does give those which are considered to be the enzymatically directly reactive forms. For example, N^{10}-formyltetrahydrofolic acid may be converted by acidification to the bridge compound (N^5,N^{10}-methenyl FAH_4). For the purpose of this discussion we will consider the N^5,N^{10}-methenyltetrahydrofolate to be the active formyl donor.

At the level of formaldehyde, the active derivative is N^5,N^{10}-methylene FAH_4 and is formed in at least three reactions.

$$(1)\quad FAH_4 + HCHO \xrightarrow[\text{enzyme}]{\text{activating}} N^5,N^{10}\text{-Methylene } FAH_4$$

$$(2)\quad \underset{\text{Serine}}{\begin{array}{l} CH_2{-}CH{-}COOH \\ | \qquad\; | \\ OH \quad NH_2 \end{array}} + FAH_4 \xleftrightarrow[\text{hydroxymethylase, pyridoxal phosphate}]{\text{serine}} \underset{\text{Glycine}}{\begin{array}{l} CH_2{-}COOH \\ | \\ NH_2 \end{array}} + N^5,N^{10}\text{-Methylene } FAH_4$$

$$(3)\quad N^5,N^{10}\text{-Methenyl } FAH_4 + NADPH + H^+ \xleftrightarrow[\text{dehydrogenase}]{\text{tetrahydrofolate}} N^5,N^{10}\text{-Methylene } FAH_4 + NADP^+$$

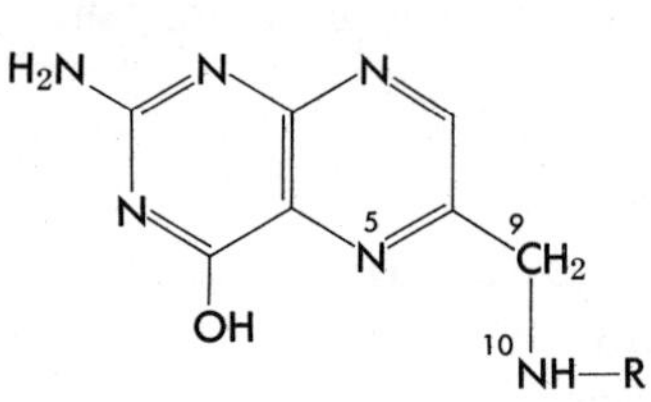

Folic acid (FA)

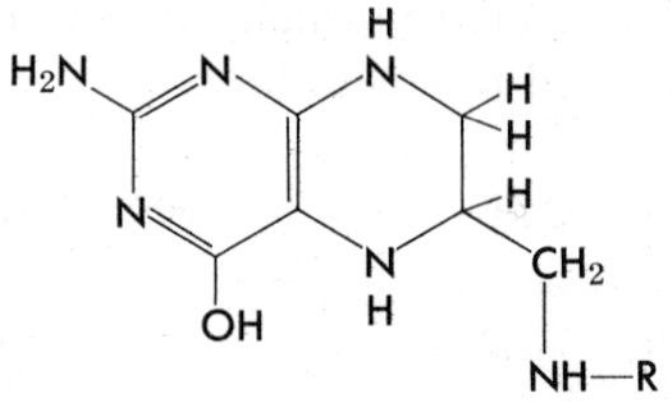

5,6,7,8-Tetrahydrofolic acid (FAH_4)

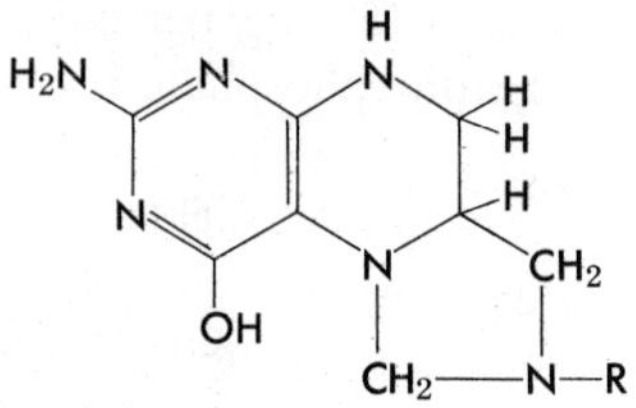

N^5,N^{10}-Methylenetetrahydrofolic acid (N^5,N^{10}-methylene FAH_4)

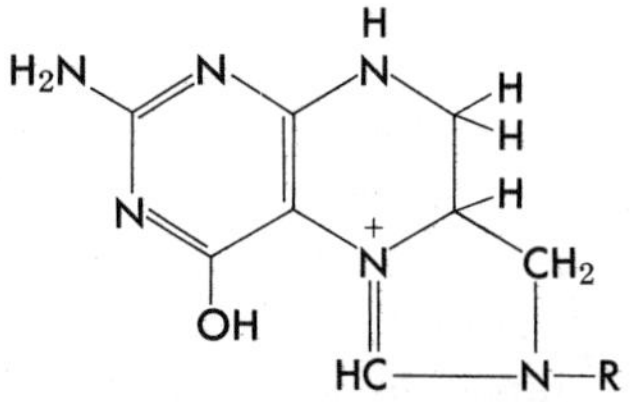

N^5,N^{10}-Methenyltetrahydrofolic acid (N^5,N^{10}-methenyl FAH_4)

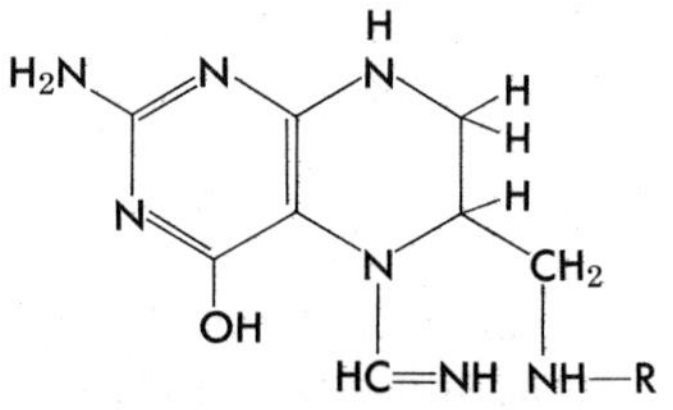

N^5-Formiminotetrahydrofolic acid (N^5-formimino FAH_4)

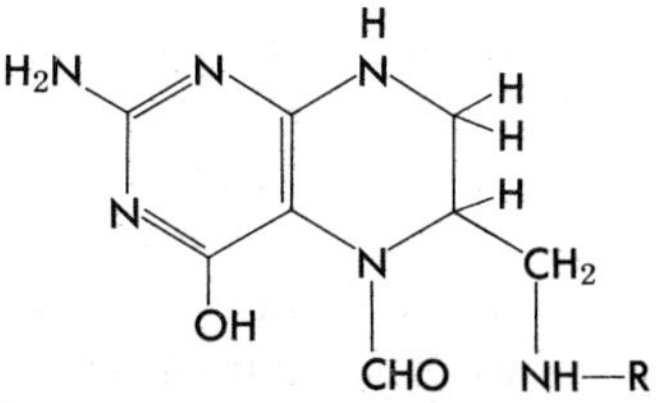

N^5-Formyltetrahydrofolic acid (N^5-formyl FAH_4)

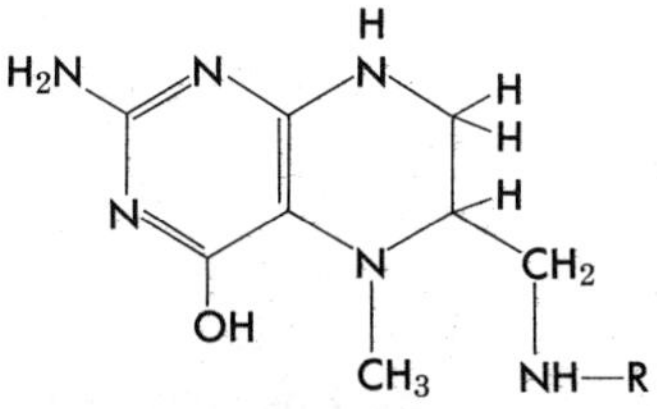

N^5-Methyltetrahydrofolic acid (N^5-methyl FAH_4)

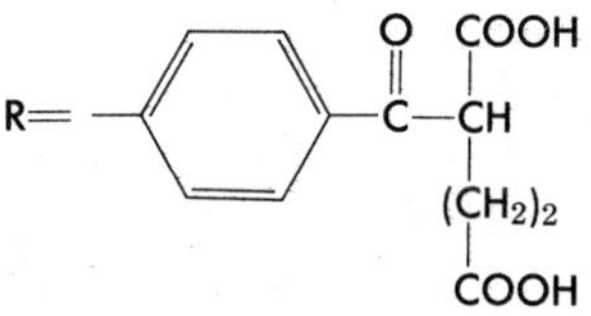

Figure 7.5. Folic acid and its functional forms.

Reaction 3 serves to intraconvert the active formaldehyde and formate derivatives.

N^5,N^{10}-Methenyl FAH_4, the primary formate donor, is formed via the following reactions

(1) From N^5,N^{10}-methylene FAH by the reversible tetrahydrofolate dehydrogenase reaction

(2) $FAH_4 + HCOOH + ATP \xrightarrow[\text{enzyme, } Mg^{++}]{\text{formate activating}} N^{10}\text{-Formyl } FAH_4 + ADP + P_i$

$$N^{10}\text{-Formyl FAH} \xleftrightarrow[H_2O]{H^+} N^5,N^{10}\text{-Methenyl } FAH_4$$

(3) N^5-Formyl FAH_4 + ATP $\xleftrightarrow{\text{isomerase}}$ N^{10}-Formyl FAH_4 + ADP + P_i

N^{10}-Formyl FAH_4 $\xleftrightarrow{H^+}$ N^5,N^{10}-Methenyl FAH_4 (− H_2O)

There are two additional folic acid compounds at the level of oxidation of formate, N^5-formyl FAH_4 and N^5-formimino FAH_4. The first enters into a reversible reaction with glutamate.

Glutamate + N^5-Formyl FAH_4 $\longleftrightarrow$ N-Formylglutamate + FAH_4

N^5-Formimino FAH_4 is involved in a number of formimino-transfer reactions of which the following is representative.

CH_2—COOH (with NH_2), Glycine + N^5-Formimino FAH_4 $\xrightarrow{\text{transferase}}$ CH_2—COOH (with HN—CH=NH), Formiminoglycine + FAH_4

The metabolic interrelations of the various folic acid derivatives are summarized in Figure 7.6.

The methyl group of thymidylic acid is formed via the following reaction.

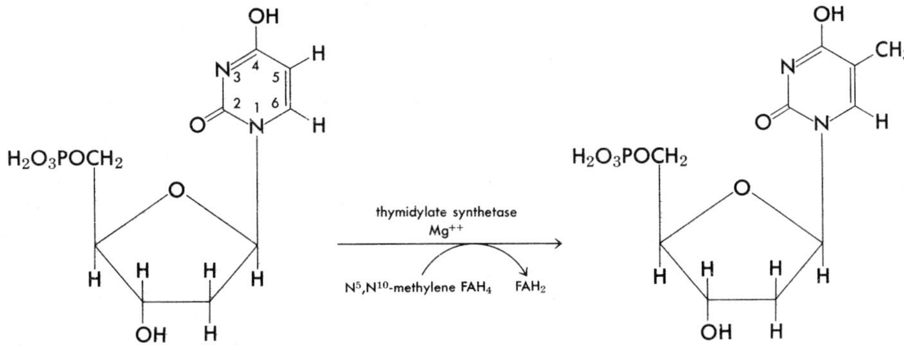

Deoxyuridylate ($H_2O_3POCH_2$; ring positions N3, 4, 5, 6, 1, 2; OH, H, H, O, N) $\xrightarrow[\text{Mg}^{++}]{\text{thymidylate synthetase}}$ Thymidylate ($H_2O_3POCH_2$; OH, CH_3, H)

N^5,N^{10}-methylene FAH_4 → FAH_2

This reaction is presumed to proceed over an intermediate in which the methylene group of N^5,N^{10}-methylene FAH_4 is bridged between the N^5 position of the tetrahydrofolate and the C^5 position of the pyrimidine. An intramolecular oxidation-reduction reaction follows in which the bond between the N^5-nitrogen and the methylene bridge is cleaved to give FAH_2 and thymidylate, the hydrogen from the

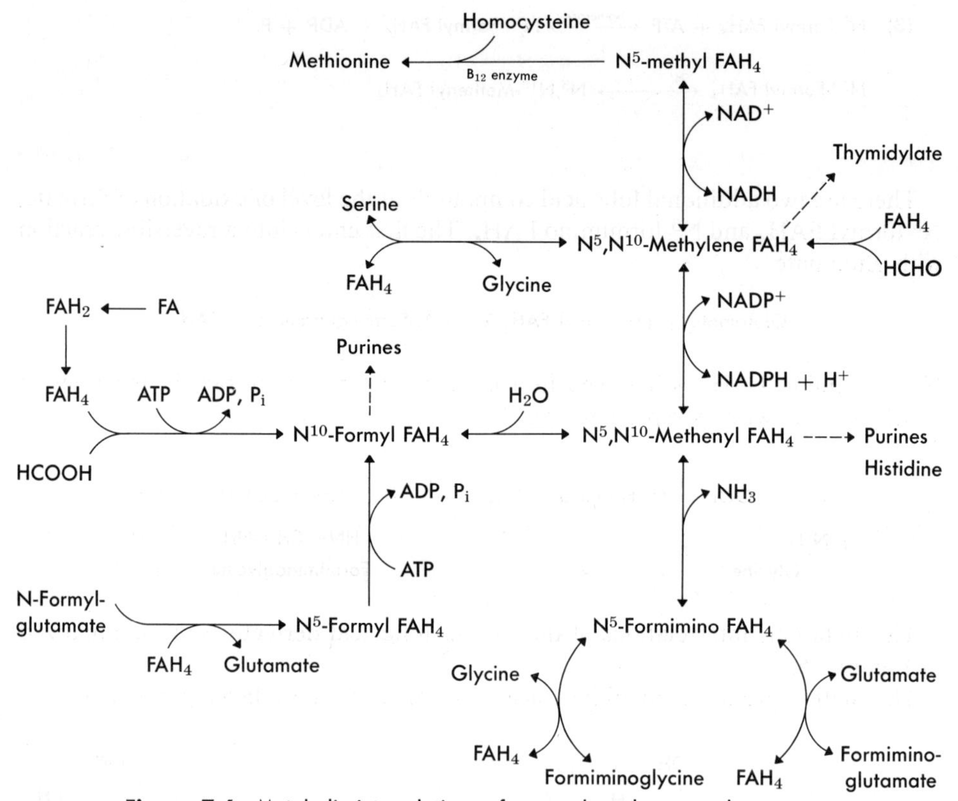

Figure 7.6. Metabolic interrelations of one-carbon donors and acceptors.

C^6 position of the N^5,N^{10}-methylene FAH_4 being transferred to the methyl group of thymidylate.

The transfer of a methyl group from a folic acid coenzyme to homocysteine for the synthesis of methionine is brought about by a pathway quite different from that involved in the formation of thymidylate. Homocysteine is methylated by N^5-methyl FAH_4 in a complicated reaction requiring a reducing system, S-adenosylmethionine, and a cobamide-dependent enzyme,

$$\underset{\text{Homocysteine}}{\begin{array}{c} CH_2{-}SH \\ | \\ CH_2 \\ | \\ H{-}C{-}NH_2 \\ | \\ COOH \end{array}} + N^5\text{-Methyl } FAH_4 \xrightarrow[\text{S-adenosylmethionine, reducing system}]{B_{12}\text{ enzyme}} \underset{\text{Methionine}}{\begin{array}{c} CH_2{-}S{-}CH_3 \\ | \\ CH_2 \\ | \\ H{-}C{-}NH_2 \\ | \\ COOH \end{array}} + FAH_4$$

Methyl-B_{12} vitamin B_{12} in which the cobalt ligand cyanide has been replaced by a methyl group, appears to be an intermediate in the reaction. One current scheme visualizes the reaction as follows.

$$\text{N}^5\text{-Methyl FAH}_4 + \text{Reduced cobamide enzyme} \xrightarrow[\text{reducing system}]{\text{S-adenosylmethionine}} \text{Methyl-cobamide enzyme} + \text{FAH}_4$$

$$\text{Methyl-cobamide enzyme} + \text{Homocysteine} \longrightarrow \text{Methionine} + \text{Reduced cobamide enzyme}$$

The reducing system ($FADH_2$) keeps the enzyme-bound cobamide in a reduced state. The role of S-adenosylmethionine is not clear, but it may act as an allosteric effector.

Transmethylation reactions and some of the above reactions will be considered in detail in subsequent sections. It is clear that one-carbon metabolism is extensive and involves amino acids, purines, and pyrimidines.

SPECIFIC METABOLISM OF INDIVIDUAL AMINO ACIDS

Consideration has been given in an earlier section to the metabolic fate of the amino group of the amino acids. The metabolism of the keto acids produced by deamination follows two generalized pathways. These are the glucogenic pathway which forms carbohydrate metabolites and the ketogenic pathway which leads to the production of ketone bodies. Amino acids which can form pyruvate or components of the tricarboxylic acid cycle which can be converted to glucose or glycogen are referred to as glucogenic. Thus, the glucogenic nature of alanine, aspartic acid, cysteine, glutamic acid, and serine, which all give rise to α-keto acid intermediates of carbohydrate metabolism, is explained. Amino acids which give rise to acetyl-CoA or acetoacetic acid are said to be ketogenic. Certain of the amino acids are metabolized to yield both glucose and ketone bodies. Table 7.5 classifies the common amino acids with regard to their metabolic fates.

Metabolism of Glycine

Although glycine engages in diverse metabolic reactions involving other amino acids, purines, lipids, and carbohydrates, its most characteristic reaction is condensation with and incorporation into other molecules. In many of these biochemical reactions, the glycine structure remains essentially intact. The formation from glycine of serine, creatine, purines, glycocholic acid, hippuric acid, and glutathione are appropriate examples of such reactions. Some of the metabolic transformations of glycine are summarized in Figure 7.7 and are discussed in detail below.

Glycine-Serine Interconversions

As noted previously (page 229), glycine and serine are interconverted via the liver serine hydroxymethylase reaction which requires N^5, N^{10}-methylenetetrahydrofolic acid as the cofactor.

Table 7.5
Glucogenic and Ketogenic Amino Acids

Amino Acid	Glucogenic	Ketogenic
Alanine	+	−
Arginine	+	−
Aspartic acid	+	−
Cysteine (cystine)	+	−
Glutamic acid	+	−
Glycine	+	−
Histidine	+	−
Hydroxyproline	+	−
Isoleucine	+	+
Leucine	−	+
Lysine	−	+
Methionine	+	−
Phenylalanine	+	+
Proline	+	−
Serine	+	−
Threonine	+	−
Tryptophan	+	−
Tyrosine	+	+
Valine	+	−

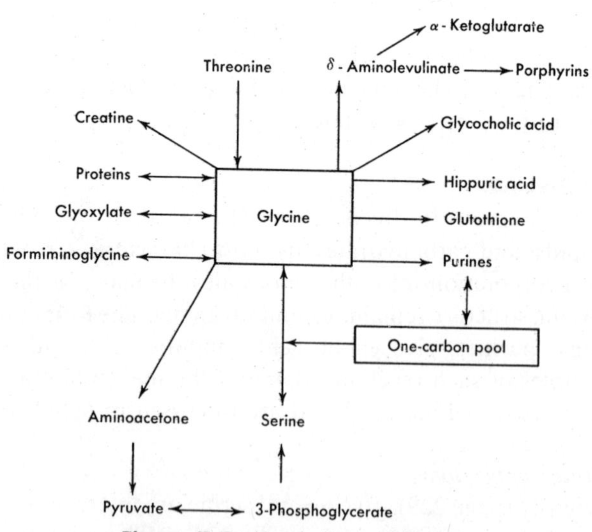

Figure 7.7. Metabolism of glycine.

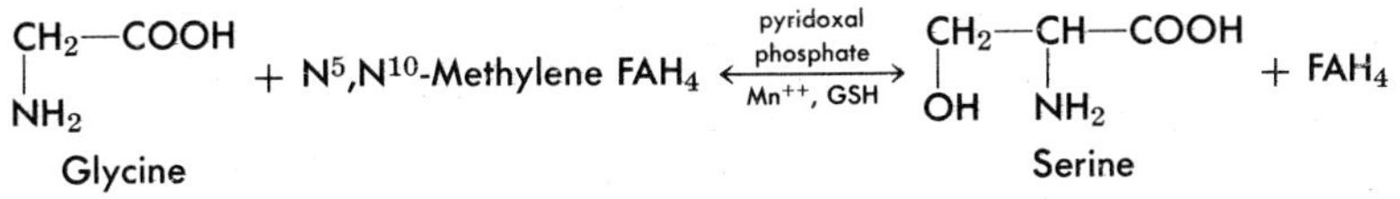

$$\underset{\text{Glycine}}{\underset{NH_2}{\overset{|}{CH_2}}\text{—COOH}} + N^5,N^{10}\text{-Methylene } FAH_4 \xrightleftharpoons[\text{Mn}^{++},\ \text{GSH}]{\text{pyridoxal phosphate}} \underset{\text{Serine}}{\underset{OH}{\overset{|}{CH_2}}\text{—}\underset{NH_2}{\overset{|}{CH}}\text{—COOH}} + FAH_4$$

It appears that this reaction may operate for the most part in the direction of glycine synthesis, serine being largely formed from glucose via 3-phosphoglycerate and other intermediates (page 242). Glycine in excess of need is presumably disposed of by conversion to serine which may be transformed to pyruvate and metabolized as such.

Deamination and Amination

Glycine may be converted to glyoxylic acid by the action of D-amino acid oxidase (page 214) or by transamination, the latter reaction providing a pathway for glycine amination. Glyoxylic acid is oxidized to formic acid and carbon dioxide, and to a smaller extent to oxalic acid. Formic acid is oxidized to carbon dioxide or enters the pool of one-carbon units (page 230) where it may be used for the synthesis of other compounds, for example, serine.

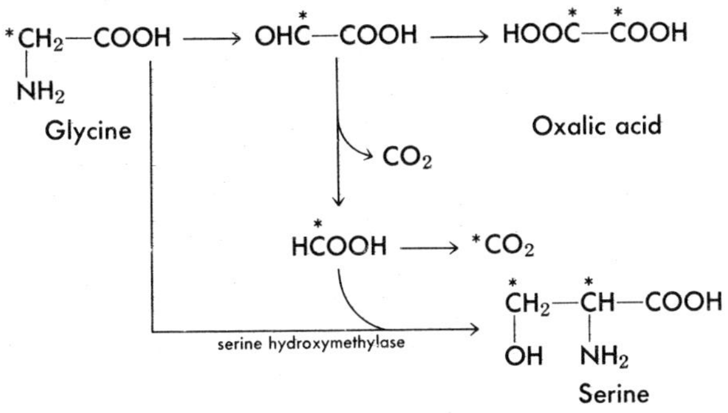

$$\underset{\text{Glycine}}{{}^*CH_2(NH_2)\text{—COOH}} \longrightarrow OH\overset{*}{C}\text{—COOH} \longrightarrow \underset{\text{Oxalic acid}}{HOO\overset{*}{C}\text{—}\overset{*}{C}OOH}$$

$$OH\overset{*}{C}\text{—COOH} \longrightarrow CO_2 + H\overset{*}{C}OOH; \quad H\overset{*}{C}OOH \longrightarrow {}^*CO_2$$

$$\text{Glycine} + H\overset{*}{C}OOH \xrightarrow{\text{serine hydroxymethylase}} \underset{\text{Serine}}{\underset{OH}{\overset{|}{\overset{*}{C}H_2}}\text{—}\underset{NH_2}{\overset{|}{\overset{*}{C}H}}\text{—COOH}}$$

In this manner glycine can provide a one-carbon unit for condensation with itself to form serine or engage in other reactions involving the one-carbon pool.

A quantitatively significant pathway for glycine catabolism appears to be its conversion in liver to lactic acid via aminoacetone. This is summarized by the following reactions.

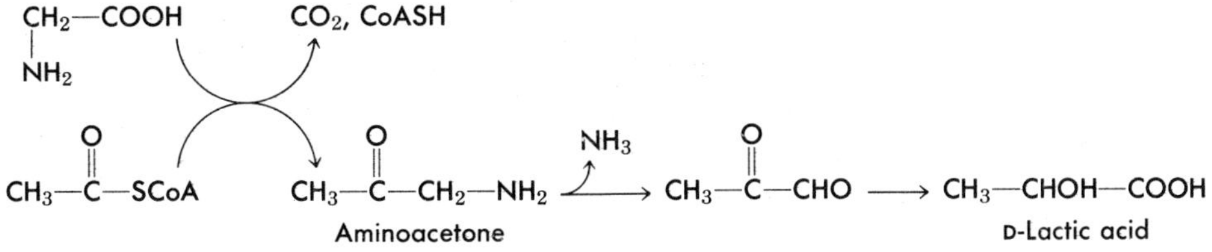

$$\underset{NH_2}{\overset{|}{CH_2}}\text{—COOH} + CH_3\text{—}\overset{O}{\overset{\|}{C}}\text{—SCoA} \longrightarrow CO_2,\ CoASH + \underset{\text{Aminoacetone}}{CH_3\text{—}\overset{O}{\overset{\|}{C}}\text{—}CH_2\text{—}NH_2} \xrightarrow{-NH_3} CH_3\text{—}\overset{O}{\overset{\|}{C}}\text{—CHO} \longrightarrow \underset{\text{D-Lactic acid}}{CH_3\text{—CHOH—COOH}}$$

Heme Biosynthesis

The two most prominent pigments of living organisms are the red pigment heme, or iron protoporphyrin, and the green pigment chlorophyll which contains magnesium. Although these compounds have the same basic tetrapyrrole structure, they differ in the side chains attached to the pyrrole units. The porphyrins are substituted derivatives of a cyclic ring system composed of four pyrrole units linked by methene carbons. Uroporphyrin is taken as the parent compound as it has the largest number of carbon atoms in the side chains.

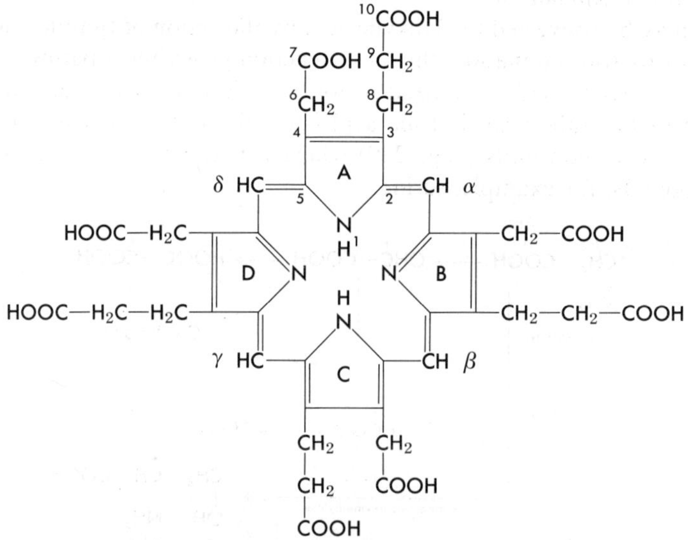

Uroporphyrin III

The pyrrole rings are designated A, B, C, D and the methene carbons as α, β, γ, and δ. Ring carbons are so designated that similar side chains are on the same numbered carbon atoms of each pyrrole ring. Of the four possible isomers of uroporphyrin, only two types are found in nature. These are type III, the far more abundant which is shown above, and type I in which there is a symmetrical arrangement of the substituent side chains.

The vast majority of organisms are able to synthesize porphyrins, and our knowledge of the biosynthetic pathway is quite extensive. The synthesis of the porphyrin-like moiety of vitamin B_{12} (cyanocobalamin) follows a similar biosynthetic pathway. The nucleated red cell of the duck and mammalian reticulocytes and enzyme preparations derived from these sources were the principle systems used to elucidate the synthetic pathway. These studies first established that

1. The four nitrogen atoms are derived from glycine.
2. The four methene bridge carbon atoms and the four carbon atoms numbered 2 are derived from the α-carbon of glycine; the carboxyl carbon of glycine is not used for porphyrin synthesis.
3. All remaining carbon atoms arise from succinyl-SCoA.

The first reaction is the condensation of succinyl-SCoA with glycine to produce δ-aminolevulinic acid (δAL) and carbon dioxide. It is catalyzed by δ-aminolevulinic acid synthetase.

$$\begin{array}{c} COOH \\ | \\ CH_2 \\ | \\ CH_2 \\ | \\ {}^{\#}C\text{—}SCoA \\ \| \\ O \end{array} + \begin{array}{c} {}^{*}NH_2 \\ | \\ {}^{\bullet}CH_2 \\ | \\ \boxed{COOH} \end{array} \xrightarrow[\text{pyridoxal phosphate}]{\text{CoASH}} \begin{array}{c} COOH \\ | \\ CH_2 \\ | \\ CH_2 \\ | \\ {}^{\#}C{=}O \\ | \\ {}^{\bullet}CHN^{*}H_2 \\ | \\ \boxed{COOH} \end{array} \xrightarrow{\boxed{CO_2}} \begin{array}{c} COOH \\ | \\ CH_2 \\ | \\ CH_2 \\ | \\ {}^{\#}C{=}O \\ | \\ {}^{\bullet}CH_2 \\ | \\ {}^{*}NH_2 \end{array}$$

Succinyl-SCoA Glycine α-Amino-β-Ketoadipic acid δ-Aminolevulinic acid

In the second step two molecules of δAL are condensed to form porphobilinogen (PBG).

$$\begin{array}{cc} & COOH \\ & | \\ COOH & CH_2 \\ | & | \\ CH_2 & CH_2 \\ | & | \\ CH_2 \cdots O & {=}C^{\#} \\ | & | \\ {}^{\#}C{=}O & {}^{\bullet}CH_2 \\ | & | \\ {}^{\bullet}CH_2 & N^{*}H_2 \\ | & \\ {}^{*}NH_2 & \end{array} \xrightarrow[\delta\text{AL dehydrase}]{2\ H_2O} \begin{array}{ccc} & & COOH \\ & & | \\ COOH & & CH_2 \\ | & & | \\ CH_2 & & CH_2 \\ | & & | \\ C & \text{———} & C^{\#} \\ \| & & \| \\ {}^{\#}C & & CH^{\bullet} \\ & \diagdown \; N^{*} \; \diagup & \\ {}^{\bullet}CH_2 & H & \\ | & & \\ {}^{*}NH_2 & & \end{array}$$

2 δAL Porphobilinogen (PBG)

In acute intermittent porphyria, an inborn error of metabolism, there is the excretion of excessive amounts of δ-aminolevulinic and porphobilinogen in the urine.

The level of δ-aminolevulinic acid synthetase is increased in the liver of these patients and is thought to be the cause of overproduction of porphyrin precursors.

The next and more complicated reaction involves the condensation of four molecules of porphobilinogen to yield uroporphyrinogen III. This reaction appears to require two kinds of enzymic reactions, one a deaminase reaction which eliminates ammonia, the second being an isomerase reaction which inverts one or three of the porphobilinogen molecules to give the isomer of the III type. This transformation is complicated and the mechanism is not known. After heating to 60°C, the enzyme systems form only uroporphyrinogen I. This is of interest, as there is a well-defined hereditary disease of porphyrin metabolism, *porphyria erythropoietica,* in which the major biochemical lesion is an enzymatic defect in the developing red blood cell which leads to the increased synthesis of type I porphyrins.

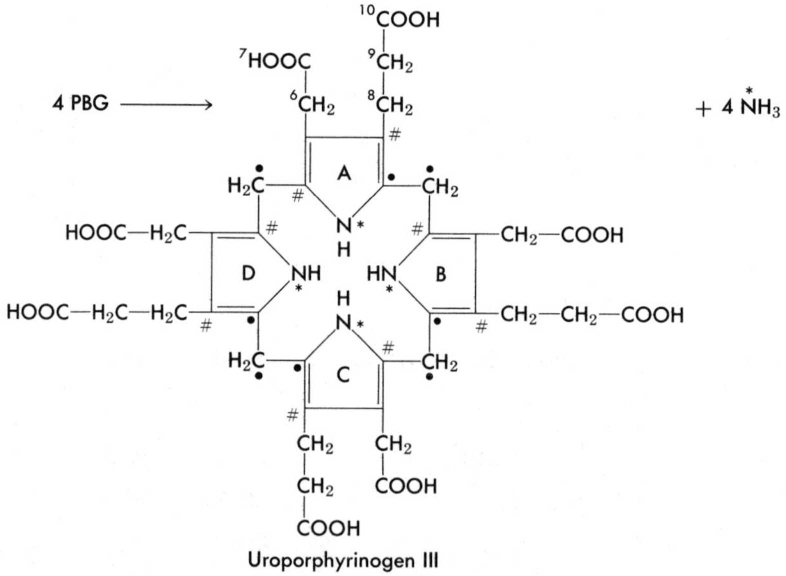

4 PBG ⟶ Uroporphyrinogen III + 4 $\overset{*}{N}H_3$

Uroporphyrinogen III

Uroporphyrinogen III is converted to coproporphyrinogen III by enzymatic decarboxylation of the four acetic acid side chains to produce methyl groups at positions numbered 6.

Uroporphyrinogen III ⟶ Coproporphyrinogen III + 4 CO_2 (from positions 7)

Coproporphyrinogen III is converted by oxidative decarboxylation and autooxidation to protoporphyrin 9. The two propionic acid side chains on rings A and B are converted to vinyl groups, and six hydrogens are removed from the molecule. Protoporphyrinogen 9 is probably an intermediate in the reaction.

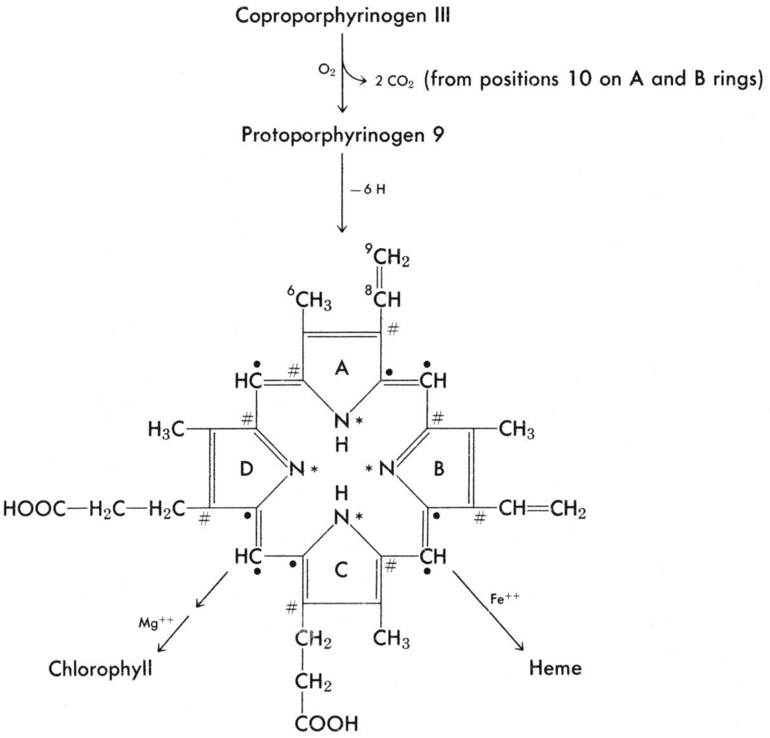

Coproporphyrinogen III

O_2 → $2\ CO_2$ (from positions 10 on A and B rings)

Protoporphyrinogen 9

$-6\ H$

Protoporphyrin 9

Mg^{++} → Chlorophyll

Fe^{++} → Heme

Protoporphyrin 9 is one of 15 possible isomers, and all hemes are either iron-protoporphyrin 9 or derived from it without rearrangement of the side chains. Enzymes are known in both the reticulocyte and liver that catalyze the incorporation of Fe^{++} into protoporphyrin 9. Iron protoporphyrin is the prosthetic group of hemoglobin, myoglobin, peroxidase, catalase, certain cytochrome b proteins, cytochrome c, and cytochrome oxidase. Further aspects of iron metabolism are considered on page 348.

Uroporphyrin and coproporphyrin are not intermediates in the biosynthetic pathway but are derived from the corresponding colorless porphyrinogen by autooxidation.

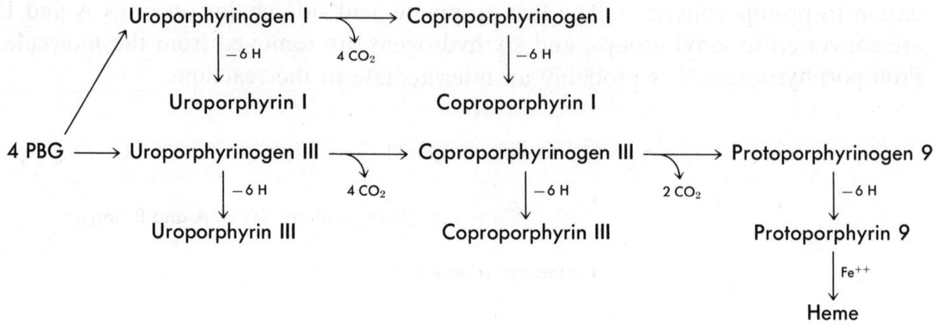

Uroporphyrinogen I ⟶ ($4 CO_2$) Coproporphyrinogen I

Uroporphyrinogen I ↓ $-6 H$ Uroporphyrin I

Coproporphyrinogen I ↓ $-6 H$ Coproporphyrin I

4 PBG ⟶ Uroporphyrinogen I

4 PBG ⟶ Uroporphyrinogen III ⟶ ($4 CO_2$) Coproporphyrinogen III ⟶ ($2 CO_2$) Protoporphyrinogen 9

Uroporphyrinogen III ↓ $-6 H$ Uroporphyrin III

Coproporphyrinogen III ↓ $-6 H$ Coproporphyrin III

Protoporphyrinogen 9 ↓ $-6 H$ Protoporphyrin 9 ↓ Fe^{++} Heme

Neither coproporphyrinogen I nor coproporphyrin I can be decarboxylated further, and no protoporphyrin or heme of the I series has been found.

The conversion of protoporphyrin to bile pigments is discussed on page 346.

The structure of vitamin B_{12} is shown in Figure 7.8. The carbon atoms marked * are derived from the methyl group of methionine and those marked • are derived from δ-aminolevulinic acid-1,4-C^{14}.

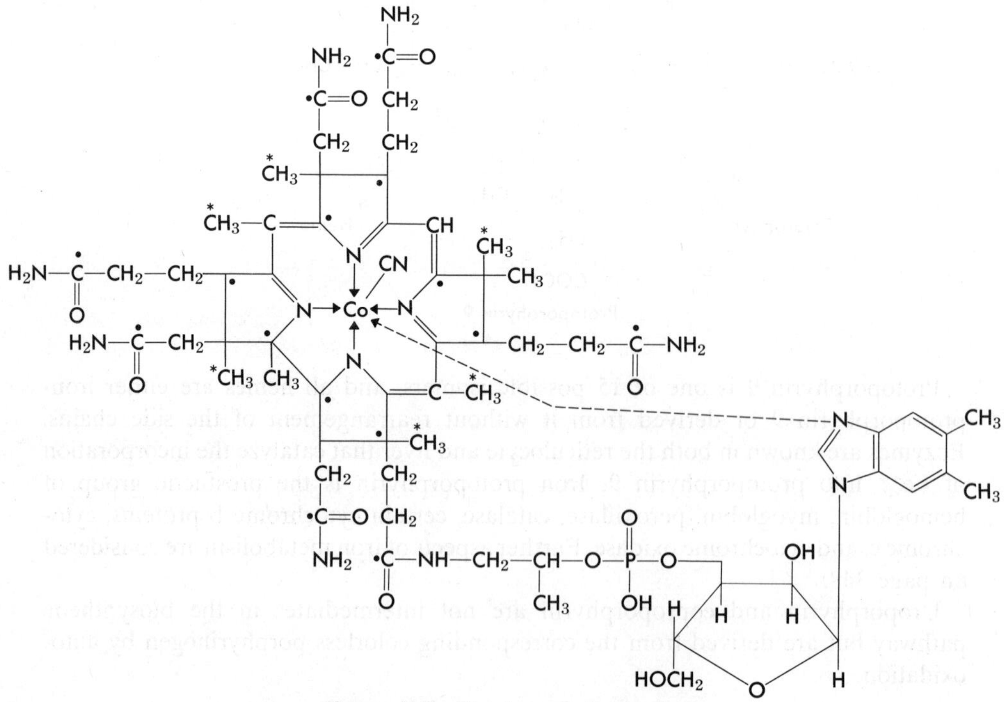

NH_2
NH_2 •C=O
•C=O CH_2
CH_2 CH_2
$\overset{*}{C}H_3$
$\overset{*}{C}H_3$—C
CH
$\overset{*}{C}H_3$
H_2N—C—CH_2—CH_2
CN
CH_3
O
N→Co←N
H_2N—C—CH_2
CH_2—CH_2—C—NH_2
N
*CH_3 CH_3
C—$\overset{*}{C}H_3$
O O
N
CH_3
$\overset{*}{C}H_3$
N
CH_3
CH_2 CH_2
•C=O CH_2
O
NH_2 •C—NH—CH_2—CH—O—P—O
OH
O
CH_3
OH H H H
$HOCH_2$ O H

Figure 7.8. Structure of vitamin B_{12}.

Other Reactions of Glycine

GLUTATHIONE. This tripeptide is synthesized by the following reactions.

$$\text{L-Glutamate} + \text{L-Cysteine} + \text{ATP} \longrightarrow \text{L-}\gamma\text{-Glutamylcysteine} + \text{ADP} + P_i$$

$$\text{L-}\gamma\text{-Glutamylcysteine} + \text{Glycine} + \text{ATP} \longrightarrow \gamma\text{-Glutamylcysteinylglycine (Glutathione)} + \text{ADP} + P_i$$

Glutathione is widely distributed in animal cells and is present for the most part in the reduced form. It serves as a coenzyme in certain enzymatic reactions and appears to be an important agent for the maintenance of the structural integrity of the erythrocyte.

CREATINE AND PURINES. Creatine synthesis has been discussed in the section on end-products of nitrogen metabolism (page 228). The conversion of glycine to purines will be considered in detail under nucleic acid metabolism (page 228).

HIPPURIC ACID. Hippuric acid, a detoxication product of benzoic acid, is synthesized in the liver and excreted in the urine (0.1 to 1.0 g per day).

$$\text{Benzoic acid} + \text{ATP} \xrightarrow{\text{activating enzyme}} \text{Adenylbenzoate} + PP_i$$

$$\text{Adenylbenzoate} + \text{CoASH} \longrightarrow \text{Benzoyl-S-CoA} + \text{AMP}$$

$$\text{Benzoyl-S-CoA} + \text{Glycine} \xrightarrow[\text{enzyme}]{\text{condensing}} C_6H_5\text{—}\overset{O}{\overset{\|}{C}}\text{—}\overset{H}{N}\text{—}CH_2\text{—COOH} + \text{CoASH}$$

Hippuric acid
(Benzoylglycine)

The role of glycine in the biosynthesis of the purine ring is discussed in Chapter 8.

Metabolism of Serine

Serine is a dispensable amino acid and need not be included in the diet. The glucogenic nature of serine and its interconversion with glycine have been discussed in earlier sections (page 235). A summary of the metabolic reactions of serine is shown in Figure 7.9.

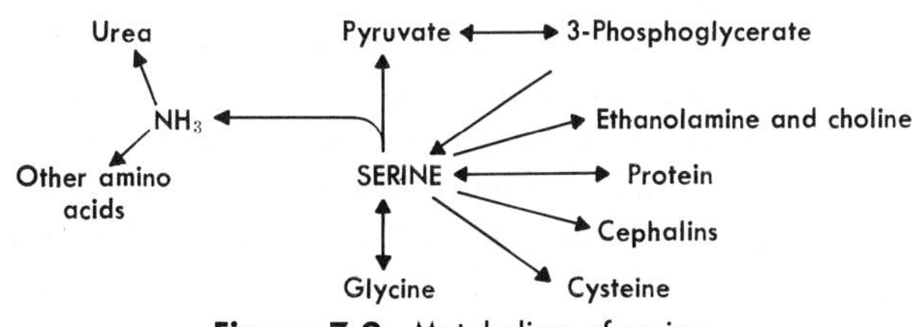

Figure 7.9. Metabolism of serine.

Biosynthesis

The major source of serine is from glucose via 3-phosphoglyceric acid.

$$\text{Glucose} \longrightarrow \longrightarrow \underset{\text{3-Phosphoglyceric acid}}{\begin{array}{c} COOH \\ | \\ H{-}C{-}OH \\ | \\ CH_2{-}O{-}PO_3H_2 \end{array}} \xrightleftharpoons[NAD^+ \quad NADH + H^+]{\text{phosphoglycerate dehydrogenase}} \underset{\text{3-Phosphohydroxypyruvic acid}}{\begin{array}{c} COOH \\ | \\ C{=}O \\ | \\ CH_2{-}O{-}PO_3H_2 \end{array}}$$

$$\text{3-Phosphohydroxypyruvic acid} \xrightleftharpoons[\text{transaminase}]{\text{Glutamate} \rightarrow \alpha\text{-Ketoglutarate}} \text{3-Phosphoserine}$$

$$\underset{\text{L-Serine}}{\begin{array}{c} COOH \\ | \\ H_2N{-}C{-}H \\ | \\ CH_2OH \end{array}} \xleftarrow[P_i \quad H_2O]{\text{phosphoserine phosphatase}} \underset{\text{3-Phosphoserine}}{\begin{array}{c} COOH \\ | \\ H_2N{-}C{-}H \\ | \\ CH_2{-}O{-}PO_3H_2 \end{array}}$$

Conversion to Ethanolamine and Choline

Very recent studies have established the following sequence of reactions in liver which does not decarboxylate free serine.

$$\text{Phosphatidylethanolamine} + \text{L-Serine} \rightleftharpoons \text{Phosphatidylserine} + \text{Ethanolamine}$$

$$\text{Phosphatidylserine} \longrightarrow \text{Phosphatidylethanolamine} + CO_2$$

$$\text{Net reaction: L-Serine} \longrightarrow \text{Ethanolamine} + CO_2$$

These reactions constitute a metabolic cycle in which L-serine is decarboxylated to ethanolamine and carbon dioxide, the phosphatidyl compounds acting catalytically. This cycle is shown in Figure 7.10. It should be noted that the ethanolamine is derived from phosphatidylethanolamine and not directly from L-serine.

The phosphatidylethanolamine is then transformed by sequential methylations with S-adenosylmethionine to form phosphatidylcholine (lecithin). Free choline is produced by hydrolysis of the phosphatidylcholine.

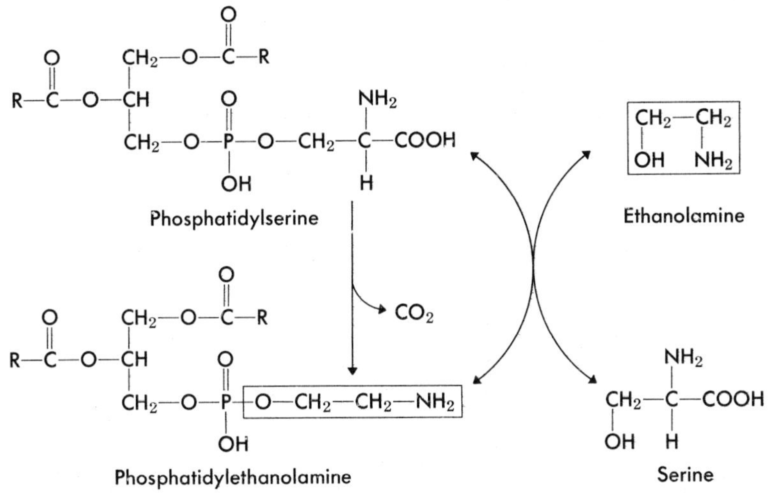

Phosphatidylserine → (− CO_2) → Phosphatidylethanolamine

Phosphatidylserine + Ethanolamine ($CH_2(OH)—CH_2(NH_2)$) ⇌ Phosphatidylethanolamine + Serine ($CH_2(OH)—CH(NH_2)—COOH$)

Phosphatidylserine

Ethanolamine

Phosphatidylethanolamine

Serine

Figure 7.10. Serine decarboxylation.

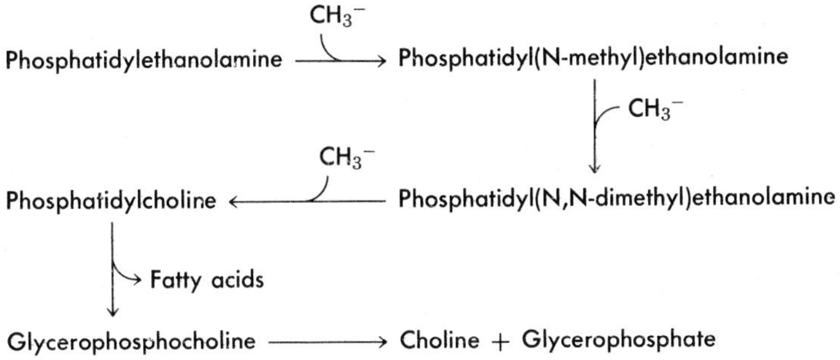

Phosphatidylethanolamine —(CH_3^-)→ Phosphatidyl(N-methyl)ethanolamine

Phosphatidyl(N-methyl)ethanolamine —(CH_3^-)→ Phosphatidyl(N,N-dimethyl)ethanolamine

Phosphatidyl(N,N-dimethyl)ethanolamine —(CH_3^-)→ Phosphatidylcholine

Phosphatidylcholine —(→ Fatty acids)→ Glycerophosphocholine

Glycerophosphocholine ⟶ Choline + Glycerophosphate

Thus, the pathway for the *de novo* synthesis of choline from serine now appears to involve phosphatide derivatives as substrates. Further aspects of choline metabolism will be considered in conjunction with methionine metabolism and transmethylation reactions (page 270).

Other Reactions of Serine

The conversion of serine to dihydrosphingosine is discussed on page 202 and that of serine to cysteine on page 269.

Serine may be converted to pyruvic acid and ammonia via the action of the pyridoxal phosphate-requiring enzyme L-serine dehydratase (page 218).

$$\begin{array}{c} CH_2OH \\ | \\ H-C-NH_2 \\ | \\ COOH \end{array} \xrightarrow{H_2O} \begin{array}{c} CH_2 \\ \| \\ C-NH_2 \\ | \\ COOH \end{array} \rightleftharpoons \begin{array}{c} CH_3 \\ | \\ C=NH \\ | \\ COOH \end{array} \xrightarrow[H_2O \quad NH_3]{} \begin{array}{c} CH_3 \\ | \\ C=O \\ | \\ COOH \end{array}$$

Metabolism of Threonine

L-Threonine is indispensable in the diet of animals. There are three pathways known for its catabolism in liver. It may be deaminated by a dehydratase reaction to produce α-ketobutyric acid which is converted to carbohydrate via propionic acid (page 187).

$$\underset{\text{L-Threonine}}{\begin{array}{c} CH_3 \\ | \\ HO-C-H \\ | \\ H-C-NH_2 \\ | \\ COOH \end{array}} \xrightarrow[H_2O]{\text{threonine dehydratase}} \begin{array}{c} CH_3 \\ | \\ H-C \\ \| \\ C-NH_2 \\ | \\ COOH \end{array} \longleftrightarrow \begin{array}{c} CH_3 \\ | \\ H-C-H \\ | \\ C=NH \\ | \\ COOH \end{array} \xrightarrow[NH_3]{H_2O} \underset{\alpha\text{-Ketobutyric acid}}{\begin{array}{c} CH_3 \\ | \\ H-C-H \\ | \\ C=O \\ | \\ COOH \end{array}}$$

$$\alpha\text{-Ketobutyric acid} \longrightarrow \underset{\text{Propionic acid}}{\begin{array}{c} CH_3 \\ | \\ H-C-H \\ | \\ COOH \end{array}} \longrightarrow \text{Succinyl CoA} \longrightarrow \text{Glucose}; \quad \text{Succinyl CoA} \longrightarrow CO_2 + H_2O$$

The second reaction is the cleavage of threonine to glycine and acetaldehyde.

$$\underset{\text{L-Threonine}}{\begin{array}{c} CH_3 \\ | \\ HO-C-H \\ | \\ H-C-NH_2 \\ | \\ COOH \end{array}} \xrightarrow{\text{hydroxamino-aldolase}} \begin{array}{c} CH_3 \\ | \\ CHO \end{array} \xrightarrow{O_2} \begin{array}{c} CH_3 \\ | \\ COOH \end{array} \quad + \quad \begin{array}{c} CH_2-NH_2 \\ | \\ COOH \end{array}$$

L-Threonine is neither deaminated at an appreciable rate by the L-amino acid oxidases nor does it participate in reversible transamination reactions. L-Threonine may be converted to aminoacetone by L-threonine dehydrogenase.

$$\text{L-Threonine} \xrightarrow{-2\,H} \alpha\text{-amino-}\beta\text{-ketobutyric acid} \xrightarrow[-CO_2]{\text{spontaneous}} \text{Aminoacetone}$$

Aminoacetone may be metabolized to lactic acid as previously discussed (page 235).

Metabolism of Alanine, Aspartic Acid, and Glutamic Acid

Since alanine, aspartic acid, and glutamic acid are related directly to the tricarboxylic acid cycle by reversible transamination reactions, it is convenient to consider their metabolism together. Figure 7.11 summarizes the metabolism of these amino acids.

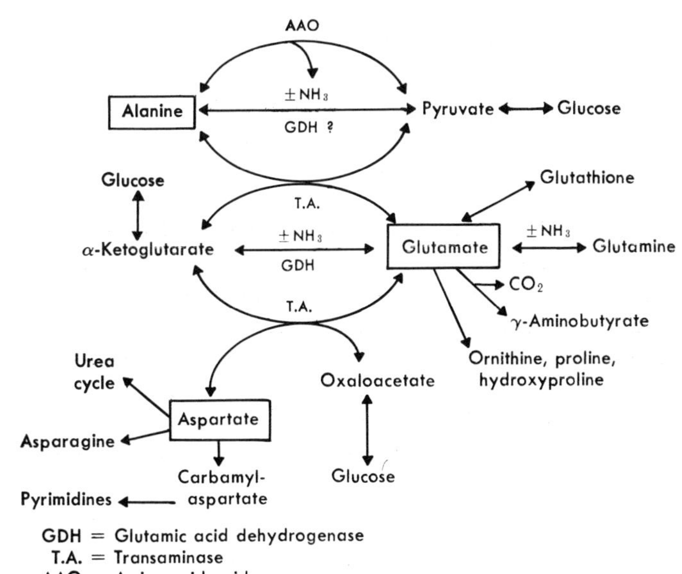

GDH = Glutamic acid dehydrogenase
T.A. = Transaminase
AAO = Amino acid oxidase

Figure 7.11. Metabolism of alanine, aspartate, and glutamate.

Most of these reactions have been considered in earlier sections. One additional reaction of glutamate, its decarboxylation to γ-aminobutyric acid (GABA) is worth noting here.

$$\underset{\text{L-Glutamic acid}}{\boxed{COOH}-CH(NH_2)-CH_2-CH_2-COOH} \xrightarrow[\text{pyridoxal phosphate}]{\alpha\text{-decarboxylase}} \boxed{CO_2} + \underset{\gamma\text{-Aminobutyric acid (GABA)}}{H_2N-CH_2-CH_2-CH_2-COOH}$$

Brain tissue contains a considerable amount of GABA, and the enzyme which produces it from glutamate. GABA apparently serves a role in regulating the generation of impulses in the central nervous system. GABA may enter into a transamination reaction to form succinic semialdehyde which can be converted to succinic acid and metabolized via the tricarboxylic acid cycle.

$$\underset{\gamma\text{-Aminobutyric acid}}{\begin{array}{l} CH_2{-}NH_2 \\ | \\ CH_2 \\ | \\ CH_2 \\ | \\ COOH \end{array}} + \alpha\text{-Ketoglutarate} \rightleftharpoons \underset{\text{Succinic semialdehyde}}{\begin{array}{l} CHO \\ | \\ CH_2 \\ | \\ CH_2 \\ | \\ COOH \end{array}} + \text{L-Glutamate}$$

Metabolism of Ornithine, Arginine, Proline, and Hydroxyproline

The metabolic interrelations of these amino acids are shown in Figure 7.12.

The central role of glutamate in these interconversions was evident from the first nutritional studies. Animals on diets free of proline, hydroxyproline, or ornithine but rich in glutamic acid grew well. Glutamic acid, proline, and ornithine were shown to be glucogenic and strengthened the proposal that these amino acids were metabolically related. Among the dietary indispensable amino acids, the position of arginine is equivocal. If arginine is omitted from the diet of the young rat, its growth

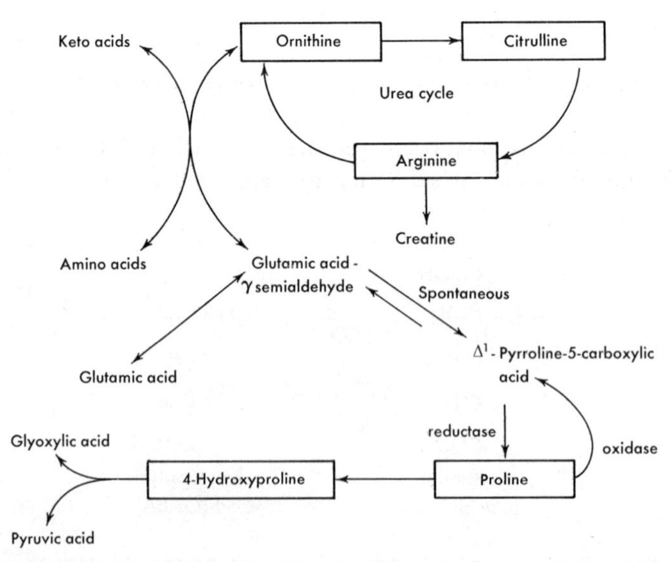

Figure 7.12. Metabolic interrelations of arginine, ornithine, citrulline, proline, and 4-hydroxyproline.

is restricted but not abolished. As previously noted, adult human males can be maintained in nitrogen balance in the absence of arginine, but their ability to produce spermatozoa is reduced. It must be concluded that although arginine may be synthesized by the rat and man and other species, the rate of synthesis is not adequate to support growth or maintain normal functions.

The key compound in the interconversions of these amino acids is glutamic acid-γ-semialdehyde. It is produced from ornithine by a reversible transamination reaction and arises reversibly from glutamate by reactions not as yet clearly understood.

$$\underset{\text{L-Ornithine}}{\begin{array}{c} CH_2-NH_2 \\ | \\ CH_2 \\ | \\ CH_2 \\ | \\ H-C-NH_2 \\ | \\ COOH \end{array}} \xrightleftharpoons{\text{transamination}} \underset{\text{L-Glutamic acid-}\gamma\text{-semialdehyde}}{\begin{array}{c} CHO \\ | \\ CH_2 \\ | \\ CH_2 \\ | \\ H-C-NH_2 \\ | \\ COOH \end{array}} \xrightleftharpoons[\text{ATP, kinase, dehydrogenase?}]{\text{glutamic semialdehyde dehydrogenase}} \underset{\text{L-Glutamic acid}}{\begin{array}{c} COOH \\ | \\ CH_2 \\ | \\ CH_2 \\ | \\ H-C-NH_2 \\ | \\ COOH \end{array}}$$

Glutamic acid semialdehyde is in equilibrium with Δ¹-pyrroline-5-carboxylic acid, the major component at pH 7.0.

$$\underset{\text{Glutamic acid-}\gamma\text{-semialdehyde}}{\begin{array}{ccc} CH_2 & - & CH_2 \\ | & & | \\ HC & & CH-COOH \\ \| & & | \\ O & & NH_2 \end{array}} \xrightleftharpoons[+H_2O]{-H_2O} \underset{\Delta^1\text{-Pyrroline-5-carboxylic acid}}{\begin{array}{ccc} CH_2 & - & CH_2 \\ | & & | \\ HC & & CH-COOH \\ & \diagdown\!\!\diagdown N \diagup & \end{array}}$$

A liver Δ¹-pyrroline-5-carboxylate reductase catalyzes the irreversible reduction of the pyrroline carboxylic acid to proline.

$$\underset{\Delta^1\text{-Pyrroline-5-carboxylic acid}}{\begin{array}{ccc} CH_2 & - & CH_2 \\ | & & | \\ HC & & CH-COOH \\ & \diagdown\!\!\diagdown N \diagup & \end{array}} \xrightarrow[\text{NAD(P)H + H}^+ \ \to \ \text{NAD(P)}^+]{\text{reductase}} \underset{\text{L-Proline}}{\begin{array}{ccc} CH_2 & - & CH_2 \\ | & & | \\ CH_2 & & CH-COOH \\ & \diagdown \underset{H}{N} \diagup & \end{array}}$$

The conversion of proline to glutamic acid appears to involve the action of a proline oxidase which is distinct from the reductase. The reaction requires oxygen and cytochrome c and is inhibited by cyanide and azide. The enzyme appears to be a dehydrogenase linked to a cytochrome-containing system.

L-Proline $\xrightarrow[O_2]{\text{Proline oxidase}}$ Δ^1-Pyrroline-5-carboxylic acid $\rightleftharpoons$ Glutamic acid-γ-semialdehyde

Glutamic acid-γ-semialdehyde $\xrightarrow{\text{glutamic semialdehyde dehydrogenase}}$ L-Glutamic acid

Δ^1-Pyrroline-5-carboxylic acid $\xrightarrow{\text{NAD-dehydrogenase}}$ L-Glutamic acid

The Δ^1-pyrroline-5-carboxylate formed may be converted via glutamic acid semialdehyde to glutamate. An alternate possibility is the direct oxidation of the pyrroline carboxylate to glutamate by a NAD-dehydrogenase reported to be present in liver. These reactions may be of considerable importance for the catabolism of proline, since the reactions from glutamate to pyrroline carboxylate and proline described above appear to be essentially irreversible.

Although hydroxyproline biosynthesis has not been separable from collagen synthesis, a cell-free system has been obtained which is capable of converting free proline to collagen hydroxyproline. The results with this system show that a ribosome-bound peptidyl-prolyl-tRNA (page 315) is the substrate for hydroxylation and not free proline or prolyl-tRNA. Proline is hydroxylated at a step subsequent to its incorporation into peptide linkage. Isotope studies have shown that the hydroxyl moiety of collagen hydroxyproline is derived from molecular oxygen and not from water. Hydroxylation is therefore catalyzed by an oxygenase enzyme.

The catabolism of 4-hydroxyproline in liver is summarized in Figure 7.13.

Degradation of Branched-Chain Amino Acids

L-Valine, L-isoleucine, and L-leucine are indispensable amino acids for all higher animals. The corresponding α-keto acids are able to replace the amino acids in the diet. The pathways for the degradation of valine, isoleucine, and leucine are presented in Figures 7.14, 7.15, and 7.16. It may be seen that the pathways of metabolism of these amino acids have several features in common and features in common with the oxidation of fatty acids.

The initial step in all three cases is transamination to yield the corresponding α-keto acids. The α-keto acids are subjected to oxidative decarboxylation with concomitant formation of the CoA esters. The three keto acids appear to be de-

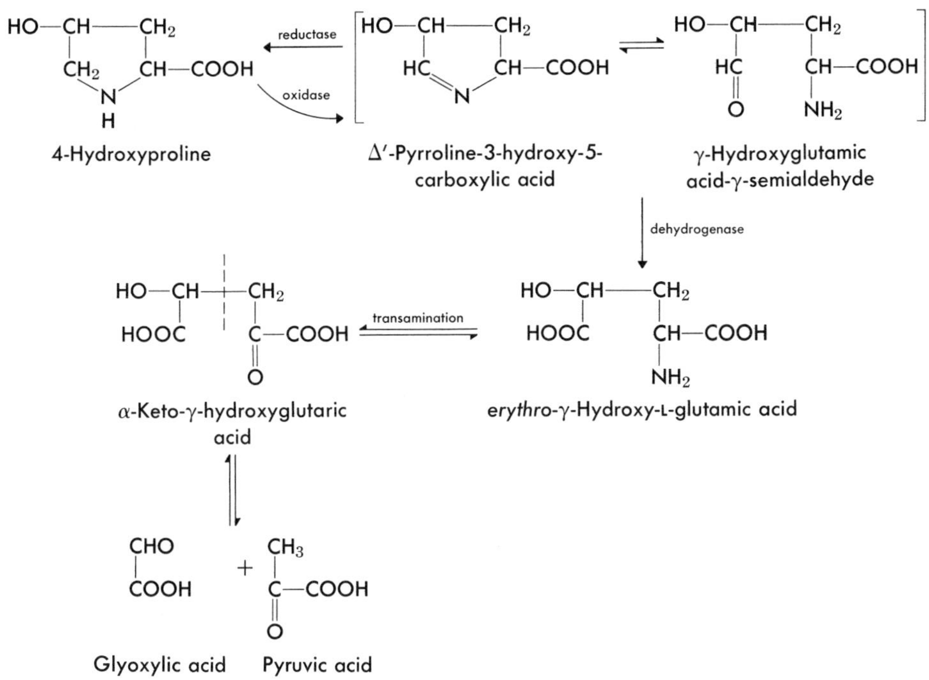

4-Hydroxyproline — reductase / oxidase — Δ′-Pyrroline-3-hydroxy-5-carboxylic acid ⇌ γ-Hydroxyglutamic acid-γ-semialdehyde

dehydrogenase

α-Keto-γ-hydroxyglutaric acid ⇌ (transamination) erythro-γ-Hydroxy-L-glutamic acid

Glyoxylic acid + Pyruvic acid

Figure 7.13. Catabolism of 4-hydroxyproline.

carboxylated by a single enzyme. This is followed by an α,β dehydrogenation to yield α,β unsaturated intermediates. With valine and isoleucine there is the α,β addition of water, followed by a β-oxidation step. With leucine the α,β addition of water follows a CO_2-fixation step.

Valine yields methylmalonyl-CoA which can be converted to succinyl-CoA (page 250) and is therefore glucogenic. Isoleucine is degraded to acetyl-CoA and to propionyl-CoA which can be converted via methylmalonyl-CoA to succinyl-CoA. Isoleucine is thus both ketogenic and glucogenic. Leucine yields upon degradation both acetoacetate and acetyl-CoA and is strongly ketogenic. Of interest is the fact that leucine gives rise, as an intermediate in its degradation, to β-hydroxy-β-methyglutaryl-CoA an intermediate in the synthesis of cholesterol from acetyl-CoA (page 205).

There is an interesting disorder of the metabolism of the branched-chain amino acids in which the blood and urinary levels are high and the keto acid analogs are present in the urine. This familial cerebral degenerative disease is called maple syrup urine disease because the urine has an odor resembling that of maple syrup. The metabolic block appears to be in the oxidative decarboxylation of the α-keto acids. As noted above, a single enzyme is thought to decarboxylate all three keto acids.

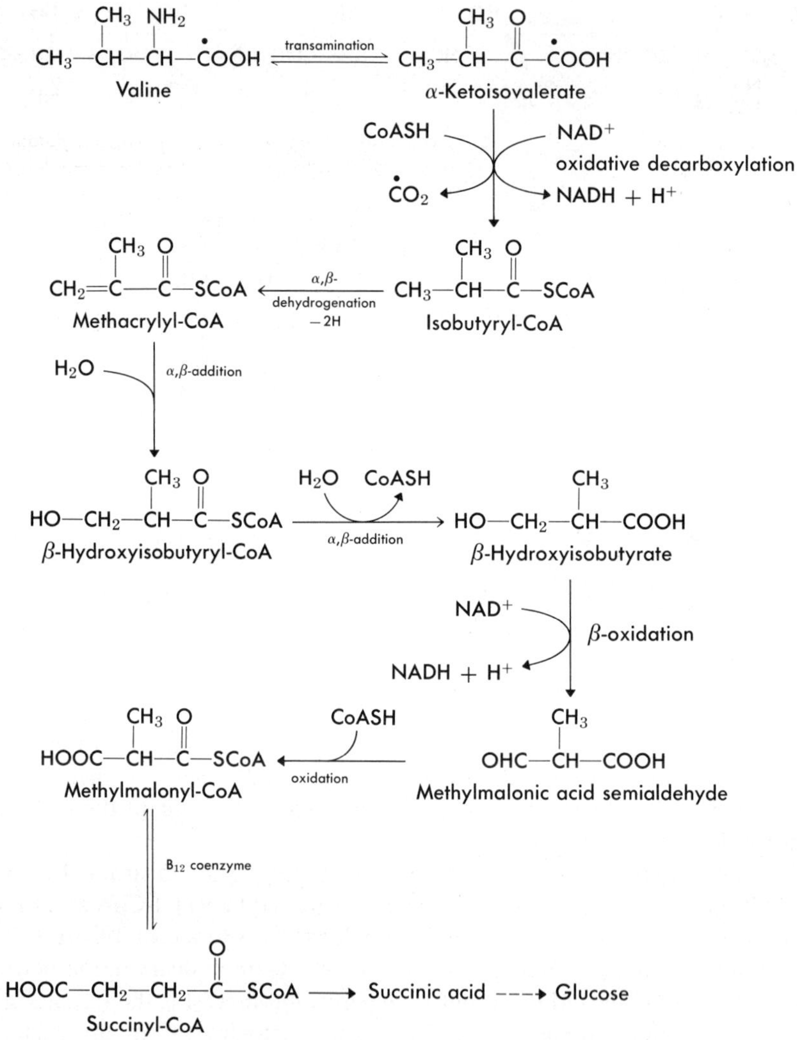

$CH_3—CH(CH_3)—CH(NH_2)—\dot{C}OOH$ (Valine) $\xrightleftharpoons{\text{transamination}}$ $CH_3—CH(CH_3)—C(=O)—\dot{C}OOH$ (α-Ketoisovalerate)

α-Ketoisovalerate → (CoASH, NAD^+; oxidative decarboxylation; $\dot{C}O_2$, $NADH + H^+$) → $CH_3—CH(CH_3)—C(=O)—SCoA$ (Isobutyryl-CoA)

Isobutyryl-CoA → (α,β-dehydrogenation, −2H) → $CH_2{=}C(CH_3)—C(=O)—SCoA$ (Methacrylyl-CoA)

Methacrylyl-CoA → (H_2O; α,β-addition) → $HO—CH_2—CH(CH_3)—C(=O)—SCoA$ (β-Hydroxyisobutyryl-CoA)

β-Hydroxyisobutyryl-CoA → (H_2O, CoASH; α,β-addition) → $HO—CH_2—CH(CH_3)—COOH$ (β-Hydroxyisobutyrate)

β-Hydroxyisobutyrate → (NAD^+, $NADH + H^+$; β-oxidation) → $OHC—CH(CH_3)—COOH$ (Methylmalonic acid semialdehyde)

Methylmalonic acid semialdehyde → (CoASH; oxidation) → $HOOC—CH(CH_3)—C(=O)—SCoA$ (Methylmalonyl-CoA)

Methylmalonyl-CoA $\rightleftharpoons$ (B_{12} coenzyme) $HOOC—CH_2—CH_2—C(=O)—SCoA$ (Succinyl-CoA) → Succinic acid ---→ Glucose

Figure 7.14. Catabolism of valine.

Lysine Metabolism

L-Lysine is an indispensable dietary amino acid for all animals studied. While it can be deaminated, the corresponding α-keto acid cannot replace lysine. Lysine does not partake appreciably in reversible transamination reactions (page 210). The scheme for the degradation of lysine shown in Figure 7.17 is not clearly established.

A great deal of the investigations have been done with liver homogenates and the intact animal and several of the steps have not been demonstrated with isolated enzyme preparations. It is possible that ϵ-N-acetyl-lysine and α-keto-ϵ-acetamidocaproate are intermediates in the degradation of lysine, the latter compound being deacetylated before conversion to Δ^1-piperidine-2-carboxylate. Lysine is ketogenic by the pathway shown.

Metabolism of Phenylalanine and Tyrosine

Phenylalanine is a dietary indispensable amino acid in animals, whereas tyrosine,

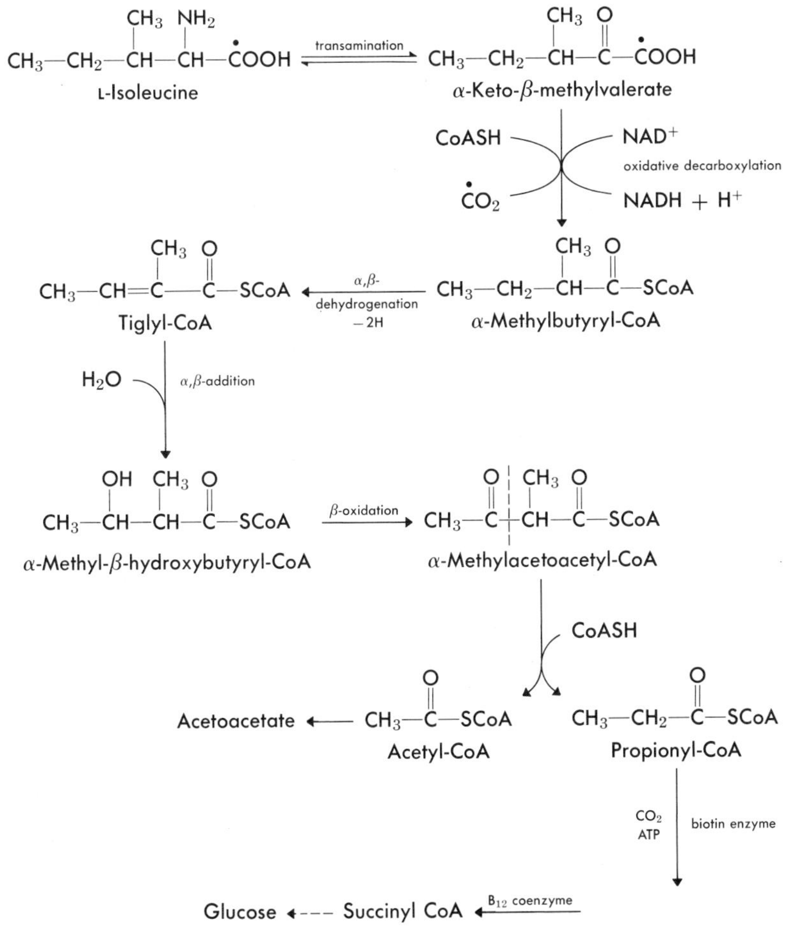

$CH_3—CH_2—CH(CH_3)—CH(NH_2)—\dot{C}OOH$ (L-Isoleucine) $\xrightleftharpoons{\text{transamination}}$ $CH_3—CH_2—CH(CH_3)—C(=O)—\dot{C}OOH$ (α-Keto-β-methylvalerate)

α-Keto-β-methylvalerate + CoASH + NAD^+ → (oxidative decarboxylation) → α-Methylbutyryl-CoA + $\dot{C}O_2$ + $NADH + H^+$

$CH_3—CH_2—CH(CH_3)—C(=O)—SCoA$ (α-Methylbutyryl-CoA) $\xrightarrow[-2H]{\alpha,\beta\text{-dehydrogenation}}$ $CH_3—CH{=}C(CH_3)—C(=O)—SCoA$ (Tiglyl-CoA)

Tiglyl-CoA + H_2O → (α,β-addition) → $CH_3—CH(OH)—CH(CH_3)—C(=O)—SCoA$ (α-Methyl-β-hydroxybutyryl-CoA)

α-Methyl-β-hydroxybutyryl-CoA $\xrightarrow{\beta\text{-oxidation}}$ $CH_3—C(=O)—CH(CH_3)—C(=O)—SCoA$ (α-Methylacetoacetyl-CoA)

α-Methylacetoacetyl-CoA + CoASH → $CH_3—C(=O)—SCoA$ (Acetyl-CoA) + $CH_3—CH_2—C(=O)—SCoA$ (Propionyl-CoA)

Acetyl-CoA → Acetoacetate

Propionyl-CoA + CO_2 + ATP → (biotin enzyme) → → (B_{12} coenzyme) → Succinyl CoA - - -> Glucose

Figure 7.15. Catabolism of isoleucine.

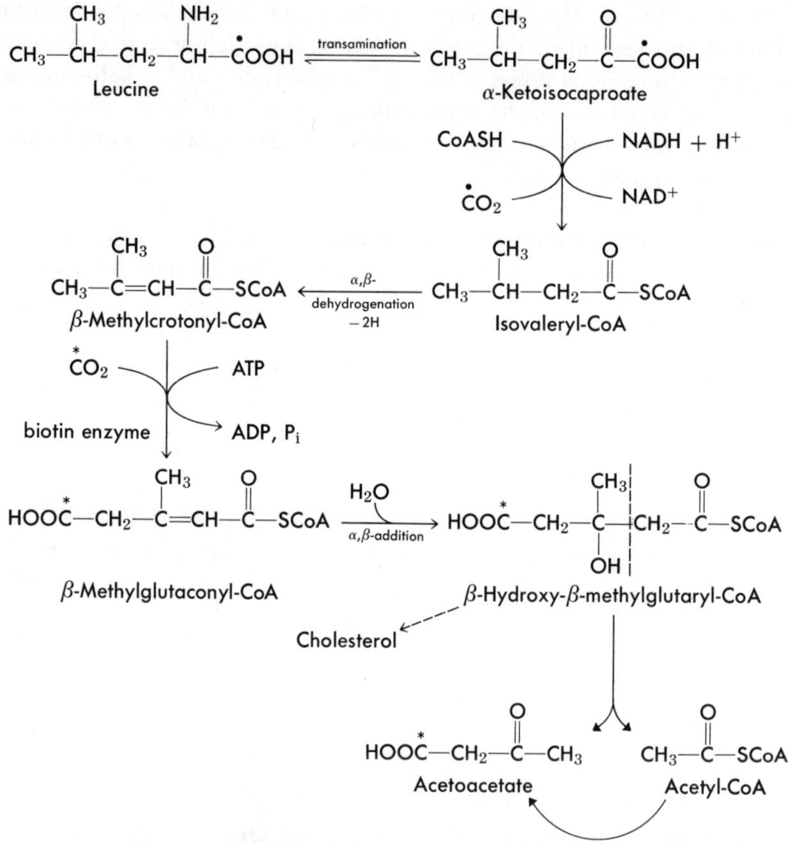

CH_3—$CH(CH_3)$—CH_2—$CH(NH_2)$—$\overset{\bullet}{C}OOH$ (Leucine) ⇌ transamination ⇌ CH_3—$CH(CH_3)$—CH_2—$C(=O)$—$\overset{\bullet}{C}OOH$ (α-Ketoisocaproate)

α-Ketoisocaproate → (CoASH, NAD^+ in; $\overset{\bullet}{C}O_2$, $NADH + H^+$ out) → CH_3—$CH(CH_3)$—CH_2—$C(=O)$—SCoA (Isovaleryl-CoA)

Isovaleryl-CoA → α,β-dehydrogenation, $-2H$ → CH_3—$C(CH_3)$=CH—$C(=O)$—SCoA (β-Methylcrotonyl-CoA)

β-Methylcrotonyl-CoA → ($\overset{*}{C}O_2$, ATP; biotin enzyme; ADP, P_i) → $HOO\overset{*}{C}$—CH_2—$C(CH_3)$=CH—$C(=O)$—SCoA (β-Methylglutaconyl-CoA)

β-Methylglutaconyl-CoA → H_2O, α,β-addition → $HOO\overset{*}{C}$—CH_2—$C(CH_3)(OH)$—CH_2—$C(=O)$—SCoA (β-Hydroxy-β-methylglutaryl-CoA) ⇢ Cholesterol

β-Hydroxy-β-methylglutaryl-CoA → $HOO\overset{*}{C}$—CH_2—$C(=O)$—CH_3 (Acetoacetate) + CH_3—$C(=O)$—SCoA (Acetyl-CoA)

Acetyl-CoA → Acetoacetate

Figure 7.16. Catabolism of leucine.

which can be formed from phenylalanine, is dispensable providing adequate amounts of phenylalanine are present. Provision of tyrosine in the diet lowers the requirement for phenylalanine by decreasing the amount of phenylalanine converted to tyrosine. In addition to the major pathway of catabolism of phenylalanine via tyrosine to acetoacetate and fumarate, there are a number of other metabolic pathways leading to the formation of several important substances. The overall metabolism is summarized in Figure 7.18.

Catabolism of Phenylalanine and Tyrosine

A great deal of our general knowledge of the fate of phenylalanine and tyrosine was derived from the study of patients with hereditary inborn errors of metabolism.

In these conditions, various metabolic products are excreted in the urine. These metabolic diseases will be discussed at the appropriate step in the catabolic pathway.

The initial step in the catabolism of phenylalanine is its irreversible conversion

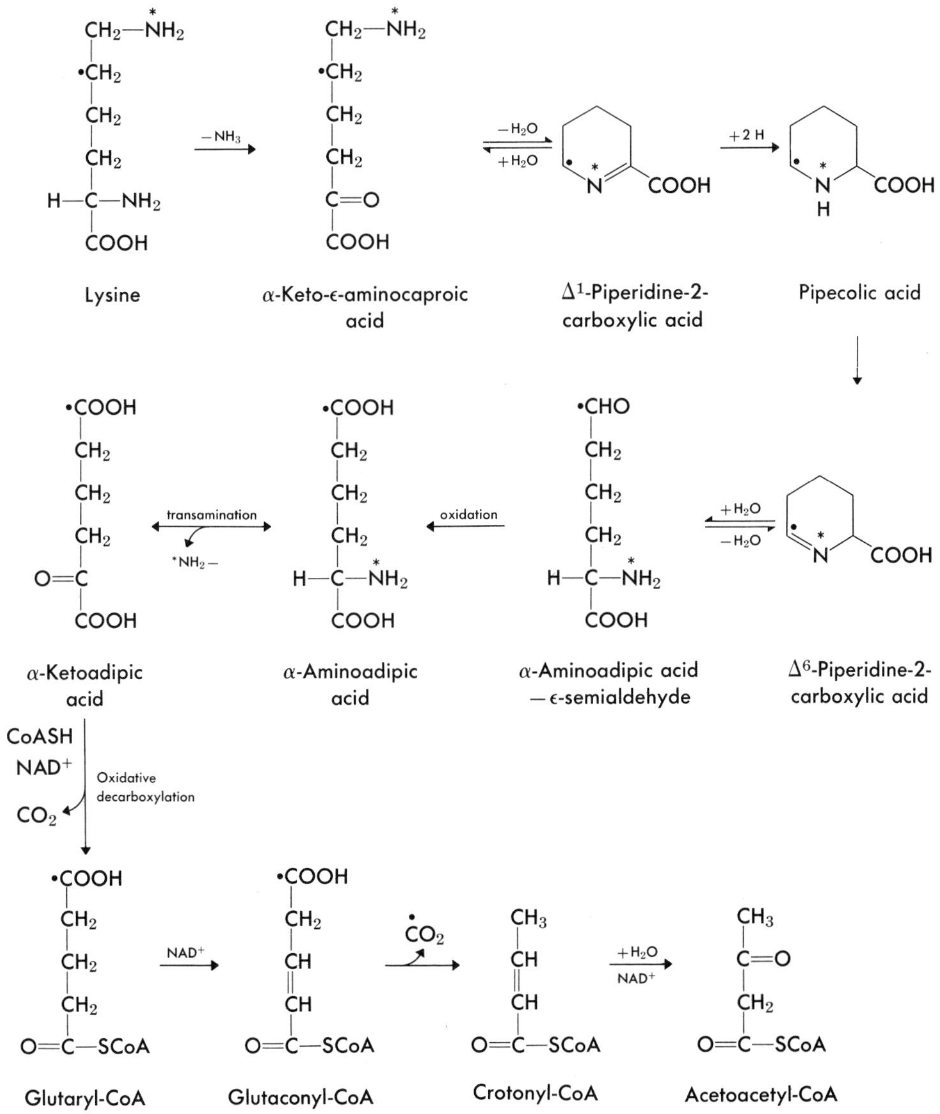

Lysine

$-NH_3$

α-Keto-ε-aminocaproic acid

$-H_2O$ / $+H_2O$

Δ^1-Piperidine-2-carboxylic acid

$+2\,H$

Pipecolic acid

Δ^6-Piperidine-2-carboxylic acid

$+H_2O$ / $-H_2O$

α-Aminoadipic acid — ε-semialdehyde

oxidation

α-Aminoadipic acid

transamination

$^{*}NH_2-$

α-Ketoadipic acid

CoASH

NAD^+

CO_2

Oxidative decarboxylation

Glutaryl-CoA

NAD^+

Glutaconyl-CoA

CO_2

Crotonyl-CoA

$+H_2O$

NAD^+

Acetoacetyl-CoA

Figure 7.17. Catabolism of lysine.

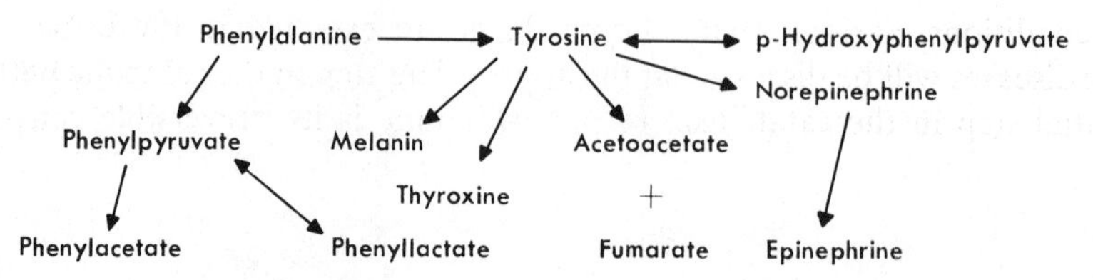

Figure 7.18. Metabolism of phenylalanine and tyrosine.

to tyrosine in liver. The conversion has been formulated as follows

(1) Phenylalanine + O_2 + Tetrahydropteridine $\xrightarrow{\text{phenylalanine hydroxylase}}$

Tyrosine + H_2O + Quinonoid dihydropteridine

(2) Quinonoid dihydropteridine + NADPH + H^+ $\xrightarrow{\text{dihydropteridine reductase}}$

Tetrahydropteridine + $NADP^+$

---

(3) Phenylalanine + O_2 + NADPH + H^+ $\xrightarrow[\text{pteridine}]{\text{tetrahydro-}}$ Tyrosine + H_2O + $NADP^+$

The first enzyme catalyzes the hydroxylation reaction (1), and the second enzyme (2) regenerates the oxidized cofactor. The cofactor has been shown to be dihydrobiopterin which must be reduced to the tetrahydro level to be active in reaction 1.

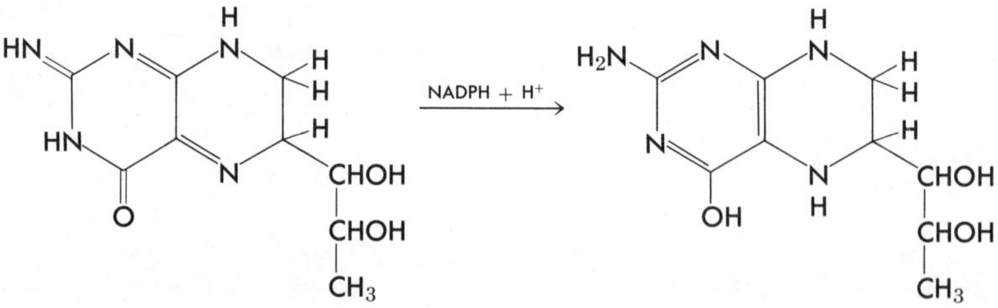

$\xrightarrow{\text{NADPH + H}^+}$

Quinonoid dihydro form

Tetrahydro form

In a hereditary disease known as phenylketonuria, phenylpyruvic acid, large amounts of phenylalanine, phenyllactate, and other products are excreted in the urine. Severe mental retardation is the most prominent clinical characteristic. These individuals lack or are deficient in phenylalanine hydroxylase. As a consequence, they have little or no ability to convert phenylalanine to tyrosine, and phenylalanine and its normal metabolites accumulate in large amounts.

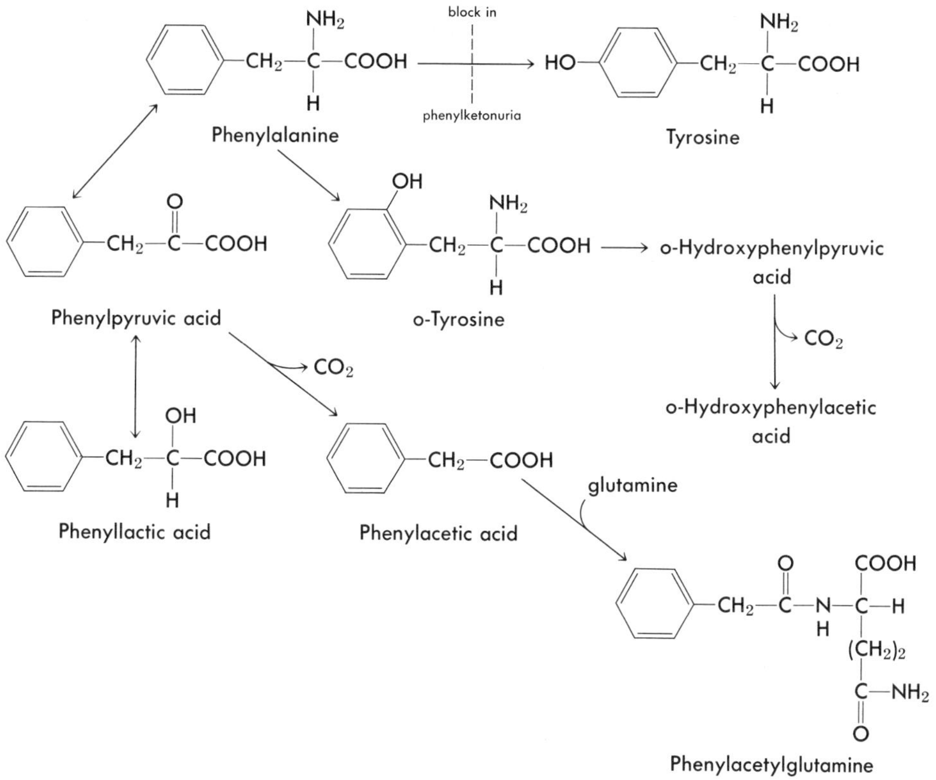

Phenylalanine —(block in phenylketonuria)→ Tyrosine

Phenylalanine ↔ Phenylpyruvic acid

Phenylalanine → o-Tyrosine → o-Hydroxyphenylpyruvic acid → (CO_2) → o-Hydroxyphenylacetic acid

Phenylpyruvic acid ↔ Phenyllactic acid

Phenylpyruvic acid → (CO_2) → Phenylacetic acid

Phenylacetic acid + glutamine → Phenylacetylglutamine

It seems that the biochemical environment of the phenylketonuric patient produces the clinical manifestations of the disease, since restriction of the dietary intake of phenylalanine in the infant prevents the development of the clinical aspects of the disease.

In the next reaction tyrosine is converted by transamination to *p*-hydroxyphenylpyruvic acid.

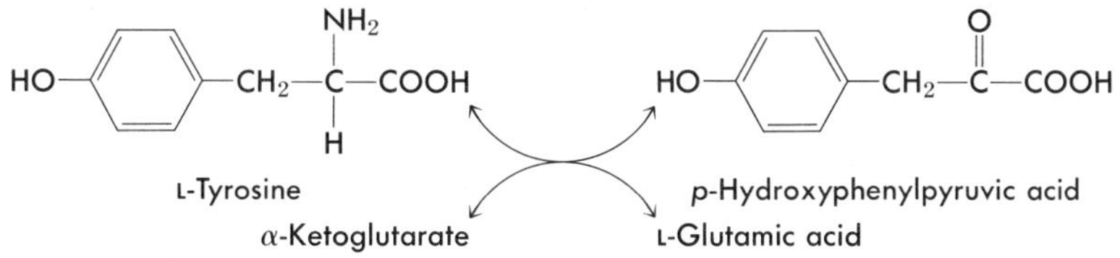

L-Tyrosine + α-Ketoglutarate ⇌ *p*-Hydroxyphenylpyruvic acid + L-Glutamic acid

p-Hydroxyphenylpyruvic acid is oxidized to homogentisic acid (2,5-dihydroxyphenylacetic acid) by a liver oxidase system. The mechanism of the reaction is a complex one, and appears to involve simultaneously a hydroxylation, a shift of the side chain, and decarboxylation.

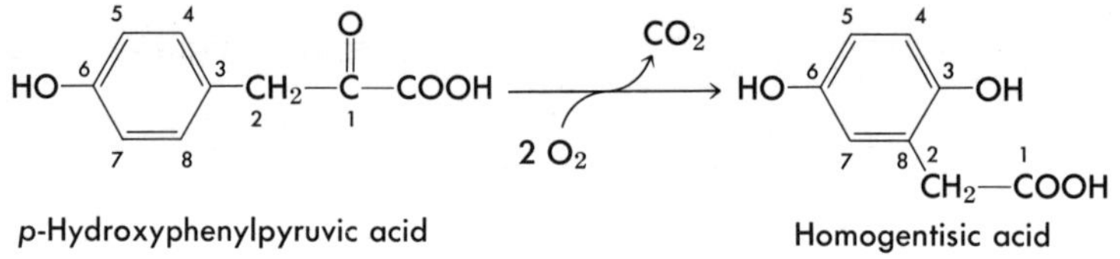

CO_2

$2\ O_2$

p-Hydroxyphenylpyruvic acid

Homogentisic acid

Homogentisic acid oxidase catalyzes the oxidation of homogentisic acid to the open-chain diketone-dicarboxylic acid compound, 4-maleylacetoacetic acid.

O_2

liver oxidase

block in alcaptonuria

4-Maleylacetoacetic acid

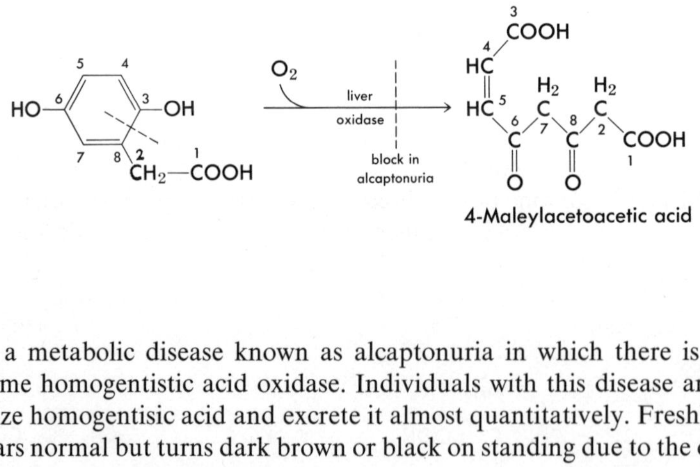

There is a metabolic disease known as alcaptonuria in which there is a defect in the enzyme homogentistic acid oxidase. Individuals with this disease are unable to metabolize homogentisic acid and excrete it almost quantitatively. Freshly voided urine appears normal but turns dark brown or black on standing due to the oxidation of homogentisic acid. Persons with this inborn error of metabolism are usually asymptomatic until late in life when a severe arthritis may develop due to deposition of pigment in bones and fibrous tissues.

Maleylacetoacetic acid is isomerized to fumarylacetoacetic acid which is cleaved to fumaric and acetoacetic acids.

$$\underset{\text{4-Maleylacetoacetic acid}}{HOOC{-}CH{=}CH{-}CO{-}CH_2{-}CO{-}CH_2{-}COOH} \xrightarrow[\text{GSH}]{\text{liver isomerase}} \underset{\text{4-Fumarylacetoacetic acid}}{HOOC{-}CH{=}CH{-}CO{-}CH_2{-}CO{-}CH_2{-}COOH}$$

$$\text{4-Fumarylacetoacetic acid} \xrightarrow{\text{liver hydrolase}} \underset{\text{Fumaric acid}}{HOOC{-}CH{=}CH{-}COOH} + \underset{\text{Acetoacetic acid}}{CH_3{-}CO{-}CH_2{-}COOH}$$

$$\text{Glucose} \longleftarrow \text{Fumaric acid}$$

Thus, phenylalanine and tyrosine are both glucogenic and ketogenic.

Melanin Formation

Melanins are the dark pigments of hair and skin and are formed via the reactions outlined in Figure 7.19. In albinism, an inherited disease, there is a failure of the melanocyte to produce normal amounts of melanin. The basic metabolic defect appears to be a reduction in the melanocyte of the enzyme tyrosinase. This copper-containing enzyme catalyzes the hydroxylation of the melanin precursor, tyrosine, to dihydroxyphenylalanine (DOPA) and DOPA quinone.

Formation of Thyroxine

The pathways for the synthesis and liberation of the thyroid hormone are illustrated in Figure 7.20. At least six enzymatic steps are thought to be involved; these are indicated by number in the appropriate step in the metabolic scheme.

Iodide Trapping (Enzyme 1)

The trapping of iodide by the thyroid gland is oxygen-dependent and inhibited by respiratory inhibitors such as cyanide and azide. The mechanism of transport of iodide against a concentration gradient is unknown but presumed to be enzymic in nature.

Oxidation of Iodide to Iodine (Enzyme 2) and Organification (Enzyme 3)

This oxidation appears to involve a peroxidase

$$H_2O_2 + 2\ {}^{-} + 2\,H^+ \xrightarrow[\text{peroxidase}]{\text{iodine}} 2\ \text{"Active I"} + 2\,H_2O$$

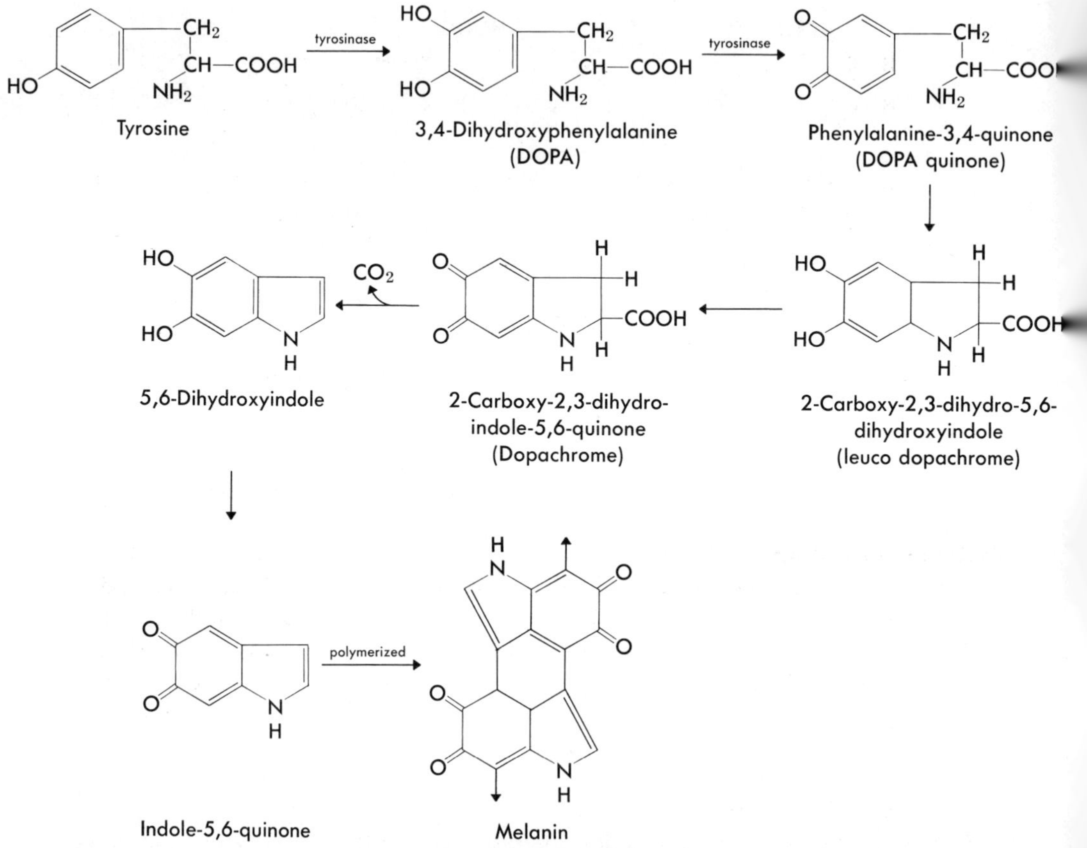

tyrosinase
tyrosinase
Tyrosine
3,4-Dihydroxyphenylalanine (DOPA)
Phenylalanine-3,4-quinone (DOPA quinone)
2-Carboxy-2,3-dihydro-5,6-dihydroxyindole (leuco dopachrome)
2-Carboxy-2,3-dihydro-indole-5,6-quinone (Dopachrome)
CO_2
5,6-Dihydroxyindole
Indole-5,6-quinone
polymerized
Melanin

Figure 7.19. Formation of melanin.

G.I. tract
I^-, I_2
Plasma iodine
Thyroid iodine
Thyroid iodine
Other tissues
Plasma Protein-Bound triodothyronine and thyroxine
Thyroglobulin Triiodothyronine Thyroxine
Thyroglobulin Mono-and diiodotyrosine
I^-
1
2
3
4
5
6

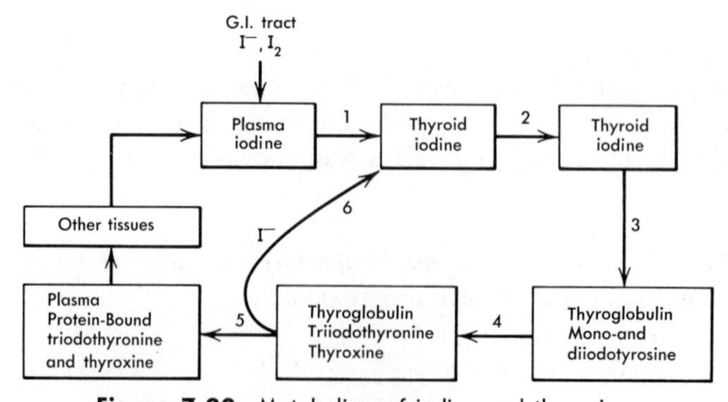

Figure 7.20. Metabolism of iodine and thyroxine.

which is coupled to a tyrosine iodinase. The "active I" is probably the iodinium ion (I^+).

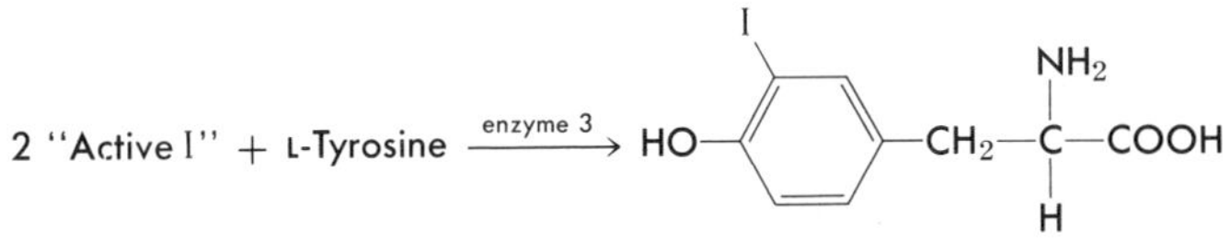

2 "Active I" + L-Tyrosine $\xrightarrow{\text{enzyme 3}}$

L-3-Iodotyrosine (MIT)

+

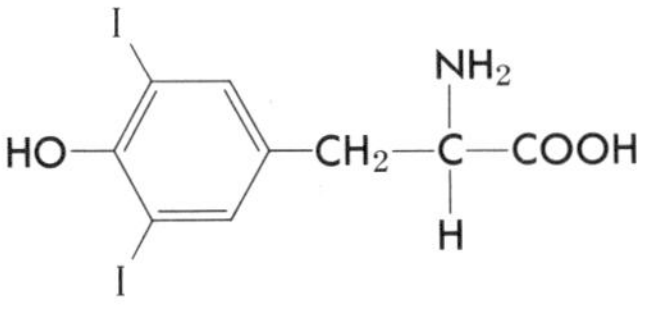

L-3,5-Diiodotyrosine (DIT)

Both MIT and DIT are formed and are stored in the glycoprotein thyroglobulin.

Coupling Reaction (*Enzyme System 4*)

Thyroxine (T_4) and triiodothyronine (T_3) are presumably formed by the enzymic coupling of two molecules of iodotyrosine and extrusion of a side chain.

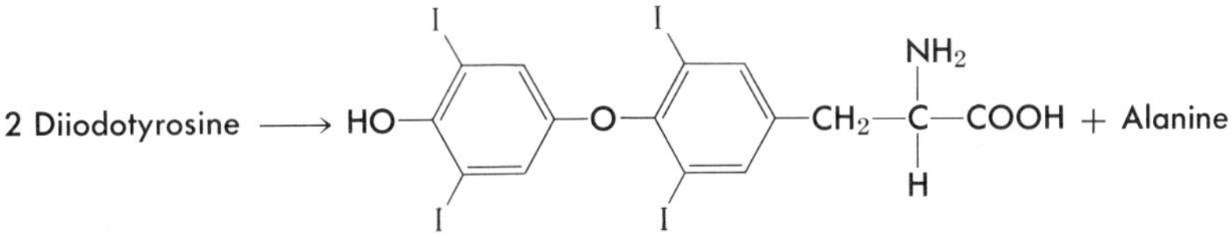

2 Diiodotyrosine $\longrightarrow$ + Alanine

3,5,3′,5′-Tetraiodotyrosine (T_4)
thyroxine

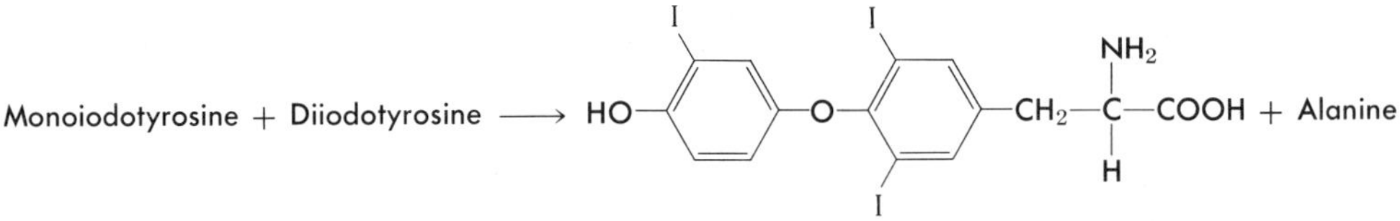

Monoiodotyrosine + Diiodotyrosine $\longrightarrow$ + Alanine

3,5,3′-Triiodothyronine (T_3)

This coupling is thought to occur while the iodotyrosine derivatives are in peptide linkages and not in the free form. T_4 and T_3 are stored in the thyroid as peptide linked residues in thyroglobulin.

Release of T_3 and T_4 (Enzyme System 5)

T_3 and T_4 are released from thyroglobulin by proteolysis and diffuse into the blood where they are bound reversibly to two main thyroxine-binding proteins, an α-globulin and a prealbumin. This is what is measured in the protein-bound iodine (PBI) determination. The normal serum level of PBI is usually accepted as being from 4.0 to 8.0 μg/100 ml. According to current concepts it is the small amount of free thyroxine and not the protein-bound thyroxine that is metabolically active. Since the level of free thyroxine is dependent on the concentration of thyroxine-binding proteins, care must be taken in interpreting PBI results at the lower limit of the range. Whether an individual is normal or hypothyroid may depend on the concentration of thyroxine-binding proteins.

At the time of release of T_3 and T_4, MIT and DIT are also released. These are deiodinated by a NADP-enzyme (enzyme 6) found in the microsomes of thyroid parenchymal cells. As a result, MIT and DIT are not found in blood leaving the thyroid. Similar enzymes are found in the liver, kidney, and other tissues. A lack of this enzyme results in a loss of hormone precursors sufficient to produce hypothyroidism.

Hypothyroid disease may result from blockage at any of the metabolic steps. Five defects have so far been established.

1. Failure to concentrate iodide
2. Failure to convert iodide to organic form
3. Inability to couple iodotyrosines
4. Inability to deiodinate iodotyrosines
5. Presence of abnormal iodoprotein in serum

The metabolic effects of T_4 and T_3 are qualitatively similar and diverse in nature. The rates of oxygen consumption and heat production are stimulated in all cells except thyroid by administration of these hormones. This is believed to be brought about by the uncoupling of a rate-limiting step in the chain of energy-yielding reactions (page 93). Subsequent steps proceed more rapidly and increase the overall yield of high-energy phosphate bonds despite a decreased coupling and lower efficiency at the rate-limiting step.

Synthesis and Metabolism of Norepinephrine and Epinephrine

The catecholamines (norepinephrine and epinephrine) are synthesized in brain, sympathetic ganglia, and sympathetic nerve endings, and in chromaffin cells of peripheral tissues, especially the adrenal medulla. Norepinephrine and epinephrine are secreted into the circulation from the adrenal medulla, and norepinephrine is released locally as a neurotransmitter by sympathetic nerve endings.

The major pathway for the biosynthesis of norepinephrine and epinephrine is shown in Figure 7.21. Alternate pathways for their formation exist, but they are not quantitatively significant. A specific tyrosine hydroxylase catalyzes the conversion of tyrosine to 3,4-dihydroxyphenylalanine (DOPA); the enzyme is similar to phenylalanine hydroxylase (page 224) in having a specific requirement for a tetrahydropteridine. Tyrosine hydroxylation is the rate-limiting step in norepinephrine biosynthesis. DOPA decarboxylase, which catalyzes the decarboxylation of DOPA can also convert histidine (page 267), phenylalanine, tyrosine, tryptophan, and 5-hydroxytryptophan (page 266) to their corresponding amines.

The specificity of the β-hydroxylase is low, almost any phenylethylamine derivative being oxidized to the corresponding ethanolamine derivative. The N-methyltransferase, which converts norepinephrine to epinephrine, is also active with other phenylethanolamines. The lack of strict specificity of the enzymes involved in catecholamine synthesis undoubtedly accounts for the alternate synthetic pathways found.

Almost all of the norepinephrine or epinephrine formed in the body is metabolized and excreted as O-methylated products. The liver is the main site for the O-methylation of circulating catecholamines. The major metabolites found in the urine are 3-methoxy-4-hydroxymandelic acid (VMA), 3-methoxy-4-hydroxyphenylglycol (free and conjugated), metanephrine (free and conjugated), and normetanephrine (free and conjugated). Small amounts of norepinephrine, epinephrine, and other metabolites are also excreted in the urine. The determination of the urinary levels of 3-methoxy-4-hydroxymandelic acid (VMA), normetanephrine, metanephrine, and the catecholamines is of considerable value for the diagnosis of the adrenal medullary tumor, pheochromocytoma, and for the diagnosis of neuroblastoma.

Metabolism of Tryptophan

Tryptophan was discovered in 1901 by Hopkins and Cole and was the first amino acid shown to be indispensable in the diet of the animal. Both the D isomer and the corresponding keto acid, indole-3-pyruvic acid, will replace L-tryptophan in the rat. The pathways for the dissimilation of tryptophan are given in Figure 7.22.

The degradation of tryptophan has been of considerable interest because it is converted to the vitamin nicotinic acid and to the neurohumoral substance 5-hydroxytryptamine.

Conversion to Nicotinic Acid

Liver contains the iron porphyrin enzyme, tryptophan pyrrolase, which catalyzes the conversion of tryptophan to N-formylkynurenine. Molecular oxygen is incorporated into the product as shown.

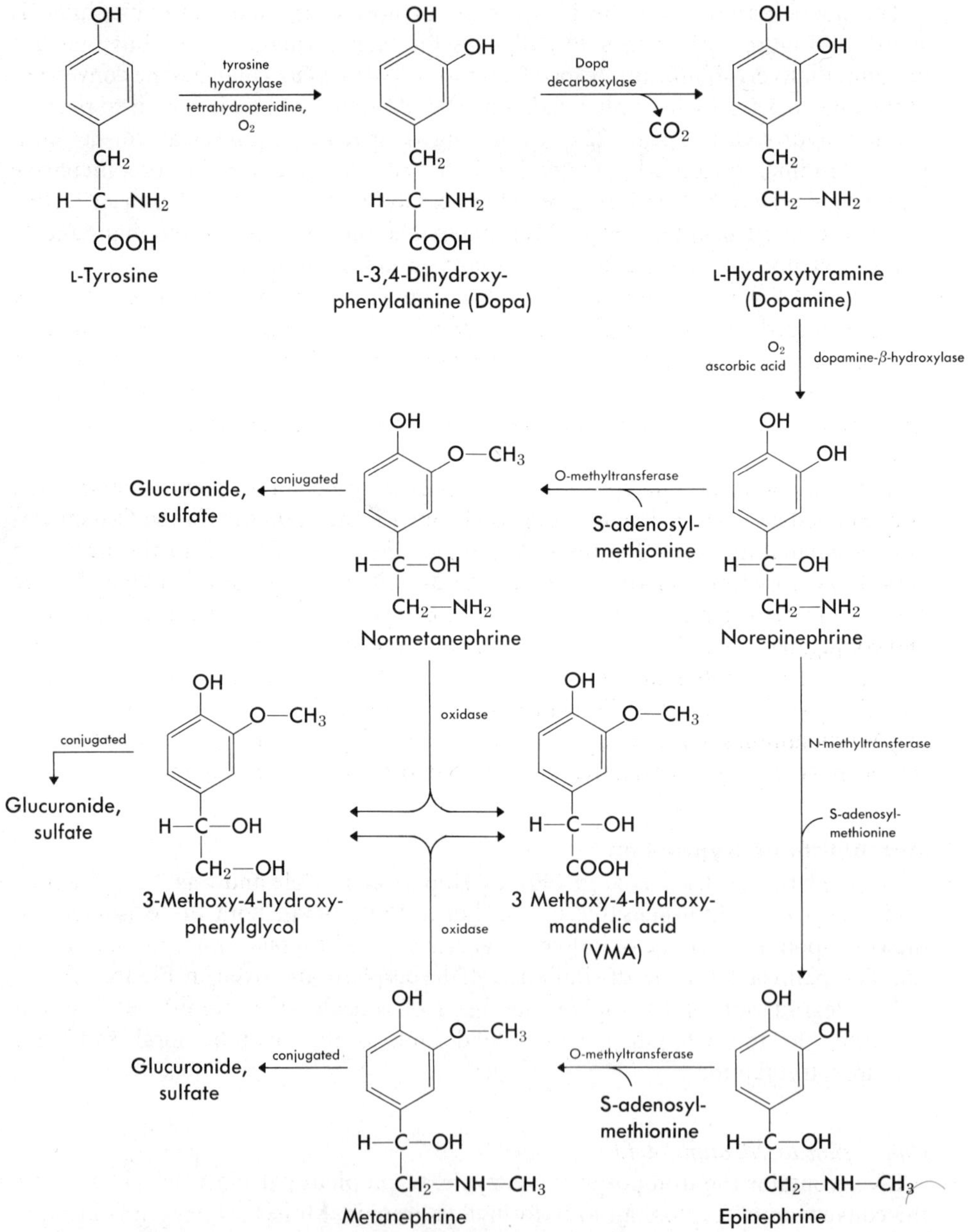

L-Tyrosine

tyrosine hydroxylase

tetrahydropteridine, O_2

L-3,4-Dihydroxyphenylalanine (Dopa)

Dopa decarboxylase

CO_2

L-Hydroxytyramine (Dopamine)

O_2 ascorbic acid

dopamine-β-hydroxylase

Norepinephrine

O-methyltransferase

S-adenosylmethionine

Normetanephrine

conjugated

Glucuronide, sulfate

oxidase

3-Methoxy-4-hydroxyphenylglycol

conjugated

Glucuronide, sulfate

3 Methoxy-4-hydroxymandelic acid (VMA)

oxidase

N-methyltransferase

S-adenosylmethionine

Epinephrine

O-methyltransferase

S-adenosylmethionine

Metanephrine

conjugated

Glucuronide, sulfate

Figure 7.21. Biosynthesis and metabolism of epinephrine and norepinephrine.

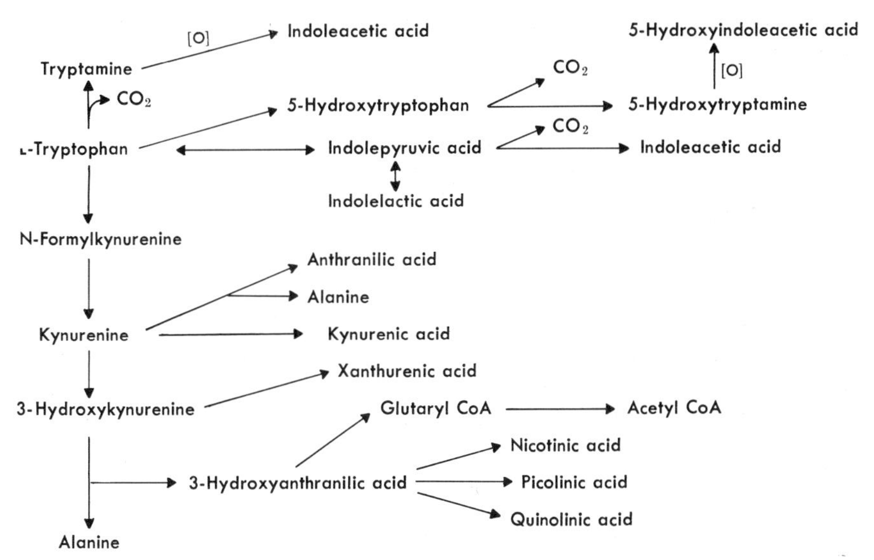

Figure 7.22. Metabolism of tryptophan.

L-Tryptophan $\xrightarrow{O_2^*}$ N-Formylkynurenine

In the next reaction, formylkynurenine is hydrolyzed to kynurenine and formic acid.

N-Formylkynurenine $\xrightarrow[H_2O]{\text{formamidase}}$ Kynurenine + HC^*O^*OH

The formic acid produced in this reaction may enter the one-carbon pool and give rise to those carbon atoms of amino acids and purines derived from the functional forms of tetrahydrofolic acid.

Kynurenine is hydroxylated to 3-hydroxykynurenine in a reaction with molecular oxygen and NADPH.

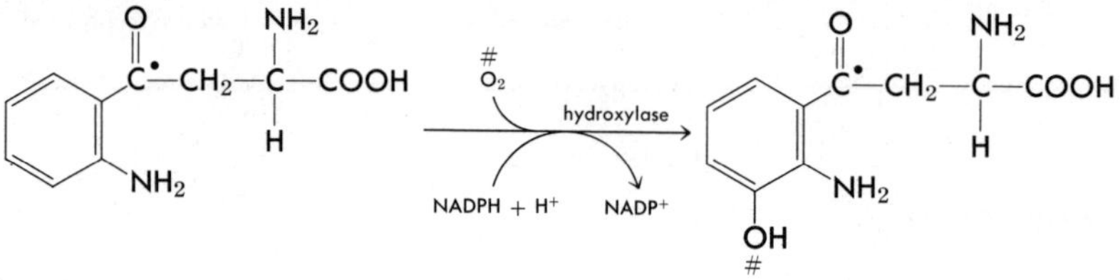

Kynurenine

O_2

hydroxylase

NADPH + H^+ NADP$^+$

3-Hydroxykynurenine

Both kynurenine and 3-hydroxykynurenine can participate in transamination reactions leading to the production of the corresponding diketo acids. These diketo acids cyclize readily to yield quinoline derivatives which are excreted in the urine.

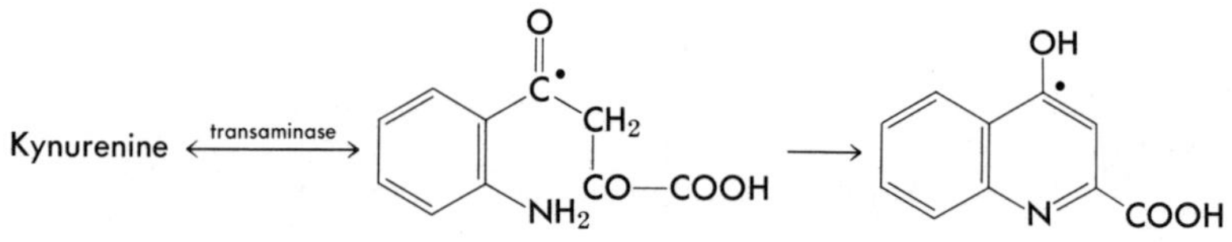

Kynurenine ⟷ transaminase

Kynurenic Acid

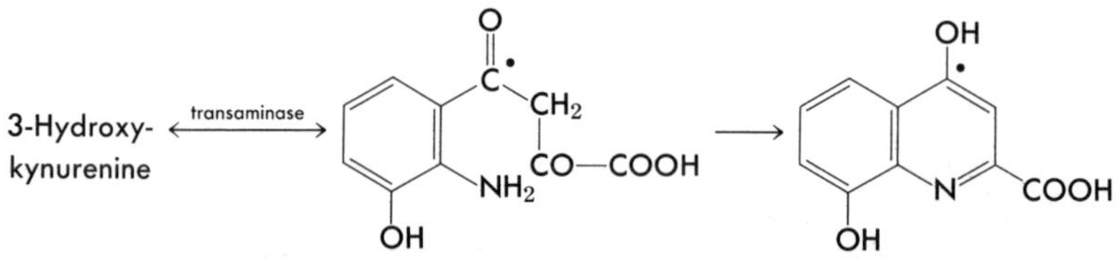

3-Hydroxy-kynurenine ⟷ transaminase

Xanthurenic Acid

The excretion of these compounds is increased in B_6-deficiency. Since both the transaminase and kyureninase (see below) are B_6-enzymes, it appears that the kynureninase reaction is more sensitive to B_6-deficiency than the transaminase.

Kynureninase catalyzes the hydrolysis of both kynurenine and 3-hydroxykynurenine.

pyridoxal phosphate

H_2O

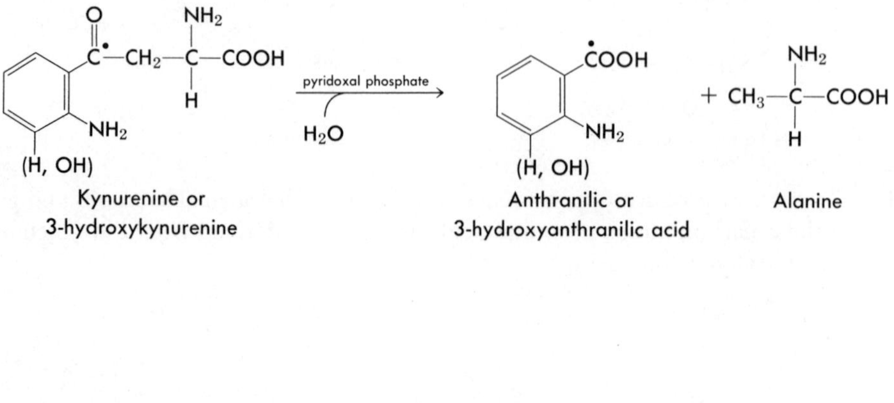

Kynurenine or 3-hydroxykynurenine

Anthranilic or 3-hydroxyanthranilic acid

Alanine

3-Hydroxyanthranilic acid is converted to NAD by the following series of reactions

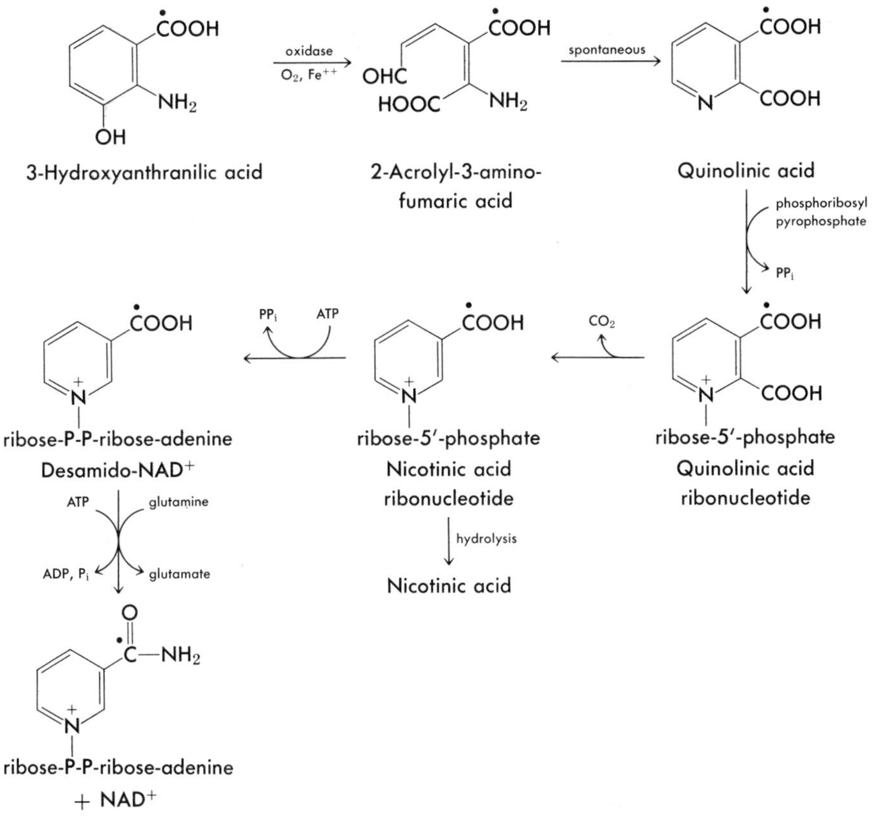

3-Hydroxyanthranilic acid → (oxidase; O_2, Fe^{++}) → 2-Acrolyl-3-amino-fumaric acid → (spontaneous) → Quinolinic acid

Quinolinic acid → (phosphoribosyl pyrophosphate; PP_i) → Quinolinic acid ribonucleotide (ribose-5′-phosphate)

Quinolinic acid ribonucleotide → (CO_2) → Nicotinic acid ribonucleotide (ribose-5′-phosphate)

Nicotinic acid ribonucleotide → (hydrolysis) → Nicotinic acid

Nicotinic acid ribonucleotide → (ATP; PP_i) → Desamido-NAD$^+$ (ribose-P-P-ribose-adenine)

Desamido-NAD$^+$ → (ATP, glutamine; ADP, P_i, glutamate) → NAD$^+$ (ribose-P-P-ribose-adenine)

It has been estimated in the human that 60 mg of dietary tryptophan is equivalent to 1 mg of nicotinic acid.

Conversion to Tryptamine and 5-Hydroxytryptamine

The conversion of tryptophan to tryptamine and 5-hydroxytryptophan is summarized in Figure 7.23.

In the normal individual the conversion of tryptophan to tryptamine and 5-hydroxytryptamine (serotonin) and the corresponding indole acetic acid is minor. Excessively large amounts of serotonin are produced in patients with malignant carcinoid. This is converted in large part to 5-hydroxyindole-3-acetic acid and excreted in the urine in amounts many times that of normal.

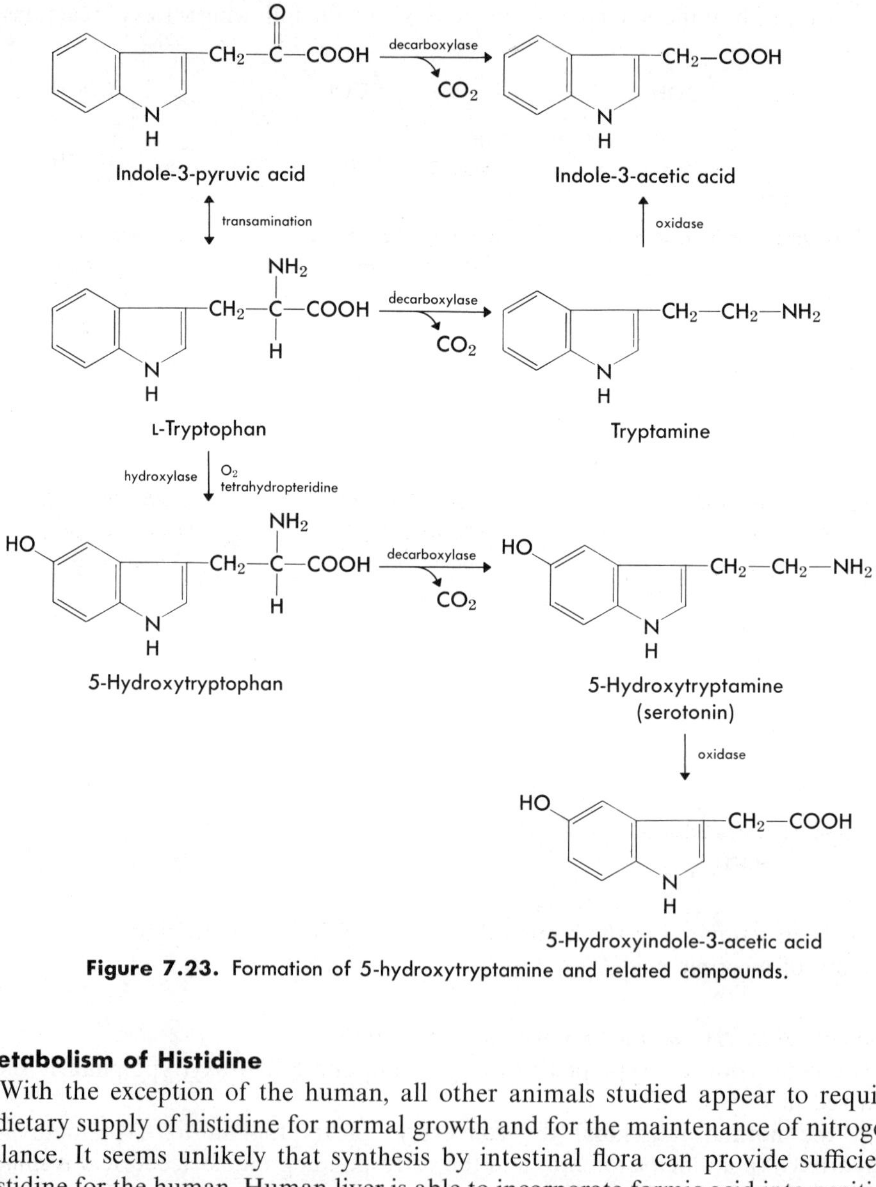

Indole-3-pyruvic acid

Indole-3-acetic acid

L-Tryptophan

Tryptamine

5-Hydroxytryptophan

5-Hydroxytryptamine (serotonin)

5-Hydroxyindole-3-acetic acid

Figure 7.23. Formation of 5-hydroxytryptamine and related compounds.

Metabolism of Histidine

With the exception of the human, all other animals studied appear to require a dietary supply of histidine for normal growth and for the maintenance of nitrogen balance. It seems unlikely that synthesis by intestinal flora can provide sufficient histidine for the human. Human liver is able to incorporate formic acid into position 2 of the imidazole ring of histidine; beyond this little is known. The α-keto and α-hydroxy acids and the D isomer of histidine are utilizable by the rat.

Metabolic Breakdown of Histidine

L-Histidine is converted to L-glutamic acid, 2 NH_3, and a one-carbon unit in liver as shown in Figure 7.24.

As would be predicted from the above pathway, folic acid-deficient animals excreted large amounts of formiminoglutamic acid in their urine.

An important product of histidine metabolism is the pharmacologically active amine, histamine, which is formed by the action of histidine decarboxylase. Histaminase (diamine oxidase) oxidizes histamine to imidazole acetaldehyde which is in turn oxidized to imidazole acetic acid by aldehyde oxidase.

Histidine —(decarboxylase, CO_2)→ Histamine —(diamine oxidase, O_2, H_2O)→ Imidazoleacetaldehyde

Histidine ↔ (transamination) Imidazolepyruvic acid

Imidazoleacetaldehyde —(aldehyde oxidase, NAD^+)→ Imidazoleacetic acid

Imidazolepyruvic acid —(decarboxylase, CO_2)→ Imidazoleacetic acid → Imidazoleacetic acid riboside

Imidazoleacetic acid and its riboside are end products of histamine metabolism and are excreted in the urine. In addition 3-methylhistamine and 3-methylimidazoleacetic acid are formed and excreted in the urine in significant amounts.

Metabolism of Sulfur-Containing Amino Acids: Methionine, Cysteine, and Related Compounds

Methionine is an indispensable amino acid, whereas cysteine and cystine are dispensable providing sufficient methionine is present. D-Methionine and the α-keto acid can replace methionine in the diet. The major biochemical functions of methio-

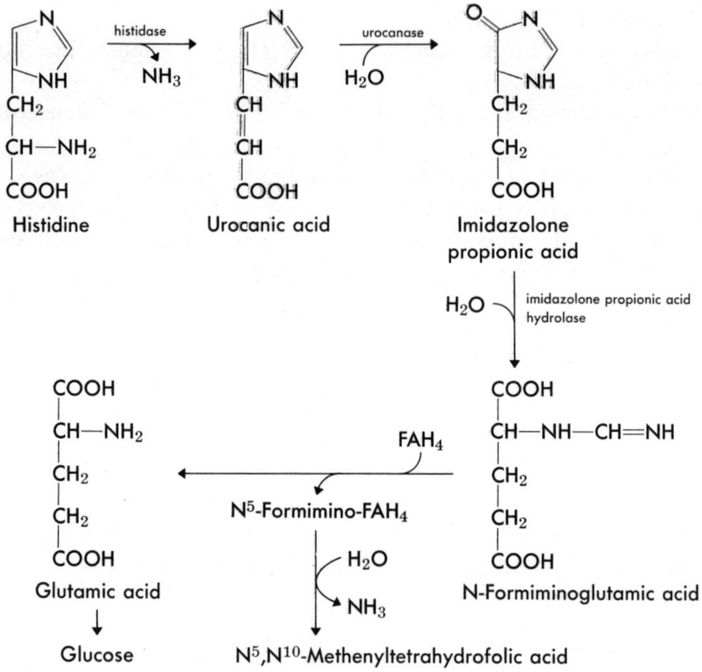

Figure 7.24. Breakdown of histidine.

nine, in addition to being utilized for protein synthesis, are its roles as a methyl donor and as a precursor of cysteine. The metabolism of methionine is outlined in Figure 7.25.

Conversion of Methionine to Cysteine

It seems likely that the metabolism of L-methionine is initiated by its conversion to S-adenosylmethionine (page 228).

S-Adenosylmethionine can enter into a number of enzyme catalyzed trans-methylation reactions with several methyl acceptors. These lead to the production of S-adenosylhomocysteine plus the methylated acceptor.

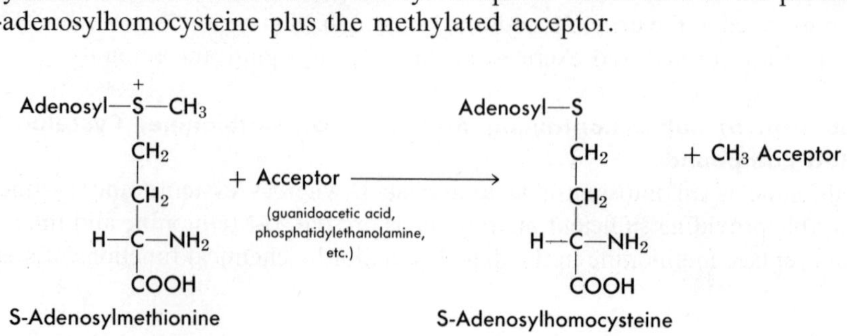

S-Adenosylhomocysteine may be cleaved by a liver enzyme to adenosine and homocysteine; homocysteine can then condense with L-serine to form cystathionine.

S-Adenosylhomocysteine $\xrightleftharpoons{\text{enzyme}}$ Adenosine + HS—CH₂—CH₂—CH(NH₂)—COOH (L-Homocysteine)

L-Homocysteine + L-Serine (CH₂OH—CH(NH₂)—COOH) $\xrightarrow[\text{pyridoxal phosphate}]{\text{cystathionine synthetase}}$ L-Cystathionine + H_2O

L-Cystathionine: CH₂—S—CH₂—CH₂—CH(NH₂)—COOH, with H—C—NH₂ and COOH on the first CH₂

The cleavage of cystathionine by the splitting enzyme produces L-cysteine, α-ketobutyrate, and NH_3.

L-Cystathionine $\xrightarrow{\text{pyridoxal phosphate}}$ L-Cysteine (CH₂—SH, H—C—NH₂, COOH) + [CH₂=C(H)—C(H)(NH₂)—COOH ⇌ CH₃—C(H)=C(NH₂)—COOH ⇌ CH₃—CH₂—C(=NH)—COOH]

$\xrightarrow{H_2O}$ CH₃—CH₂—C(=O)—COOH (α-Ketobutyrate) + NH_3

Glucose ⟵ Succinate ⟵ ⟵ α-Ketobutyrate

It will be noted that only the sulfur atom of methionine is used for the formation of cysteine, the carbon chain and amino group being derived from serine. Homoserine was originally thought to be the other product of the cleavage reaction. The product is, however, α-ketobutyrate which can be converted to carbohydrate, thus explaining the glucogenic nature of methionine. Cystathionine can replace cysteine in the diet but not methionine. It can be cleaved to form cysteine but not homocysteine.

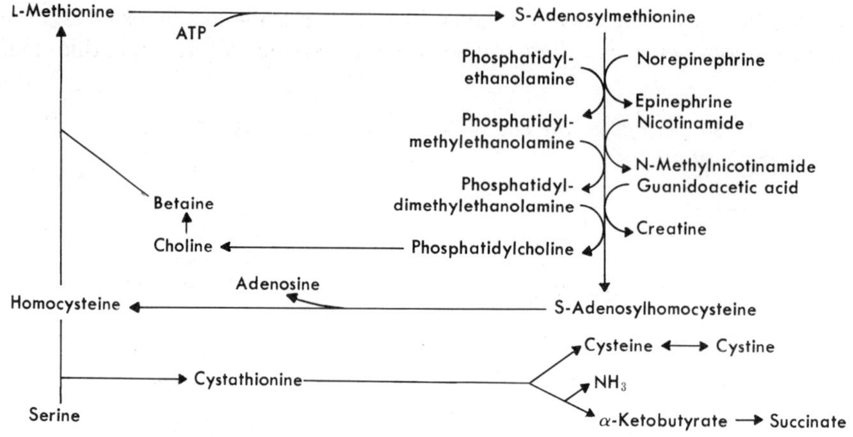

Figure 7.25. Metabolism of methionine.

Transmethylation Reactions

Brief reference has been made to the reactions of S-adenosylmethionine as a methyl donor for the synthesis of creatine, epinephrine, phosphatidylcholine, and other methylated compounds. Nutritional experiments had also established that choline plus homocysteine (or homocystine) could replace methionine in the diet of the growing rat. It seemed apparent from these results that biologic transmethylation reactions had occurred. Isotope studies subsequently established a number of transmethylation reactions. It was found in addition that the intact rat and various rat liver preparations were able to synthesize methyl groups from formate. Since formate can arise from a number of sources such as the β-carbon atom of serine, carbon 2 of glycine, carbon 2 of histidine, and so forth, it is understandable that these groups also can give rise to methyl groups by their relationships to the active folic acid one-carbon donors and acceptors (see page 232). Since normal development does not take place in animals on a diet free from methyl groups although some synthesis does take place, it would appear that methyl groups are in the same position as arginine. Arginine is synthesized by the young rat, but not at a rate sufficient to meet all the demands of the growing animal. Thus, methyl groups may be regarded as a dietary essential.

We have the situation of having a "reversible" pathway between certain methyl groups and the active one-carbon compounds. This idea of a methyl cycle is opposed to the older concept of a methyl pool in which methyl groups supplied in the diet were transferred only by transmethylation reactions. There is also a definite specificity prevailing among methyl donors and methyl acceptors. A given methyl compound may be related to another only through a series of methyl transfer reactions. Figure 7.26 summarizes the methyl group cycle.

It may be seen that choline must be oxidized to betaine before a methyl group can be transferred to homocysteine to form methionine. Methionine in turn must be activated by conversion to S-adenosylmethionine prior to the transfer of its methyl group to specific acceptors. The methyl groups of dimethylglycine and sarcosine are removed by conversion to the active formaldehyde derivative (N^5,N^{10}-methylene-tetrahydrofolic acid) and thence to the β-carbon of serine, carbon dioxide, or to other compounds related to the active folic acid derivatives.

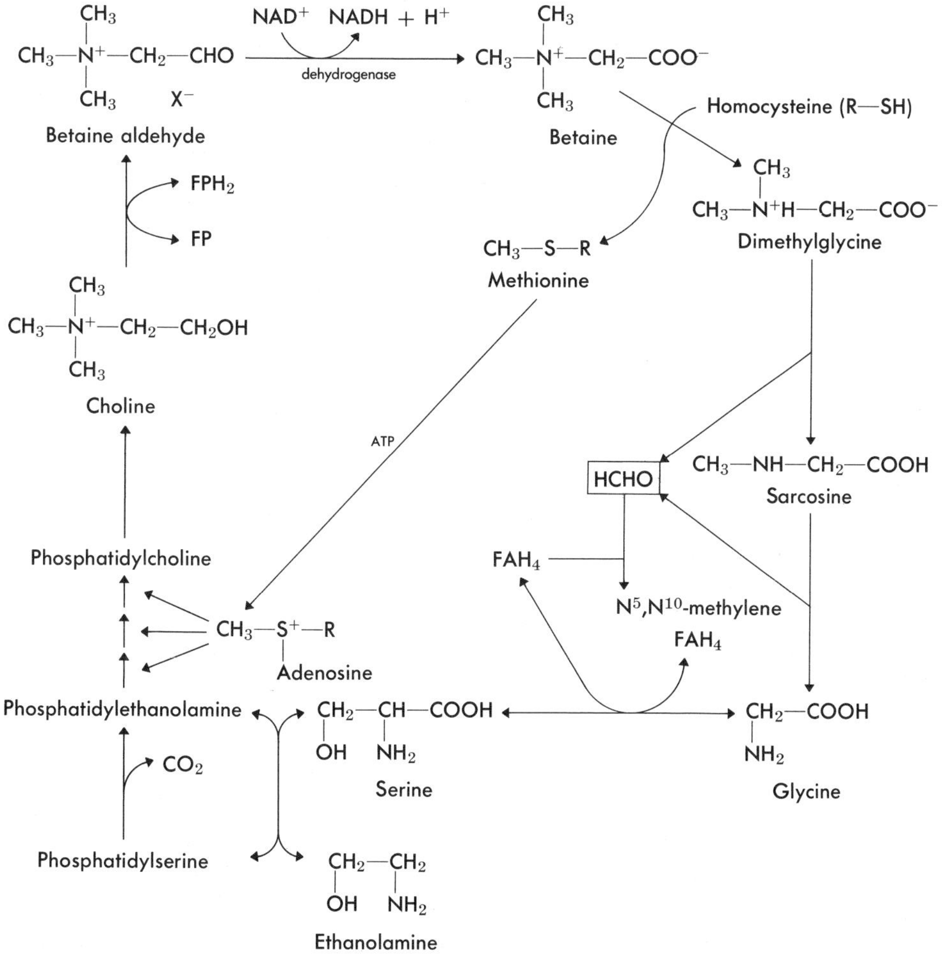

Figure 7.26. Methyl group cycle.

It was pointed out earlier (page 242) that the decarboxylation of serine and the conversion of ethanolamine to choline apparently take place with phosphatidyl-bound compounds and not the free compounds as originally proposed.

It will be noted that the two active methyl donors are "onium" compounds. Betaine is a quaternary ammonium compound, and S-adenosylmethionine is a sulfonium compound.

The concept has developed that sulfonium compounds are energy-rich compounds and can serve as sources of chemical energy for biosynthetic reactions. The sulfonium compounds have free energies of hydrolysis of about -8 kcal. Quaternary am-

$$CH_3-\overset{+}{N}(CH_3)_2-CH_2-COO^-$$

Betaine

$$CH_3-\overset{+}{S}(CH_2\text{-adenosine})-(CH_2)_2-CH(NH_2)-COO^-$$

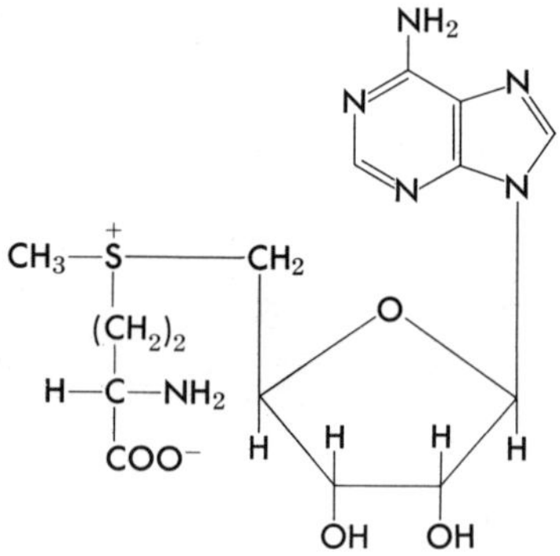

S-Adenosylmethionine

monium compounds do not appear to be "high energy" but can transfer a methyl group to thiols because the thiols are at a low energy level.

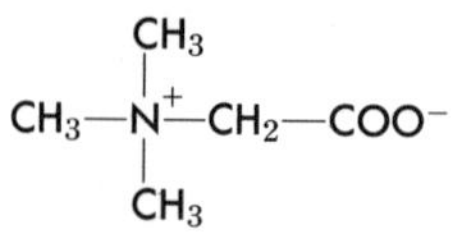

Metabolism of Cysteine (Cystine)

The synthesis of cysteine from methionine and serine was discussed on page 269.

Cystine reductase is widely distributed in tissues and catalyzes the reversible reduction of cystine to cysteine.

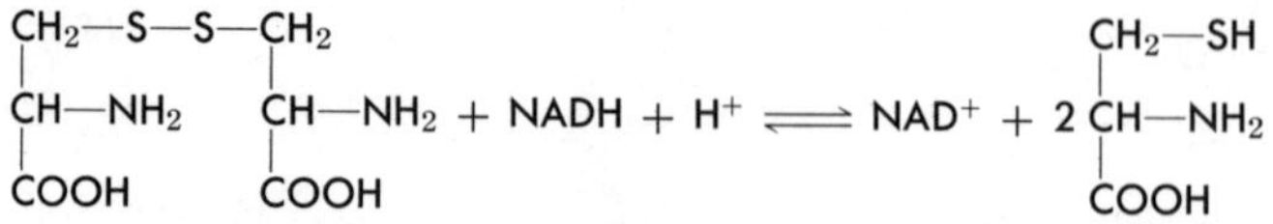

$$\underset{\text{Cystine}}{HOOC-CH(NH_2)-CH_2-S-S-CH_2-CH(NH_2)-COOH} + NADH + H^+ \rightleftharpoons NAD^+ + 2\,\underset{\text{Cysteine}}{HS-CH_2-CH(NH_2)-COOH}$$

Cysteine is then metabolized via several pathways, some of which are outlined in Figure 7.27.

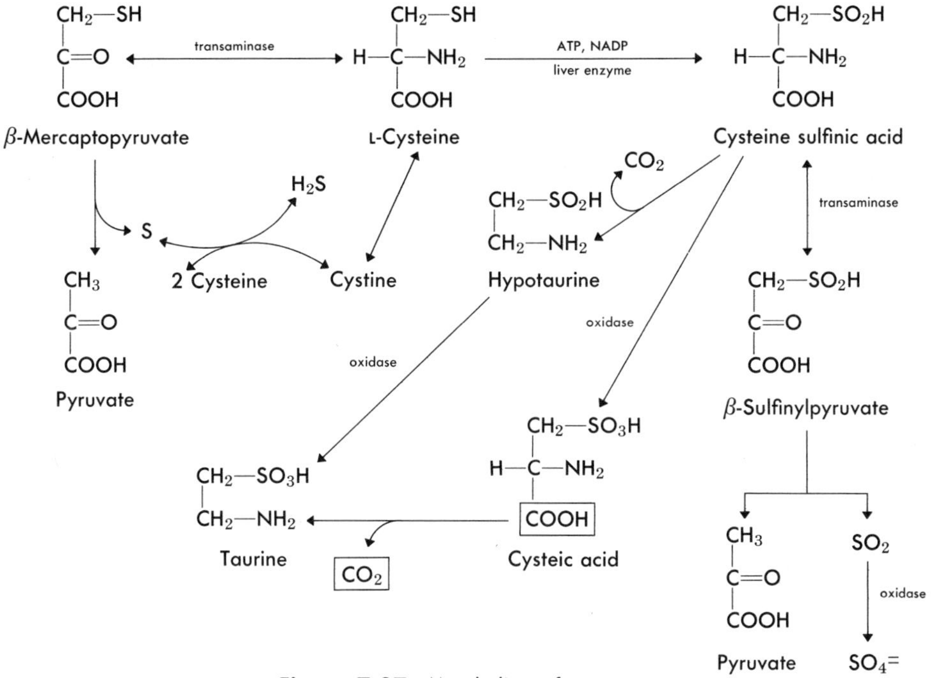

$CH_2—SH$, $C=O$, $COOH$ — β-Mercaptopyruvate ⟷ (transaminase) ⟷ $CH_2—SH$, $H—C—NH_2$, $COOH$ — L-Cysteine → (ATP, NADP; liver enzyme) → $CH_2—SO_2H$, $H—C—NH_2$, $COOH$ — Cysteine sulfinic acid

β-Mercaptopyruvate → S → CH_3, $C=O$, $COOH$ — Pyruvate

H_2S; S; 2 Cysteine; Cystine

Cysteine sulfinic acid → CO_2 + $CH_2—SO_2H$, $CH_2—NH_2$ — Hypotaurine

Cysteine sulfinic acid ⟷ (transaminase) ⟷ $CH_2—SO_2H$, $C=O$, $COOH$ — β-Sulfinylpyruvate

Hypotaurine → (oxidase) → $CH_2—SO_3H$, $CH_2—NH_2$ — Taurine

Cysteine sulfinic acid → (oxidase) → $CH_2—SO_3H$, $H—C—NH_2$, COOH — Cysteic acid → Taurine + CO_2

β-Sulfinylpyruvate → CH_3, $C=O$, $COOH$ — Pyruvate + SO_2 → (oxidase) → $SO_4^{=}$

Figure 7.27. Metabolism of cysteine.

Sulfate Activation

Urinary sulfur is derived primarily from the catabolism of sulfur-containing amino acids. It is partitioned between ethereal sulfate sulfur (5 per cent), organic sulfide sulfur (15 per cent), and inorganic sulfate sulfur (80 per cent). Although the excretion varies directly with protein intake, the usual 24 hour output is 0.6 to 1.4 g sulfate as sulfur. The ethereal sulfates represent detoxication compounds of phenols and are formed by the liver. These compounds as well as compounds like chondroitin sulfate are formed in reactions employing an "active" sulfate. The active sulfate, 3′-phosphoadenosine-5′-phosphosulfate, is formed and utilized as shown in Figure 7.28.

Disorders of Amino Acid Metabolism

From his observations on alcaptonuria, albinism, cystinuria, and pentosuria, Garrod in 1908 formulated the concept that certain diseases of lifelong duration were due to the loss of or decrease in activity of an enzyme catalyzing a single metabolic step. Garrod called these diseases "inborn errors of metabolism." A vast number of these defects have been recognized that involve all areas of metabolism.

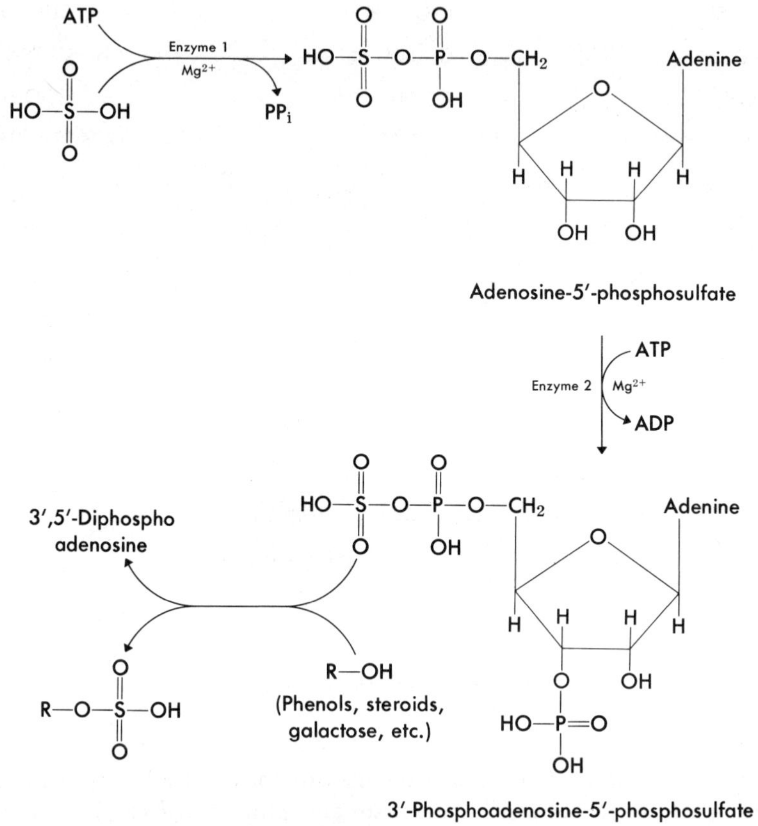
ATP
Enzyme 1
Mg2+
HO—S—OH
PPi
HO—S—O—P—O—CH2
OH
Adenine
H
OH
Adenosine-5′-phosphosulfate
ATP
Enzyme 2
Mg2+
ADP
3′,5′-Diphospho adenosine
R—O—S—OH
R—OH
(Phenols, steroids, galactose, etc.)
HO—P=O
3′-Phosphoadenosine-5′-phosphosulfate

Figure 7.28. Formation and utilization of active sulfate.

Some 50-odd amino acid-opathies alone are known involving both transport and postabsorptive metabolism. Some of these have been discussed elsewhere in this chapter.

Table 7.6 presents a list of inborn errors of amino acid metabolism in which the specific enzyme defect has been identified. At present it is not possible to describe each defect in terms comparable to the classification of the hemoglobins (page 330).

The abundance of enzymatic mutations in man has provided an unusual opportunity to study metabolic pathways in a complex, highly differentiated organism. Indeed, a great deal of basic and universal information has been derived from the study of these metabolic defects.

Table 7.6
Inborn Errors of Amino Acid Metabolism with Defined Enzymatic Defects

Disease	Defective Enzyme
Alcaptonuria	Homogentisic acid oxidase (page 256)
Phenylketonuria	Phenylalanine hydroxylase (page 254)
Tyrosinemia	*p*-Hydroxyphenylpyruvic oxidase (page 256)
Maple Syrup	α-keto decarboxylase (page 249)
Hyperammonemia	Ornithine transcarbamylase (page 223)
Arginosuccinicaciduria	Arginosuccinase (page 224)
Histidenemia	Histidase (page 268)
Hyperprolinemia	Proline oxidase (page 248)
Cystathioninuria	Cystathionase (page 269)
Homocystinuria	Cystathionine synthetase (page 269)
Isovalericacidemia	Isovaleryl-CoA dehydrogenase (page 252)
Methylmalonicaciduria	Methylmalonyl-CoA isomerase (page 250)
Citrullinemia	Arginosuccinic synthetase (page 224)
Hypervalinemia	Valine transaminase (page 250)
Hydroxyprolinemia	Hydroxyproline oxidase (page 249)

REFERENCES

Greenberg, D. M. (ed): *Metabolic Pathways,* Vol. 3, 3rd ed., Chaps. 14, 15, 16, 17, and 18. Academic Press, New York, 1969.

Mahler, H. R., and Cordes, E. H., *Biological Chemistry,* Chap. 16. Harper and Row, New York, 1966.

Meister, A.: *Biochemistry of the Amino Acids,* Vol. I, II, 2nd ed. Academic Press, New York, 1965.

Nyhan, W. L. (ed.): *Amino Acid Metabolism and Genetic Variation.* McGraw-Hill, New York, 1967.

Stanbury, J. B., Wyngaarden, J. B., and Fredrickson, D. S. (eds.): *The Metabolic Basis of Inherited Disease,* 2nd ed. McGraw-Hill, New York, 1966.

West, E. S., Todd, W. R., Mason, H. S., and Van Bruggen, J. T.: *Textbook of Biochemistry,* 4th ed., Chap. 25. The Macmillan Co., New York, 1966.

White, A., Handler, P., and Smith E. L.: *Principles of Biochemistry,* 4th ed., Chaps. 23, 24, 25, and 26. McGraw-Hill, New York, 1966.

Chapter 8

CHEMISTRY AND METABOLISM OF NUCLEIC ACIDS AND NUCLEOPROTEINS

Chemistry

BIOLOGICAL SYSTEMS employ large molecules, the nucleic acids, for the essential functions of storage and transmission of genetic information. With the exception of certain viruses, it is deoxyribonucleic acid (DNA) that serves as the primary genetic material, the gene. Various ribonucleic acid (RNA) species are involved in the transcription of genetic information from DNA and its translation to primary protein structure. This chapter presents the chemistry of these macromolecules, as well as some aspects of their origins and metabolic fates. The roles of nucleic acids in protein synthesis are discussed more fully in the following chapter.

Nucleoproteins

As both DNA and RNA are strong anions at physiologic pH, they are found associated in the cell with basic proteins or other cations. In higher organisms the DNA of the cell nucleus is frequently combined with histones. Histones (M.W. 10,000 to 20,000) are the basic chromosomal proteins containing a high content of arginine and lysine and lacking tryptophan. In the sperm cells of some birds and fish, histones are replaced by protamines, which are smaller (M.W. 3000 to 5000) and have an even higher content of basic amino acids. There is also evidence to indicate that RNA can also associate with histones.

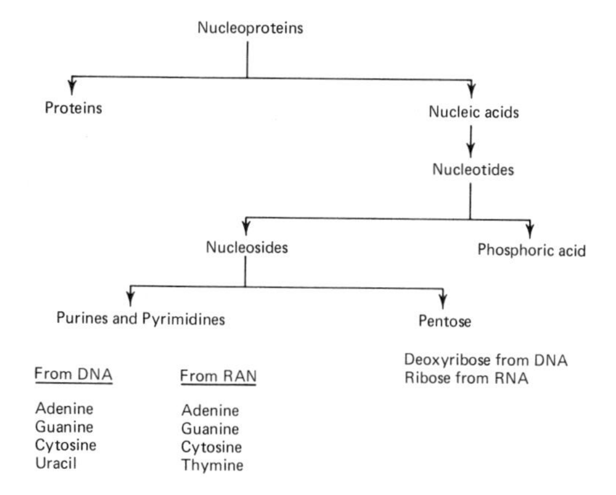

Figure 8.1. Hydrolytic products of nucleic acids.

Much of the interest in chromosomal proteins was elicited by the observation that histones decrease the ability of DNA to act as primer for the DNA-dependent RNA polymerase (page 307). Thus, these proteins might act as repressors and block the transcription (page 306) of a given bit of genetic information.

The nucleoproteins can be dissociated into protein and nucleic acid by a variety of methods such as treatment with high ionic strength salt or acid, detergent or phenol extraction, or by repeated emulsification of a salt solution of the nucleoprotein with chloroform-octanol.

Complete chemical hydrolysis of nucleic acids gives a mixture of purine and pyrimidine bases, a pentose or deoxypentose sugar, and phosphoric acid. A partial hydrolysis of nucleic acids yields nucleotides (base-sugar-phosphate) and nucleosides (base-sugar). The sugar is either 2-deoxy-D-ribose (in DNA) or D-ribose (in RNA). In DNAs the common bases are the purines adenine (A) and guanine (G) and the pyrimidines cytosine (C) and thymine (T). RNAs contain the purines adenine and guanine and the pyrimidines uracil (U) and cytosine. Figure 8.1 shows the relationship of these compounds.

Pyrimidines

Pyrimidine Ring | Cytosine (C) (RNA and DNA) | Uracil (RNA) (U) | Thymine (DNA) (T)

In addition to the three major pyrimidines a number of minor components have been isolated from various sources. In transfer ribonucleic acids (page 289) various methylated bases such as 5-methylcytosine, thymine, and others replace some of the normal constituents. Some 5-hydroxymethyluracil and 5-methylcytosine occur in DNA from bacteria and bacterial viruses and 5-hydroxymethyl cytosine replaces cytosine in the T-even phages of *Escherichia coli.*

The pyrimidines, like the purines, are capable of lactam-lactim tautomerism and may occur in two forms.

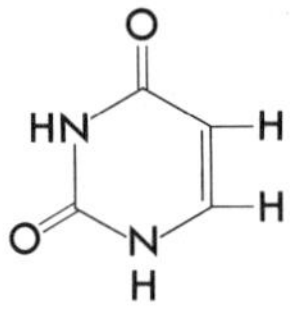

Uracil
(lactam or keto form)

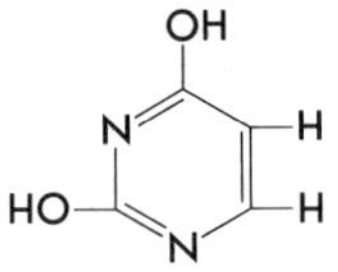

Uracil
(lactim or enol form)

Due to the presence of the conjugated ring system, both pyrimidines and purines and their derivatives exhibit characteristic absorption in the ultraviolet range of the spectrum, with absorption maxima near 260 mμ. The absorption depends on the substituents present and the pH at which it is measured. The dependence of absorption on pH is due to the degree of ionization. These spectral properties are useful for the detection and estimation of these compounds.

A characteristic property of nucleic acids is that their quantitative absorption of ultraviolet light is less than would be predicted on the basis of their nucleotide content. This *hypochromisim* depends on the A + T content; it is greater for DNA than RNA and greater for double-stranded than single-stranded polymers. When a nucleic acid is converted from the ordered native form to a more disordered conformation, there is an increase in absorption (*hyperchromic effect*). Thus this provides a technique for the study of the secondary structure of nucleic acids.

Purines

Purines contain a six-membered pyrimidine ring fused to a five-membered imidazole ring.

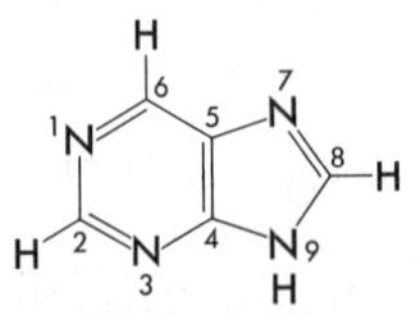

Purine

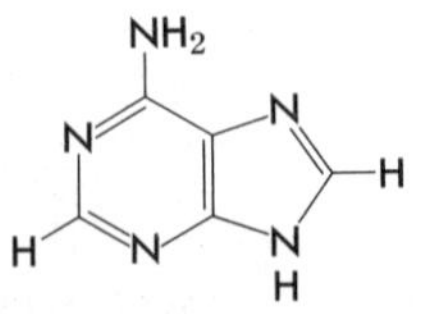

Adenine (A)

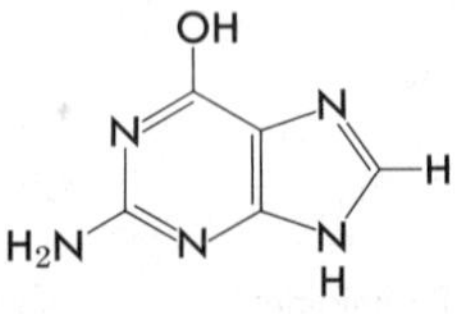

Guanine (G)

Adenine and guanine are the major purines isolated from nucleic acids and are constituents of both DNA and RNA. Other purines occur in living organisms; these include hypoxanthine, xanthine, and uric acid which are catabolic products of adenine and guanine metabolism. Caffeine (1,3,7-trimethylxanthine) and theobro-

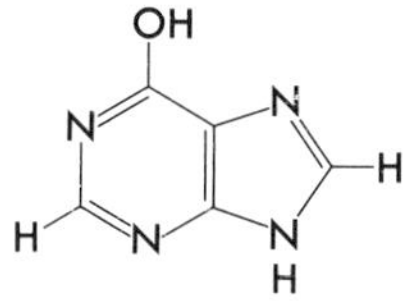

Hypoxanthine 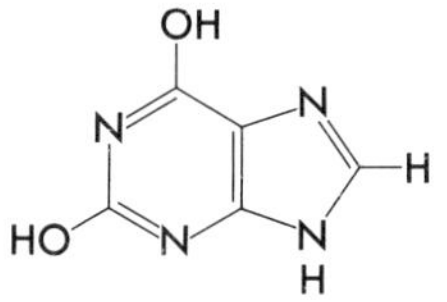　　Xanthine 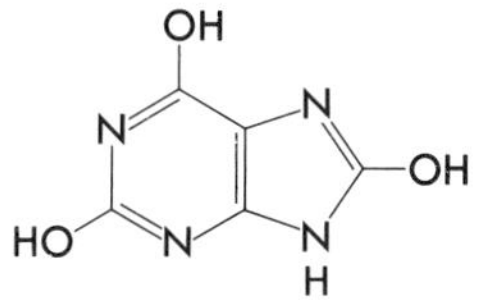　　Uric acid

mine (3,7-dimethylxanthine) are constituents of plants.

Sugars of Nucleic Acids

The classification of the nucleic acids is based on the type of sugar present—deoxyribose in DNA and ribose in RNA.

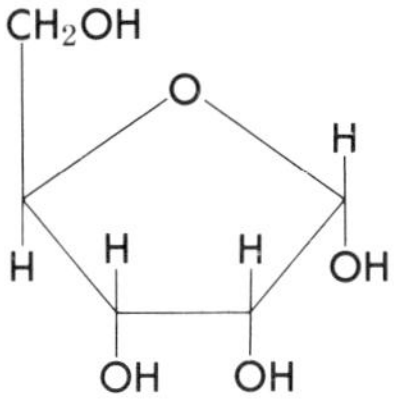

Ribose (α-D-ribofuranose)

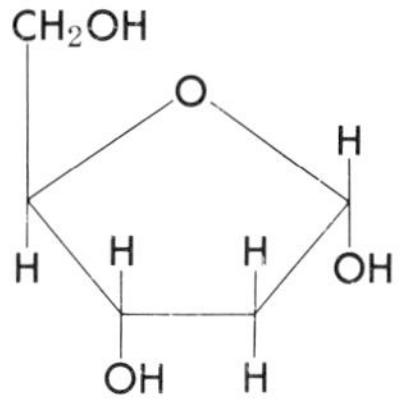

Deoxyribose (α-D-2-deoxyribofuranose)

Assays of nucleic acids are frequently based on their sugar content. The reaction with orcinol and $FeCl_3$ in HCl is used to quantitate RNA on the basis of its ribose content. Similarly the reaction of diphenylamine or indole in HCl with deoxyribose is used for DNA.

Nucleosides

Nucleosides are purines or pyrimidines that have either ribose or deoxyribose attached through a N-β-glycosidic linkage. In the pyrimidines the attachment is to the nitrogen in position 1; in the purines the attachment is to the nitrogen in position

9. Pseudouridine (Ψ.U.), an unusual nucleoside of uracil with a C-glycosidic bond, has been found in transfer RNA (page 289), along with a number of nucleosides of methylated bases.

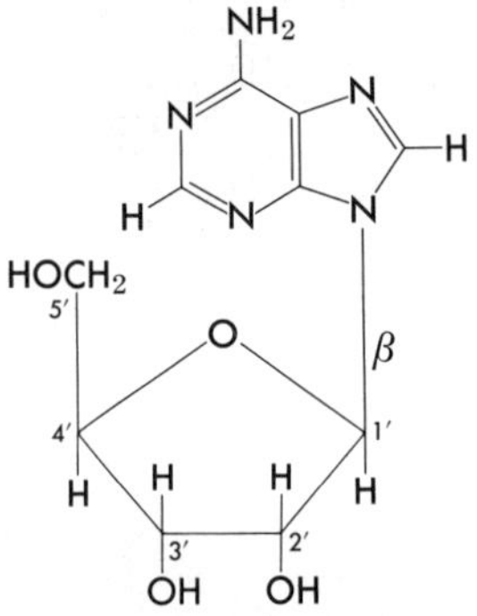

Adenosine
(adenine riboside)

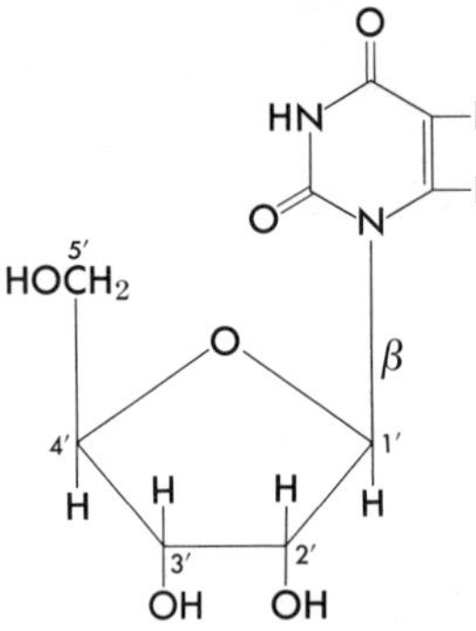

Uridine
(uracil riboside)

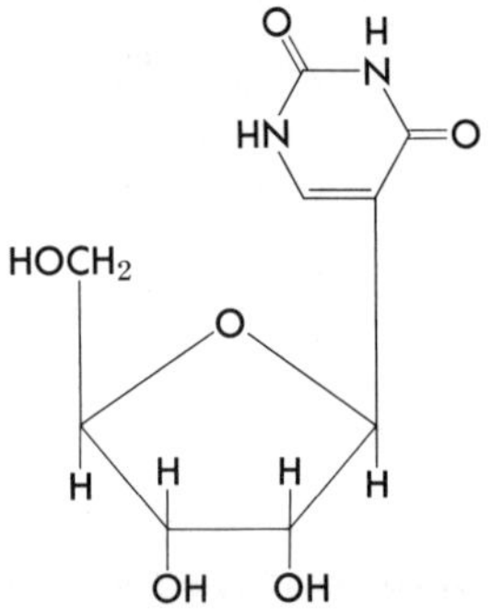

Pseudouridine
(5-β-D-ribofuranosyluridine)

Other nucleosides present in nucleic acids are

Guanosine (guanine riboside)
Cytidine (cytosine riboside)
Deoxycytidine (cytosine deoxyriboside)
Deoxyadenosine (adenine deoxyriboside)
Thymidine (thymine deoxyriboside)

Nucleotides

Nucleotides are sugar-O-phosphate esters of nucleosides. There are three positions in the ribose moiety for esterification, the 2′-, 3′-, and 5′-hydroxyls, whereas deoxyribose has only two positions available, the 3′- and 5′-hydroxyls. All these isomers are found in nature. Adenylic and thymidylic acid are typical examples of this group of compounds.

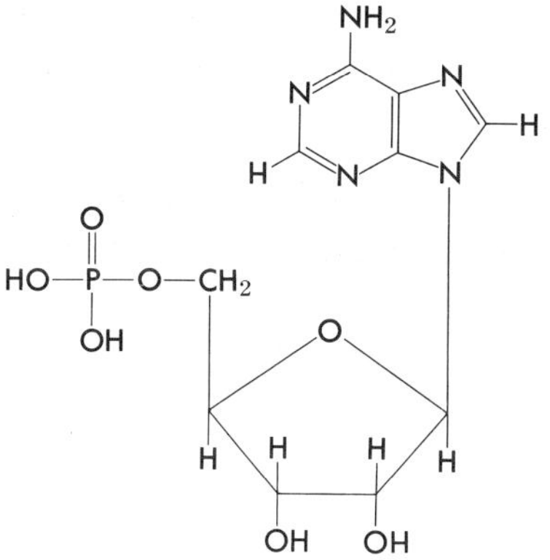

Adenosine-5′-monophosphate
Adenylic acid (AMP)

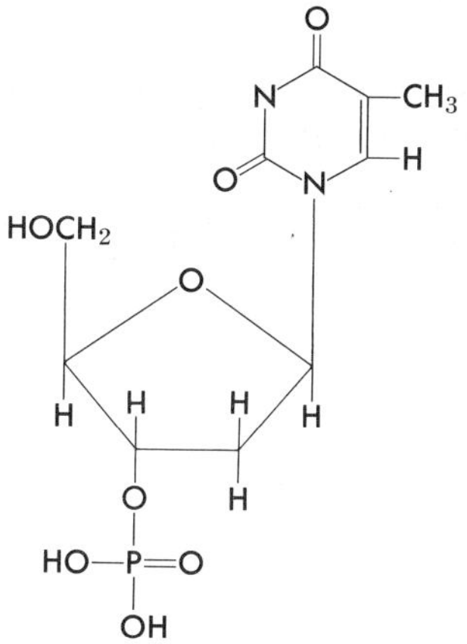

Thymidine-3′-monophosphate

The nucleotides are usually named as the nucleoside phosphates or with reference to the base, for example, guanosine monophosphate (GMP) or guanylic acid, thymidine monophosphate or thymidylic acid, and so on.

In addition cyclic 2′-,3′-phosphates are produced by alkaline hydrolysis of RNA. Adenosine-3′,5′-cyclic phosphate (page 121) is an activator of phosphorylase, and adenosine-2′,5′-diphosphate and adenosine-3′,5′-diphosphate are components respectively of NADP (page 76) and coenzyme A (pages 74–75).

Nucleotides of particular importance in metabolism and transfer of energy are adenosine-5′-diphosphate (ADP) and adenosine-5′-triphosphate (ATP).

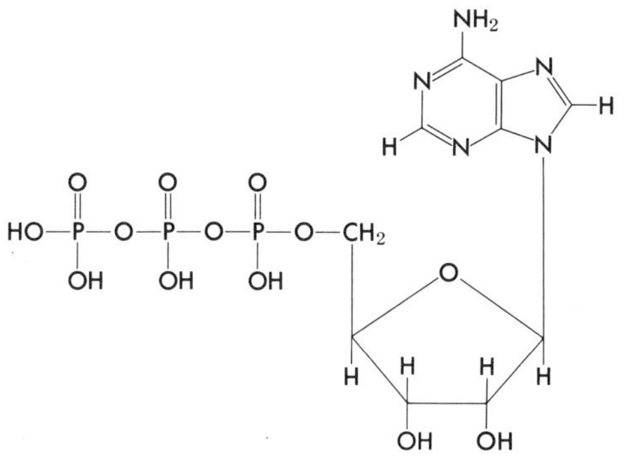

Adenosine-5′-triphosphate (ATP)

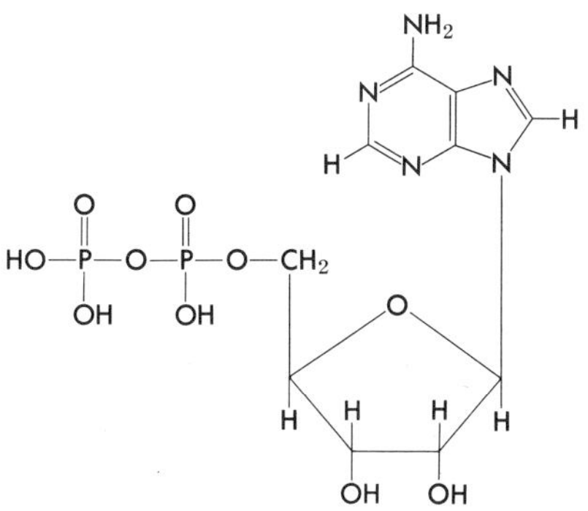

Adenosine-5′-diphosphate (ADP)

Similar 5′-nucleoside di- and triphosphates of other pyrimidine and purines occur in tissues. However, the di- and triphosphates are not components of nucleic acids.

Nucleic Acids—Primary Structure

Both DNA and RNA are linear polymers in which successive nucleotides are joined through a phosphodiester from the 5′ position of one nucleoside to the 3′ position of the next nucleotide. A portion of a RNA molecule is as follows

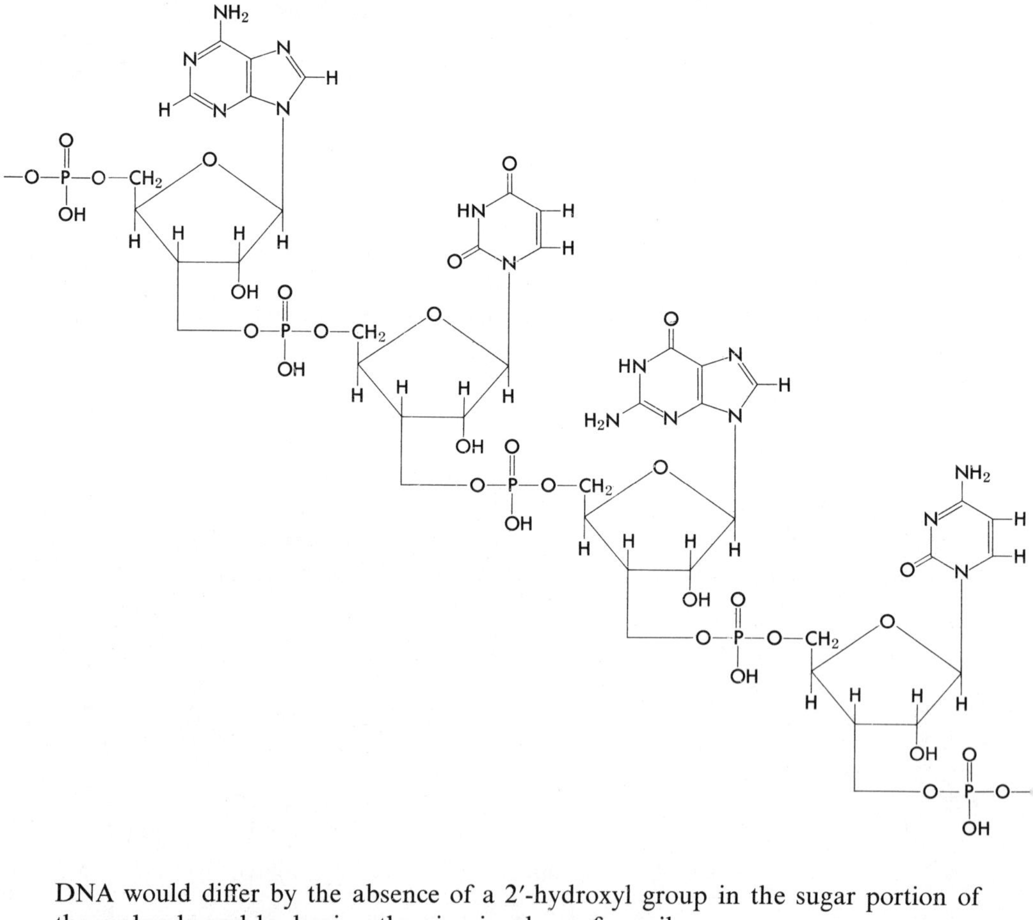

DNA would differ by the absence of a 2′-hydroxyl group in the sugar portion of the molecule and by having thymine in place of uracil.

A convenient shorthand method is used to represent the structure of these linear

polymers, for example

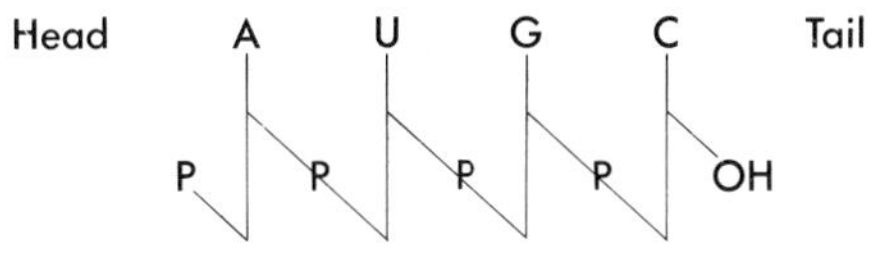

Head A U G C Tail

P P P P OH

where the left end of the chain is the 5′-phosphate terminus and the right end of the chain shows the 3′-hydroxyl group.

Polyribonucleotides are degraded by alkali to yield a mixture of 2′- and 3′-phosphate ribonucleotides. Cyclic 2′,3′-phosphates are formed as intermediates and either of the C—O—P bonds is cleaved by OH^- to produce the mixture of isomers.

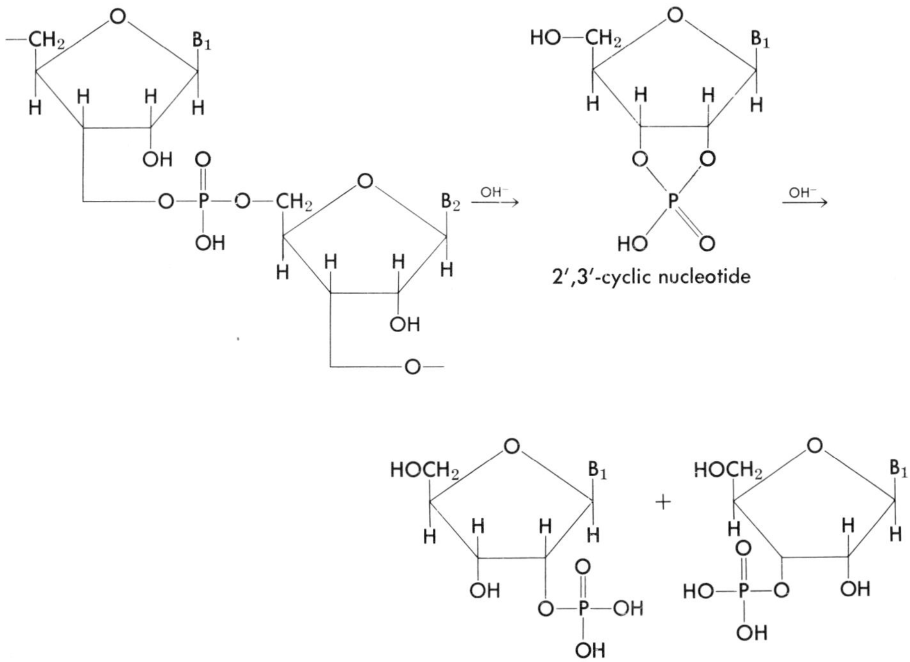

2′,3′-cyclic nucleotide

The fact that the polydeoxyribonucleotides are devoid of a 2′-hydroxyl group for participation in the alkaline hydrolysis explains their stability to alkali. The net effect of this hydrolysis is the transfer of a phosphate group from a 5′-nucleotide to its nearest neighbor to the left.

A consideration of the enzymes that can break down RNA is difficult because of the multiplicity of enzymes and their varying specificities. In general two main groups can be recognized. Group I consists of those enzymes specific for RNA (ribonucleases) in which nucleoside-2′,3′-cyclic phosphates are obligate intermediates. They produce 2′- or 3′-ribonucleotides and have certain specificities towards the base adjacent to the bond cleaved and towards the macromolecular form of RNA (single- or double-stranded). The second group consists of phosphodiesterases and nucleases (sugar-nonspecific enzymes) that do not employ the cyclic phosphate esters as an intermediate in the hydrolysis. They produce 3′- or 5′-phosphate monoesters and have certain specificities towards bases. They may be ribose-specific, sugar-nonspecific, or deoxyribose-specific. The relative activities of an enzyme for RNA and DNA depends on the medium employed and the nature and chain length of the nucleic acids used. Various of these enzymes were employed successfully to determine the nucleotide sequence in yeast alanine and tryosine transfer RNA (pages 289–290).

Although DNA is stable to the action of alkali, it may be degraded by a variety of deoxyribonucleases and nucleases. Bovine pancreas deoxyribonuclease (DNase I) is an endonuclease that cleaves intranucleotide bonds at the 3′ position. Snake venom phosphodiesterase, an exonuclease, liberates 5′-deoxyribonucleotides from the 3′-OH end (tail) of DNA oligodeoxyribonucleotides. The combined action of these two enzymes is to convert DNA almost completely to 5′-deoxyribonucleoside monophosphates. Similarly, the combined action of DNase II (from spleen, thymus, or bacterial sources) and spleen phosphodiesterase yields 3′-deoxyribonucleotide monophosphates.

Structure of DNA

With the exception of a small amount of DNA in mitochondria, the DNA of cells of higher organisms is concentrated in the nucleus. It is estimated that the nuclei of animal cells contain about 4 to 8×10^{-12} g of DNA per nucleus, bacteria about one thousandth (0.001×10^{-12}) of this amount, and DNA-containing viruses from one ten-thousandth (0.0001×10^{-12}) down to one-millionth ($0.000{,}001 \times 10^{-12}$) of this amount. The sizes of DNA molecules remain uncertain within limits due to unknown shearing damage incurred during isolation. If shearing forces are too high, the molecular weight will be low; and if protein removal is not complete, aggregation will produce results that are too high. Within these considerations the molecular weights for the DNAs from higher organisms range from 5 to 12×10^{6}.

E. Chargaff and his collaborators in the period 1949–1953 carried out extensive and careful analyses of the base composition of DNAs from diverse sources. A number of important generalizations about the base composition of DNA were drawn.

1. Regardless of the source, DNA contains equivalent amounts of purines and pyrimidines, that is, A + G = T + C.
2. A = T, G = C, and A + C = G + T. That is, the molar quantity of adenine is equal to that of thymine and that of guanine is equal to cytosine, and the number of 6(4)-amino bases is equal to the number of 6(4)-keto bases.
3. As A + G + C + T = 1, any DNA that deviates from this must have some unusual structural features.
4. The base composition of a DNA is characteristic for the species from which it is derived, and closely related organisms exhibit similar base composition.
5. DNA from different tissues and organs of the same host appears to have identical or very similar base composition.
6. The ratio (A + T)/(G + C) exhibits a wide variation from organism to organism reflecting a variation in base composition. DNA from animal sources always appears to have a ratio greater than 1, whereas microorganisms appear to have an almost continuous spectrum of composition with ratios from 0.35 to 2.70 being obtained.

These chemical analyses suggested some unusual relationship existed between specific bases to account for the observed ratios. Earlier data had shown that DNA gave regular x-ray diffraction patterns which were interpreted as being due to the flat purine and pyrimidine bases being stacked one above the other and arranged at right angles to the long axis of the polynucleotide chain. On the basis of the base-pairing (A = T and G = C) and x-ray diffraction data, J. D. Watson and F. H. C. Crick proposed in 1953 their new model for DNA structure.

The Watson-Crick model (Figure 8.2) for DNA consists of two right-handed helical polynucleotide chains of opposite polarity wrapped around each other in a regular fashion and around the same axis to form a double helix. The purine and pyrimidine bases are on the inside of the helix with the sugar and phosphate backbone on the outside. The chains are held together by hydrogen-bonding between specific base pairs. Only base-pairing between A and T and between C and G can be accomodated spatially. Thus the chains are *complementary* and *antiparallel* so that the sequence in one chain determines the sequence of the other. These considerations are important for the replication of DNA. The following structures illustrate the hydrogen bonding between the bases and the opposite polarity of the two DNA chains.

Thymine

Cytosine

Adenine

Guanine

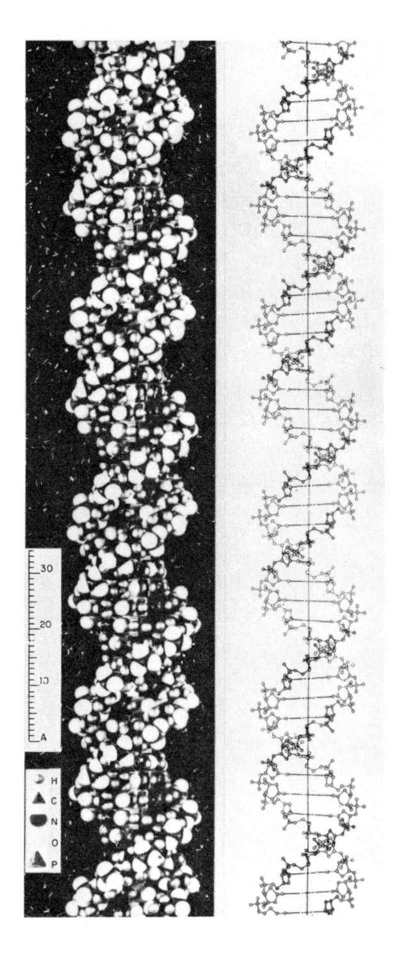

Figure 8.2. Schematic diagram of Watson and Crick structure. *Left:* Space-filling model. *Right:* Projection of model. (From the Nobel Lecture by M. H. F. Wilkins: *Science,* **140:**941, 1963.)

The DNA molecule is extremely extended in neutral salt solutions, having an intrinsic viscosity higher in proportion to its molecular weight than any other known substance. Although the DNA double helix is relatively stable, it can be denatured (helix to coil transition) by heat and extremes of pH. In the complete denaturation of DNA the two strands are separated, the molecular weight is halved, the viscosity and optical rotation are markedly decreased, and the absorbance at 260 mμ is increased. Reversion (renaturation) to the double helical form can be effected under suitable conditions. This reconstitution occurs only between DNA strands which are derived from the same or closely related organisms and apparently requires near perfect complementarity of bases. Such a technique (hybridization) has been used not only to detect complementary nucleotide sequence between single-stranded DNA samples, but also to establish complementary nucleotide sequences between DNA and various RNA molecules.

Structure of RNAs

RNA is found in all fractions of the cell and at least three main types are known. Some 60 to 80 per cent of the RNA in a cell is accounted for by ribosomal RNA (rRNA). Ribosomes may exist free or in aggregates called polysomes which in turn may be attached to the endoplasmic reticulum. The ribosomes are necessary for protein synthesis and in this process move along a strand of a second type of RNA that provides the template for specific protein synthesis. This is called messenger RNA and is synthesized on and complementary to one of the strands of a specific segment of DNA. *Transfer RNA* (the third type) recognizes a specific site on the messenger RNA, and it transfers a specific amino acid to the growing peptide chain.

Ribosomal RNAs (rRNA)

Ribosomes from different cells vary somewhat in both composition and size. They usually contain about 40 per cent protein and 60 per cent RNA. Bacterial ribosomes have a *sedimentation coefficient(s)* (page 42) of 70S and may be dissociated into two functionally different subunits (page 316) a larger 50S subunit and a smaller 30S subunit. The 50S ribosome subunit contains a single 23S RNA molecule, a single 5S RNA molecule, and some 30 protein molecules. The 30S ribosome subunit contains one 16S RNA molecule and 19 or 20 different protein molecules. Animal ribosomes exist as 80S monosomes and form 60S and 35S subunits upon dissociation.

The 23S and 16S ribosomal RNAs, although similar in base composition, appear to have different nucleotide sequences. rRNA has various minor components such as methylated bases and pseudouridylic acid. These molecules are single-stranded and have regions containing hydrogen-bonded AU and GC pairs and regions where no such structures are possible. As a result the molecule has some helical content and structural stability.

The 5S RNA appears to be a structural component of the 50S subunit and may be involved in the assembly of that subunit.

It has been suggested that rRNA codes for ribosomal proteins. Although it is certain that rRNA is synthesized on a DNA template (page 310), the available evidence does not prove that ribosomal proteins are coded by rRNA.

Messenger RNA (mRNA)

mRNAs are found in both the nucleus and cytoplasm of the cell. They are single-stranded molecules having a base composition complementary to a portion of one DNA strand. Relatively little is known about their secondary structure, although it is likely that the strand is folded on itself in some places and thus has some regions of double helical structure. The role of mRNA in protein synthesis is discussed in Chapter 9.

Transfer RNA (tRNA)

This family of RNAs serves the function of transferring the activated amino acid to the ribosome for assembly into polypeptide chains. There is at least one tRNA for each of the commonly occurring amino acids. The tRNAs are small polyribonucleotides with molecular weights of about 25,000. Among the 75 to 85 nucleotides making up the tRNAs are a number of unusual bases such as pseudouridylic acid, 5,6-dihydrouridine, N^2-dimethylguanosine, and others. The nucleotide sequences of six tRNA molecules are available at present. All the known tRNA sequences can be arranged in a cloverleaf structure as shown for the structures of yeast alanine and tyrosine tRNAs in Figure 8.3. The C—C—AOH sequence at the tail of the chain is common to all tRNAs and carries the amino acid as an acyl ester to the 3′ position of adenosine. The anticodons are believed to be I—G—C for alanine and G—Ψ—A for tyrosine and are found in the same positions in the bottom loops of the molecules. It is usual to refer to the code groups in mRNA as the primary coding units. Thus, the complementary group or the tRNA is an anticodon. Since mRNA is complementary to DNA this makes tRNA codons identical with DNA codons.

Metabolism

It has been known for many years that nucleic acids are synthesized by animal tissues from smaller metabolic precursors and that purines and pyrimidines need not be supplied in the diet. Man, for example, excretes a far greater quantity of purines than is ingested in the diet. Cogent proof for the synthesis of nucleic acids and their constituents from simple dietary components was provided by nutritional experiments in which animals achieved a normal growth rate on synthetic diets lacking purines and pyrimidines. We consider in this section the following main topics: (1) the biosynthesis of purine and pyrimidine nucleotides, (2) the polymerization of these nucleotides into nucleic acids, (3) the catabolism of purines and pyrimidines, and (4) some aspects of the overall metabolism and rolls of nucleic acids. The roles of nucleic acids in protein synthesis are discussed in Chapter 9.

DE NOVO BIOSYNTHESIS OF NUCLEOTIDES

Purine Ribonucleotides

Although most of our knowledge of the mechanism of the *de novo* synthesis of purines has been derived from studies on pigeon liver, a similar pathway appears

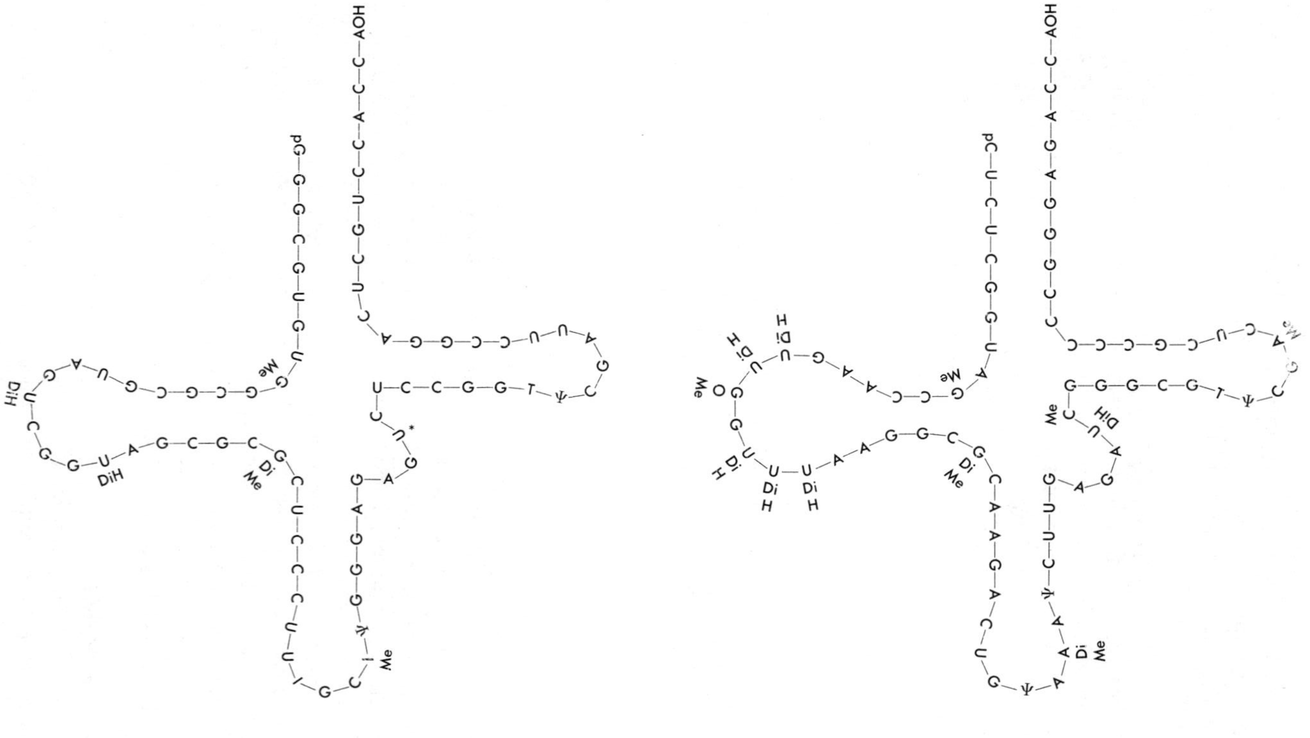

Ala t-RNA

Tyr t-RNA

Figure 8.3. Proposed cloverleaf structure for yeast alanine and tyrosine tRNAs.

to operate in all animal cells. The determination of the metabolic origin of most of the carbon and nitrogen atoms of the purine ring preceded the studies with isolated enzyme systems. The origin of the various atoms is shown in Figure 8.4. It will be noted that the carboxyl, α-carbon and nitrogen atoms of glycine give rise respectively to carbon atoms 4 and 5 and nitrogen atom 7 of the purine ring. Active formate (N^5,N^{10}-methenyltetrahydrofolic acid or N^{10}-formyltetrahydrofolic acid) is the precursor of carbon atoms 2 and 8, and carbon dioxide (or bicarbonate) is the precursor of carbon atom 6 of the purine ring. The amide nitrogen of glutamine is the source of nitrogen atoms 3 and 9. Nitrogen atom 1 is derived from the amino nitrogen of aspartic acid. The detailed enzymic pathway for the synthesis of purine ribonucleotides was elucidated primarily by J. Buchanan and G. R. Greenberg. In the representation of the structures to follow, Ⓟ is used to represent the phosphoryl group ($-PO_3H_2$).

The initial step in the biosynthesis of purine ribonucleotides is the transfer of a pyrophosphate group from ATP to ribose-5-phosphate to yield 5-phosphoribosyl-1-pyrophosphate (PRPP).

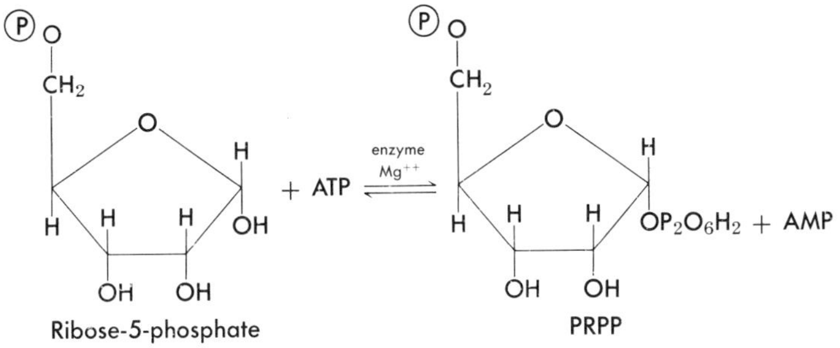

$$\text{Ribose-5-phosphate} + \text{ATP} \xrightleftharpoons[]{\text{enzyme, } Mg^{++}} \text{PRPP } (C_1\text{-}OP_2O_6H_2) + \text{AMP} \qquad (1)$$

The second reaction is the formation of *5-phosphoribosyl-1-amine* from PRPP and glutamine.

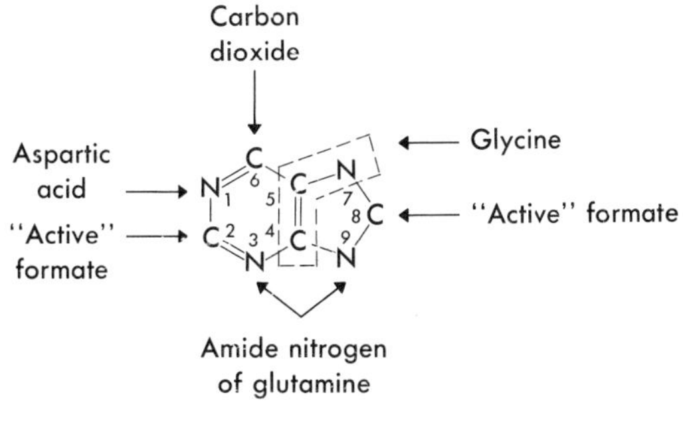

Figure 8.4. Precursors of purine ring atoms.

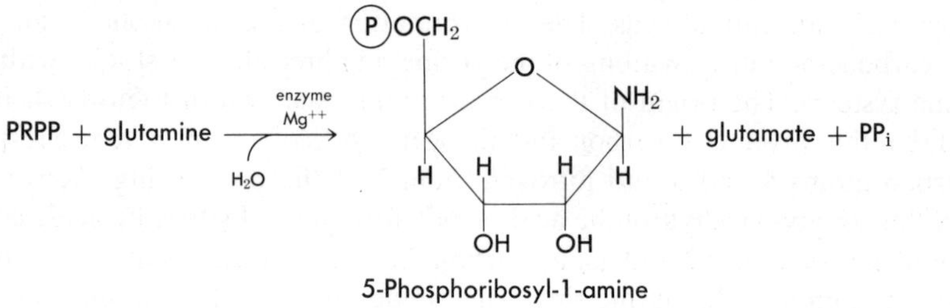

PRPP + glutamine —(enzyme, Mg^{++}; H_2O)→ [ⓅOCH_2 ribose ring: O; NH_2; H, H, H, H; OH, OH] + glutamate + PP_i (2)

5-Phosphoribosyl-1-amine

It should be noted that an inversion has occurred in this reaction and that the amino sugar has the β-configuration.

In the next reaction, 5-phosphoribosyl-1-amine is conjugated with glycine to give *glycinamide ribonucleotide.*

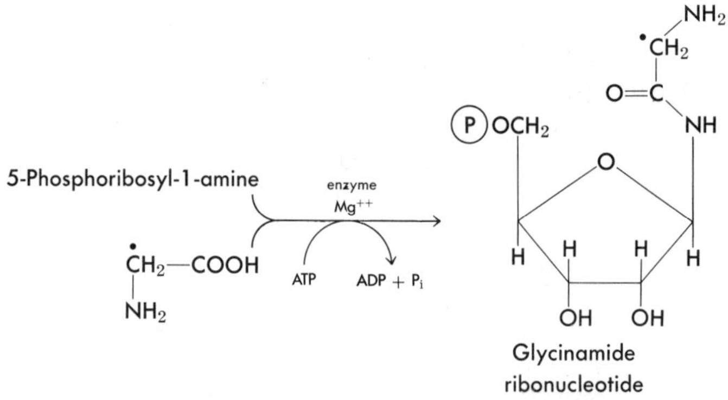

5-Phosphoribosyl-1-amine + $\overset{\bullet}{C}H_2$—COOH (with NH_2) —(enzyme, Mg^{++}; ATP → ADP + P_i)→ [NH_2—$\overset{\bullet}{C}H_2$—C(=O)—NH— on ⓅOCH_2 ribose ring: O; H, H, H, H; OH, OH] (3)

Glycinamide ribonucleotide

In the fourth reaction, N^5,N^{10}-methenyltetrahydrofolic acid transfers a formyl group to glycinamide ribonucleotide to form *α-N-formylglycinamide ribonucleotide.*

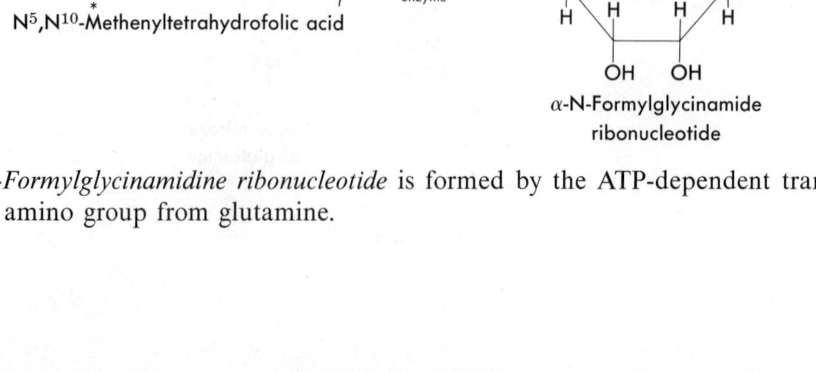

Glycinamide ribonucleotide + N^5,N^{10}-$\overset{*}{M}$ethenyltetrahydrofolic acid —(enzyme; H_2O → FAH_4)→ [$\overset{*}{C}HO$—NH—$\overset{\bullet}{C}H_2$—C(=O)—NH— on ⓅOCH_2 ribose ring: O; H, H, H, H; OH, OH] (4)

α-N-Formylglycinamide ribonucleotide

N-Formylglycinamidine ribonucleotide is formed by the ATP-dependent transfer of a amino group from glutamine.

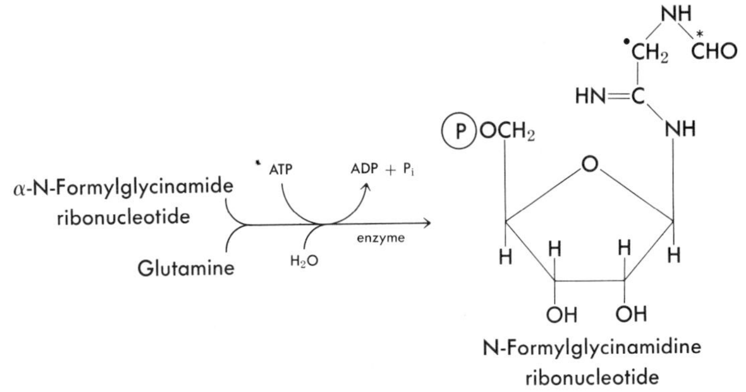

α-N-Formylglycinamide ribonucleotide
Glutamine
ATP
ADP + P_i
H_2O
enzyme

NH
$\cdot CH_2$ $\overset{*}{C}HO$
HN=C
NH
(P)OCH_2
O
H H H H
OH OH

N-Formylglycinamidine ribonucleotide

(5)

Ring closure to *5-aminoimidazole ribonucleotide* is accomplished by an ATP-dependent dehydration.

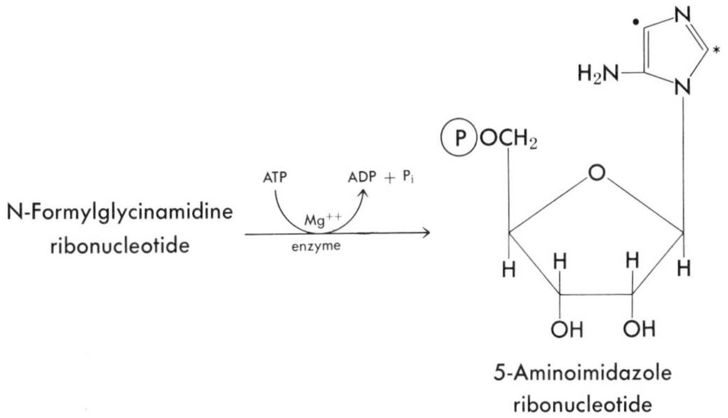

N-Formylglycinamidine ribonucleotide
ATP
ADP + P_i
Mg^{++}
enzyme

N
H_2N
N
(P)OCH_2
O
H H H H
OH OH

5-Aminoimidazole ribonucleotide

(6)

5-Aminoimidazole-4-carboxylic acid ribonucleotide is formed from bicarbonate and 5-aminoimidazole ribonucleotide.

5-Aminoimidazole ribonucleotide
enzyme
$\overset{\Delta}{CO_2}$

$\overset{\Delta}{HOOC}$
N
H_2N
N
(P)OCH_2
O
H H H H
OH OH

5-Aminoimidazole-4-carboxylic acid ribonucleotide

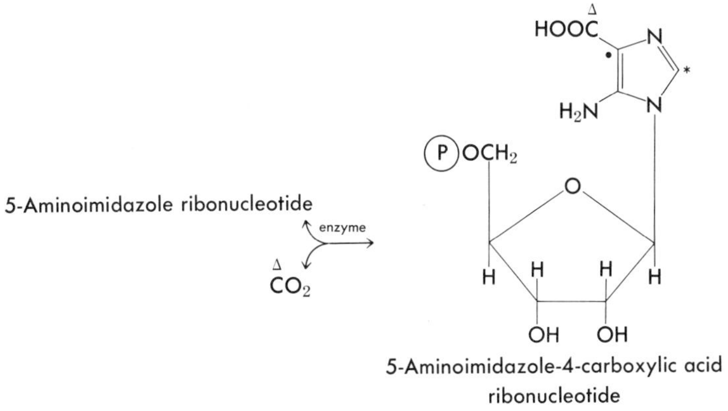

(7)

Aspartic acid is added to the aminoimidazole carboxylic acid derivative in an ATP-dependent process to form *5-aminoimidazole-4-N-succinocarboxamide ribonucleotide.*

5-Aminoimidazole-4-carboxylic acid ribonucleotide + aspartic acid + ATP

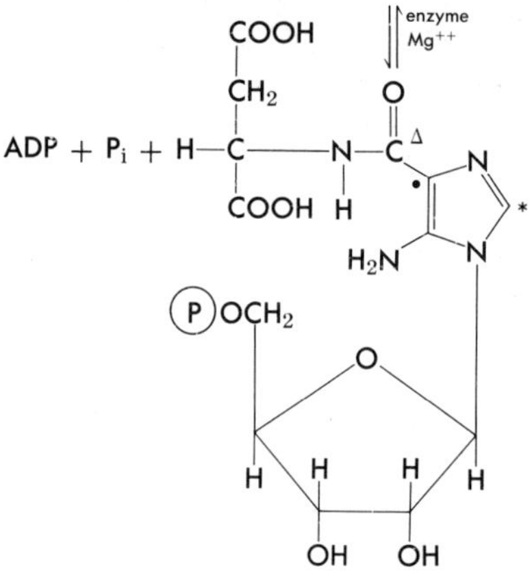

⇅ enzyme, Mg^{++}

$ADP + P_i + H{-}C(CH_2COOH)(COOH){-}N(H){-}C(=O)$– [imidazole ring with H_2N; ribose 5′-phosphate: ⓅOCH_2, H, H, H, H, OH, OH] (8)

5-Aminoimidazole-4-N-succinocarboxamide ribonucleotide

The conversion to *5-aminoimidazole-4-carboxamide ribonucleotide* is achieved by the elimination of fumaric acid.

5-Aminoimidazole-4-N-succinocarboxamide ribonucleotide $\overset{\text{enzyme}}{\rightleftharpoons}$

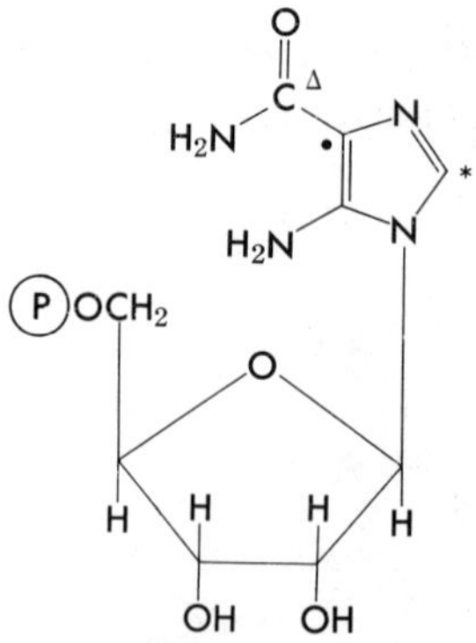

$H_2N{-}C(=O)$– [imidazole ring with H_2N; ribose 5′-phosphate: ⓅOCH_2, H, H, H, H, OH, OH] + Fumarate (9)

5-Aminoimidazole-4-carboxamide ribonucleotide

The last carbon atom is introduced by the reaction of N^{10}-formyltetrahydrofolic acid and aminoimidazole carboxamide ribonucleotide.

5-Aminoimidazole-4-carboxamide ribonucleotide + N^{10}-formyltetrahydrofolic acid $\xrightarrow[\text{enzyme}]{K^+}$

tetrahydrofolic acid + [structure: O, H_2N, CHC, N, H, N, N, (P)OCH_2, O, H, H, H, H, OH, OH] (10)

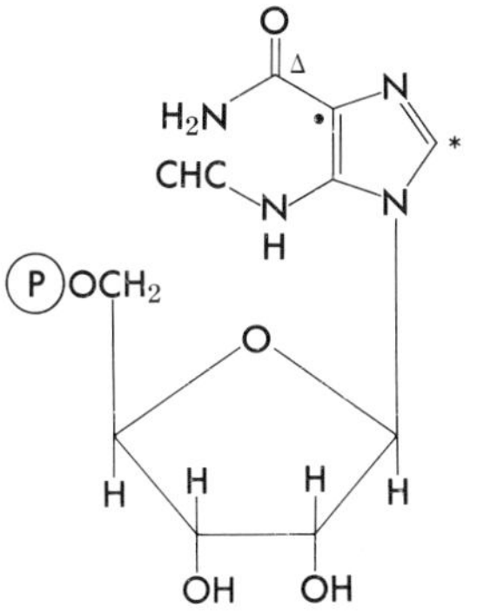

5-Formamidoimidazole-4-carboxamide ribonucleotide

Formation of the purine ring is completed by dehydration and ring closure.

5-Formamidoimidazole-4-carboxamide ribonucleotide $\overset{\text{enzyme}}{\rightleftharpoons}$ [structure: O, HN, N, N, N, (P)OCH_2, O, H, H, H, H, OH, OH] + H_2O (11)

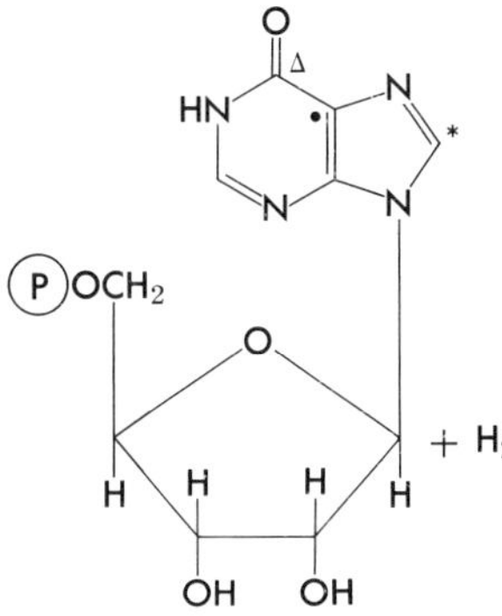

Inosinic acid

There are eleven enzymatic steps in the biosynthetic pathway from ribose-5-phosphate to inosinic acid, each catalyzed by a specific enzyme. Five ATPs participate directly (steps 1, 3, 5, 6, and 8), and four ATPs are involved indirectly, two for the synthesis of the nitrogen donor glutamine (steps 2 and 5) and two for the conversion of formate to N^5, N^{10}-methenyltetrahydrofolate or N^{10}-formyltetrahydrofolate (page

232) which are used respectively in steps 4 and 10. Thus a minimum of nine equivalents of ATP are used to drive the synthetic reaction.

It is important to note that in no stage in the pathway of *de novo* synthesis is a free purine base involved. All reactions utilize phosphoribosyl derivatives, the purine ring being completed only as the purine ribonucleotide. The biosynthesis of the purine ring is not separable from purine nucleotide synthesis, there being no systems known for the biosynthesis of a free purine as such. This is in contrast to the synthesis of the pyrimidine ring in which a free pyrimidine base is first formed and subsequently converted to the nucleotide.

Inosinic acid is the key purine nucleotide from which are derived adenylic and guanylic acids. These conversions are shown in Figure 8.5.

The conversion of inosinic acid to adenylic acid is similar to the conversion of 5-aminoimidazole-4-carboxylic acid ribonucleotide to 5-aminoimidazole-4-carboxamide ribonucleotide (reactions 8 and 9, page 294). The enzyme (adenylosuccinase) catalyzing the elimination of fumarate from adenylosuccinate is probably the same enzyme that catalyzes the elimination of fumarate from 5-aminoimidazole-4-N-succinocarboxamide ribonucleotide (reaction 9, page 294).

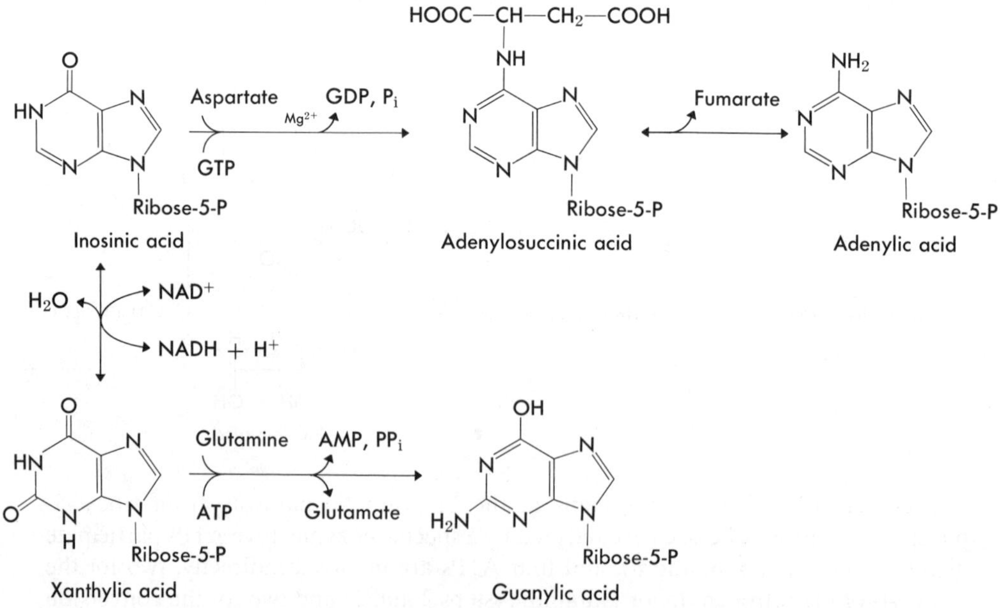

Aspartate, GTP → (Mg^{2+}) GDP, P_i
Fumarate
Inosinic acid
Adenylosuccinic acid
Adenylic acid
H_2O, NAD^+ → $NADH + H^+$
Glutamine, ATP → AMP, PP_i, Glutamate
Xanthylic acid
Guanylic acid

Figure 8.5. Conversion of inosinic acid to adenylic and guanylic acids.

Pyrimidine Ribonucleotides

The initial step in the biosynthesis of pyrimidines is the formation of carbamyl phosphate, which is also the first intermediate in the formation of urea (page 223).

$$CO_2 + NH_3 + 2\,ATP \xrightarrow[Mg^{++}]{\text{enzyme, acetylglutamate}} H_2N{-}\overset{\displaystyle O}{\overset{\|}{C}}{-}OPO_3H_2 + 2\,ADP + P_i$$

In the second step, catalyzed by aspartate-carbamyl transferase, the carbamyl moiety of carbamyl phosphate is transferred to the amino group of aspartate.

$$H_2N{-}\overset{\displaystyle O}{\overset{\|}{C}}{-}OPO_3H_2 + HOOC{-}CH_2{-}CH(NH_2){-}COOH \underset{}{\overset{\text{enzyme}}{\rightleftharpoons}} H_2N{-}C({=}O){-}NH{-}CH(COOH){-}CH_2{-}COOH + P_i \qquad (1)$$

Carbamyl phosphate — Aspartic acid — Carbamyl aspartic acid

Dihydroorotic acid is formed in a reversible reaction catalyzed by dihydroorotase.

$$H_2N{-}C({=}O){-}NH{-}CH(COOH){-}CH_2{-}COOH \overset{\text{enzyme}}{\rightleftharpoons} \text{Dihydroorotic acid} + H_2O \qquad (2)$$

Carbamyl aspartic acid — Dihydroorotic acid

Dihydroorotic acid is oxidized to orotic acid by dihydroorotic dehydrogenase. The enzyme is a flavoprotein and in linked to NAD.

$$\text{Dihydroorotic acid} + NAD^+ \overset{\text{enzyme}}{\rightleftharpoons} \text{Orotic acid} + NADH + H^+ \qquad (3)$$

Dihydroorotic acid — Orotic acid

Ribose-5′-phosphate is now attached to the pyrimidine ring in the reaction catalyzed by orotidine-5′-phosphate pyrophosphorylase.

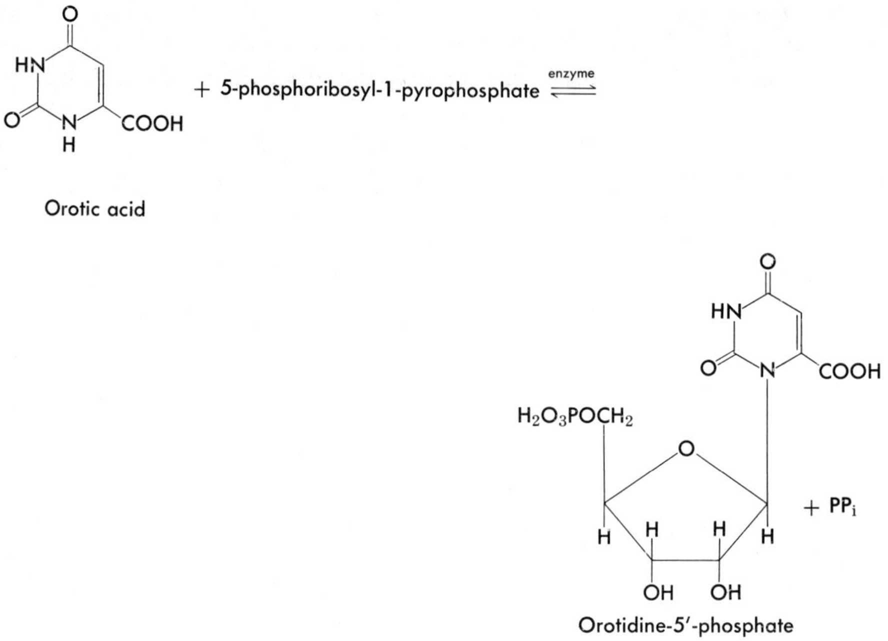

Orotic acid + 5-phosphoribosyl-1-pyrophosphate $\xrightleftharpoons{\text{enzyme}}$ Orotidine-5′-phosphate + PP_i (4)

An irreversible decarboxylation forms uridylic acid.

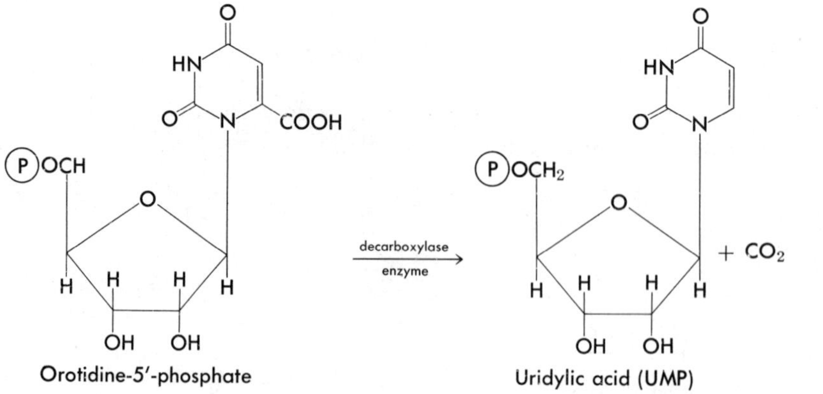

Orotidine-5′-phosphate $\xrightarrow[\text{enzyme}]{\text{decarboxylase}}$ Uridylic acid (UMP) + CO_2 (5)

The biosynthesis of cytidine ribonucleotides proceeds via the amination of uridine-5′-triphosphate.

$$(1)\ \text{UMP} + \text{ATP} \xrightleftharpoons{\text{kinase}} \text{UDP} + \text{ADP}$$

$$(2)\ \text{UDP} + \text{ATP} \xrightleftharpoons{\text{kinase}} \text{UTP} + \text{ADP}$$

$$(3)\ \text{UTP} + \text{ATP} + \text{NH}_3 \xrightarrow[\text{Mg}^{++}]{\text{GTP}} \text{ADP} + \text{P}_i + \quad (6)$$

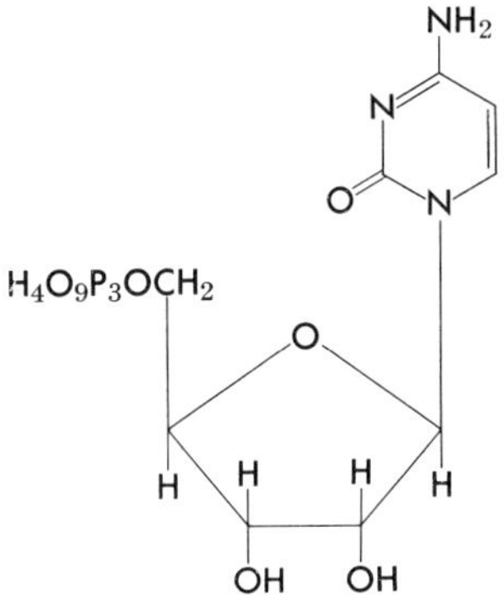

NH_2, N, O, N, $H_4O_9P_3OCH_2$, O, H, H, H, H, OH, OH

Cytidine triphosphate (CTP)

Biosynthesis of Purine and Pyrimidine Deoxyribonucleotides

Although the details are not as yet known, the preponderance of evidence indicates that the deoxyribonucleotides of adenine, guanine, cytosine, and uracil are derived by reduction from the corresponding ribonucleotides. Evidence is available which supports the following reactions

$$\text{GMP} \xrightarrow{+2\text{H}} \text{deoxy GMP}$$

$$\text{AMP} \xrightarrow{+2\text{H}} \text{deoxy AMP}$$

$$\text{CMP} \xrightarrow{+2\text{H}} \text{deoxy CMP}$$

$$\text{UMP} \xrightarrow{+2\text{H}} \text{deoxy UMP}$$

The conversion of deoxy UMP to thymidine-5′-phosphate is discussed on page 231.

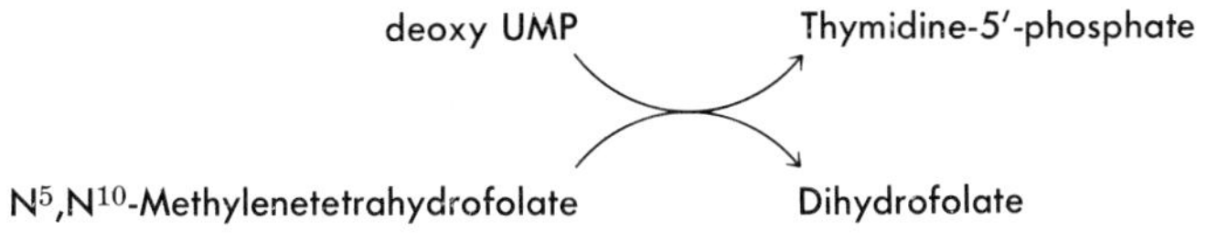

deoxy UMP → Thymidine-5′-phosphate

N^5,N^{10}-Methylenetetrahydrofolate → Dihydrofolate

Formation of Nucleoside Di- and Triphosphates

There are nucleoside monophosphate kinases present in mitochondria and the cytoplasm of the cell which catalyze the formation of nucleoside diphosphates according to the following general reaction in which Z can be adenosine, guanosine, cytidine, uridine, the corresponding deoxy compounds, and thymidine.

$$\text{ATP} + \text{ZMP} \rightleftharpoons \text{ADP} + \text{ZDP}$$

The nucleoside diphosphates can in turn be further phosphorylated to the corresponding triphosphates by nucleoside diphosphate kinases.

$$\text{ZDP} + \text{ATP} \rightleftharpoons \text{ADP} + \text{ZTP}$$

These reactions provide for the synthesis of deoxyribonucleoside triphosphates which are the immediate precursors of DNA and for the synthesis of ribonucleoside triphosphates which are precursors of RNA.

Utilization of Preformed Purines and Pyrimidines

There are in addition to the *de novo* pathways, pathways that utilize intact bases for nucleotide synthesis. These pathways appear to be of special importance in certain tissues, for example, brain. For the purines, the following enzymic pathways are available

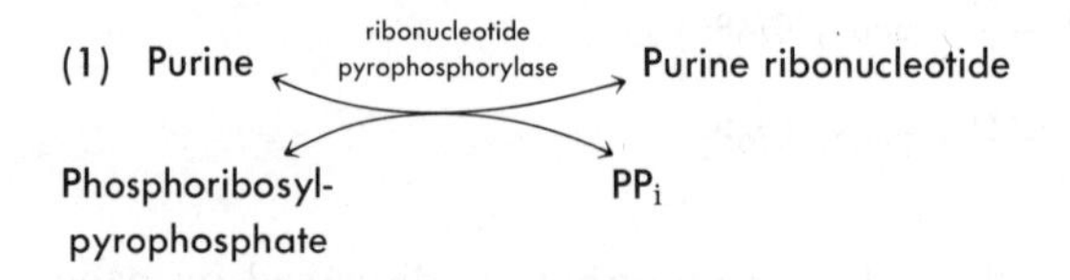

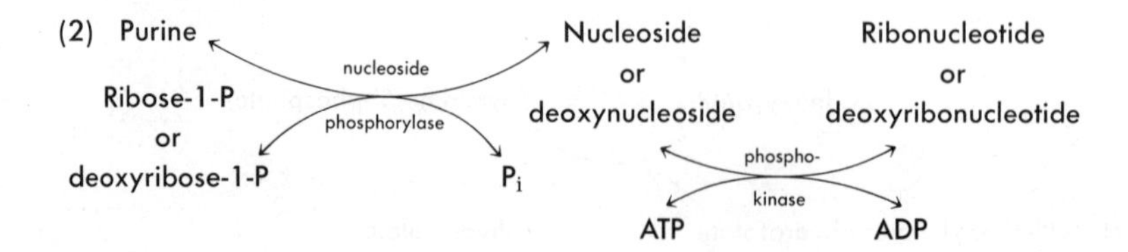

In the case of intact pyrimidines the reactions known to occur are

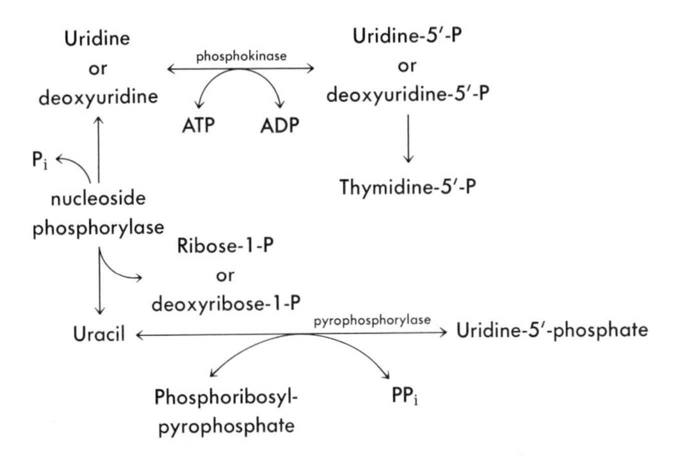

Relatively little is known about the enzymic reactions of cytosine. It is apparently not reactive in either the nucleoside phosphorylase reaction or in the pyrophosphorylase reaction.

Catabolism of Purines and Pyrimidines

Purines

In man and other primates, the purine nucleotides are degraded stepwise to uric acid which is excreted largely in the urine. The pathway in liver leading to uric acid is shown in Figure 8.6. Similar reactions are available for deoxyderivatives. A very active guanase hydrolyzes guanine to xanthine. In the case of adenine an active adenase does not seem to exist and the deamination occurs at the levels of adenylic acid and adenosine. Hypoxanthine and xanthine are converted to uric acid by the action of the flavin enzyme, xanthine oxidase. Man and other primates do not have an active uricase and do not metabolize uric acid further. A portion of the uric acid in man may be excreted into the intestine where it can be broken down by the bacterial flora to allantoin, allantoic acid, and urea plus glyoxylate.

The normal serum level of uric acid is 3 to 6.5 mg per 100 ml for men and 2.5 to 5.5 mg per 100 ml for women. Some 200 to 400 mg of uric acid is excreted daily in the urine of the normal individual. These levels are elevated in gout, a disease in which there is hyperuricemia (elevated blood uric acid) and the deposition of crystalline monosodium urate in cartilage and the kidney. The metabolic defect in

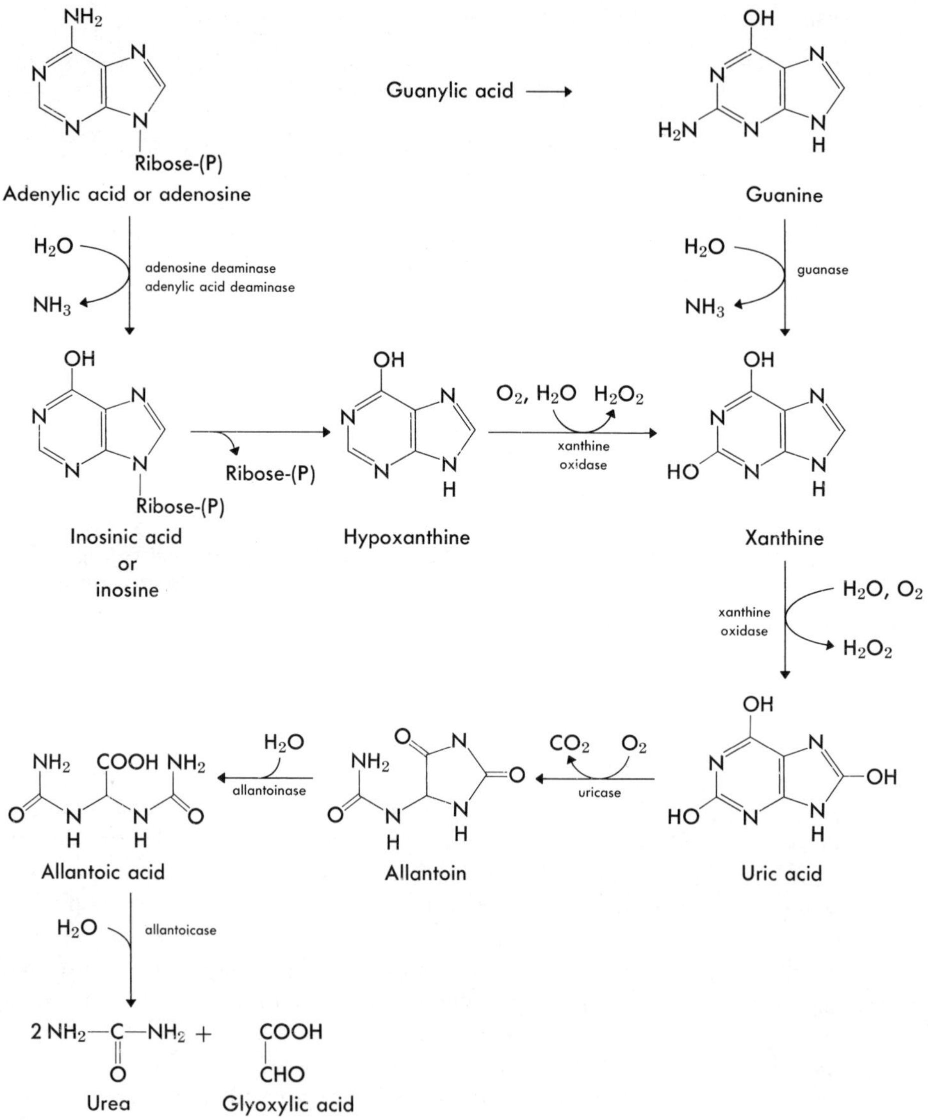

Adenylic acid or adenosine

H_2O → NH_3 (adenosine deaminase, adenylic acid deaminase)

Guanylic acid ⟶ Guanine

H_2O → NH_3 (guanase)

Inosinic acid or inosine → Ribose-(P) + Hypoxanthine

Hypoxanthine → Xanthine (O_2, H_2O → H_2O_2; xanthine oxidase)

Xanthine → Uric acid (xanthine oxidase; H_2O, O_2 → H_2O_2)

Uric acid → Allantoin (uricase; O_2 → CO_2)

Allantoin → Allantoic acid (H_2O; allantoinase)

Allantoic acid → (H_2O; allantoicase)

$2\ NH_2—C(=O)—NH_2$ + COOH–CHO

Urea   Glyoxylic acid

Figure 8.6. Catabolism of purines.

gout is thought to be an overproduction of uric acid, possibly as a consequence of a failure in the feedback control of the synthesis of phosphoribosyl-1-amine (page 292).

Other species excrete various end products: allantoin by mammals other than primates, allantoic acid by teleost fish, and urea + glyoxylate by amphibia and most fish.

Pyrimidines

In the case of the pyrimidines, little is known about the conversion of pyrimidine nucleotides to free pyrimidine bases. Reactions similar to those for the purine nucleotides presumably effect the same types of transformations. The free pyrimidine bases are metabolized in liver according to the scheme shown in Figure 8.7. The malonic semialdehyde derived from β-alanine may be converted to propionyl-SCoA and to succinyl-CoA via methylmalonyl-SCoA (page 187). The methylmalonyl semialdehyde may be oxidized to methylmalonyl-SCoA and converted to succinate.

Control of Purine and Pyrimidine Biosynthesis

The first irreversible reaction involved directly in purine synthesis is that concerned with the formation of 5-phosphoribosyl-1-amine. This reaction, catalyzed by glutamine phosphoribosylpyrophosphate amidotransferase, is inhibited by various adenine and guanine nucleotides. Thus a feedback control of *de novo* purine synthesis of purine ribonucleotides is quite likely, and the formation of phosphoribosylamine may be the rate-limiting step in purine synthesis. In addition, the formation of GMP from IMP (Figure 8.5) is inhibited by GMP, suggesting the possibility of a feedback regulation of GMP synthesis from IMP. Since ATP is required for the synthesis of GMP and GTP is required for the synthesis of AMP (Figure 8.6), a balance between GMP and AMP synthesis may be achieved by positive feedback control. These relationships are shown in the following diagram.

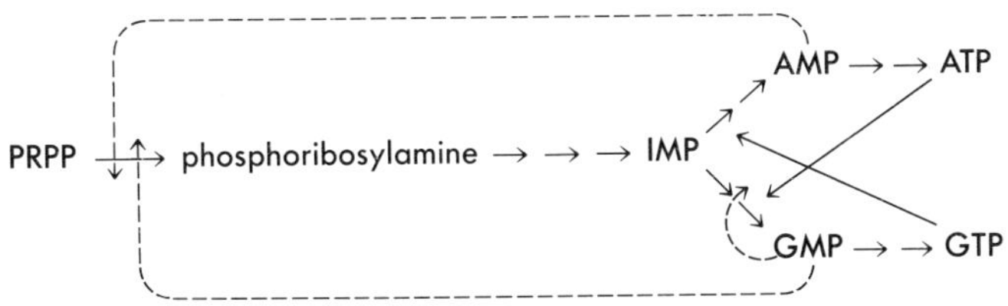

AMP → → ATP

PRPP → phosphoribosylamine → → → IMP

GMP → → GTP

In *de novo* pyrimidine biosynthesis, the most likely point for exercise of control is the aspartate carbamyl transferase reaction (page 297). This reaction is favored because of its essential irreversibility and because it is the first reaction specific for pyrimidine biosynthesis (see also page 68). Although this enzyme has been shown in bacteria to be under feedback control by various nucleotides, such a mechanism can only be inferred in mammalian systems.

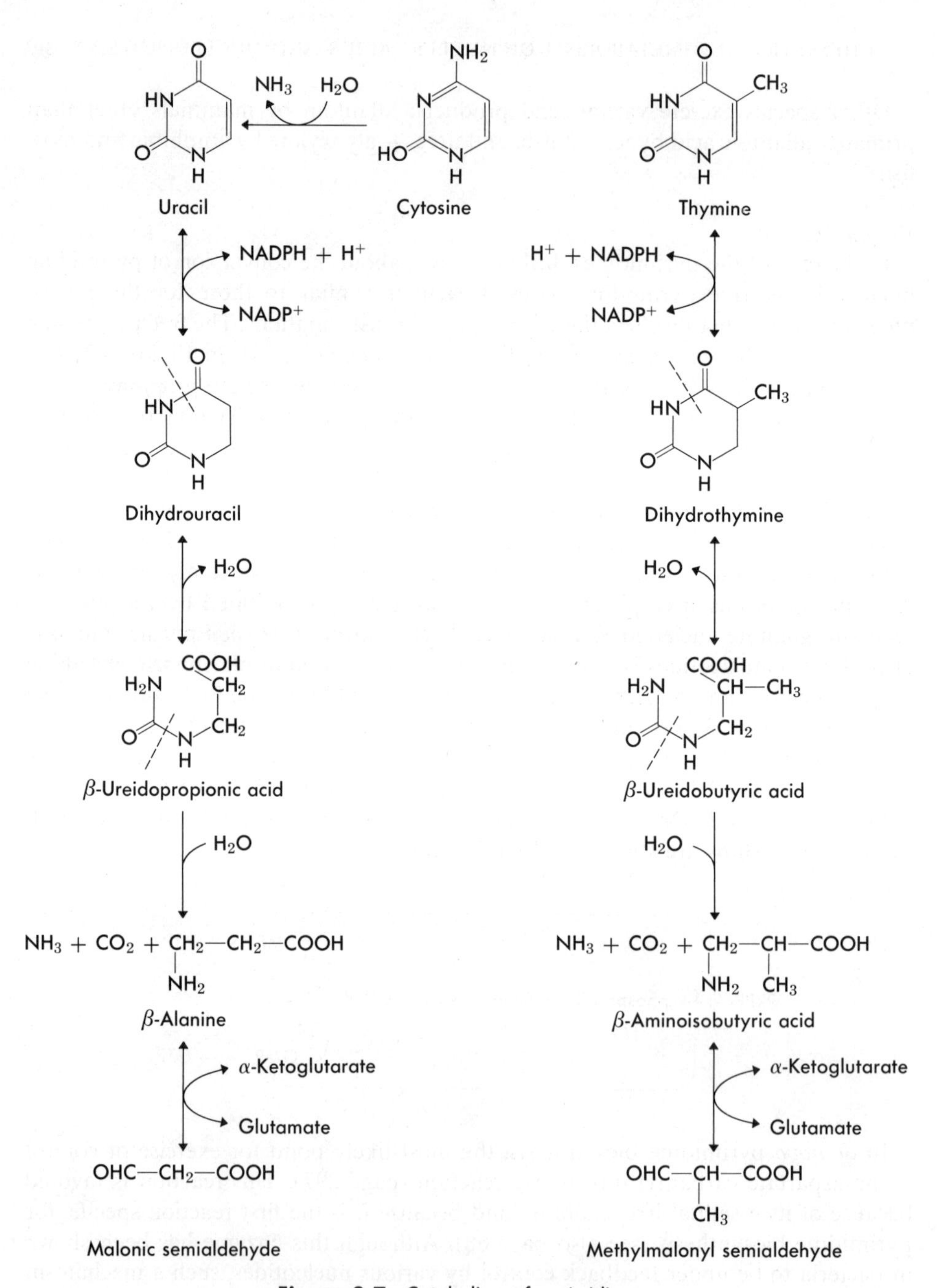

Figure 8.7. Catabolism of pyrimidines.

SOME ASPECTS OF THE OVERALL METABOLISM AND ROLES OF NUCLEIC ACIDS

DNA

Many of the earlier findings of DNA metabolism may be summarized by stating that DNA is metabolically relatively stable and is normally produced in rather strict relationship to cell division. Intact purines and pyrimidines, their derivatives, and their precursors (glycine, formate, NH_3, CO_2, and so on) are not incorporated as readily into tissue DNA as into RNA. It is reasonably established that an isotopically labeled precursor is incorporated into DNA mainly when there is mitotic division, as in growing tissue or regenerating liver. When once incorporated, the labeled atoms disappear more slowly from DNA than from RNA. This "metabolic stability" may be of importance for a substance which is to serve as the source of genetic information.

Three main lines of evidence—cytologic, genetic, and biochemical—have merged to provide us with our current ideas on the genetic significance of DNA. Feulgen developed a specific reaction for DNA and was able to show that the acid component of the chromatin was a DNA-protein. DNA was shown to be responsible for the characteristic chromogenicity of chromatin and virtually all of the DNA of the animal cell was shown to be located in the chromatin of the nucleus. The behavior of the sperm nucleus provided evidence that the nucleus was the site of hereditary factors. T. H. Morgan had pointed out the parallels between organized chromosome movements and the movements of the genetic determinants deduced from mendelian genetics. The chromosome was accepted as a "gene carrier." Subsequent data of various kinds such as the genetic results accompanying translocation of chromosome fragments or breakage of chromosomes by x-ray made it clear that mendelian genes were linearly arranged on the chromosome.

An extremely important contribution was the demonstration that DNA was constant in amount in the somatic nuclei (diploid) of various tissues within a species and was present in half the amount in germ cells (haploid). DNA thus satisfied the quantitative requirements for a specific genetic substance. Subsequently, the genetic action of isolated DNA was shown by the demonstration that DNA extracted from an encapsulated (smooth) strain of pneumococcus would transform an unencapsulated (rough) strain into encapsulated cells. The smooth cells produced by transformation propagated indefinitely as a smooth strain without further exposure to the DNA. The transformed cells produced a DNA with the same ability to produce transformation of rough to smooth. The pneumococcal DNA preparations had induced a specific inheritable function and had initiated its own replication, functions usually attributable to genes.

RNA

The RNA of the cell is distributed throughout all cell fractions, whereas most of the DNA is concentrated in the nucleus, small amounts being found in mitochondria. The following are some representative figures for the intracellular distribution of RNA

RNA Content of Mouse Liver	
Whole homogenate	100%
Nucleus	10
Mitochondria	16
Microsomes	55
Soluble fraction	19

Various studies had shown that these RNA fractions had different turnover rates as measured by the incorporation of $P^{32}O_4$, glycine, or orotic acid. The nuclear RNA always exhibited greater turnover activity than the other fractions. This led to numerous speculations that the nuclear RNA was the precursor of the RNA of the other fractions. In addition, it became apparent that the RNA of these fractions was not homogeneous and that there was more than one RNA in any given fraction. It was noted that cells that had a high rate of protein synthesis had a high content of RNA which is localized in the nucleolus and in the cytoplasm near the nuclear membrane. This was interpreted as indicating the nuclear origin of cytoplasmic RNA and indicating a role for RNA in protein synthesis. The localization of protein synthesis in the ribosome which contains some 50 to 60 per cent RNA substantiated the relationship between RNA and protein synthesis. These studies pointed the direction for present day experimentation and led to the discovery of the various RNAs now known to be involved intimately with protein biosynthesis.

BIOSYNTHESIS OF NUCLEIC ACIDS

It is now agreed that the genome of any cell is composed of an unique set of DNA molecules. This unique set of molecules is exactly duplicated during cell division and one set distributed to each daughter cell. In addition to this function of insuring the transmission of heritable characteristics, the encoded genetic information in DNA must be transcribed and translated into specific protein structures needed for metabolic function and control during the life span of a cell. The overall pattern of information flow may be summarized as follows

$$\text{DNA Replication} \xrightarrow{\text{transcription}} \text{RNA} \xrightarrow{\text{translation}} \text{Protein}$$

We consider in this section some details of what is known about the replication of DNA and gene expression transcription or RNA synthesis at both the cellular and enzymatic level. The problems of transport of RNA and amino acids to the sites of protein synthesis and translation (or protein synthesis) in ribosomes on mRNA templates are discussed in the following chapter.

Replication of DNA and DNA Polymerase

Enzymes have been isolated from a number of bacterial and animal sources that fulfill many of the requirements for the replication of DNA. These enzymes are known as DNA polymerases and catalyze the *net* synthesis of DNA via the reaction

$$\begin{matrix} n\text{dATP} \\ n\text{dGTP} \\ n\text{dCTP} \\ n\text{dTTP} \end{matrix} + \text{DNA} \underset{\text{Mg}^{++}}{\overset{\text{polymerase}}{\rightleftharpoons}} \text{DNA}\text{—}\begin{bmatrix} \text{dAMP} \\ \text{dGMP} \\ \text{dCMP} \\ \text{dTMP} \end{bmatrix}_n$$

In this reaction mononucleotide units are added to the 3′-hydroxyl terminus of a DNA chain in the direction of 5′ to 3′. There is an absolute requirement for all four deoxyribonucleoside triphosphates (symbolized dATP, dGTP, and so on), magnesium ion, and a DNA *template* and *primer*. A *primer* is regarded as polynucleotide that is required for the new synthesis of polynucleotide by a process of extension or elongation. A *template* can be regarded as a molecule which is required for the syntheses of new polynucleotides by a process in which the constituent nucleotides of the new strand are aligned by complementary base-pairing to those in the preexisting strand. Thus, the nature and sequence of the nucleotides are determined by those of the template.

It is now clear that initiation of synthesis by the polymerase requires both a 3′-hydroxyl terminated primer and a template. For example, treatment of a bihelical DNA primer with exonuclease III removes nucleotide residues progressively from the 3′-hydroxyl ends and produces a DNA molecule that has a central double-stranded region with extensive single-stranded peripheral regions, DNA polymerase acts on such molecules to restore the portion of the molecule which had been removed by the exonuclease. The partially degraded strand acts as a primer, whereas the non-degraded strand acts as a template. The newly formed material is linked covalently to the primer, and the fully repaired DNA is undistinguishable from the original undegraded DNA.

All known DNA polymerases can use denatured (single-stranded) DNA as a template. This appears to involve a mechanism in which the 3′ terminus of the template loops back upon itself and acts as the priming end for replication of the remainder of the template. In this case the 5′ end of the product is linked covalently to the 3′ end of the template.

Circular single-stranded DNA (see below) can be replicated by DNA polymerase.

There is the requirement for an oligonucleotide which apparently attaches to the template by base-pairing. The oligonucleotide then serves as a primer for the new DNA chain in a manner similar to that described above for repair synthesis.

When a native, bihelical DNA is used as the template for DNA polymerase, the synthesized product has two properties that differ from those of the native DNA. First, electron micrographs show the product to have a branched structure. Second, permanent denaturation cannot be achieved; separation of the strands occurs on heating or exposure to alkali but, unlike native DNA, helix structure is reformed promptly when alkali or heat is removed. It appears likely that the branching is caused by the enzyme leaving the 3′ strand which is serving as the template for the synthesis of the new strand in the 5′ to 3′ direction and beginning to copy the 5′ strand, thus producing a hairpin loop.

This failure of DNA polymerases to catalyze a linear, sequential replication of a DNA duplex cast doubt on DNA polymerase being solely responsible for the *in vivo* replication of a duplex chromosome. All available evidence shows a linear sequential replication of the chromosome starting from one point on the helix. Replication apparently occurs at a single fork where both strands are laid down in the same direction. Since the two chains are of opposite polarity, one chain must be synthesized in the 5′ to 3′ direction while the other proceeds in the 3′ to 5′ direction. However, DNA polymerases cannot replicate the 5′ strands of the template.

In order to explain the replication of the 5′ strand, a number of mechanisms have been proposed. One of these, the model of discontinuous synthesis, has some experimental support. This model suggests the possibility that one or both strands are synthesized in a discontinuous manner in the 5′ to 3′ direction and that only short segments of DNA are synthesized at a time. These short segments are subsequently covalently linked, the net result being the unidirectional growth of both strands. It has been found that newly synthesized DNA consists mainly of small units of DNA which are later incorporated into large DNA molecules. In addition, an enzyme has been discovered which could join these short segments of DNA. This enzyme is *DNA ligase,* and it catalyzes the formation of a phosphodiester bond by esterification of the 5′-phosphoryl group to the 3′-hydroxyl group in a single chain if the complementary chain is intact and holds the 5′-phosphoryl and 3′-hydroxyl end-groups in juxtaposition.

Although there are many unanswered questions pertaining to DNA replication and it is possible that enzymes other than polymerase and ligase may be involved, the enzymatic synthesis of a biologically active DNA has recently been accomplished. Arthur Kornberg and associates, using DNA polymerase, DNA ligase, an oligonucleotide primer, and the single-stranded DNA of phage $\phi \times 174$ DNA as the template, synthesized a covalent duplex circular DNA. The product is indistinguishable from the duplex circular $\phi \times 174$ DNA (replicative form). The synthetic negative strand was separated from the positive template strand and its infectivity tested *in vitro*. The isolated negative circles were infectious. In further studies the negative strand was used as a template and the synthesized positive strand (which is a true replica of the natural virus) was shown to be infective. The ability to synthesize infectious $\phi \times 174$ molecules with the DNA polymerase and ligase support the view that these enzymes are capable of replicating DNA for normal cell division. The fidelity of copying is very precise since a single error in placement of the 5500 deoxyribonucleotides could have resulted in a loss of infectivity. Among yet unresolved questions is why the purified polymerase under certain circumstances makes branched DNA with nonseparable strands? There is evidence that the growing region of DNA in certain bacteria is attached to the cell membrane. This might provide a way to obtain some special tertiary structure of the template. It may be extremely difficult if not impossible in all cases to simulate *in vitro* the special arrangements that exist *in vivo*. Thus, the difficulty in obtaining a DNA template with the appropriate tertiary structure may explain the atypical behavior of the polymerase *in vitro*.

It has recently been shown that purified DNA polymerases can hydrolyze native or denatured DNA chains from either end. This hydrolytic activity, termed exonuclease II is inseparable from polymerase activity and considered an integral part of the enzyme. Its biologic role is unknown. Given chain breaks, the enzyme might trim down individual chains by exonuclease action until only perfectly fitting hydrogen bonded base pairs are left, and then built them back by polymerase action.

The model of DNA structure proposed by Watson and Crick (page 285) consists of two helical strands wound about each other and held together by hydrogen bonding between complementary base pairs. It was further proposed that replication of DNA occurred by separation of the two complementary chains followed by formation of a new complementary chain on each of the parental chains. This mechanism was comfirmed *in vivo* at the molecular level. *Escherichia coli* were grown for several generations in a medium containing N^{15} ammonium chloride. It is possible by density gradient centrifugation to separate N^{15}-DNA (heavy) from N^{14}-DNA (light), and from DNA of intermediate densities. Fully N^{15}-labeled cells were transferred to a medium containing N^{14}-ammonium chloride, and the DNA was isolated and examined at various generation times. After one cycle of replication, all the DNA molecules were half labeled. After two generation periods, equal amounts of half labeled and unlabeled DNA molecules were present. These results indicate that the DNA after one replication consists of one heavy and one light chain. After two replications there are two DNA molecules containing only light chains and two DNA molecules each containing one heavy and one light chain. This mechanism is referred to as semiconservative replication since one strand of the parental DNA molecule is conserved in each daughter molecule. This is shown schematically in Figure 8.8.

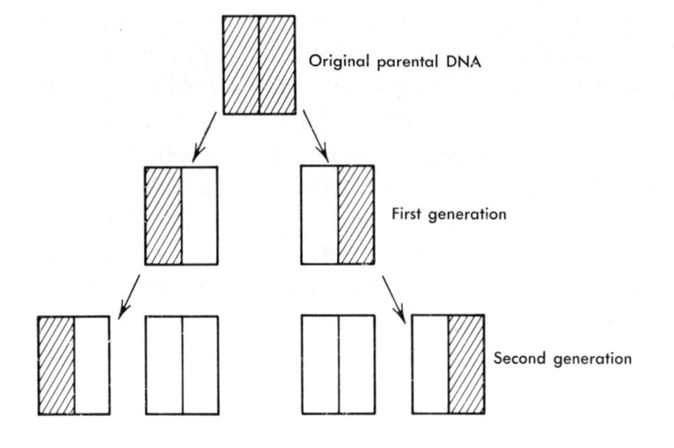

Figure 8.8. Semiconservative replication of DNA.

Biosynthesis of RNA

It appears likely that all types of normal cellular RNA are synthesized on a DNA template. Hybridization experiments have shown that mRNA, tRNA, and both of the larger sizes of ribosomal RNA form stable hybrids with denatured DNA from the same source and presumably are complementary copies of a portion of this DNA.

Enzymes known as RNA polymerases (transcriptases) are widely distributed and catalyze the DNA-dependent synthesis of RNA from ribonucleoside-5′-triphosphates. There is a strict requirement for all four ribonucleoside-5′-triphosphates, a DNA template, and a bivalent cation. The reaction may be formulated similarly to that catalyzed by the DNA polymerase.

$$\begin{matrix} n\text{ATP} \\ n\text{GTP} \\ n\text{CTP} \\ n\text{UTP} \end{matrix} \underset{\text{Mg}^{++}}{\overset{\text{DNA}}{\rightleftharpoons}} \text{RNA}-\begin{bmatrix} \text{AMP} \\ \text{GMP} \\ \text{CMP} \\ \text{UMP} \end{bmatrix}_n + 4\ (n)\ \text{PP}_i$$

The product of the reaction is a polyribonucleotide of high molecular weight, containing all four ribonucleosides in 3′,5′-phosphodiester bonds. The DNA acts as a template and the base sequence of a portion of the DNA is transcribed into a corresponding sequence in the RNA. This conclusion is supported by the following observations.

1. The base composition of the RNA formed is the same (with uracil replacing thymine) as that of the template when a linear bihelical DNA is used.

2. When the single-stranded DNA of bacteriophage $\phi\times 174$ is used, the base composition of the product is *complementary* to that of the DNA template.
3. When the synthetic, alternating copolymer dAd-T is used as the template, only UMP and AMP are incorporated into the product.
4. Nearest neighbor analysis indicates that the frequencies of the various dinucleotides in the RNA product are very similar to that in the DNA template.
5. The newly synthesized RNA can form hybrids with the DNA template, indicating substantial regions of complementary sequences of bases.

The results with the linear bihelical DNA template imply that both strands of the DNA were transcribed (symmetric transcription) into RNA. This in fact is so. Two complementary RNA strands are produced which can form RNA-DNA hybrids with both strands of the template DNA and which can be annealed to form double-stranded RNA. This would seem to be paradoxical since *in vivo* studies had shown the only one strand of the DNA was transcribed (asymmetric transcription) and, as noted above, a single-stranded DNA may function as the template. The reason for this discrepancy is not entirely clear, but it may be related to the need for some special structural feature of the template. For example, if the bihelical, circular, replicative form of $\phi\times 174$ DNA is employed as the template, the product RNA can form hybrids with only one of the two DNA strands. If opened circles of DNA are used as the template, the RNA product forms hybrids with both strands of DNA. The situation is perhaps analagous to that observed with DNA polymerase (page 307). There is evidence that the synthesis of RNA by RNA polymerase separates the strands of a bihelical template product helix and ejects one of the template strands.

In addition to the DNA-dependent RNA polymerases, there are RNA polymerases that require a RNA template. These enzymes, called RNA-dependent RNA polymerases or replicases, are found in certain microorganisms and especially in virus-infected cells. These enzymes require all four ribonucleoside-5′-triphosphates, Mg^{++}, and RNA which acts as a template. Such an enzyme has not been detected in normal animal cells, but may be induced by infection with a RNA-containing virus and is responsible for the replication of the viral nucleic acid.

Inhibitors of Nucleic Acid Synthesis

A number of inhibitors of nucleic acid synthesis are known. Some are structural analogs of intermediates and cofactors involved in nucleic acid biosyntheses and act as competitive inhibitors at one or more of the enzymatic steps. Examples of this group are 6-mercaptopurine (a purine analog), 5-fluorodeoxyuridine (a pyrimidine analog), and azaserine (a glutamine analog). The structures of these compounds and that of amethopterin are given on the next page.

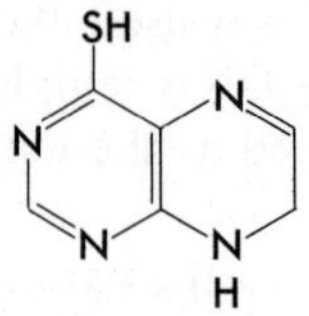

6-Mercaptopurine

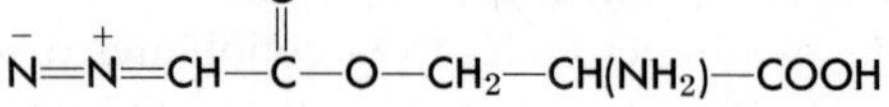

$$\overset{-}{N}=\overset{+}{N}=CH-\overset{O}{\overset{\|}{C}}-O-CH_2-CH(NH_2)-COOH$$

Azaserine

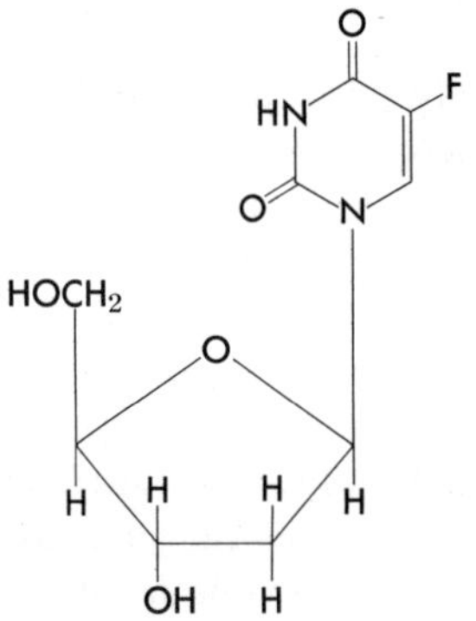

5-Fluorodeoxyuridine

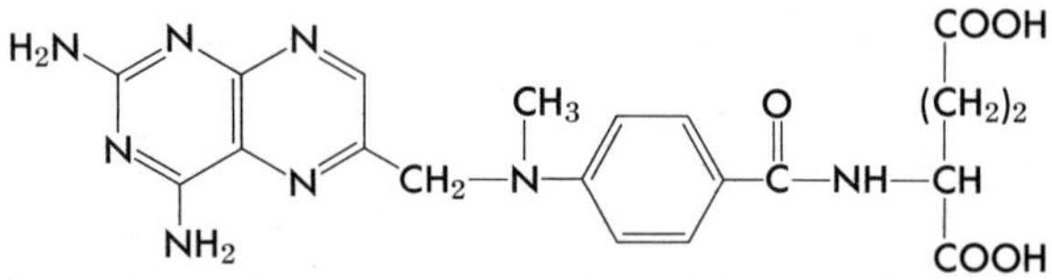

Amethopterin

6-Mercaptopurine is converted metabolically to its ribonucleotide which is an inhibitor of both steps in the conversion of inosinic acid to adenylic acid (Figure 8.5). 5-Fluoro-deoxyuridine inhibits the thymidylate synthetase reaction (page 231), and the antibiotic azaserine inhibits the conversion of formylglycinamide ribonucleotide to formylglycinamidine ribonucleotide (page 292).

Amethopterin inhibits the conversion of folic acid to tetrahydrofolic acid and thus interferes with the production of coenzymes required in two steps in the *de novo* synthesis of purines [reactions 4 (page 292) and 10 (page 295)]. This inhibition cannot be reversed with folic acid.

The antibiotic actinomycin D (page 324) inhibits the synthesis of RNA but not that of DNA. It apparently forms a complex with DNA that makes it unsuitable as a template for DNA-dependent RNA polymerases. Another antibiotic, mitomycin C, after metabolic modification inhibits DNA synthesis by introducing covalent linkages between the DNA strands.

Mitochondrial DNA

It has become clear in recent years that mitochondria contain DNA and a protein synthesizing system distinct from the system of the extramitochondrial cytoplasm.

The most common conformation of mitochondrial DNA from animal cells is a bihelical circle with a linear dimension of about 5 μ. Unlike the nuclear DNA, the mitochondrial DNA is apparently not associated with protein and resembles the "chromosomes" of *Escherichia coli*. The combined DNA content of all mitochondria per cell is perhaps 0.1 to 0.2 per cent of the DNA of the nucleus. It appears that there is both a mitochondrial DNA polymerase which differs from the nuclear DNA polymerase and a DNA ligase. The mitochondrial DNA is reported to be replicated by a semiconservative mechanism.

Mitochondrial protein is formed in two ways: (1) by synthesis within the mitochondria, and (2) by synthesis outside the mitochondria in association with the usual ribosomal system of the cytoplasm (Chapter 9). Mitochondria are also capable of synthesizing RNA and they contain ribosomes and ribosomal RNAs which differ from those of the cytoplasm in size and functions. They also contain specific species of tRNA and aminoacyl-tRNA synthetases (page 305). The biogenesis of mitochondria seems to involve, first, a synthesis of certain structural proteins by the system within the mitochondria and then the integration of soluble proteins synthesized in the cytoplasm. Thus, the interaction of the nuclear genetic system with that of the mitochondrial genetic system appears essential in the biosynthesis of these organelles. Mitochondria are then molecular symbionts with their own DNA-RNA-protein-synthesizing system. This confers upon them a degree of genetic continuity and mutability independent of the nuclear system.

REFERENCES

Buchanan, J. M.: The Enzymatic Synthesis of the Purine Nucleotides, *Harvey Lectures*, **55**:104 (1959–1960).

Chargaff, E., and Davidson, J. N. (eds): *The Nucleic Acids: Chemistry and Biology*, 3 vols. Academic Press, New York, 1955, 1960.

Cold Spring Harbor Symposia on Quantitative Biology: *Replication of DNA in Micro-Organisms*, Vol. XXXIII, 1968.

Khorana, H. G.: Polynucleotide Synthesis and the Genetic Code, *Harvey Lectures*, **62**:79 (1966–1967).

Kornberg, A.: *Enzymatic Synthesis of DNA*. Wiley, New York, 1962.

Mahler, H. R., and Cordes, E. H.: *Biological Chemistry*, Chapters 4 and 17. Harper and Row, New York, 1966.

West, E. S., Todd, W. R., Mason, H. S., and Van Brugger, J. T.: *Textbook of Biochemistry*, 4th ed., Chaps. 9 and 26. The Macmillan Co., New York, 1966.

White, A., Handler, P., and Smith, E. L.: *Principles of Biochemistry*, 4th ed., Chaps. 9 and 27. McGraw-Hill, New York, 1968.

Chapter 9

PROTEIN BIOSYNTHESIS

THE CURRENTLY accepted model for protein biosynthesis may be summarized as follows.

The primary site of protein biosynthesis is the ribosome. Genetic information in one of the strands of the DNA is transcribed in the form of a complementary polyribonucleotide chain, the messenger RNA (mRNA), which specifies the sequence of amino acids in the polypeptide chain. Each amino acid is activated by an amino acid-specific enzyme (aminoacyl-tRNA synthetases) in an ATP-dependent reaction to form enzyme-bound aminoacyl adenylates. The aminoacyl moiety of the enzyme-bound complex is transferred by the synthetases to amino acid-specific RNA molecules known as transfer RNAs (tRNAs). The aminoacyl group is attached by ester linkage to the ribose portion of a terminal adenosine residue of the tRNA to form aminoacyl-tRNA. The aminoacyl-tRNAs link up with mRNA-bound ribosomes at specific sites determined by the nucleotide sequence of the mRNA. A peptide bond is formed between an incoming aminoacyl-tRNA and the growing peptide chain with the release of the tRNA that was attached to the carboxyl end of the growing polypeptide chain. Several factors are required in this complex process. Following peptide bond formation, the ribosome and mRNA move one coding unit over in relation to each other thus positioning the next mRNA coding unit for base-pairing with the next aminoacyl-tRNA. As this process is repeated, the polypeptide chain grows stepwise from the amino-terminal amino acid to the carboxy-terminal amino acid. When the translation is complete, the newly synthesized polypeptide chain, the ribosome, and the terminal tRNA are released from the mRNA. The ribosome is free to start a new chain. A ribosome makes one polypeptide at a time, but the mRNA may be simultaneously employed by more than one ribosome. These and other considerations are discussed in more detail in the following sections.

ACTIVATION OF AMINO ACIDS AND FORMATION OF AMINOACYL-tRNAs

The enzymes activating the amino acids and forming aminoacyl-tRNAs are called *aminoacyl-tRNA synthetases*. In the first reaction the amino acids are converted to transient enzyme-bound aminoacyl adenylates in which the carboxyl group of the amino acid is linked to the phosphate group of the adenylate moiety by an anhydride bond. The aminoacyl group is transferred in a second reaction to the amino acid-specific acceptors, the transfer RNAs (tRNAs).

$$1)\ \text{Amino acid}_1 + \text{Synthetase}_1 + \text{ATP} \overset{Mg^{2+}}{\rightleftharpoons} \text{Aminoacyl}_1\text{-AMP-synthetase}_1 + PP_i$$

$$2)\ \text{Aminoacyl}_1\text{-AMP-synthetase}_1 + \text{tRNA} \overset{Mg^{2+}}{\rightleftharpoons} \text{Aminoacyl}_1\text{-tRNA}_1 + \text{AMP} + \text{synthetase}_1$$

$$3)\ \text{Amino acid}_1 + \text{ATP} + \text{tRNA}_1 \rightleftharpoons \text{Aminoacyl}_1\text{-tRNA} + \text{AMP} + PP_i$$

Each synthetase is specific for a single amino acid. Once an amino acid is attached to its tRNA it no longer controls its own fate. This was shown by converting enzymatically formed cysteinyl-$tRNA_{cys}$ to alanyl-$tRNA_{cys}$ by chemical reduction and following the incorporation of the aminoacyl moiety of the hybrid aminoacyl-tRNA into polypeptide linkages in positions normally occupied by cysteine. Thus, the amino acid was selected by virtue of the base sequence in the tRNA adaptor. It is believed that a specific sequence of nucleotides (anticodon) in a tRNA molecule interacts by base-pairing with the codon on the mRNA chain to select the "correct" amino acid. The specificity of the aminoacylation of the tRNAs is crucial, since no subsequent mechanism is known for the rejection of "wrong" amino acids.

The tRNAs are a group of polyribonucleotides containing 75 to 85 nucleotide residues per molecule. They contain a relatively high proportion of unusual bases in addition to the four major bases (page 289). The individual tRNAs appear to have certain structural features in common: (1) they have the same terminal sequence (cytidylate-cytidylate-adenosine) at the amino acid acceptor end; (2) they may possess a common pentanucleotide sequence within the molecule; (3) all known tRNA sequences can be arranged in a cloverleaf structure (page 290) with the presumed anticodons located in the bottom loops of the structure; and (4) most have a guanylic acid as the 5′-nucleotide residue. Since the majority of studies have been done with unfractionated tRNA and the primary sequence of only six specific tRNAs has been established, these generalizations should be accepted with caution.

Apart from these possible common features, the base sequence of each tRNA is different and confers the structural specificity for the specific interaction with the appropriate aminoacyl-tRNA synthetase. There is at least one specific tRNA for each amino acid. Since the code is degenerate and more than one triplet codes for each amino acid (see below), there is probably a tRNA for each codon.

Attachment of the amino acid to tRNA occurs by formation of an acyl ester linkage between the carboxyl group of the amino acid and the 3′- or 2′-hydroxyl groups of the terminal adenosine residue. Although most of the evidence favors the 3′-ester as the requisite structural feature, the possibility of the involvement of the 2′-ester cannot be excluded.

tRNAs have a codon recognition site (anticodon) and a site for interaction with the appropriate aminoacyl-tRNA synthetase (acceptor recognition site). These two sites probably involve separate areas of the molecule. This conclusion is based on the observations that there are species specificity differences in the acceptor and transfer reactions, and that various chemical modifications of tRNAs affect the acceptor and transfer functions to different extents.

FUNCTION OF RIBOSOMES

Ribosomes from different cells vary somewhat in both composition and size (page 288). It is the ribosome at which mRNA, aminoacyl-tRNA, and various other factors are convened to carry out protein synthesis. The process at the ribosome may be divided into three phases: *initiation, elongation,* and *termination.*

In bacteria, synthesis is initiated by the binding of the mRNA to the 30S ribosomal subunit and the binding of the chain initiator formylmethionyl-tRNA to its codon AUG. At the same time the 50S ribosomal subunit joins the 30S subunit to produce the 70S ribosomal unit. Three factors are apparently necessary at this stage, one for the binding of mRNA and two for the binding of the chain initiator. Following a full cycle of initiation, elongation and termination, the ribosome separates again into 50S and 30S subunits.

The *elongation* cycle may be divided into four phases as follows

1. In the starting phase of peptidyl-tRNA occupies a donor site on the 50S subunit, and the aminoacyl site (*acceptor site*) on the 50S subunit is open.
2. The *acceptor site* becomes occupied by an aminoacyl-tRNA whose anticodon matches the RNA codon next to the one occupied by the peptidyl-tRNA. This binding requires GTP and elongation factor T and initiates an enzyme catalyzed transpeptidation from peptidyl-tRNA to the free amino group of aminoacyl-tRNA and frees the tRNA originally attached to peptidyl-tRNA.

3. The newly elongated peptidyl-tRNA now situated on the acceptor site is now translocated to the donor site with the displacement of the free tRNA.
4. In the last phase the peptidyl-tRNA has returned to its starting position but now extended by one amino acid. It has also carried the mRNA one codon over and exposed under the acceptor site a new codon for attachment of the next aminoacyl-tRNA. These translocations require an elongation factor G and GTP-derived energy.

Some of these relationships are shown in Figure 9.1.

Termination takes place when one of the termination triplets (UAA, UAG, or UGA) appears on the mRNA next to the peptidyl-tRNA. The final product is released from the tRNA in a manner not yet understood.

The tRNAs recycle through the system being discharged by transpeptidation. They are recharged with amino acids and, after reaction with the appropriate factors, they are carried back to the ribosome-bound mRNA. The successive addition of new ribosomes to the mRNA to produce a polysome can occur. Each ribosome travels independently along the same mRNA progressively completing its own protein copy.

MESSENGER RNA (mRNA)

The idea that some species of RNA might serve as a template and specify the arrangement of amino acids in a polypeptide was considered for several years. This was suggested by earlier experiments (page 306) in which it was noted that cells with a high rate of protein synthesis had a high content of RNA localized in the nucleolus and in the cytoplasm near the nuclear membrane. Protein synthesis also was localized in the ribosome which contains some 60 per cent RNA. Ribosomal RNA (rRNA) was the original candidate for the informational molecule. However, a number of subsequent facts and considerations excluded rRNA as being the coding agent. Since the base composition of rRNA is remarkably homogeneous even in species whose DNA compositions differ considerably as do the composition and size of their cellular proteins, it is difficult to assign to rRNA the role of the template when it does not reflect either the composition of the DNA or the cellular proteins.

In an attempt to unify the interpretation of experiments involving the kinetics of enzyme induction and repression, and the synthesis of RNA in bacteria infected with phages, Jacob and Monod postulated the existence of a messenger RNA (mRNA). The mRNA was postulated to be the primary gene product and to be complementary to the DNA on which it was synthesized. It is transported from the nucleus to the cytoplasm where it interacts with nonspecialized ribosomes to dictate the synthesis of specific proteins. During the translation process the mRNA is destroyed and the released ribosome is then available for the same or another messenger; the messenger is thus unstable. Considerable evidence has accumulated

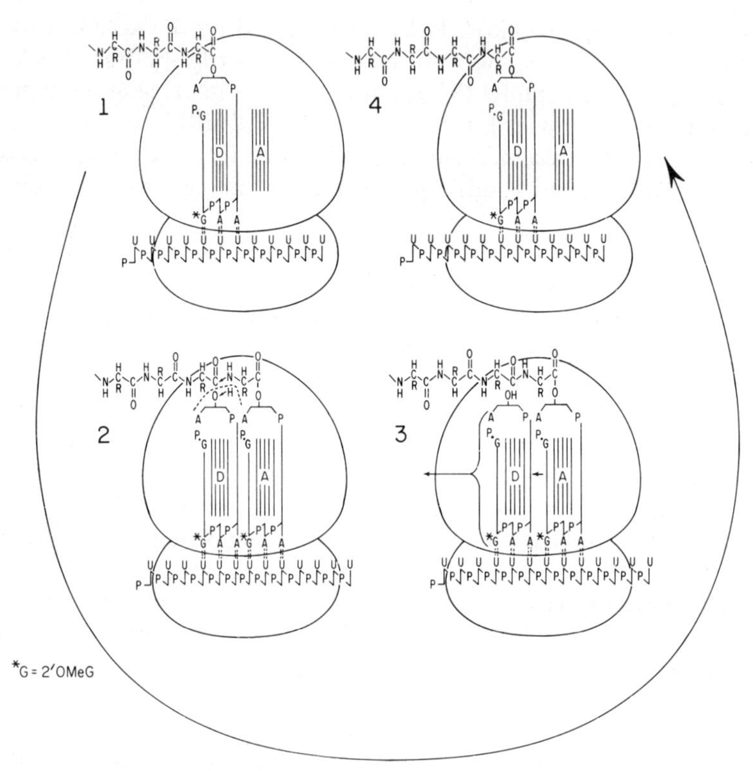

Figure 9.1. Peptide condensation cycle. (1) In the starting phase (upper left) a peptidyl-tRNA has just been transferred to the donor site (D) on the ribosomes; (2) (lower left), the acceptor site (A) becomes occupied by an aminoacyl-tRNA, whose anticodon matches mRNAs codon triplet next to the one occupied by the peptidyl-tRNA. The binding is a function of factor T (T_u and T_s) + GTP; it initiates transpeptidation from peptidyl-tRNA to the free amino group of aminoacyl-tRNA and frees the tRNA formerly linked to peptidyl-tRNA. To open the donor site for translocation of the newly extended peptidyl-tRNA, now situated on the acceptor site, factor G + GTP simultaneously promote in (3) a displacement of free tRNA from, and transfer of extended peptidyl-tRNA to, the donor site. With translocation completed in (4), the peptidyl-tRNA returns to the starting position of (1) after it has been extended by one amino acid; it carries mRNA to the left by the length of one codon, and exposes under the donor site a new codon for attachment of the next aminoacyl-tRNA. (From F. Lippman, *Science,* **164:**1024, 1969. Copyright 1969 by the American Association for the Advancement of Science.)

to support the concept of mRNA, although some modifications in the original concept are necessitated by new experimental evidence and by the extension of the idea to nonbacterial systems. The most cogent support for the concept of mRNA has come from cell-free systems in which synthetic polyribonucleotides can function as templates for the synthesis of specific polypeptides (page 320).

A number of criteria have been used for identifying a RNA fraction as mRNA.

These include base composition and sequence, stimulation of amino acid incorporation, stability, and size.

Base Composition and Sequence

One of the predictions of the messenger model was that mRNA would reflect the base composition and sequence of the DNA of the organism. The equivalence of base composition is of limited significance since only one strand of DNA is transcribed into mRNA (page 311). Unless the average base ratio of a single strand of DNA is the same as that of the whole, there is no need to assume a base equivalence between mRNA and DNA. In addition only a fraction of DNA may be functional in any given system at a given time. Thus, there is no reason to expect that the base composition of a limited area of DNA will bear any resemblance in its base composition to the gross composition of the helical DNA of the organism. As to the sequence of nucleotides, it now appears that all cellular RNAs, including tRNA and rRNA, form hybrids with homologous DNA. Hybridization is also not an unique property of mRNA.

Stimulation of Amino Acid Incorporation

Another prediction of the original messenger model was that mRNA alone would stimulate amino acid incorporation in a cell-free protein synthesizing system. Subsequent studies have shown this property is not unique to mRNA. Almost any RNA and even DNA, under appropriate conditions, can stimulate the incorporation of amino acids into protein. The main requirement for a nucleic acid to stimulate amino acid incorporation is the lack of a secondary structure.

Stability

In the original formulation of the messenger model for bacterial systems, the messenger model was regarded as highly unstable. It now seems clear that the stability of mRNA molecules varies considerably between microorganisms and higher organisms. mRNAs may also vary in stability within the same cell, and it seems likely that all cells produce both stable and unstable mRNA. Mammalian erythrocytes and platelets continue to synthesize proteins for days in the absence of nuclear DNA and any detectable RNA synthesis. Instability is not an obligate requirement for the characterization of a mRNA.

Size

In the original formulation the messenger was thought to be relatively small, since the sedimentation coefficients of bacterial mRNAs were found to be in the range of 6S to 14S. It was thought that while the genes comprising one operon (page 326) were transcribed together, a separate mRNA molecule was transcribed for each polypeptide. mRNA was thus considered to be monocistronic and of relatively small molecular weight. However, newer isolation techniques which avoid degradation of

RNA give values up to 30S. Some RNAs are now believed to be polycistronic, that is, they contain the information of several genes. Although a specific mRNA would be expected to have a unique molecular weight, heterogeneity in size of the total mRNA of the cell is expected. The size of the mRNA will depend on the size of the unit to be transcribed.

It is apparent that no one of the above criteria by itself is unique for mRNA. The definitive experiment for the characterization of a mRNA would be the isolation of a specific RNA that had been made on a DNA template and the demonstration that this RNA could direct the synthesis of a specific protein. The base sequence of the mRNA should be consistent with the amino acid sequence of the polypeptide as predicted by the codon dictionary (see below).

The demonstration that synthetic polynucleotides can function as templates for the synthesis of specific polypeptides in cell-free systems provides the most important support for the messenger concept. Such experiments have also provided one of the means for the deciphering of the genetic code.

THE GENETIC CODE

The current codon assignments for the genetic code are shown in Table 9.1. Since more than one technique was used in assigning these codons, there is reasonable confidence in their correctness.

Nirenberg and coworkers were the first to use synthetic polynucleotides as messengers in a cell-free system from *E. coli*. The system consisted of ribosomes and soluble enzymes which had been depleted of mRNA by preincubation and of DNA by treatment with deoxyribonuclease. Such a system has minimal endogenous capacity to synthesize protein. The addition of polyribonucleotides to this system promoted amino acid incorporation, the specific amino acids incorporated depending on the nucleotide content of the added polyribonucleotide. Using a variety of homopolymers and random heteropolymers and assuming a nonoverlapping triplet code, Nirenberg and Ochoa and their coworkers were able to establish the nucleotide composition but not sequence of some 50 triplets coding for amino acids.

The sequences of nucleotides within codons was established by two additional techniques: the polyribonucleotide-directed specific binding of aminoacyl-tRNA to ribosomes by nucleotide triplets of known sequence, and the use of long-chain synthetic ribonucleotide polymers also of known structure. An examination of the "dictionary" reveals that the code is highly degenerate in a semisystematic way, and there are assignments for all of the possible 64 trinucleotides. Codons representing the same amino acid are called *synonyms*. Several patterns of degeneracy are apparent. First, for a given amino acid, the first two bases of degenerate codons (synonyms) are the same whereas the third base (3′-hydroxyl terminal) may be any of the four bases (for example, the codons for valine, alanine, and glycine). Second, for a given amino acid the first two bases of synonyms are the same, whereas the third base may be either of the two purines, or either of the two pyrimidines (for example,

Table 9.1
Nucleotide Sequences of RNA Codons in E. coli

5′-Hydroxyl Terminal Base	Second Base				3′-Hydroxyl Terminal Base
	U	C	A	G	
	Phe	Ser	Tyr	Cys	U
U	Phe	Ser	Tyr	Cys	C
	Leu	Ser	CT	NONS	A
	Leu	Ser	CT	Try	G
	Leu	Pro	His	Arg	U
C	Leu	Pro	His	Arg	C
	Leu	Pro	Gln	Arg	A
	Leu	Pro	Gln	Arg	G
	Ileu	Thr	Asn	Ser	U
A	Ileu	Thr	Asn	Ser	C
	Ileu	Thr	Lys	Arg	A
	Met (CI)	Thr	Lys	Arg	G
	Val	Ala	Asp	Gly	U
G	Val	Ala	Asp	Gly	C
	Val	Ala	Glu	Gly	A
	Val	Ala	Glu	Gly	A
	Val (CI)	Ala	Glu	Gly	G

Abbreviations: Ala = alanine, Arg = arginine, Asn = asparagine, Asp = aspartate, Cys = cysteine, Glu = glutamate, Gly = glysine, His = histidine, Ileu = isoleucine, Leu = lecuine, Lys = lysine, Met = methionine, Phe = phenylalanine, Pro = proline, Ser = serine, Thr = threonine, Try = tryosine, Val = valine, NONS = nonsense codons, CT = chain termination, CI = chain initiation.

the codons for lysine, glutamate, phenylalanine, and tyrosine). Third, the first base (5′-hydroxy terminal) may vary while the second two remain constant (for example, the codons for leucine and argenine).

In addition, two codons (AUG and GUG) have two meanings and stand for both insertion of amino acids and for chain initiation signals. When polypeptide synthesis is initiated, the first aminoacyl-tRNA must directly enter the peptide site on the ribosome. This is apparently accomplished by a specific tRNA which carries N-formylmethionine. This tRNA recognizes the codons AUG and GUG and inserts N-formylmethionine at the beginning of the chain, but not in the middle. Methionine in the middle of a chain is inserted by another tRNA that recognizes only AUG. AUG and GUG are the codons for initiation in *E. coli* in which some 40 per cent of the polypeptides have methionine as the N-terminal amino acid. Other proteins may also start with formylmethionine which may be subsequently removed by special enzymes. No information of this type is yet available for chain initiation in mammalian systems.

There are three triplets (UAA, UAG, and UGA) which are not involved in amino acid insertion. UAA and UAG appear to be involved in chain termination. It is not known whether UGA is a chain terminator or is a nonsense triplet.

It is generally believed that recognition between the codon on mRNA and the anticodon triplet on the tRNA occurs by standard base-pairing. It appears that pairing may be more important in the first two positions of the triplet and somewhat more relaxed in the pairing at the third position. Crick has proposed what he called the "wobble" hypothesis in which the third base may recognize in an antiparallel manner codons with U or C. U might recognize A and G and I (inosine) if present might recognize U, C, and A. The anticodon of yeast alanine-tRNA (page 290) is thought to be IpGpC and is antiparallel to the codons for alanine (GpCpC[U] and GpCpA[G]). Pairing could take place in an antiparallel manner as follows

pGpCpU pGpCpC pGpCpA
CpGpIp CpGpIp CpGpIp

Alanine-tRNA recognizes the above codons but not GpCpG to any appreciable extent.

The direction of translation of mRNA is the same as its direction of synthesis, that is, from the 5′ end to the 3′ end. Ribosomes attach to the 5′ end of the mRNA and the codon for the N-terminal amino acid is at or near this end. The order of the codons in the mRNA is colinear with the amino acids in the corresponding polypeptide as is the information in the DNA genome. The RNA code is nonoverlapping and contiguous triplets are read sequentially and consecutively. This was clearly shown in studies with polyribonucleotides containing alternate triplets. For example, the alternating copolymer AC which contains the two alternating trinucleotide sequences ApCpA and CpApC directed the synthesis of a polypeptide containing alternate threonine (codon ACA) and histidine (codon CAC) residues.

Further support for the validity of the cell-free technique for assigning the code has come from the study of known mutations in which a single amino acid in a protein molecule is known to be changed. For example, in sickle cell hemoglobin (page 328) there is a single change from glutamic acid in normal hemoglobin to valine in sickle cell hemoglobin. Since glutamic acid is coded by GAA and GAG (Table 9.1) and valine by GUA and GUG, a single base change from A to U could account for the substitution. In hemoglobin C the change is from glutamate (GAA, GAG) to lysine (AAA, AAG). Thus, the substitution of one amino acid for another may be related in a logical way to the change of a single base in the coding triplet. The study of a large number of mutational changes has revealed a general although not absolute agreement with the results of the cell-free systems.

The code appears to be universal to the extent that synthetic polynucleotides appear to code in the same way in both mammalian and bacterial systems and components of the protein synthesizing system of some species can operate with components from certain others. There are, however, certain species-specific features of the aminoacyl-tRNA synthetases, the tRNAs, and the aminoacyl transferases.

INHIBITORS OF PROTEIN BIOSYNTHESIS

We consider in this section several antibiotics and other substances that have been used in the study of protein biosynthesis. For the most part antibiotics that inhibit protein biosynthesis are far more active against bacteria and *in vitro* bacterial systems than against higher organisms and their *in vitro* systems. The difference may be due to differences in the properties of the ribosomes.

Tetracyclines

These antibiotics, comprising a family of related structures, are inhibitors of protein synthesis. Recent evidence indicates they inhibit specifically the binding of aminoacyl-tRNAs. Tetracycline $R_1R_2 = H$; Chlorotetracycline $R_1 = Cl$, $R_2 = H$; Oxytetracycline $R_1 = H$, $R_2 = OH$.

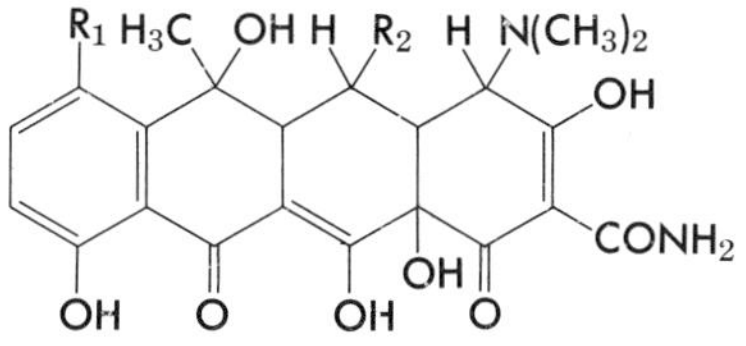

R_1 H_3C OH H R_2 H $N(CH_3)_2$
OH
$CONH_2$
OH O OH OH O

Tetracyclines

Puromycin

This antibiotic interferes with protein synthesis both *in vivo* and *in vitro*. The structure shown below bears a resemblance to that of the terminal aminoacyl adenosine of tRNA. The isomer of puromycin analogous to the 2′ isomer of aminoacyl-tRNA is inactive.

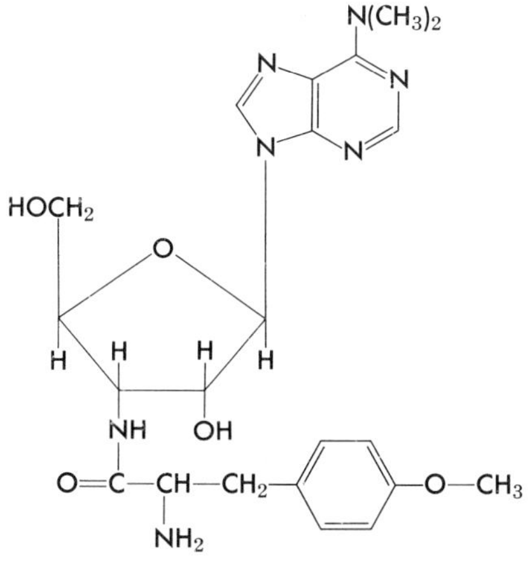

$N(CH_3)_2$
N N N N
$HOCH_2$
O
H H H H
NH OH
$O{=}C{-}CH{-}CH_2{-}C_6H_4{-}O{-}CH_3$
NH_2

Puromycin

NH_2
N N N N
$R'OCH_2$
O
H H H H
O OH
$O{=}C{-}CH{-}R$
NH_2

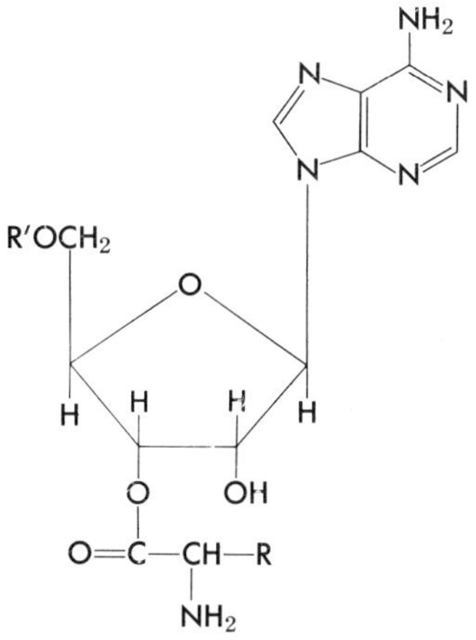

Amino acyl adenosine terminus of tRNA molecule

Puromycin acts by replacing aminoacyl-tRNA and forming a peptide bond so that puromycin becomes the carboxy terminus of the peptide chain. The peptidyl puromycins are not capable of further chain growth; peptides of varying length and bearing puromycin at the carboxy terminus are released from the ribosome.

Chloramphenicol

The site of action of this antibiotic is not known with certainty. It does not inhibit the amino acid-activating enzymes or interfere with the binding of mRNA or aminoacyl-tRNA to ribosomes. The effect would appear to be localized at the peptide bond synthesis stage, the transfer of the peptidyl-tRNA to the aminoacyl-tRNA.

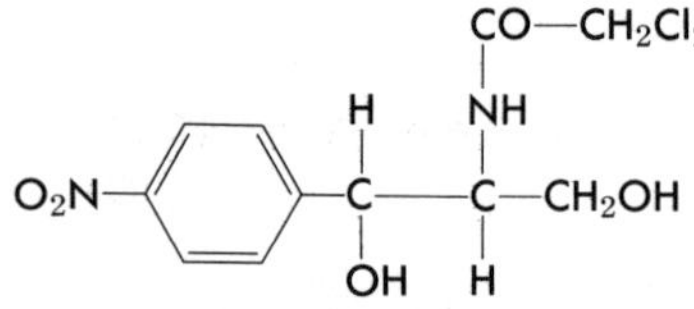

Chloramphenicol

Streptomycin

This antibiotic (page 115) appears to be bound to ribosomes in such a way as to distort the normal mRNA-ribosome-tRNA interaction and allow incorrect base pairing to occur.

Cycloheximide

Cycloheximide (actidione) is not effective in bacteria but is active in mammalian and fungal systems. It appears to block the peptide bond formation step.

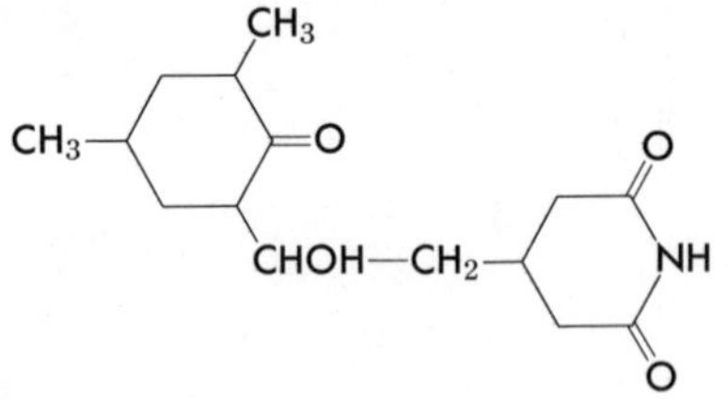

Cycloheximide

Actinomycin D

As noted on page 312, this antibiotic inhibits DNA-dependent RNA syntheses. Thus, it indirectly inhibits protein synthesis by diverting the flow of information from DNA to mRNA.

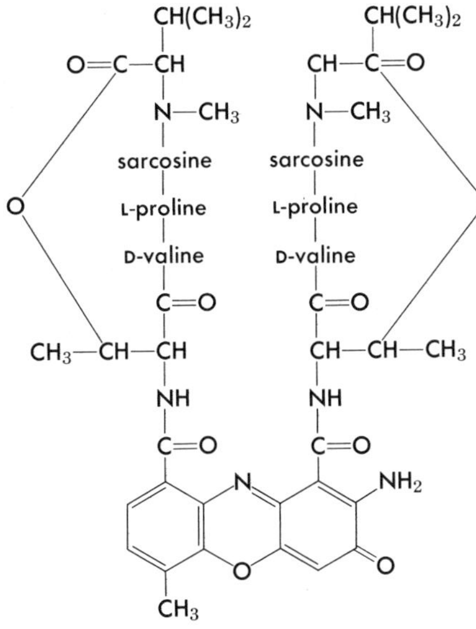

Actinomycin D

REGULATION OF PROTEIN BIOSYNTHESIS

The diverse biochemical processes within the cell must be precisely regulated in order to insure the integrated and orderly function of the cell. Numerous factors will govern whether an enzyme once present will function or not. The availability of substrates and cofactors, the optimal pH, the effects of mass action, and the competition of various enzyme systems for a specific substrate in specific areas within the cell are some of the factors that will regulate biochemical processes. It has been known for some time that bacteria can alter their enzyme composition in response to changes in their environment. In general the synthesis of *catabolic* enzymes is induced by the substrate of the enzymes, whereas the production of *anabolic* enzymes concerned with the synthesis of cellular components is repressed by their final products. Thus, there are some other kinds of controls operating at the level of enzyme (protein) synthesis. Genetic mechanisms must exist which can either switch on or switch off the synthesis of specific proteins. In 1961 F. Jacob and J. Monod proposed the first theory to explain the regulation of bacterial enzyme synthesis. According to this model, called the operon theory, there are two basic types of genes: *structural genes* whose primary product is messenger RNA (mRNA) which carries structural information from cistrons (genes) to the sites of protein synthesis (ribosomes); and *control genes* which regulate the action of the structural genes. Control genes were considered to be of two types, operator genes and regulator genes. The *operator genes* lie adjacent to the structural genes and may represent the initial segment of the structural genes. They are thought to initiate the transcription of the information (mRNA synthesis) contained in the structural genes. A single operator gene may control several structural genes, as in the case of histidine biosynthesis in microorganisms in which the series of ten enzymes involved is under coordinate functional control by a single operator. The unit containing the operator gene and the structural genes coding for related proteins is called an *operon*. The operator acts only on the adjacent structural genes and has no action on genes located on the other chromosome in diploid organisms.

The activity of the operator gene is controlled by the *regulator gene* which synthesizes a specific protein—the *repressor* substance. The *repressor* can combine reversibly with the operator. This combination blocks the action of the operator gene, preventing transcription of the operon, and thereby protein synthesis. The repressor acts on the operator genes of both homologous chromosomes. In the case of *enzyme induction* the specific substrate for an enzyme may combine with the repressor to inactivate it. This would release the operon from inhibition and allow the initiation of protein synthesis. For *enzyme repression* the repressor of the regulator gene may be inactive (aporepressor) until activated by the end product of a metabolic sequence. The activated repressor can then combine with the operator gene to block synthesis at the operon. The regulation of protein synthesis in this model occurs at transcription by altering the rate of synthesis of mRNA from DNA. Some of these relationships are shown in Figure 9.2.

The unit of transcription is the operon, a single polycistronic mRNA corresponding to the entire operon being synthesized. This may be referred to as the one operon–one messenger theory. An alternate possibility, now thought unlikely, is the one gene–one messenger theory in which a messenger corresponding to each gene is synthesized. Whether or not the operator is transcribed or translated in not known. The synthesis of certain enzymes may be induced by some critical concentration of substrate which hinders the interaction of the repressor and operator or converts the repressor to an inactive form. In the case of enzyme repression, the accumulation of an end product enhances the interaction of the repressor with the operator or modifies the repressor to a form that is active. It has been proposed that the repressor, like enzymes subject to feedback control, has two specific allosteric sites. One of these sites has an affinity for the operator and the other to a specific metabolic effector. The combination of repressor and effector at the second allosteric site reduces the affinity of the first site for the operator and inactivates the repressor, thus allowing the operator to remain open and the synthesis of inducible enzymes to take place. The combination of the repressor with the effector increases the affinity of the first site for the operator, and the operator is closed and enzyme synthesis is repressed.

Regulation need not occur only at the transcriptional level. Alternate theories have been proposed in which the stability of the messenger or its attachment to ribosomes

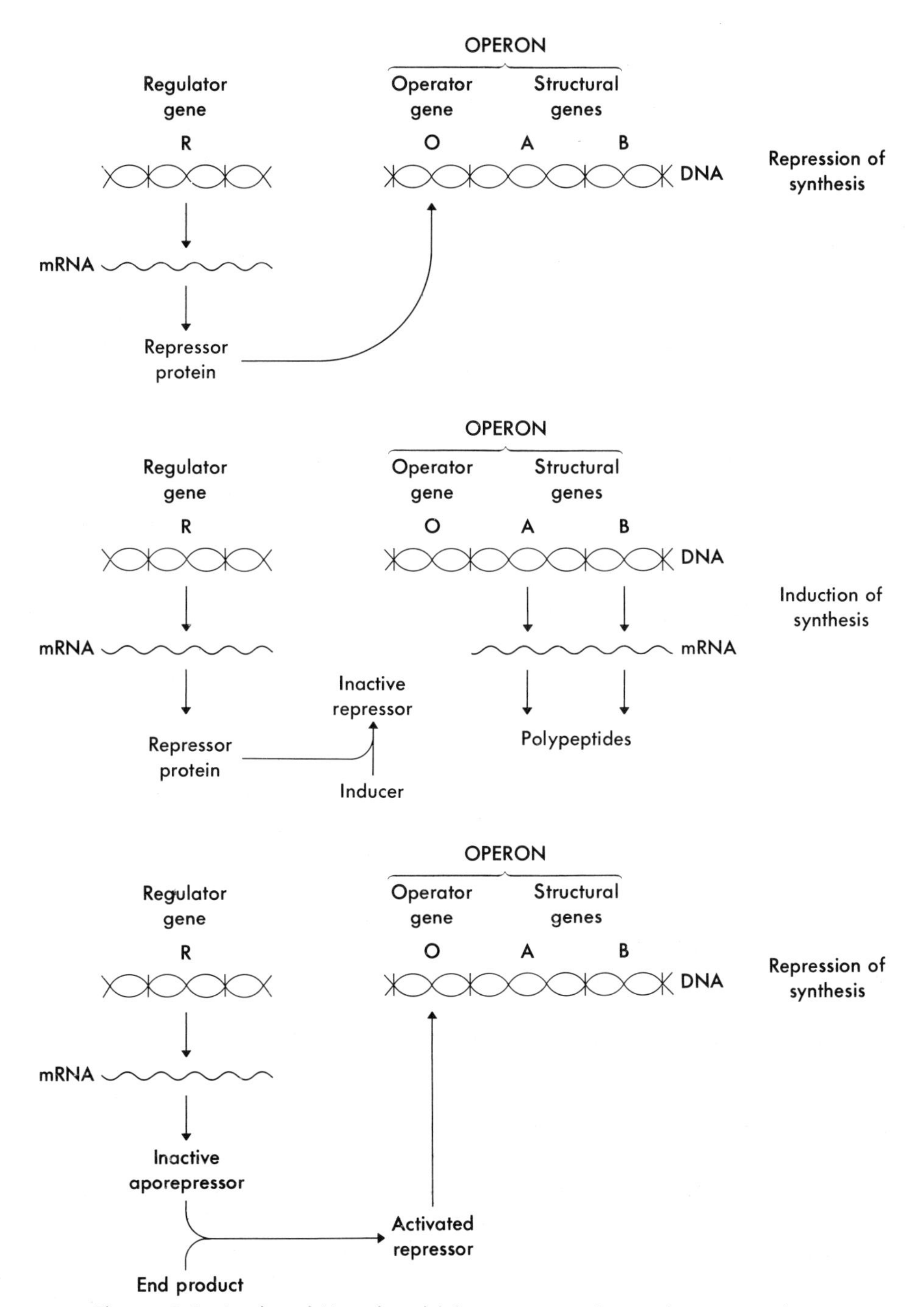

Figure 9.2. Jacob and Monod model for genetic regulation of protein synthesis.

is determined by the repressor. It seems likely that some translational control must operate in the cells of higher organisms whose structure and function differ from those of bacteria. Bacterial chromosomes consist of naked, double-stranded DNA molecules, whereas the chromosomes of animal cells with a nucleus contain not only DNA but RNA and a variety of proteins. Also a great variety of RNA molecules representing transcription of a large portion of the animal genome are synthesized but do not function as messenger RNAs, although they have the potential to do so.

Whether or not the proposals of Jacob and Monod or those of other investigators turn out to be correct, these investigators have brought some order to what was a chaotic area and have stimulated much thought and research on this difficult and complex problem.

APPLICATION TO HUMAN GENETICS

Although our concepts of genetic regulation have been derived primarily from the study of microorganisms, it seems likely that such mechanisms operate in principle in the cells of higher organisms including man. There are a number of conditions in the human which can be attributed to mutations of structural genes. These mutations result presumably in the production of an altered mRNA and hence a qualitatively different protein. Although some of these alterations produce clinical disease, the preponderance of such instances probably represents normal biochemical variation.

The most completely studied example of the effect of gene mutations on the primary structure of a mammalian protein is afforded by the hemoglobin variants. Hemoglobin is a conjugated protein composed of four molecules of ferroprotoporphyrin IX attached to a single globulin molecule. Human hemoglobins have a molecular weight of 68,000, and the globin moiety is composed of two identical half-molecules, each half-molecule consisting of two different polypeptide chains. In the normal adult hemoglobin, HbA, these are the α and β chains. The normal fetal hemoglobin (HbF) and the normal minor component (HbA_2) have respectively α, γ and α, δ chains. The synthesis of these four peptide chains is controlled by four distinct genes designated α, β, γ and δ. The hemoglobin variants either contain altered combinations of these four chains, or chains with an altered sequence of amino acids.

Pauling and coworkers made the important discovery in 1949 that the hemoglobin from patients with sickle-cell anemia differed in electrophoretic mobility from the hemoglobin of normal individuals. Furthermore, it could be shown that the clinical manifestations of the disease could be attributed to the effect of a gene mutation on the structure of a protein. In HbS there is the replacement of a glutamic acid by a valine residue at position 6 of the β-chain. Following the discovery of HbS, many other abnormal hemoglobins were discovered on the basis of differential electrophoretic behavior. Certain hemoglobin variants of identical electrophoretic behavior were detected by digesting with enzymes and examining the resulting peptides with the "finger-printing" technique. In this technique, the denatured hemoglobin is subjected to tryptic digestion. The peptides obtained are separated by paper electrophoresis in one direction, followed by paper chromatography in a perpendicular direction. A tracing of the fingerprints of HbA and HbS is shown in Figure 9.3. Peptide four, which was shown to contain eight amino acids, occupies a different position in the two fingerprints and is the one which has the amino acid substitution (valine for a glutamate residue in HbS). In this manner minor differences in chemical composition could be detected.

Subsequently, the amino acid sequence of the α and β chains of HbA and the γ-chain of HbF were elucidated. In Table 9.2 are shown some of the abnormal variants of HbA in which the position and nature of the amino acid substitution are known. All the abnormal hemoglobin variants so far studied differ from HbA in a single amino acid.

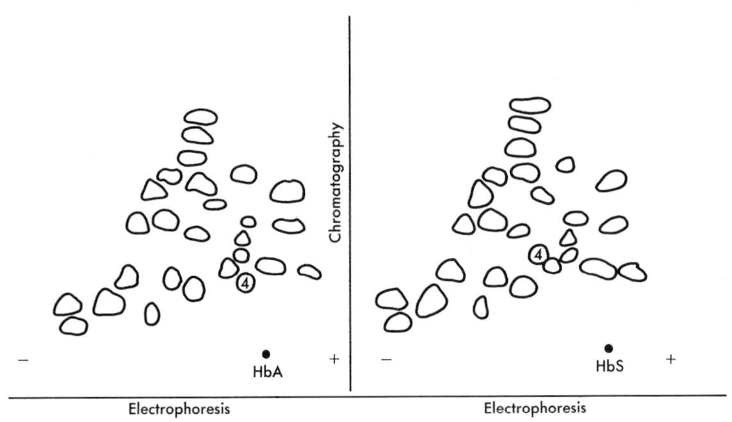

Figure 9.3. Fingerprints of tryptic digests of hemoglobin (A) and (S).

These alterations are the smallest ones possible in the primary structure of a protein and as noted previously (page 322) provided support for the genetic code as elucidated by other techniques.

Many other instances of biochemical variation are known. The polymorphism of the transferrins is another example of genetic heterogenity referrable to a single mutational event. In other instances, such as the blood group substances, more than one mutation must have been involved as the structural differences are too extensive to be the result of a single-step mutation. Other human proteins which occur in structurally altered forms are, among many others, the haptoglobins, gamma globulins, glucose-6-phosphate dehydrogenase, and cholinesterase.

Table 9.2
Some of the Known Amino Acid Substitutions in Abnormal Hemoglobins

Hemoglobin	Chain and Position	Amino Acid Substitution*	Codon Alteration
G Honolulu	α,30	Glu ⟶ Gln	GAA(G) ⟶ CAA(G)
Norfolk	α,57	Gly ⟶ Asp	GGU(C) ⟶ GAU(C)
M Boston	α,58	His ⟶ Tyr	CAC(U) ⟶ UAC(U)
G. Philadelphia	α,68	Asn ⟶ Lys	AAU(C) ⟶ AAA(G)
O Indonesia	α,116	Glu ⟶ Lys	GAA(G) ⟶ AAA(G)
S	β,6	Glu ⟶ Val	GAA(G) ⟶ GUA(G)
C	β,6	Glu ⟶ Lys	GAA(G) ⟶ AAA(G)
G San Jose	β,7	Glu ⟶ Gly	GAA(G) ⟶ GGA(G)
E	β,26	Glu ⟶ Lys	GAA(G) ⟶ AAA(G)
Zurich	β,63	His ⟶ Arg	CAC(U) ⟶ CGC(U)

* See Table 9.1 (page 321) for abbreviations.

In certain human hereditary conditions, there may be a considerable decrease in the amount of a protein synthesized, although some is usually present. Such conditions as agammaglobulinemia, analbuminemia, hypophosphatasia, galactosemia, alcaptonuria, phenlyketonuria, and many others may be ascribed to the mutation of a control gene. Here, it is presumably only the amount of the protein and not the nature of the protein that is affected. In the case of hypergammaglobulinemia, the amount of γ-globulin is strikingly increased and is due presumably to changes in control mechanisms.

A gene mutation may then result either in the production of an altered protein or in an alteration in the amount of the protein synthesized.

REFERENCES

Ames, B. N., and Martin, R. G.: Biochemical Aspects of Genetics: The Operon, *Ann. Rev. Biochem.*, **33**:235 (1964).
Campbell, P. N.: The Biosynthesis of Proteins, *Progr. Biophys. Mol. Biol.*, **15**:3 (1965).
Cold Spring Harbor Symposia on Quantitative Biology: *The Genetic Code,* Vol. XXXI, 1966.
Ingram, V.: *The Biosynthesis of Macromolecules.* W. A. Benjamin, Inc., New York, 1965.
Jacob, F., and J. Monod: Genetic Regulatory Mechanisms in the Synthesis of Proteins, *J. Mol. Biol.*, **3**:318 (1961).

Lippman, F.: Polypeptide Chain Elongation in Protein Biosynthesis: *Science*, **164**, 1024 (1969).
Mahler, H. R., and Cordes, E. H.: *Biological Chemistry*, Chap. 18. Harper and Row, New York, 1966.
Nirenberg, M.: The Genetic Code II, *Sci. Am.*, March, 1963, p. 80.
Novelli, G. D.: Amino Acid Activation for Protein Synthesis, *Ann. Rev. Biochem.*, **36**(II):449 (1967).
Steiner, R. F.: *The Chemical Foundations of Molecular Biology*. Van Nostrand, Princeton, N. J., 1965.
Watson, J. D.: *Molecular Biology of the Gene*. W. A. Benjamin, Inc., New York, 1965.
West, E. S.; Todd, W. R.; Mason, H. S.; and Van Bruggen, J. T.: *Textbook of Biochemistry*, 4th ed., Chaps. 25 and 26. The Macmillan Co., New York, 1966.
White, A.; Handler, P.; and Smith, E. L.: *Principles of Biochemistry*, 4th ed., Chaps. 29, and 30. McGraw-Hill, New York, 1968.

Chapter 10

BLOOD

In chapter 1 it was pointed out that the tissue cells are connected with the outside environment through the circulation, and it is through the medium of the blood that nutriments are transported to cells and waste products removed. The general composition of the blood remains remarkably constant despite numerous biochemical substances and cellular elements constantly leaving and entering the blood stream. This *steady state* is maintained by biochemical and physiologic regulatory processes that result in equal rates of destruction and formation, or output and input, of cellular elements and biochemical substances.

Cells in the higher organism are bathed in interstitial fluid to which waste products are added and from which substances are removed by processes of diffusion. Regulation of the composition of interstitial fluid is achieved by diffusion through the capillary walls to and from the circulating blood. Through these mechanisms cells are able to maintain their cellular environment within certain limits and are protected temporarily from large external physical and chemical variations.

The primary functions of the blood may be considered in a broad sense to be

1. *Metabolic regulation*—transport of O_2, CO_2, metabolites, hormones, and so forth.
2. *Physical chemical regulation*—Temperature, acid-base balance, and osmotic pressure and fluid balance.

3. *Regulation of body defenses*—protection against infection by action of antibodies, leukocytes, and other mechanisms, and prevention of hemorrhage.

Some of these functions are discussed in this and the following chapter.

GENERAL PROPERTIES OF BLOOD

Normal values for total blood volume vary depending on the methods used for determination and the basis of reference (sex, age, weight, height, and surface area). In a healthy adult of normal body habitus, the circulating blood comprises 6 to 8 per cent of the body weight, or 5000 to 6000 ml.

Cells, Serum and Plasma

The cellular components of blood are the erythrocytes (red cells), leukocytes (white cells), and the platelets (thrombocytes). The blood cells make up 40 to 45 per cent of the blood volume and the hematocrit is the per cent of the blood volume occupied by the blood cells as determined by centrifugation under standard conditions.

The blood cells are chiefly *erythrocytes* (4.2 to 5.4 million per mm^3) which contain about 35 per cent solids of which some one third is hemoglobin (page 328). There is then about 15 g of hemoglobin per 100 ml of blood. This is twice the quantity of the other proteins in blood. Some details of erythrocyte metabolism are discussed on page 344.

The *leukocytes* are of several types and the normal range for the adult is 4500 to 11,000 cells mm^3 of blood. Unlike erythrocytes, the leukocytes are nucleated and possess the usual biochemical pathways.

The blood *platelets* are small discoid-shaped cells without a nucleus. The normal range is 180,000 to 360,000 per mm^3 of venous blood, and they are indispensable to the coagulation system of blood.

When blood clots (page 341), the plasma protein, fibrinogen, is converted to fibrin which forms a network of threads. Blood cells become enmeshed in this network to form the clot. The clear fluid which is formed as the clot retracts is the *serum*.

When citrate, heparin, or some other anticoagulant is added to blood to prevent clotting and the cells removed by centrifugation, the fluid remaining is the *plasma*. Plasma differs from serum in that it contains the protein, fibrinogen.

Table 10.1 lists the normal values for some of the more common constituents of blood. The quantitative determination of these constituents can provide useful information on the nature and extent of a number of pathologic conditions.

Plasma Proteins

Blood plasma is made up of 6 to 8 per cent protein, and this constitutes most of the solids (Table 10.1). The plasma proteins are many and complex and their functions diverse. The primary functions of the plasma proteins include the transport of fatty acids, steroids, hormones, metal ions, and so forth; the maintenance of colloid osmotic pressure, electrolyte balance, and pH; hemostasis and prevention of thrombosis; a nutritional source of amino acids for the tissues; and defense against infection through the action of antibodies and other factors. Since the plasma proteins are components of a metabolically active system and are often altered in disease, their determination and fractionation have been of considerable clinical interest.

Table 10.1
Normal Values of Blood Constituents (Values Are for Serum Unless Otherwise Stated)

Constituent	Normal Range
Hemoglobin	13–18 gm/100 ml blood (male)
	11–16 gm/100 ml blood (female)
Serum proteins	6–8 gm/100 ml
Glucose (fasting)	70–115 mg/100 ml
Urea nitrogen	8–20 mg/100 ml
Nonprotein nitrogen	18–30 mg/100 ml
Uric acid	3.0–6.5 mg/100 ml (male)
	2.5–5.5 mg/100 ml (female)
Creatinine	0.6–1.1 mg/100 ml
Cholesterol, total	140–280 mg/100 ml
Cholesterol esters	72–78% of total
Bilirubin, total	0.3–1.0 mg/100 ml
Bilirubin, direct	0.1–0.4 mg/100 ml
Sodium	137–148 meq/l
Chloride	98–110 meq/l
Potassium	3.5–5.2 meq/l
CO_2-content	21–29 mM/l
Calcium	8.7–10.7 mg/100 ml
Phosphorus (inorganic)	2.5–4.8 mg/100 ml (adult)
	4.0–6.5 mg/100 ml (child)
Magnesium	1.6–2.2 meq/l
Alkaline phosphatase	0.6–2.5 BLB units/l (adult)
	3.4–9.0 BLB units/l (child)
Acid phosphatase, total	0.05–0.70 BLB units/l
Acid phosphatase, prostatic	0.01–0.20 BLB units/l
Amylase	60–140 Somogyi units/100 ml
Lactic dehydrogenase	150–350 units/ml
Glutamic-oxaloacetic transaminase	10–30 units/ml
Glutamic-pyruvic transaminase	5–25 units/ml
Isocitric dehydrogenase	50–250 units/ml

Separation and Identification

The isolation of plasma proteins is based usually on exploiting the differences in one or more of the following properties: solubility under controlled conditions, net electric charge, and the size, shape, or density of the molecule. The first fractionation procedures employed salting-out or precipitation by organic solvents of low polarity to yield fractions containing various mixtures of proteins. The various factors influencing protein solubility (ionic strength, hydrogen ion concentration, temperature, dielectric constant of medium, protein concentration, and so forth) were manipulated to give the desired end. The gross separation of serum proteins into albumin and globulin fractions (A/G ratio) by sodium sulfate is an example of such a method as is the cold-ethanol procedures developed by E. J. Cohn and associates during World War II.

The introduction of the moving-boundary electrophorisis apparatus by Tiselius and the analytical ultracentrifuge by Pedersen and Svedberg materially improved the separation and characterization of plasma proteins and provided systems of nomenclature. From moving-boundary electrophoresis came the classic fractions of albumin and α-, β-, γ-globulins (page 47). Two major components with sedimentation coefficient values (page 42) of about 4S and 7S and a minor component of about 19S are seen in the classic ultracentrifuge patterns of serum. All of these fractions were subsequently shown to be heterogeneous by the newer techniques of zone electrophoreses on various supporting media, immunoelectrophoresis, and chromatography.

Figure 10.1 shows the separation patterns obtained by electrophoresis of human sera on cellulose acetate strips. The strips were stained with Ponceau S dye after an electrophoretic run of $1\frac{1}{2}$ hours in a veronal buffer of pH 8.6.

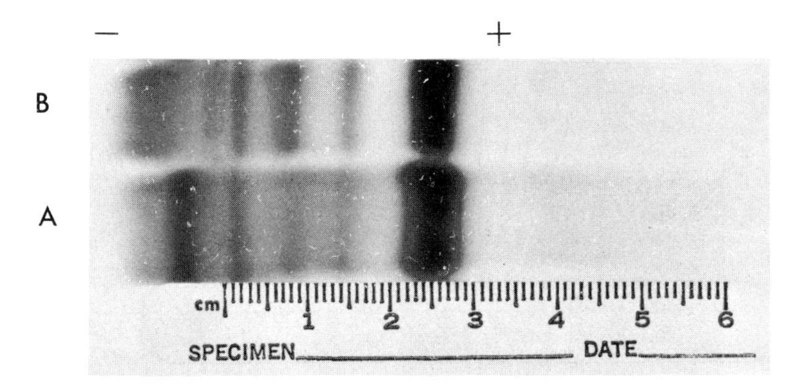

Figure 10.1. Cellulose acetate electrophoresis of human serum. (A, serum of patient with multiple myeloma; B, normal serum. Courtesy of Miss Pat Robinson, Rush–Presbyterian–St. Luke's Medical Center, Chicago.)

The fastest migrating fraction, albumin, is on the right followed by α_1-globulins, α_2-globulins, β_1-globulins, β_2-globulins, and γ-globulins. These fractions do not represent a single homogeneous protein but a group of unrelated proteins with similar electrophoretic mobilities under the conditions used. For example, the α_2 fraction is known to contain haptoglobin, ceruloplasmin, α_2-macroglobulins, and other components (see immunoelectrophoresis below). Strips A and B of Figure 10.1 show respectively the serum electrophoretic patterns for an individual with multiple myeloma (not unusual protein in γ region) and the normal individual. The most important use of serum protein electrophoresis is as an aid in the diagnoses of diseases in which abnormal proteins appear in the circulating blood (for example, multiple myeloma and macroglobulinemia) or those in which a component is missing or its content markedly decreased (analbuminemia and agammaglobulinemia, for example). Analysis of human sera by starch-gel electrophoresis reveals some 22 zones and led to the discovery of genetic variants of transferrin and haptoglobin.

A precise and discriminating technique has been developed for identifying proteins. This technique, called *immunoelectrophoresis,* utilizes both the electrophoretic and immunologic properties of proteins for their identification. The protein mixture, for example serum, is first subjected to electrophoretic separation, usually in agar gel, although other supporting media may be used. After electrophoretic separation, specific immune serum is added to a groove extending the length of the plate beside the separated protein fractions. Precipitin lines form after some hours at the points of contact, the antigens (separated proteins) diffusing radially outward and the antibodies (immune serum) diffusing laterally inward. The number, form, and intensity of the precipitin lines indicate the incidence, nature, and extent of the precipitin reactions that have occurred. Figure 10.2 shows the pattern obtained when

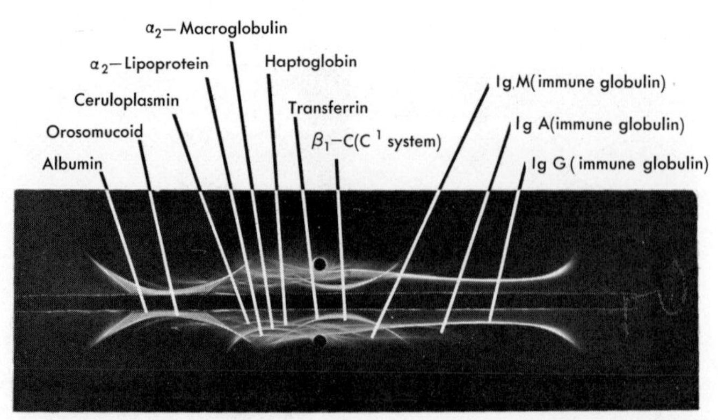

Figure 10.2. Immunoelectrophoretic pattern of normal human serum developed with goat antiserum to whole human serum. (Courtesy of Dr. C. A. Williams, Jr., Rockefeller University, New York City.)

normal human serum was separated electrophoretically and reacted with a goat antiserum to whole human serum.

Properties and Functions of Major Plasma Proteins

Table 10.2 list some of the better characterized plasma proteins.

Albumin

Albumin constitutes approximately one half of the total plasma proteins; because of this concentration and its relatively small size, it contributes 75 to 80 per cent of

Table 10.2
Some Plasma Proteins

Protein	Molecular Weight	Mg/100 ml Plasma	Function
Albumin	69,000	3500–4500	Transport, osmotic pressure regulation, storage form of amino acids
α_1-Lipoproteins (high density)	195,000 435,000	217–270 37–117	Transport of lipids
α_1-Antitrypsin	45,000	210–500	Inhibits trypsin and chymotrypsin
Haptoglobins	0.85, 2×10^5, 4×10^5	30–190	Binds hemoglobin
Ceruloplasmin	160,000	27–63	Transport of Cu? Ferro oxidase?
α_2-Lipoproteins (very low density)	5 to 20×10^6	150–230	Transport of lipids
β-Lipoproteins (low density)	2 to 3×10^6	240–440	Transport of lipids
Transferrin	90,000	200–320	Transport of Fe
β_{1A}-β_{1C}-Globulins	—	35	component of complement
Hemopexin	80,000	80–100	Binds heme
Fibrinogen	340,000	200–600	Blood coagulation
Plasminogen	140,000	50–100	Lysis of fibrin in clots
γ-G-Immuno globulin (IgG)	150,000	900–1500	Antibodies
γ-M-Immuno globulin (IgM)	$(150{,}000)_n$ $n = 6, 8 \ldots$	39–117	Antibodies
γA-Immuno globulin (IgA)	$(150{,}000)_n$ $n = 1, 2, 4$	110–180	Antibodies

the colloid osmotic pressure of plasma. It has, therefore, an important role in the regulation of intravascular volume. Albumin has a high affinity for anions and other substances and is the major protein concerned with the transport of anions (for example, bilirubin, page 346) and fatty acids, cations, drugs, and dyes.

There is a condition known as *analbuminemia* in which there is a decreased rate of albumin synthesis. The plasma level of albumin is severely reduced (1 to 25 mg per 100 ml) in this autosomal recessive disease in which edema is a common symptom.

In some healthy individuals electrophoresis reveals two distinct albumin peaks, the sum of which is equal to the albumin concentration in normal serum. This genetic polymorphism is apparently due to a structural mutation (page 330) with the substitution of a single amino acid residue.

α- and β-Lipoproteins

The concentration and distribution of serum lipoproteins are of considerable importance in the transport and metabolism of lipids. Electrophoresis and ultracentrifugation are the two standard techniques for separation and classification of lipoproteins (Table 10.2). In the ultracentrifuge three different broad density classes of lipoproteins can be obtained. These are *very low density lipoproteins* (VLDL), *low density lipoproteins* (LDL), and *high density lipoproteins* (HDL). Although each class contains cholesterol, cholesterol esters, glycerides, phospholipids, and free fatty acids, the relative content of each fraction varies greatly.

The low density lipoproteins (β-lipoproteins) is the fraction in which the age-associated increases in phospholipid and cholesterol occur and it is these proteins that are missing in a rare inherited disease known as *abetalipoproteinemea.* Among other symptoms this disease is characterized by abnormally low concentrations of cholesterol, phospholipids, and triglycerides. The primary defect may be the inability to synthesize the protein specific for LDL molecules.

In *Tangier disease* there is almost complete absence in the plasma of HDL (α_1-lipoproteins). Cholesterol is usually below 120 mg per 100 ml, phospholipids are also reduced, and triglycerides are normal or elevated. The biochemical defect is most likely the failure in synthesis of HDL protein.

There is a generic group of diseases termed *familial hyperlipoproteinemia.* Some five types are recognized in which one or a combination of two of the lipoprotein fractions are elevated.

Haptoglobins

Haptoglobins are serum glycoproteins which have the ability to form specific stable complexes with hemoglobin. Three common phenotypes exist, designated Hp 1–1 (M.W. 85,000), Hb 2–1 (M.W. 200,000), and Hp 2–2 (M.W. 400,000). Other genetic variants are also known. The presumed biologic function of haptoglobin is to bind hemoglobin to protect the kidney from damage by hemoglobin and to prevent excessive urinary loss of iron.

Ceruloplasmin

Some 98 per cent of the serum copper is bound to a specific α_2-globulin called ceruloplasmin. Ceruloplasmin (M.W. 160,000) binds eight atoms of copper per mole. In Wilson's disease there is a decreased concentration of ceruloplasmin and a deposition of copper in the liver, brain, kidneys, and cornea. There appears to be both an increased absorption and a decreased excretion through the intestinal tract in Wilson's disease. The function of ceruloplasmin is not clear. It could serve a transport function for copper and control in some way both the absorption and biliary excretion of copper. Recently it has been suggested that ceruloplasmin acts to oxidize Fe^{++} to Fe^{+++} which is requisite for the binding of iron to transferrin.

Transferrin

The major part of the plasma iron is reversibly bound to the specific iron-binding β_1-globulin, transferrin. Each transferrin molecule (M.W. 90,000) can bind two atoms of Fe^{+++}. The genetic polymorphism of transferrin is extensive. Seventeen different molecular species of transferrin have been detected by electrophoretic analysis in starch-gel. The hereditary absence of transferrin has been reported in a patient with a severe hypochromic anemia. The biologic role of transferrin is discussed under iron metabolism (page 348).

Immunoglobulins

Antibody activity is associated with three major classes of plasma proteins all of which are synthesized in plasma and lymphoid cells. These immunoglobulins are designated by IgG (γG, 7S), IgA (γA, β_2A), and IgM (γM, β_2M). A current schematic model for the structures is shown in Figure 10.3. IgG, IgA, and IgM all consist of a pair of identical light chains (κ or λ), with a molecular weight of about 25,000, and a pair of identical heavy chains (γ, α, or μ), with a molecular weight of about 50,000, all joined by disulfide bonds. Thus the three major classes may exist as six major subgroups: IgG ($\kappa_2\gamma_2$ or $\lambda_2\gamma_2$), IgA ($\kappa_2\alpha_2$ or $\lambda_2\alpha_2$), and IgM ($\kappa_2\mu_2$ or $\lambda_2\mu_2$). The heavy chains (γ, α, or μ) which differ in structure carry the antigenic determinants characteristic of the class (IgG, IgA, or IgM). The light chains will determine the antigenic types either κ or λ. The Bence Jones proteins excreted by individuals with multiple myeloma are either of the κ or λ type, although they differ in structure in each individual.

A disorder known as *agammaglobulinemia* is characterized by decreased amounts of all three immunoglobulins.

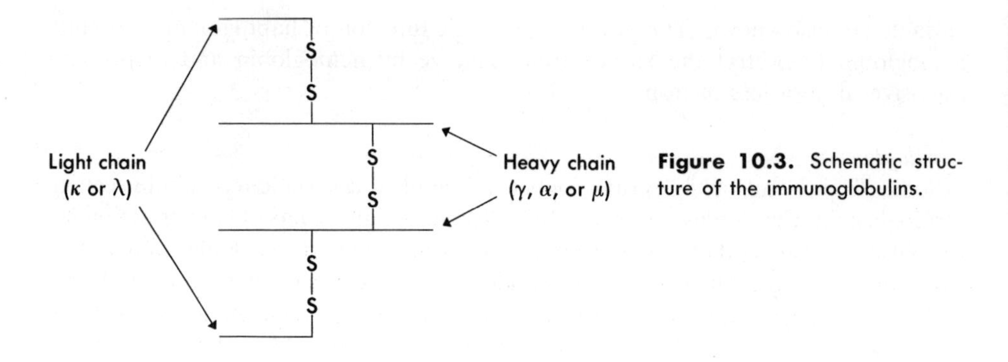

Figure 10.3. Schematic structure of the immunoglobulins.

Serum Enzymes

There are present in blood plasma (serum) a number of enzymes derived from the breakdown of tissue and blood cells. Aldolase, amylase, glutamic-oxaloacetic transaminase (SGOT), glutamic-pyruvic transaminase (SGPT), isocitric dehydrogenase (ICD), lipase, acid and alkaline phosphatases, and lactic dehydrogenase (LDH) among others are to be found in serum. The measurement of the activity of some of these enzymes in serum can provide extremely useful diagnostic and prognostic clinical information.

Certain enzymes such as amylase, lipase, and the phosphatases have a relatively limited tissue distribution, and diseases of tissues containing these enzymes are often revealed by rather specific elevations in serum enzyme activities. On the other hand, enzymes that perform generalized metabolic functions and likely to be present in substantial amounts in most tissues will be less precise as an indicator of specific tissue involvement. The transaminases (SGOT, SGPT) and lactic dehydrogenase (LDH) are examples of this type.

A further complication arises from the fact that the activity of a given serum enzyme may be due to a mixture of similar enzymes derived from different tissues in the body and even from the same tissue. Such variants are termed "isoenzymes" and were found by their differential electrophoretic migration. In the case of lactic dehydrogenase (page 69), five isoenzymes may be resolved electrophoretically from serum and most tissues and possibly seven in the red blood cells. The relative distribution of lactic dehydrogenase isoenzymes varies not only from tissue to tissue but from species to species. Figure 10.4 shows the distribution of LDH isoenzymes in serum, heart, liver, and erythrocytes.

The LDH isoenzymes differ in a number of properties such as substrate affinity, stability at elevated temperatures, and effects of inhibitors. The relative abundance of certain LDH isoenzymes in serum has been a useful aid in diagnosis: an increase of LDH-1 and LDH-2 is characteristic of myocardial infarction, whereas an increase of LDH_4 and LDH_5 is found in hepatitis (Figure 10.4).

Blood Coagulation

Blood clotting is the end result of a complex series of biochemical reactions. Some of the reactions have been shown to be enzymatic in nature, and others are presumed to be so. Despite considerable progress in the understanding of blood coagulation, there are still many uncertainties and all schemes for blood clotting must be considered tentative, including the oversimplified one presented here. Different viewpoints do exist, and the reader is directed to the appropriate literature for these.

One of the more remarkable characteristics of blood is that it normally clots when

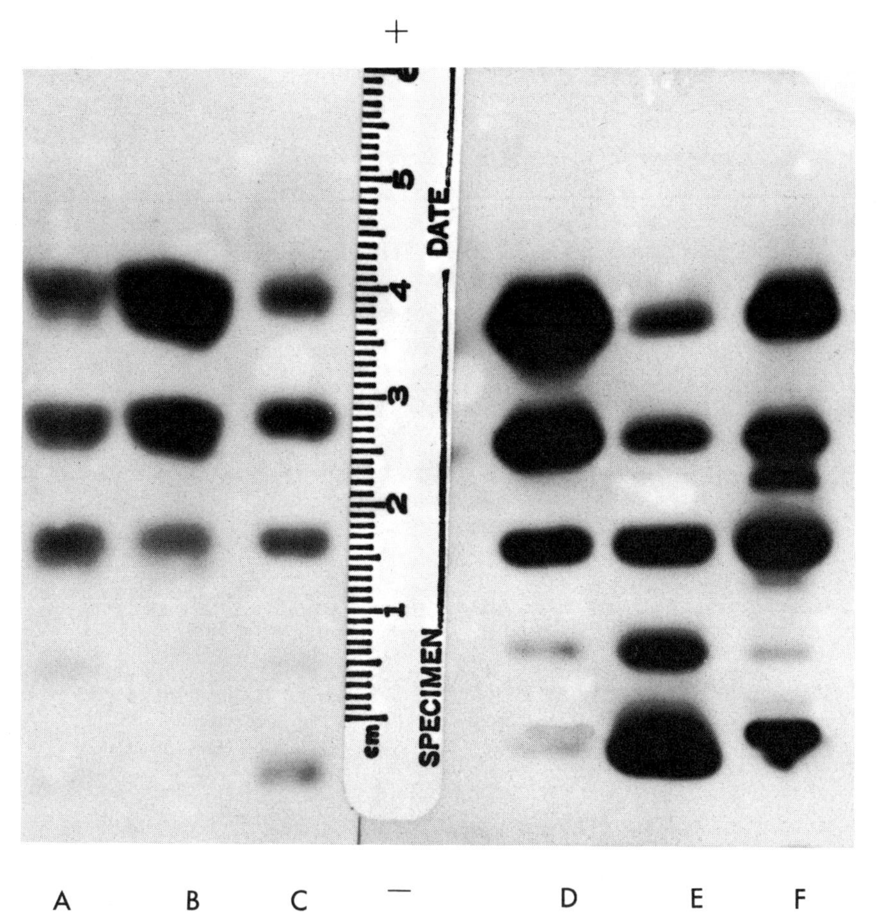

Figure 10.4. Electrophoretic separation on Cellogel of LDH isoenzymes of human serum and tissues. A, Normal human serum; B, serum from patient with myocardial infarction; C, serum from patient with infectious hepatitis; D, heart; E, liver; F, erythrocyte. (Courtesy of Dr. Margaret A. Kenny, Rush–Presbyterian–St. Luke's Medical Center, Chicago.)

necessary and usually does not when not necessary. Since all the components necessary for coagulation are present in blood, there must be some way to prevent unnecessary clotting. This is accomplished by two basic methods. First, certain of the coagulation factors are present in an inactive form and must be activated prior to functioning in the coagulation process. Second, blood contains anticoagulants which inhibit various steps in coagulation.

There are two pathways by which clotting may be initiated, the *intrinsic* system and the *extrinsic* system, both of which probably operate *in vivo*. The series of reactions in plasma leading to the formation of thrombin without the participation of tissue extracts is known as the *intrinsic* system (Figure 10.5). Those initiated by adding tissue extracts to plasma are called the *extrinsic* clotting pathway (Figure 10.6).

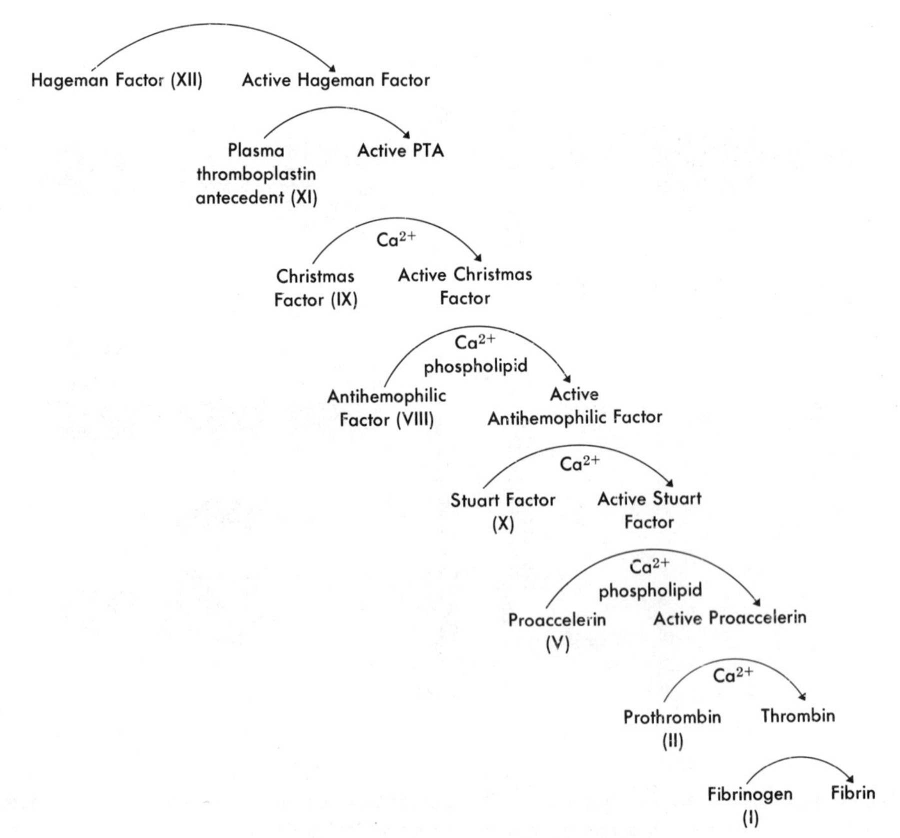

Figure 10.5. Intrinsic system for blood clotting.

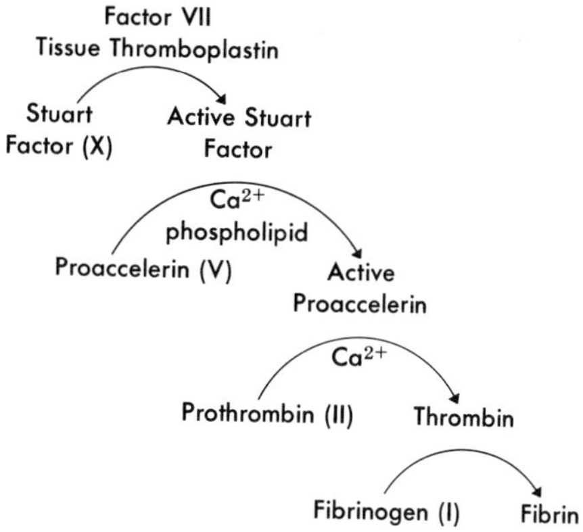

Factor VII
Tissue Thromboplastin
Stuart Factor (X) → Active Stuart Factor
Ca^{2+}
phospholipid
Proaccelerin (V) → Active Proaccelerin
Ca^{2+}
Prothrombin (II) → Thrombin
Fibrinogen (I) → Fibrin

Figure 10.6. Extrinsic system for blood clotting.

There are several important points to be made regarding the tentative "cascade" or "waterfall" sequences shown in Figures 10.5 and 10.6. These are

1. Each factor occurs in plasma in an inactive form and is shown on the left side of the cascade.
2. Upon initiation of clotting, each factor (excepting fibrinogen) is converted to an active form (shown on right) that has enzymic or presumed enzymic activity.
3. Activation of each factor occurs in a stepwise sequence with each newly formed "enzyme" reacting with its specific substrate to convert it to an active "enzyme."
4. The two pathways are probably identical after the Stuart factor.

In the intrinsic system of scheme postulates the successive participation of activated Hageman factor (Factor XII) plasma thromboplastin antecedent (PTA), Christmas factor (plasma thromboplastin component, Factor IX), antihemophilic factor (Factor VIII) and Stuart factor (Factor X) to form activated proaccelerin (Ac-globulin, Factor V) which can convert prothrombin to thrombin. Calcium ions and phospholipids probably derived from platelets are also required. The early steps of coagulation are circumvented by the addition of tissue thromboplastin (Factor VII). The successive steps in the extrinsic pathway are tissue thromboplastin, Stuart factor, and proaccelerin.

The final stages of the coagulation process involve the conversion of the soluble plasma protein, fibrinogen, into the insoluble fibers of the fibrin clot. Thrombin, which has proteolytic activity, splits two acidic peptides (A and B) from each fibrinogen molecule, producing a soluble monomer of fibrin. Peptide A has a molecular weight of 1900 and peptide B a molecular weight of 2400, and approximately 2 moles of each are released per mole of fibrinogen. The fibrin monomers are polymerized and aggregated to form a soluble fibrin clot. The soluble fibrin clot is converted to a more insoluble form by the action of a substance present in plasma and other tissues. This substance is called "*fibrin-stabilizing factor*" or "*fibrinase*" and in the presence of Ca^{++} it acts to cross link the fibrin monomers by a transamidination reaction involving the β-carboxyl of asparagine of one fibrin monomer with the N-terminal glycine residue in an adjacent fibrin monomer. The conversion of fibrinogen to fibrin may be summarized by the following reactions.

$$\text{(Fibrinogen)}_n \xrightarrow{\text{thrombin}} \text{Fibrin (monomer)}_n + \text{(Peptide A)}_{2n} + \text{(Peptide B)}_{2n}$$

$$\text{(Fibrin monomer)}_n \longrightarrow \text{Fibrin}$$

$$\text{Fibrin} \xrightarrow[\text{Ca}^{++}]{\text{fibrinase}} \text{Cross-linked fibrin}$$

The Erythrocyte

Erythrocytes are derived from primitive nucleated cells in the bone marrow by processes of mitosis and maturation. Table 10.3 summarizes the properties of erythroid cells at three principal stages of development: the nucleated erythroid cells of the bone marrow, the reticulocytes, and the mature erythrocyte.

The nucleated erythroid cells can carry out diverse metabolic reactions such as the synthesis of DNA, RNA, proteins, lipids, and carbohydrates. The bulk of the hemoglobin is synthesized at this stage.

There is little DNA in the reticulocyte which is devoid of a nucleus. RNA is present, some of which may be the messenger for globin synthesis. Mitochondria, ribosomes, and endoplasmic reticulum are visible in electron micrographs.

Table 10.3
Characteristics of Erythroid Cells

Component or Function	Nucleated Erythroid Cell	Reticulocyte	Mature Erythrocyte
RNA	Present	Present	Absent
DNA	Present	Absent	Absent
Glycolytic pathway	Active	Active	Active
Tricarboxylic acid cycle	Active	Active	Inactive
Phosphogluconate pathway	Active	Active	Active
Cytochrome system	Active	Active	Inactive
Heme synthesis	Active	Active	Inactive
Protein synthesis	Active	Active	Inactive
Lipid synthesis	Active	Active	Inactive

In contrast, the mature erythrocyte has neither mitochondria nor ribosomes and little or no RNA. The synthesis of lipid, heme, and globin does not occur. In the absence of a functioning tricarboxylic acid cycle, the erythrocyte must obtain its energy from glycolysis and from the oxidation of glucose via the phosphogluconate oxidative pathway.

The mature erythrocyte is a nonnucleated biconcave disc about 7μ in diameter. It has a surface membrane, a complex mosaic of protein and lipid, that projects irregularly into the body of the cell to form a stroma meshwork. Hemoglobin is dispersed throughout the stroma in increasing concentration from the periphery to the interior of the cell. When placed in hypotonic solutions, erythrocytes swell and the membranes rupture to release the internal components of the cell. This process, called *hemolysis,* may also be achieved in isotonic solutions by surface active agents. Virtually all of the constituents of the red cell are soluble with the exception of the stroma or ghost. The stroma contains protein, most of the lipids of the cell (cephalins, lecithins, and cholesterol), and the blood group substances.

The major component of the red cell is hemoglobin, the biosynthesis of which is discussed on page 236. The structure of hemoglobin and the abnormal hemoglobins are discussed on page 328.

In common with other cells, potassium is the predominating cation in the erythrocyte. Small amounts of sodium and magnesium are also present. The major anions are chloride, hemoglobin, 2,3-diphosphoglycerate (the coenzyme for phosphoglyceromutase, page 129), bicarbonate, ATP, inorganic phosphate, and glucose-1,6-diphosphate (the coenzyme for phosphoglucomutase, page 122). Water, small uncharged organic molecules, univalent anions such as HCO_3^-, Cl^-, and OH^- pass freely and rapidly into and out of the cell. The passage of both K^+ and Na^+ across the cell membrane and the maintenance of different concentrations in the erythrocyte and plasma are due to active transport of both of these ions.

The erythrocyte has a high concentration of glutathione which is present largely in the reduced form. Glutathione reductase the NADPH-dependent reduction of glutathione.

$$\text{GSSG} + \text{NADPH} + \text{H}^+ \rightleftharpoons 2\ \text{GSH} + \text{NADP}^+$$

The production of NADPH is dependent on the action of glucose-6-phosphate dehydrogenase (page 145). An important group of acquired hemolytic anemias is related to glutathione reductase and glucose-6-phosphate dehydrogenase. These anemias are due to sensitivity to certain drugs, for example, sulfonamides, phenacetin, primaquine, and the fava bean. The erythrocytes of sensitive individuals have a reduced ability to maintain glutathione in the reduced state. This is a consequence of a deficiency of glucose-6-phosphate dehydrogenase, upon which the production of NADPH depends. Structural integrity is not maintained and hemolysis ensues.

The number of erythrocytes in the circulating blood remains relatively constant. New cells are formed in the bone marrow at the same rate old cells are destroyed. Peripheral erythrocytes have an average life-span of about 125 days. Some 0.8 per cent of the total red cells are destroyed and formed each day. Erythrocytes are destroyed by the reticuloendothelial system mainly in the spleen, liver, and bone marrow. When the erythrocyte is destroyed, the iron is released from hemoglobin, combined with transferrin (page 339), and transported to storage depots where it is efficiently reutilized. The porphyrin moiety is converted to bile pigments and excreted from the body. The globin is degraded to amino acids and returned to the body pool of amino acids.

Bile Pigment Metabolism

The exact nature and sequence of events in the breakdown of hemoglobin to bile pigments are as yet not certain. Still a matter of some controversy is whether the protoporphyrin ring is separated from the globin prior to being cleaved at the α-methene bridge or whether the opening of the porphyrin ring occurs while the ring is still associated with the globin. Be that as it may, the first bile pigment produced in the reticuloendothelial cell appears to be *biliverdin* which is rapidly reduced to *bilirubin*. Bilirubin is transported from the reticuloendothelial cells to the liver as a bilirubin-albumin complex (2 moles of bilirubin per mole of albumin). Bilirubin is converted to bilirubin diglucuronide in the liver by the action of a microsomal enzyme, uridine diphosphate glucuronyl transferase. These reactions are shown in Figure 10.7.

For excretion into the bile the water-insoluble bilirubin must be converted to the water-soluble bilirubin diglucuronide. Because of its solubility in water, conjugated bilirubin gives a *direct* reaction with the diazotized sulfanilic acid reagent of van den Bergh, whereas the water-insoluble, unconjugated bilirubin reacts only after addition of alcohol (*indirect* van den Bergh reaction). Bile normally gives the direct reaction and serum primarily the indirect reaction. There is a small fraction of unconjugated bilirubin in serum which gives a direct reaction, probably because of the presence of other solubilizing substances. Thus, the van den Bergh reaction allows only an approximation of the amounts of conjugated and unconjugated bilirubin present.

There are three general steps in the metabolism of bilirubin

1. Transport by serum albumin into the liver
2. Conjugation by the liver microsomal enzyme system
3. Transport from hepatic cells into the bile

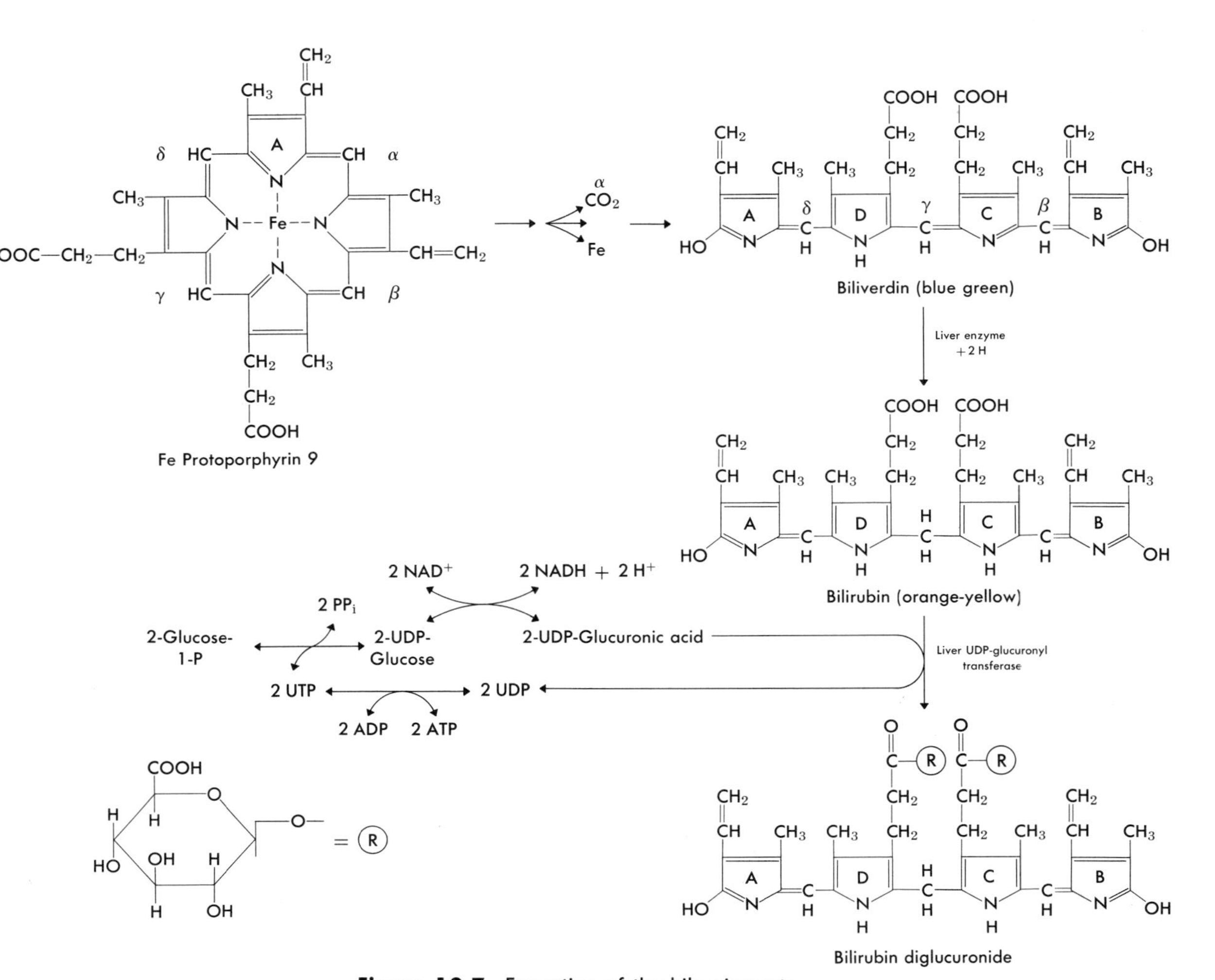

Figure 10.7. Formation of the bile pigments.

A defect or relative insufficiency in either of the first two steps would result in the decreased formation of bilirubin diglucuronide and the retention in blood of unconjugated bilirubin. Examples of such a jaundice are the physiological jaundice of the newborn, the jaundice of Crigler-Naajar syndrome, and the jaundice associated with hemolytic disease. The neonatal animal has, compared with the adult organism, low activities of both uridine diphosphate glucose dehydrogenase and glucuronyl transferase in the liver. The excessive rate of destruction of red blood cells in hemolytic disease leads to the production of bile pigment at a rate exceeding the capacity of the liver to dispose of it.

The blocking of the third step would reduce the excretion of conjugated bilirubin and lead to its regurgitation into the blood. In obstruction of the biliary passages, conjugated bilirubin appears in the blood and in the urine. The liver, in early stages of this disease, continues to excrete conjugated bilirubin, but the bile with conjugated bilirubin is regurgitated into the blood. Prolonged biliary obstruction results in liver damage and leads to a failure to conjugate bilirubin so that high levels of free bilirubin may also occur.

It is not intended to discuss in detail here the further metabolism of bilirubin. However, a brief consideration of the urobilinogens has merit because of their great clinical importance. Bilirubin in the gut is reduced by bacterial action to a series of colorless pigments known as *urobilinogens*. The urobilinogens formed in the colon are partially taken into the portal circulation and removed by the liver. There is little urobilinogen in the systemic blood and only small quantities excreted in the urine under normal conditions. With complete biliary obstruction there is little or no bilirubin reaching the colon. There is, therefore, no urobilinogen formed and none present in the urine or feces. In parenchymal hepatic lesions, urobilinogen reabsorbed in the colon is not effectively removed from the portal circulation. It gains access to the systemic circulation and is excreted in large amounts in the urine. The fecal urobilinogen is a measure of the quantity of bilirubin which reaches the gut and is not reabsorbed into the portal blood. The determination of fecal urobilinogen can be a very valuable clinical determination, and it is a measure of the degree of hemoglobin destruction in the absence of liver and biliary disease. Increased quantities are found in hemolytic disease.

Iron Metabolism

The body is economic in its conservation of iron. Iron liberated from hemoglobin amounts to 20 to 25 mg per day. Since the normal diet supplies 12 to 15 mg of iron per day, of which only 0.6 to 1.5 mg may be absorbed, a serious daily loss would result unless there was an efficient conservation of hemoglobin iron. The total daily loss of iron is difficult to assess. It is in the range of 0.5 to 1 mg per day in the urine, feces, and sweat. The average daily loss during the menstrual cycle amounts also to 0.5 to 1 mg per day. Estimated daily iron requirements are as follows: 0.5 to 1 mg for men and postmenopausal women, 1 to 2 mg for menstruating women, 1.5 to 2.5 mg for pregnant women, 1 to 1.5 mg for children, and 1.5 to 2.5 mg for the adolescent girl.

Depending on body size and hemoglobin level, the total body iron may range from 2 to 6 g. Most of the iron is found in hemoglobin (65 per cent) and in storage form (25 per cent) as ferritin and hemosiderin. Some 5 per cent is present in myoglobin, and less than 1 per cent is present in plasma and extracellular fluid. A small percentage of the iron is unaccounted for and represents either error in determination or unknown constituents.

Iron is absorbed in its ferrous form directly into the blood stream and not into the lymphatic system. Absorption can take place in any portion of the gastrointestinal tract with the duodenum being the most active. The biochemical mechanisms that control iron absorption are more efficient during periods of increased iron requirements. However, the earlier theory of a ferritin-mediated mucosal block now seems unlikely. This theory held that iron absorbed by the intestinal mucosa was converted to ferritin; when the cells were saturated with ferritin, further absorption was blocked until the iron was released from ferritin and transferred to plasma. More recent evidence indicates that an active transport system exists for the transport of iron across the mucosa to the blood plasma and that this system is influenced by the level of iron in tissues and the rate of blood cell formation. It is likely that ferritin is important and may influence the activity of the absorptive mechanism.

The small quantity (70 to 170 μg per 100 ml) of iron found in the plasma is determined by a dynamic equilibrium between the amount being added by absorption from the gastrointestinal tract, released from hemoglobin or from storage depots, and the amount being removed for storage or hemoglobin synthesis. It was found by Holmberg and Laurell that the major part of the plasma iron is bound to the specific iron-binding β_1-globulin, transferrin (page 339). The iron is transported to tissues and storage depots by this protein.

Between 500 and 1500 mg of iron are stored intracellularly as ferritin and hemosiderin in the liver, spleen, bone marrow, and other tissues in the normal adult. Ferritin is a water-soluble protein containing as high as 23 per cent protein by weight. It is composed of a protein, *apoferriten,* with a molecular weight of 455,000 and a ferric hydroxyde-phosphate complex. Hemosiderin is another storage form of iron. It is a rather ill defined and variable entity and represents colloidal iron in the form of granules much larger than ferritin molecules. Hemosiderin granules are insoluble in water and contain protein and other organic constituents. Iron in excess of ferritin storage capacity accumulates as hemosiderin. Since there is no known excretory mechanism for excess iron, continued administration of iron leads to the accumulation of hemosiderin. Such an accumulation in the liver can lead to the destruction

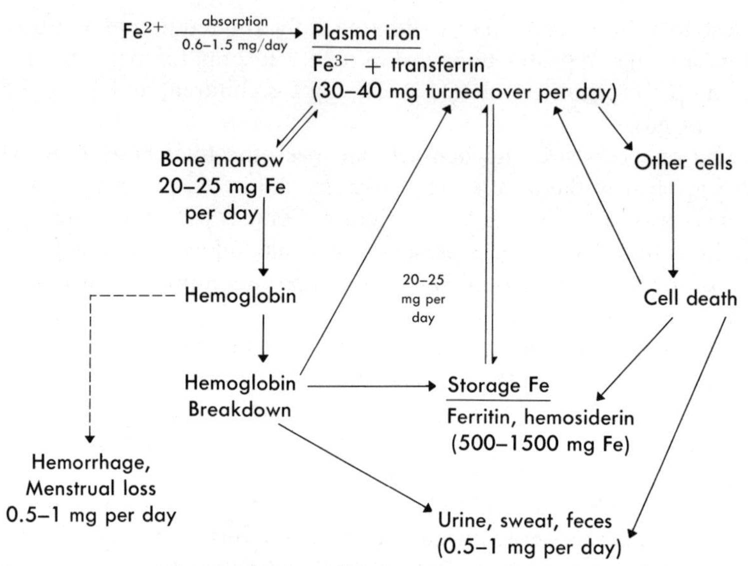

Figure 10.8. Outline of iron metabolism.

of this organ. Iron in both ferritin and hemosiderin can be mobilized when needed for hemoglobin synthesis.

The scheme shown in Figure 10.8 outlines some of these aspects of iron metabolism.

REFERENCES

London, I. M.: The Metabolism of Erythrocytes, *Harvey Lectures,* **56:**151 (1960–1961).

Moore, C. V.: Iron Metabolism and Nutrition, *Harvey Lectures,* **55:**67 (1958–1959).

Putnam, F. W.: Immunoglobulin Structure: Variability and Homology, *Science,* **163:**633 (1969).

Putnam, F. W. (ed.): *The Plasma Proteins,* 2 vols. Academic Press, New York, 1960.

Seegers, W. H. (ed.): *Blood Clotting Enzymology*. Academic Press, New York, 1967.

West, E. S.; Todd, W. R.; Mason, H. S.; and Van Bruggen, J. T.: *Textbook of Biochemistry,* 4th ed., Chap. 15. The Macmillan Company, New York, 1966.

White, A.; Handler, P.; and Smith, E. L.: *Principles of Biochemistry,* 4th ed., Chap. 31. McGraw-Hill, New York, 1968.

Wilkinson, J. H. *Isoenzymes*. J. B. Lippincott, Co., Philadelphia 1966.

Wintrobe, M. M.: *Clinical Hematology,* 5th ed. Lea & Febiger, Philadelphia, 1961.

Chapter 11

RESPIRATION AND ACID-BASE BALANCE

In complex organisms like man, the cell is maintained in an environment that changes within narrow limits. In order to accomplish this constancy of environment for the cell, man has developed efficient systems for transporting needed materials to the cell and removing waste products. The cells are bathed by interstitial fluids which, in turn, are in contact with the capillaries. It is the circulatory system that carries nutrients to the cells via the interstitial fluids and removes the waste products by the same route. Although many substances are transported to and from the cell, we will concern ourselves in this chapter with the transport of oxygen, carbon dioxide, acids, bases, and electrolytes and look at the chemical mechanism by which the job is done without changing the pH or osmotic pressure.

MAGNITUDE OF THE JOB IN AVERAGE MAN AT REST

About 15,000 liters of air per day are involved in breathing, 250 ml of O_2 per minute is consumed, 200 ml of CO_2 per minute is produced, and 5 liters of blood per minute is pumped through the lungs.

Approximately 1.5 liters of blood per minute is pumped through the kidneys and about 180 liters per day is filtered through the glomeruli. Of this, about 1.5 liters is excreted as urine and the remaining 178.5 liters are reabsorbed. About 1.5 kg of sodium chloride is filtered and reabsorbed per day. This is accomplished with about 6 liters of blood, of which 50 per cent is cells.

PHYSICAL FACTORS

The distances involved in diffusion are short, the surfaces involved in absorption are very large, and the times required to reach equilibrium are short. There is intimate contact between the cells, interstitial fluid, and capillaries.

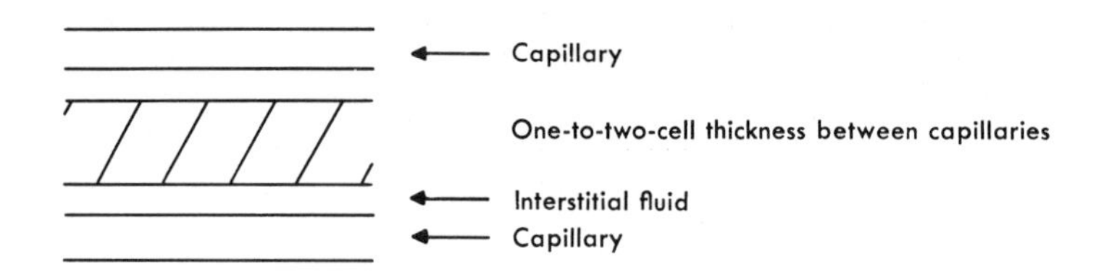

The capillaries are about 0.06 mm apart and in the alveoli, the air is separated from the blood by 1 to 2 μ. The surface area of the alveoli is 100 m^2; that of the muscle cells is about 6000 m^2. The ratio of alveolar surface to blood volume in the alveoli is such that the blood in the alveolar capillaries is in equilibrium with alveolar air within 0.7 sec. The diffusion of O_2 across capillaries requires 0.002 sec. In 0.01 sec all fluid spaces between the capillaries are brought to equilibrium.

In summary, the distances of diffusion are very short, the surfaces very large, and the times required for reaching equilibrium very short.

Several forces are used to transport fluids across capillaries and membranes. Some of these are (1) muscular propulsion by the heart, (2) muscular forces other than heart, (3) concentration gradient, (4) pressure gradient for gases, (5) osmotic pressure, and (6) biologic work.

THE FLUID COMPARTMENTS

Figure 11.1 shows the relationship between the various body fluids with respect to both volume and connection to the external environment through the gastrointestinal tract, lungs, kidneys, and skin. The body is about 70 per cent water. The intracellular fluids account for about 50 per cent of the body weight, the interstitial fluids 15 per cent, and blood plasma 5 per cent. Fluids such as gastric juice, aqueous humor of the eye, pancreatic juice, spinal fluid, and so on make up a relatively small amount of the total volume but are important for specialized functions of the organs involved.

Figure 11.2 shows the partial pressure of oxygen and carbon dioxide in the various compartments. The cell is actively using oxygen and producing large amounts of carbon dioxide. At the cell there is rapid diffusion of oxygen from the capillary into the interstitial fluid and then into the cell. The carbon dioxide diffuses from the cell where the partial pressure is high to the interstitial fluid and then into the blood. The oxygen is transported to the cell partially as dissolved oxygen and mostly in combination with hemoglobin. The carbon dioxide is transported partially as dissolved gas but primarily as bicarbonate.

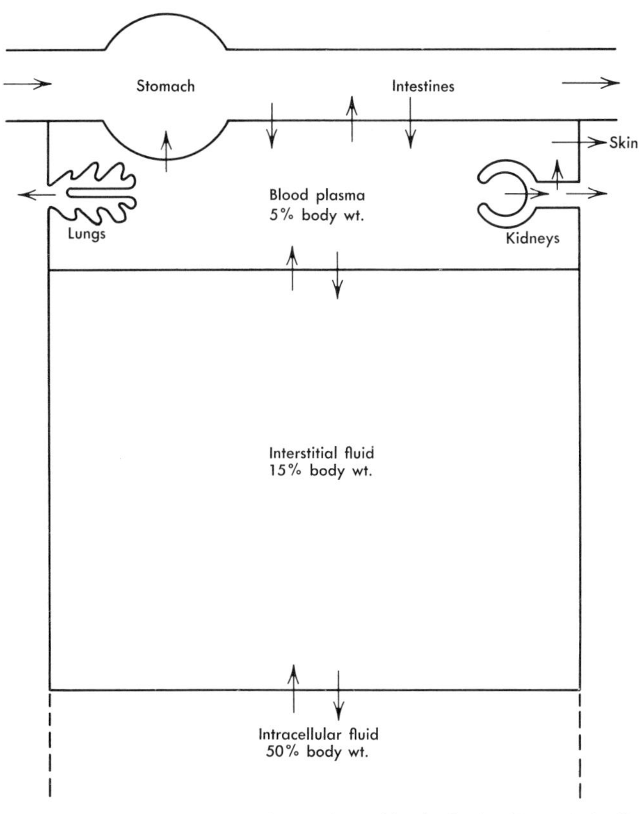

Figure 11.1. Volume and concentration relationship of body fluids. (From J. L. Gamble: *Chemical Anatomy, Physiology and Pathology of Extracellular Fluid.* Harvard University Press, Cambridge, Mass., 1958.)

The amount of gas dissolved by the plasma is dependent on the partial pressure of the gas.

$$\text{Solubility of } O_2 = K\ pO_2$$

$$\text{Solubility of } CO_2 = K\ pCO_2$$

At 38°C and 760 mm, 1 ml of plasma dissolves 0.024 ml of O_2 and 0.51 ml of CO_2. In other words, K for O_2 is 0.024 ml and the K for CO_2 is 0.51 ml. It follows that 100 ml of arterial plasma at 38° will dissolve 0.31 *ml of* O_2 (0.31 vol per cent).

$$\tfrac{100}{760} \times 100 \times 0.024 = 0.31 \text{ ml}$$

Calculations can be made for O_2 at other partial pressures and for CO_2.

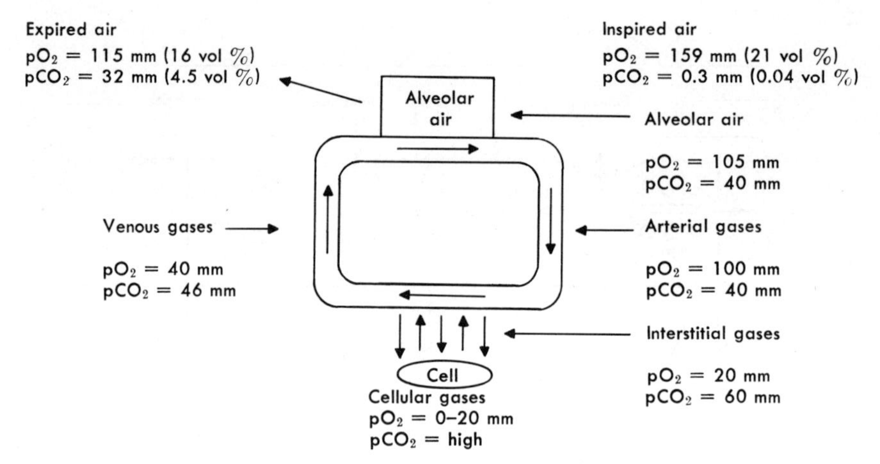

Figure 11.2. Transport of O_2 and CO_2.

TRANSPORT OF OXYGEN

A large portion of the oxygen used by cells is transported in combination with hemoglobin only a small amount of oxygen is transported as dissolved oxygen. When hemoglobin is saturated with oxygen, four moles of oxygen combine with one mole of hemoglobin (M.W. = 68,000)

$$Hb(4FeH_2O) + 4O_2 \rightleftharpoons Hb(4FeO_2) + 4H_2O$$

or more conveniently

$$Hb + O_2 \rightleftharpoons HbO_2$$

Since 1 mole of oxygen (22.4 liters, 32 g) combines with 17,000 g of hemoglobin, 1 g of hemoglobin can combine with (22,400/17,000) = 1.32 ml of O_2. There is normally about 15 g of Hb per 100 ml of blood which, when saturated, can carry $1.32 \times 15 = 19.8$ ml O_2 as oxyhemoglobin. It was seen above that 100 ml of plasma would dissolve 0.32 ml of O_2 or approximately 1/70 the amount carried by 15 g of hemoglobin.

If we convert to moles, 1 mM of Hb = 17.0 g and 1 mM of O_2 = 22.4 ml and 19.8/22.4 = 0.9 mM O_2 can combine with the hemoglobin in 100 ml of blood.

In the equation $Hb + O_2 \rightleftharpoons HbO_2$ the position of the equilibrium will depend on the pO_2. In addition to the oxygen tension in the lungs, the degree of saturation of the hemoglobin with O_2 will depend on (1) pCO_2 which influences the pH, (2) temperature, and (3) ionic strength. The amount of O_2 transported to the tissues will depend on the amount of hemoglobin in the blood and the amount of blood transported as well as the degree of saturation of the hemoglobin present.

Figure 11.3 relates the percentage saturation of hemoglobin to oxygen tension at various partial pressures for CO_2.

The curve for $pCO_2 = 40$ mm is the normal curve. The sigmoid nature of the curve is of particular importance. At pO_2 of 100 mm the hemoglobin is fully saturated and due to the flatness of the curve remains 90 per cent saturated at a pO_2 of 60 mm. At high altitudes or reduced pO_2 most of the Hb would be saturated. At a pO_2 less than 40 mm the curve drops rapidly. At the tissues where the pO_2 is very low, hemoglobin can release almost all of its oxygen to the cells. Under ordinary conditions only a portion of the O_2 is released, but under conditions of severe exercise, almost all the O_2 would be released.

The influence of CO_2 is due partially to a change in pH. As the pH is lowered, the affinity of hemoglobin for oxygen decreases. This is a desirable reaction since at the tissues, the production of CO_2 would cause the release of additional O_2. Another important factor in the effect of CO_2 is a combination with hemoglobin. Hemoglobin, which has reacted with CO_2, has less affinity for O_2. Again this reaction is in the direction to release oxygen at the cells where it is needed.

The sigmoid nature of the curve is due in part to the four sites for attachment of oxygen to the hemoglobin molecule. The first mole reacts less readily than the

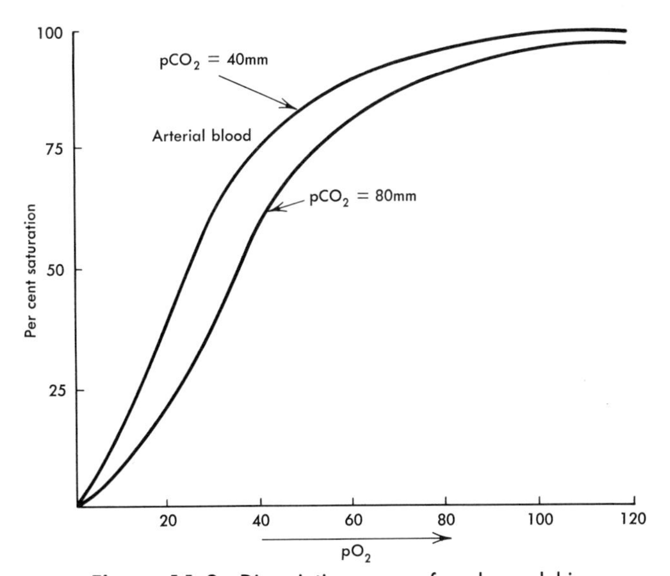

Figure 11.3. Dissociation curve of oxyhemoglobin.

second, and the fourth has the highest affinity. If we let Hb represent one equivalent of hemoglobin containing one iron atom, then

$$Hb_4 \underset{-O_2}{\overset{+O_2}{\rightleftharpoons}} Hb_4O_2 \underset{-O_2}{\overset{+O_2}{\rightleftharpoons}} Hb_4(O_2)_2 \underset{-O_2}{\overset{+O_2}{\rightleftharpoons}} Hb_4(O_2)_3 \underset{-O_2}{\overset{+O_2}{\rightleftharpoons}} Hb_4(O_2)_4$$

Hemoglobin is a molecule that has the properties needed for oxygen transport. In the next section we will see that it is also a model molecule, elegantly designed for the transport of CO_2.

Transport of CO_2

Cells carrying out active metabolism produce large amounts of CO_2. Carbon dioxide reacts slowly with water to produce carbonic acid which ionizes to a proton.

$$CO_2 + H_2O \rightleftharpoons H_2CO_3 \rightleftharpoons H^+ + HCO_3^-$$

and bicarbonate ion. The red blood cell has a zinc-containing enzyme, carbonic anhydrase, which catalyzes the hydration reaction. The neutralization of this large amount of acid (about 100 equivalents of CO_2 per day) and the transport of the CO_2 as a bicarbonate are functions which are handled by the buffers in the plasma and hemoglobin. Again, the hemoglobin molecule has all the chemical properties needed to perform these duties as well as the transport of oxygen. The organic chemist must marvel that so many functions can be built so perfectly into one molecule.

Role of Buffers

The following buffers are effective in maintaining the pH of the blood at 7.4.

(1) $\frac{B_2HPO_4}{BH_2PO_4}$ $pK_a = 6.8$

(2) $\frac{BHCO_3}{H_2CO_3}$ $pK_a = 6.1$

(3) $\frac{\text{B Protein}}{\text{H Protein}}$

(4) $\frac{BHb}{HHb}$ $pK_a = 7.93$

(5) $\frac{BHbO_2}{HBHbO_2}$ $pK_a = 6.68$

At pH 7.4 a large amount of the buffering by proteins is due to the imidazole group of histidine which occurs in large amounts in hemoglobin.

[Structural formulas: imidazole ring of histidine (H–C, N, NH, HC=C–Hb) $+ H^+ \rightleftharpoons$ protonated imidazole ring (H–C, N, NH_2^+, HC=C–Hb)]

The buffering capacity of hemoglobin is shown in Figure 11.4. For a decrease in pH of 0.1, 1 mM of hemoglobin will neutralize 0.25 mM of acid. Since hemoglobin is present in large amounts, it is the most effective buffer present.

The Isohydric Effect

The most effective means of neutralizing carbonic acid is the decrease in the acid strength of hemoglobin when it releases its oxygen.

FeO2 ⋯ N (imidazole ring: N, CH, NH, C, HC) —C— Hb Protein + H^+ $\underset{-O_2}{\overset{+O_2}{\rightleftharpoons}}$ Fe ⋯ N (imidazole ring: N, CH, NH_2^+, C, HC) —C— Hb Protein

$pK_a = 6.68$ $\qquad$ $pK_a = 7.93$

The capacity of this process is illustrated in the titration curves in Figure 11.5. One mM of hemoglobin can neutralize 0.7 mM of acid without a change in pH when it changes from oxyhemoglobin to the unoxygenated form. Since 1 mM of O_2 is released per mM of hemoglobin, it follows that 0.7 mM of H_2CO_3 can be neutralized for each mM of O_2 released. In a normal individual the ratio of CO_2 released to oxygen consumed is about 0.8. Under these conditions, the isohydric effect will handle $\frac{7}{8}$ of the acid produced.

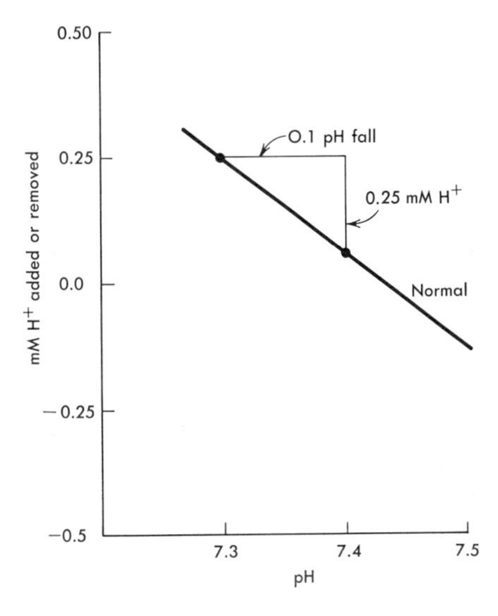

Figure 11.4. Titration curve for 1 mM oxyhemoglobin ($HHbO_2$).

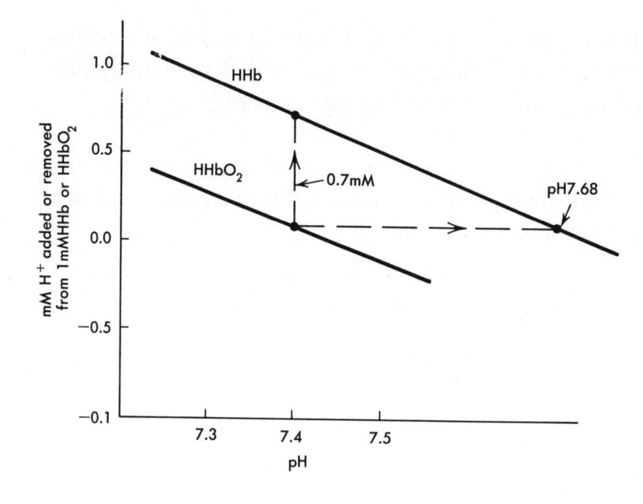

Figure 11.5. Titration curves of hemoglobin and oxyhemoglobin.

The same arguments hold, only in the opposite direction, at the lungs. Here the pO_2 is high and the pCO_2 is low. These conditions are favorable for the oxygenation of hemoglobin which leads to stronger acid and the release of H^+ which can combine with the HCO_3^- to form H_2CO_3. At the low pCO_2 the equilibrium for $H_2CO_3 \rightleftharpoons CO_2 + H_2O$ shifts to the right and CO_2 is given off.

Carbamino Compounds

Most of the CO_2 is transported as HCO_3^- in the plasma or red blood cell. About 20 per cent is transported in the form of carbamino groups on the protein portion of hemoglobin.

$$\underset{\text{Protein}}{R\text{—}NH_2} + CO_2 \rightleftharpoons \underset{\text{Protein}}{R\text{—}\overset{H}{N}\text{—}\overset{O}{\overset{\|}{C}}\text{—}O^-} + H^+$$

The proton is neutralized by buffers and the isohydric effect. The capacity of hemoglobin to transport CO_2 in this combination is shown in Figure 11.6.

At the tissues about 0.2 mM of CO_2 will be converted to carbamino groups for each mM of O_2 released. When the hemoglobin becomes oxygenated in the lungs, the same amount of CO_2 will be released from carbamino groups.

WATER SHIFT IN THE RED BLOOD CELL

The red blood cell swells at the tissue when it loses O_2 and shrinks at the lungs when the hemoglobin becomes oxygenated. This is due to an increase in the number of osmotically active particles in the red blood cell when CO_2 is taken up and a decrease in the number of osmotically active particles when CO_2 is released. In the conversion of HbO_2 to Hb, protons formed in the dissociation of H_2CO_3 are taken up by the imidazole groups of the reduced hemoglobin. This increases the number of positive charges on the hemoglobin molecule and reduces the number of net negative charges. To restore osmotic equilibrium, water passes from the plasma into the red cell. The process can be illustrated by the following scheme in which a proton is removed by the isohydric effect and a molecule of bicarbonate is formed; negative charges on the hemoglobin are not shown.

Red Cell $\qquad\qquad$ Red Cell

$$HbO_2 \xrightarrow{\searrow O_2} HbH + HCO_3^-$$

$$H_2CO_3 \uparrow \quad CO_2 \uparrow \quad CO_2 \quad (\text{Tissues}) \qquad H_2O \uparrow$$

At the lungs, the reverse reaction takes place.

Red Cells

$$HbH + HCO_3^- \xrightarrow{O_2} HbO_2 + H^+ + HCO_3^-$$

$$H_2CO_3 \downarrow \quad CO_2 \downarrow \quad CO_2 \quad (\text{Lungs}) \qquad \downarrow H_2O$$

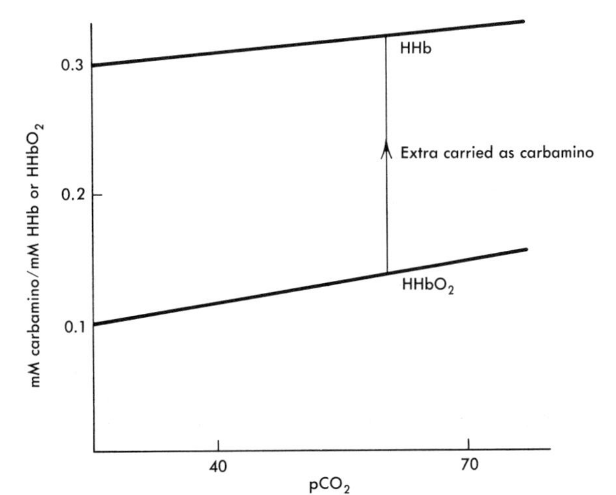

Figure 11.6. Transport of CO_2 as carbamino groups.

THE CHLORIDE SHIFT

Since there is a change in the number of particles in the red cell as a result of the isohydric effect and the membrane of the red cell is permeable to some of these particles (for example, Cl^-, HCO_3^-, OH^-) and not to others (for example, hemoglobin anion, cations), there will be diffusion of certain particles out of and into the red cell until equilibrium is reached. According to Donnan, equilibrium will be reached when the diffusible electrolytes have the following relative concentrations

$$\frac{[Na^+]_p}{[NA^+]_c} = \frac{[K^+]_p}{[K^+]_c} = \frac{[Cl^-]_c}{[Cl^-]_p} = \frac{[HCO_3^-]_c}{[HCO_3^-]_p} \quad \text{etc.}$$

The letters c and p denote respectively the concentrations within the red blood cell and the concentrations in the plasma or outside the cell. At the tissues where a large amount of HCO_3^- is formed from the CO_2, there is a large excess of HCO_3^- within the red cells. The excess bicarbonate will diffuse from the red cell into the plasma. In order to maintain electric neutrality, the chief anion of the plasma, Cl^-, will diffuse from the plasma into the red cell until equilibrium is established. Other diffusible ions will undergo similar redistributions but are present in much lower concentrations than is Cl^-. Since the red cell membrane is relatively impermeable to cations during the periods of time involved, the loss of HCO_3^- cannot be compensated for by a concomitant loss of cations from the red cell. At the tissues, then, HCO_3^- diffuses out of the red blood cells and Cl^- diffuses from the plasma into the red cell. At the lungs, the reverse process takes place. This shift of chloride is further illustrated in the general diagram below.

GENERAL DIAGRAM

The following scheme (Figure 11.7) shows a summary of the reactions which take place at the tissues when oxygen is released from hemoglobin and carbon dioxide is produced. The opposite takes place at the lungs. It should be noted that the CO_2 diffuses through the plasma into the red blood cell where it is hydrated by means of carbonic anhydrase to carbonic acid. The reactions

$$CO_2 + H_2O \underset{\text{anhydrase}}{\overset{\text{carbonic}}{\rightleftharpoons}} H_2CO_3 \rightleftharpoons H^+ + HCO_3^-,$$

proceed to the right in the erythrocytes at the tissues and in the opposite direction at the lungs.

TRANSPORT OF ACID (ACID-BASE BALANCE)

Acid-base balance is a study of the chemical mechanisms by which protons are produced and handled in the body. The pH of the blood is controlled at very near

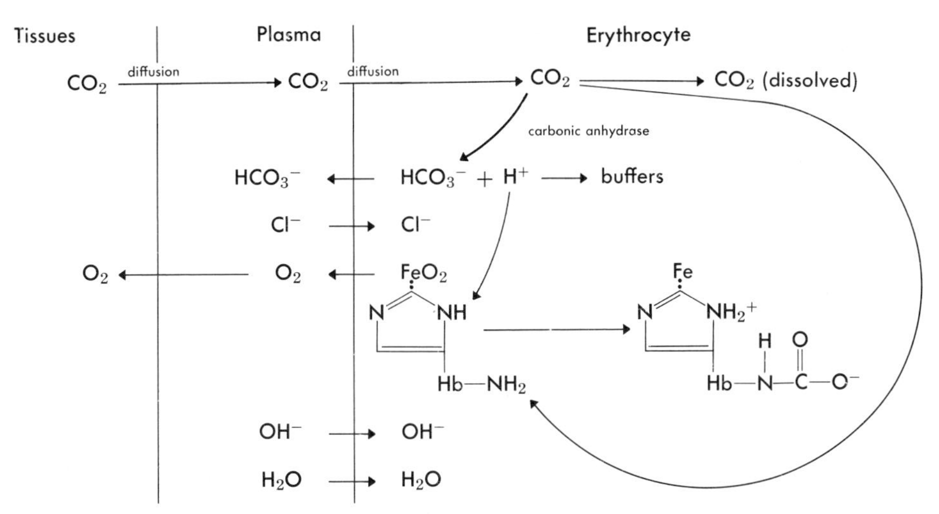

Figure 11.7. Summary of reactions occurring at tissues when O_2 is released and CO_2 is produced. (From H. W. Davenport: *ABC of Acid-Base Chemistry,* 4th ed. University of Chicago Press, Chicago, 1958.)

7.4 and the intracellular fluids are maintained at a slightly lower pH. The chemical mechanism by which large quantities of CO_2 were transported to the lungs and eliminated was discussed in a previous section. In addition to the acid arising from CO_2, there are many other acids and conditions that lead to the formation of acids which are not volatile and cannot be expired by the lungs.

Acids may arise from foods. Sulfur in proteins is oxidized to sulfuric acid and nucleoproteins contain phosphate. Salts of organic acids, for example, sodium citrate, are oxidized to $BHCO_3$ and actually increase base rather than furnish protons. The ammonia from ammonium chloride is converted to urea which is neutral leaving HCl. Drugs may cause an acidosis or alkalosis depending on the nature of the compound. Administration of $NaHCO_3$ can produce an alkalosis.

Most important are the changes taking place in acid-base balance in pathological conditions. In severe diabetes acetoacetic acid and β-hydroxybutyric acid are produced in large quantities. In kidney disease, acid cannot be excreted, leading to severe acidosis. In severe vomiting, large amounts of HCl are lost, causing alkalosis. Severe diarrhea causes a loss of $BHCO_3$ and a severe acidosis.

The mechanism for handling these conditions involves the buffers, lungs and kidneys.

The buffers are

$$\frac{NaHCO_3}{H_2CO_3} \cdot \frac{KHb}{HHb} \cdot \frac{KHbO_2}{HHbO_2} \cdot \frac{Na_2HPO_4}{NaH_2PO_4}$$

Of these, the bicarbonate system deserves special consideration. It is not only the most plentiful of the group but also differs in having an acid which is under control of the lungs. Normally the ratio of B/A is 20/1 and

$$7.4 = 6.1 + \log \frac{[NaHCO_3]}{[H_2CO_3]} = \frac{20}{1} = \frac{60 \text{ vol. \%}}{3 \text{ vol. \%}} = \frac{27 \text{ mM}}{1.35 \text{ mM}}$$

The pH can be maintained at 7.4 as long as the ratio of $[NaHCO_3]/[H_2CO_3] = 20/1$. Since $[H_2CO_3]$ is under control of the lungs, it is possible to change the ratio by hyperventilation or hypoventilation. The respiratory center is sensitive to changes in pH and CO_2 and will respond to regulate the ratio. When the bicarbonate buffer is adjusted to the 20/1 ratio, the other buffers are adjusted to the same pH.

Figure 11.8 shows the chemical structure of blood plasma, interstitial fluid, and cellular fluid. A comparison of the composition of blood plasma and interstitial fluid to seawater is of interest. The ions are the same but are more concentrated in seawater. The composition of the extracellular body fluids probably represents the composition of the seas at the time of the establishment of the internal environment.

The primary difference between the chemical structure of plasma and the interstitial fluid is the presence of protein in plasma. The importance of the protein in the transfer of fluid between the plasma and the interstitial fluid was discussed on page 338. The chief cation in the cell is potassium, whereas sodium is the chief cation in the extracellular fluids. Phosphate is the principal anion in the cell and chloride is the principal anion in the extracellular fluid. The nonelectrolytes constitute only a small portion of the fluids and are materials in transport, for example, glucose and urea.

The maintenance of the chemical structure of these fluids in face of the addition of large amounts of materials, often intermittently, is accomplished by the kidneys, the lungs, and adjustments within the systems. Actually the concentrations are not exactly maintained, and fluctuations are observed. In the presence of disease or extensive changes in the external environment, the changes may become large.

The normal hydrogen ion concentration of the plasma ranges from about pH 7.35 to 7.45 and the range compatible with life is 7.0 to 7.8. The control of the hydrogen ion concentration rests first on the bicarbonate buffering system. At pH 7.4 it can be calculated that the ratio of base/acid is 20/1 and

$$7.4 = 6.1 + \log \frac{(\text{Base})}{(\text{Acid})}$$

in normal plasma there are 27 mM H_2CO_3 and 1.35 mM HCO_3^-. This is about one fifth of the total ionic structure of the plasma. Also, CO_2 is being produced rapidly in the tissues and eliminated in the lungs. For this reason, base or acid is

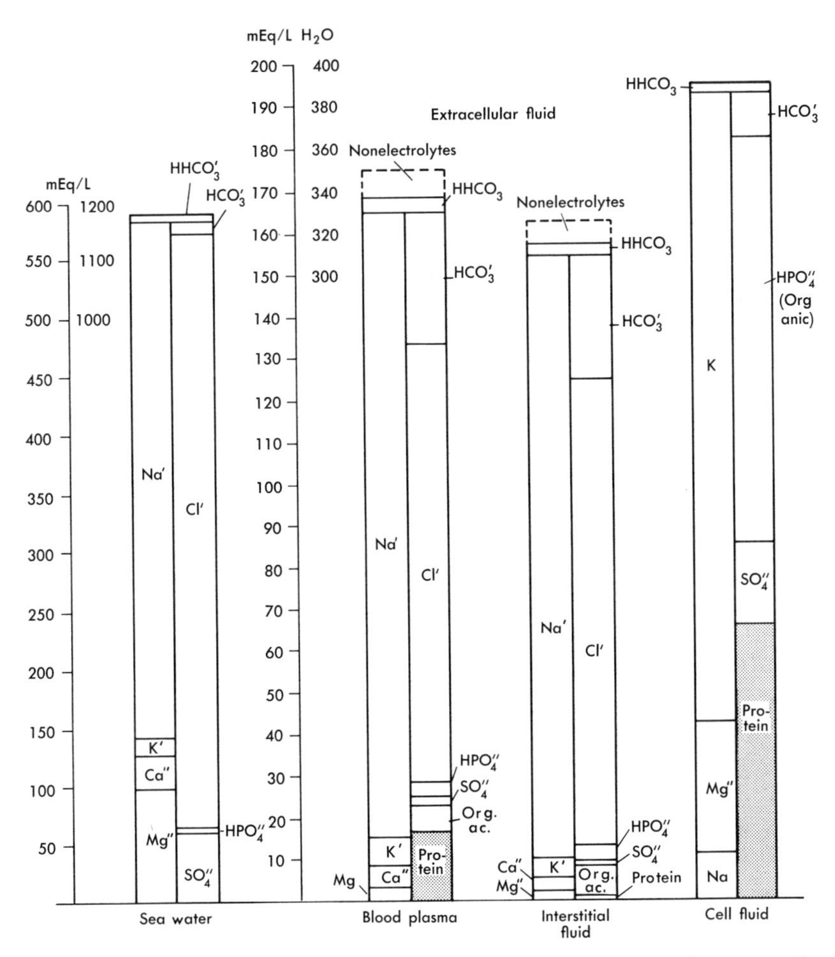

Figure 11.8. Chemical structure of body fluids. (From J. L. Gamble: *Chemical Anatomy, Physiology and Pathology of Extracellular Fluid.* Harvard University Press, Cambridge, Mass., 1958.)

immediately available for adjustment of the system. Buffering systems having a ratio of base/acid = 20/1 have a low capacity and would not be expected to be efficient in controlling the pH of a highly mobile system. However, the bicarbonate system is unique in that the acid is volatile and under control of the lungs. If acid is added to a bicarbonate system at pH 7.4, a portion of the bicarbonate will be converted to carbonic acid with a drop in pH. By removing enough H_2CO_3 to maintain a 20/1 ratio, the pH could be maintained at 7.4. The 20/1 ratio can then be controlled by the lungs.

Other buffers in the blood also play a role in maintaining the pH at 7.4. The principal ones are hemoglobin, plasma proteins, and phosphates.

Metabolic Disturbances of H^+ Concentration

Figure 11.9 shows diagrammatically the four types of disturbances encountered in the plasma bicarbonate buffering system.

The first column shows the normal situation with a ratio of $HCO_3/H_2CO_3 = 20/1$. Column A shows a situation in which HCO_3^- has been decreased from 27 mM to 16.5 mM per liter. If the decrease in HCO_3^- (10.5 mM) had appeared as H_2CO_3, the pH would be much less than 7.0. By maintaining H_2CO_3 at 1.3 mM per liter, the normal concentration, the pH is held at 7.2 which can be tolerated. By reducing the concentration of H_2CO_3 below the normal level, the pH can be brought back to 7.4. This is illustrated in column B, in which the HCO_3^- concentration was initially reduced to 10.5 mM per liter and the H_2CO_3 concentration was subsequently reduced to 0.5 mM per liter by expiring CO_2. The adjustment brings the base/acid to 20/1 and the pH to 7.4.

Conditions in which the primary defect is a reduction in HCO_3^- concentration (base deficit) are known as metabolic acidosis. The acidosis which accompanies diabetes or starvation is an example. The addition of the ketone bodies, acetoacetic acid and β-hydroxybutyric acid displaces the HCO_3^-, resulting in an increase in H_2CO_3. The H_2CO_3 concentration is reduced by hyperventilation in an attempt to bring the pH back to 7.4. The compensation is very seldom complete, and the pH is less than 7.4. Metabolic acidosis is then characterized by a decrease in (HCO_3^-) and a lowered pH.

Column C illustrates a condition in which the concentration of HCO_3^- is increased to 40.7 mM per liter and the H_2CO_3 maintained at the normal level of 1.3 mM per liter. By maintaining a normal level of H_2CO_3, the pH stays within a tolerable range. If H_2CO_3 had been permitted to fall, the pH would have been higher. Column D shows the same situation in which the (HCO_3^-) is even higher (46.8 mM per liter), but the pH has been maintained at 7.4 by increasing H_2CO_3 concentration. This is done by hypoventilation.

A condition in which the first change is an increase in HCO_3^- (base excess) is called metabolic alkalosis and can be brought about by feeding an excess of sodium bicarbonate or a loss of gastric HCl as in severe vomiting. The condition is characterized by an increase in pH and an increase in HCO_3^- concentration. There is usually partial compensation. In both types of conditions discussed, the primary change has been in the numerator of base/acid ratio. When base concentration is lowered, the defence is to lower the acid concentration to bring the ratio back to 20/1. When the base concentration is increased, the pH is increased and is compensated by an increase in the acid concentration.

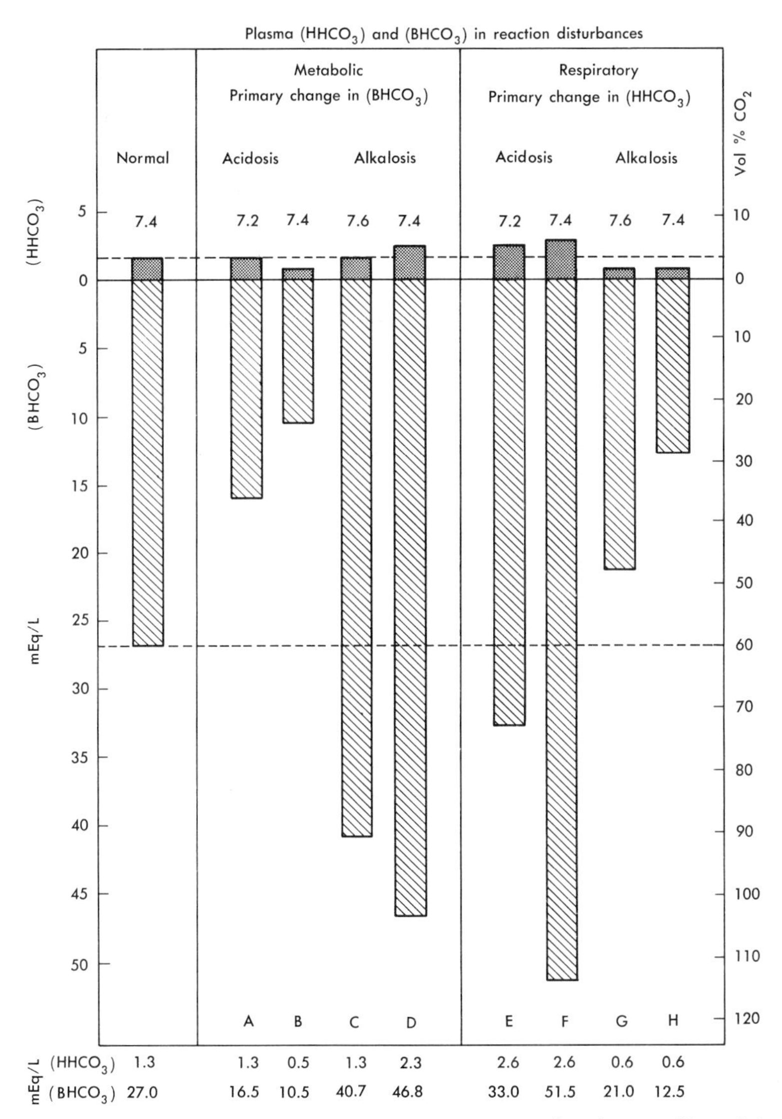

Figure 11.9. Changes in bicarbonate buffering system in various disturbances. (From J. L. Gamble: *Chemical Anatomy, Physiology and Pathology of Extracellular Fluid.* Harvard University Press, Cambridge, Mass., 1958.)

In the next two types of conditions to be discussed, the primary change is in the denominator of the base/acid ratio. The compensating changes are in the base concentration.

Column E shows a case in which the primary change is an increase in H_2CO_3 concentration to 2.6 mM per liter. There is partial compensation by increasing the HCO_3^- to 33.0 mM per liter. Column F illustrates a case in which compensation is complete. The HCO_3^- concentration has been adjusted to 51.5 mM to bring the ratio to 20/1 and the pH to 7.4. Compensation usually falls short and the pH is less than normal. A primary CO_2 excess or respiratory acidosis occurs in pneumonia and respiratory paralysis.

A condition in which the primary change is a decrease in H_2CO_3 is illustrated in column G. The decrease in H_2CO_3 to 0.6 mM per liter is partially compensated by a decrease in HCO_3^- to 21.0 mM per liter and the pH is 7.6. Column H shows a case which is completely compensated by reducing the HCO_3^- concentration to 12.5 mM per liter to restore the 20/1 ratio and a normal pH. Hyperventilation due to hypoxia, fever, and hysteria can cause a primary CO_2 deficit (respiratory alkalosis). As in the other types of conditions, compensation is usually not complete.

The Role of the Kidney in Acid-Base Balance

The kidney is the main organ which regulates the extracellular fluids. Figure 11.10 shows a comparison of the chemical structure of the extracellular fluid to that of a normal 24-hour specimen of urine. It is seen that substances which appear in very low concentration in the blood may appear in large amounts in the urine. This means that these substances are very effectively concentrated in the urine by the kidney and thereby removed from the blood. The composition of the urine from normal individuals will vary rather widely depending primarily on the composition of the diet. It should be noted that urine does not contain protein or glucose but contains ammonia which is not present in the plasma. The plasma proteins do not enter the glomerular filtrate, and glucose is completely reabsorbed in the tubule up to concentrations of about 170 mg per 100 ml. Another significant difference between plasma and urine is the acidity. In the example, the pH of the urine is 5.4 as compared to the normal of 7.4 for plasma.

Adjustment of pH by respiration is effective and immediate but very seldom complete and, by itself, would not be adequate to protect the structure of the extracellular fluids. The response of the kidney is slower than that of the lungs but very effective. Important chemical mechanisms used by the kidney to control pH are as follows.

1. Excretion of an Acid or Alkaline Urine

The kidney can excrete urine which varies in pH from about 4.8 to 8.0. By this device phosphate can be excreted as $H_2PO_4^-$. At a pH of 7.4 about 80 per cent

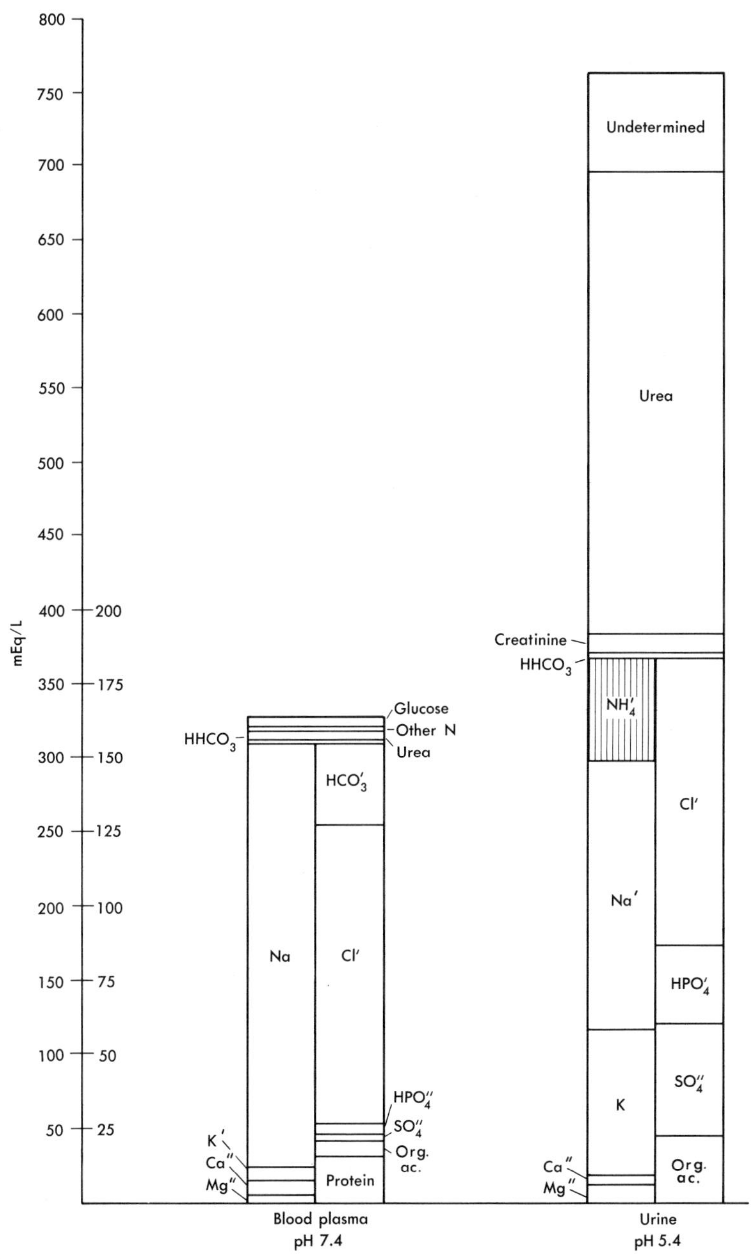

Figure 11.10. Renal regulation of extracellular fluids. (From J. L. Gamble: *Chemical Anatomy, Physiology and Pathology of Extracellular Fluid.* Harvard University Press, Cambridge, Mass., 1958.)

of the phosphate is HPO_4^-. Thus, for each millimole of phosphate excreted, 0.8 millimole of sodium would be saved. About one half of acetoacetic acid, which has a pK_a of about 4.8, could be excreted as the free acid. In contrast to the lungs which can alter the concentration of H_2CO_3, the kidney alters only the HCO_3 concentration in this buffer system. The concentration of H_2CO_3 remains constant at a pCO_2 of 40 mm. At a pH of 4.8 there is very little HCO_3^- in the urine. At a pH of 8.0 the HCO_3^- in the urine is very high.

2. Excretion of Ammonia

The kidney can neutralize acid by excreting ammonia. Ammonia is produced in the kidney from glutamine.

$$\begin{array}{l} CONH_2 \\ | \\ CH_2 \\ | \\ CH_2 \\ | \\ CH{-}NH_2 \\ | \\ COOH \end{array} \xrightarrow[H_2O]{\text{glutaminase}} \begin{array}{l} COOH \\ | \\ CH_2 \\ | \\ CH_2 \\ | \\ CH{-}NH_2 \\ | \\ COOH \end{array} + NH_3$$

This mechanism is very effective for neutralizing acids for prolonged periods of time.

By excreting acid and ammonia, the kidney is able to return an equivalent amount of Na^+ to the plasma and maintain the so-called "alkaline reserve" of the plasma.

3. Adjustment of Chloride

As the amount of HCO_3^- is increased or decreased, or organic anions are added to the plasma, the total ionic concentration of the extracellular fluids would change unless some adjustments were made. The kidney adjusts chloride ion concentration to meet these needs. If HCO_3^- concentration is decreased, chloride ion concentration is adjusted upward. In conditions where HCO_3^- concentration is high, the chloride concentration will be adjusted to lower levels.

Our discussion has centered around maintaining a 20/1 ratio in the bicarbonate buffering systems and maintaining the normal ionic structure of the extracellular fluids. It should be realized that the situation is more complicated than presented. The cells are not impassive, and there is active exchange of Na^+ and H^+ and K^+ between them and extracellular fluids. These changes are important in the overall adjustment in alkalosis or acidosis. An adjustment in ionic structure calls for an adjustment in water. In practice there is often impaired kidney function, liver function or heart function which must be considered.

Students who are interested in parenteral fluid therapy will want to acquaint themselves with some of the very good books and reviews on this subject.

REFERENCES

Christensen, H. N.: *Body Fluids and Their Neutrality*. Oxford University Press, New York, 1963.

Davenport, H. W.: *The ABC of Acid-Base Chemistry*, 4th ed. University of Chicago Press, Chicago, 1958.

Frisell, W. R.: *Acid-Base Chemistry in Medicine*. The Macmillan Co., New York, 1968.

Gamble, J. L.: *Chemical Anatomy, Physiology, and Pathology of Extracellular Fluid*. Harvard University Press, Cambridge, Mass., 1958.

Appendix

OUTLINES OF SOME SPECIAL METABOLIC PATHWAYS

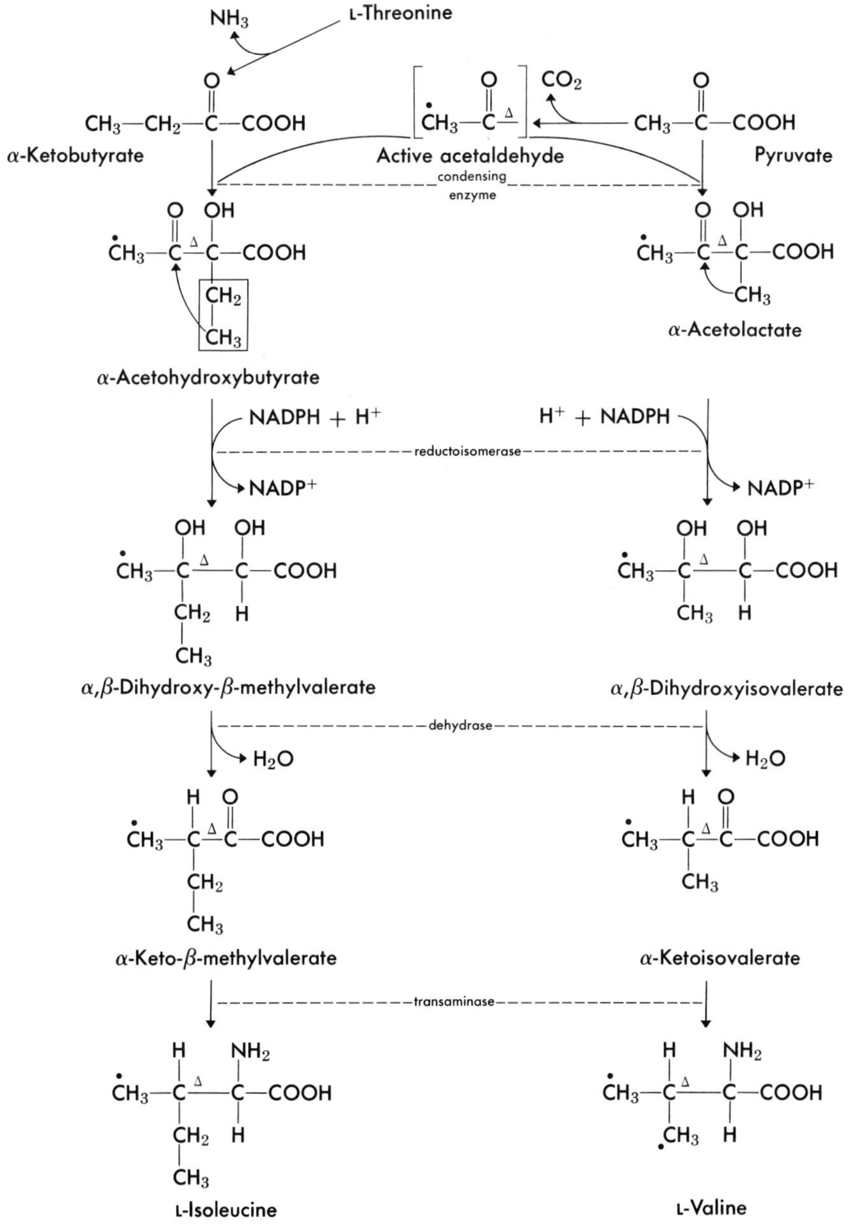

NH_3 L-Threonine

$CH_3—CH_2—C(=O)—COOH$

α-Ketobutyrate

[$\dot{C}H_3—C(=O)—^{\Delta}$] CO_2

Active acetaldehyde

$CH_3—C(=O)—COOH$

Pyruvate

condensing enzyme

$\dot{C}H_3—C(=O)\overset{\Delta}{—}C(OH)(CH_2CH_3)—COOH$

α-Acetohydroxybutyrate

$\dot{C}H_3—C(=O)\overset{\Delta}{—}C(OH)(CH_3)—COOH$

α-Acetolactate

$NADPH + H^+$ $H^+ + NADPH$

reductoisomerase

$NADP^+$ $NADP^+$

$\dot{C}H_3—C(OH)(CH_2CH_3)\overset{\Delta}{—}C(OH)(H)—COOH$

α,β-Dihydroxy-β-methylvalerate

$\dot{C}H_3—C(OH)(CH_3)\overset{\Delta}{—}C(OH)(H)—COOH$

α,β-Dihydroxyisovalerate

dehydrase

H_2O H_2O

$\dot{C}H_3—C(H)(CH_2CH_3)\overset{\Delta}{—}C(=O)—COOH$

α-Keto-β-methylvalerate

$\dot{C}H_3—C(H)(CH_3)\overset{\Delta}{—}C(=O)—COOH$

α-Ketoisovalerate

transaminase

$\dot{C}H_3—C(H)(CH_2CH_3)\overset{\Delta}{—}C(NH_2)(H)—COOH$

L-Isoleucine

$\dot{C}H_3—C(H)(\dot{C}H_3)\overset{\Delta}{—}C(NH_2)(H)—COOH$

L-Valine

Figure A.1. Biosynthesis of isoleucine and valine in bacteria.

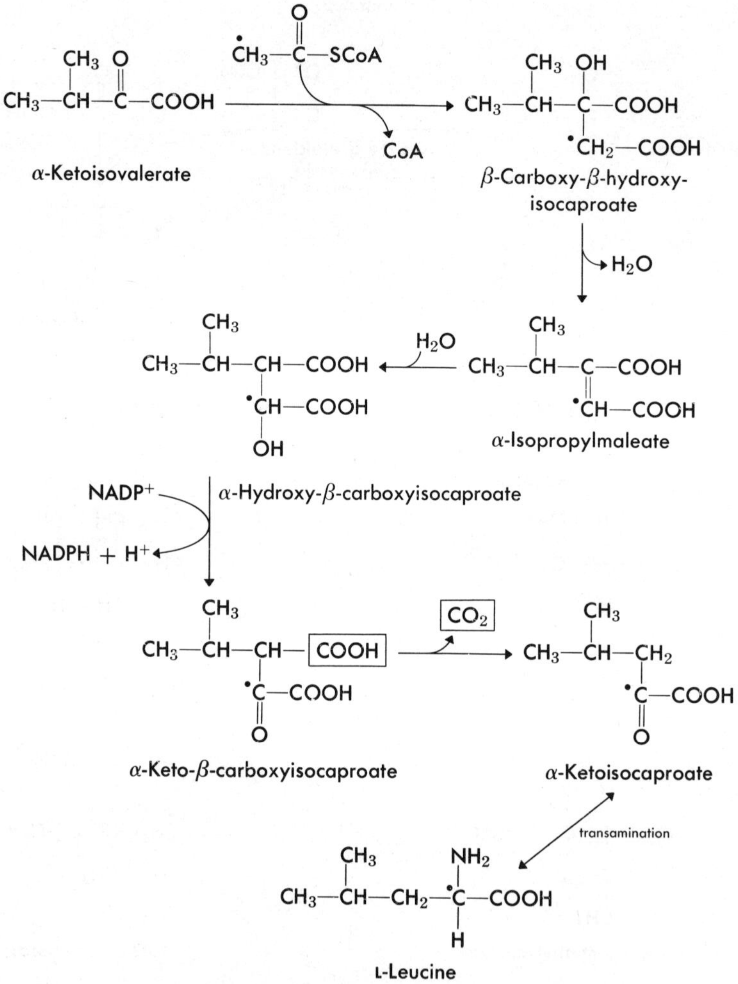

$$CH_3{-}CH(CH_3){-}CO{-}COOH \xrightarrow[\;CoA]{\;\dot{C}H_3{-}CO{-}SCoA\;} CH_3{-}CH(CH_3){-}C(OH)(COOH){-}\dot{C}H_2{-}COOH$$

α-Ketoisovalerate → β-Carboxy-β-hydroxy-isocaproate

$$\xrightarrow{-H_2O} CH_3{-}CH(CH_3){-}C(COOH){=}\dot{C}H{-}COOH$$

α-Isopropylmaleate

$$\xrightarrow{+H_2O} CH_3{-}CH(CH_3){-}CH(COOH){-}\dot{C}H(OH){-}COOH$$

α-Hydroxy-β-carboxyisocaproate

$$\xrightarrow{NADP^+ \rightarrow NADPH + H^+} CH_3{-}CH(CH_3){-}CH(\boxed{COOH}){-}\dot{C}O{-}COOH$$

α-Keto-β-carboxyisocaproate

$$\xrightarrow{-\boxed{CO_2}} CH_3{-}CH(CH_3){-}CH_2{-}\dot{C}O{-}COOH$$

α-Ketoisocaproate

$$\xleftrightarrow{\text{transamination}} CH_3{-}CH(CH_3){-}CH_2{-}\dot{C}H(NH_2){-}COOH$$

L-Leucine

Figure A.2. Biosynthesis of leucine in bacteria.

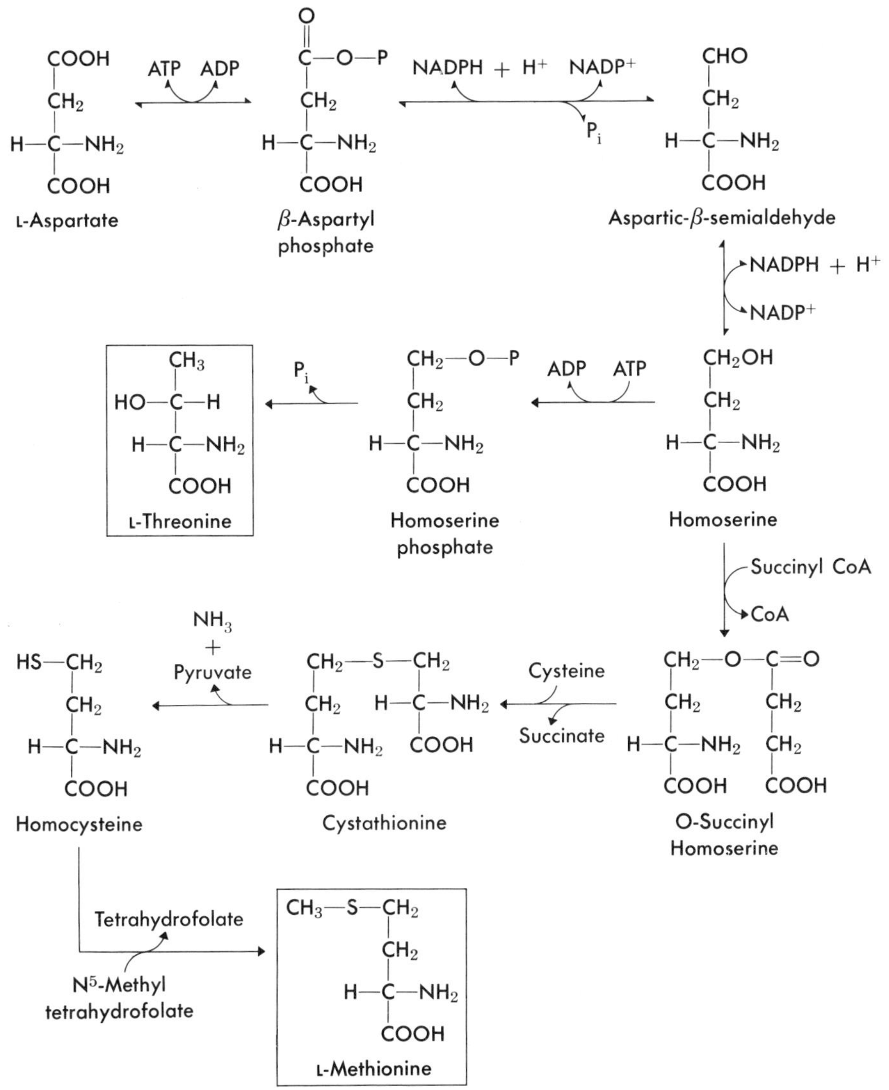

L-Aspartate $\xrightarrow{ATP \rightarrow ADP}$ β-Aspartyl phosphate $\xrightarrow{NADPH + H^+ \rightarrow NADP^+, P_i}$ Aspartic-β-semialdehyde $\xrightarrow{NADPH + H^+ \rightarrow NADP^+}$ Homoserine

Homoserine $\xrightarrow{ATP \rightarrow ADP}$ Homoserine phosphate $\xrightarrow{P_i}$ L-Threonine

Homoserine $\xrightarrow{Succinyl\ CoA \rightarrow CoA}$ O-Succinyl Homoserine $\xrightarrow{Cysteine \rightarrow Succinate}$ Cystathionine $\xrightarrow{NH_3 + Pyruvate}$ Homocysteine

Homocysteine $\xrightarrow{N^5\text{-Methyl tetrahydrofolate} \rightarrow Tetrahydrofolate}$ L-Methionine

Figure A.3. Biosynthesis of threonine and methionine in microorganisms.

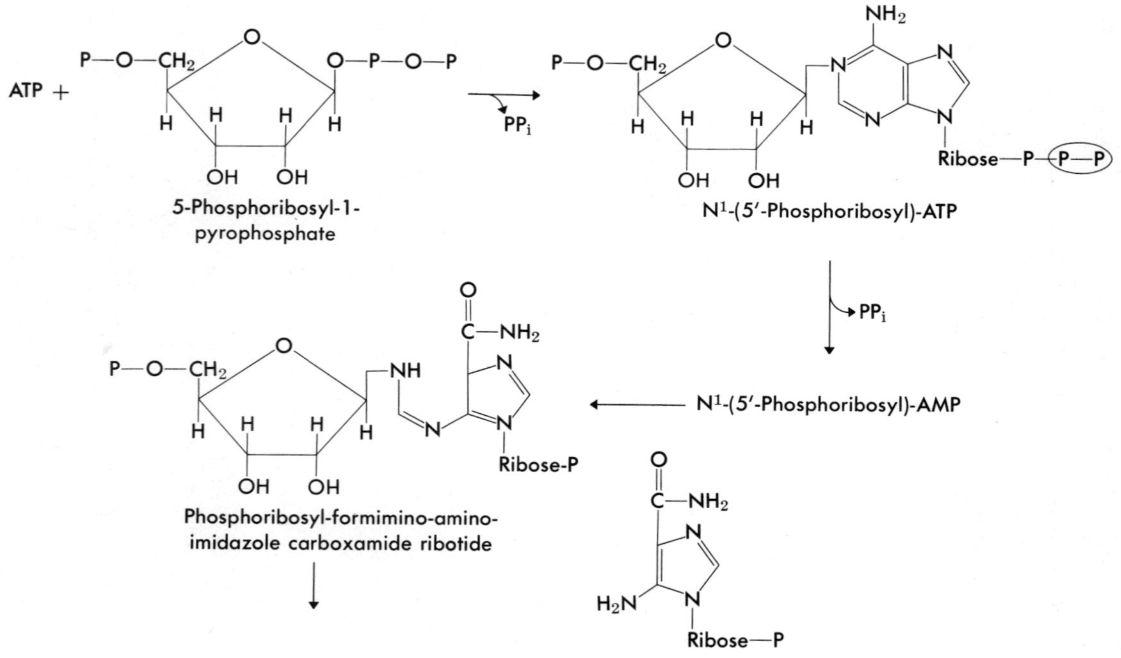

ATP + 5-Phosphoribosyl-1-pyrophosphate $\xrightarrow{-PP_i}$ N^1-(5′-Phosphoribosyl)-ATP

N^1-(5′-Phosphoribosyl)-ATP $\xrightarrow{-PP_i}$ N^1-(5′-Phosphoribosyl)-AMP

N^1-(5′-Phosphoribosyl)-AMP ⟶ Phosphoribosyl-formimino-amino-imidazole carboxamide ribotide ⟶

P—O—CH_2 O—P—O—P   H   OH NH_2 N   Ribose—P—P—P   NH   C—NH_2 O   Ribose-P H_2N Ribose—P

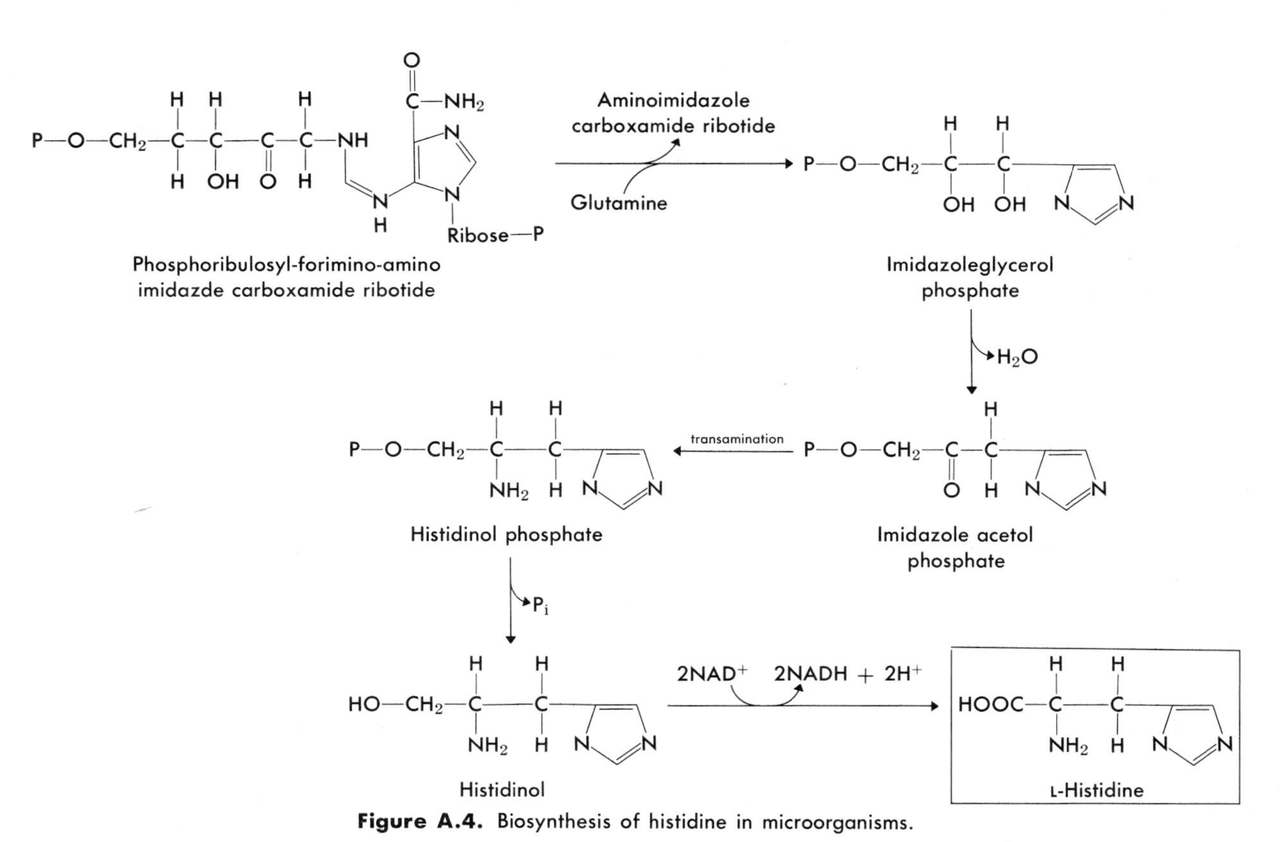

Figure A.4. Biosynthesis of histidine in microorganisms.

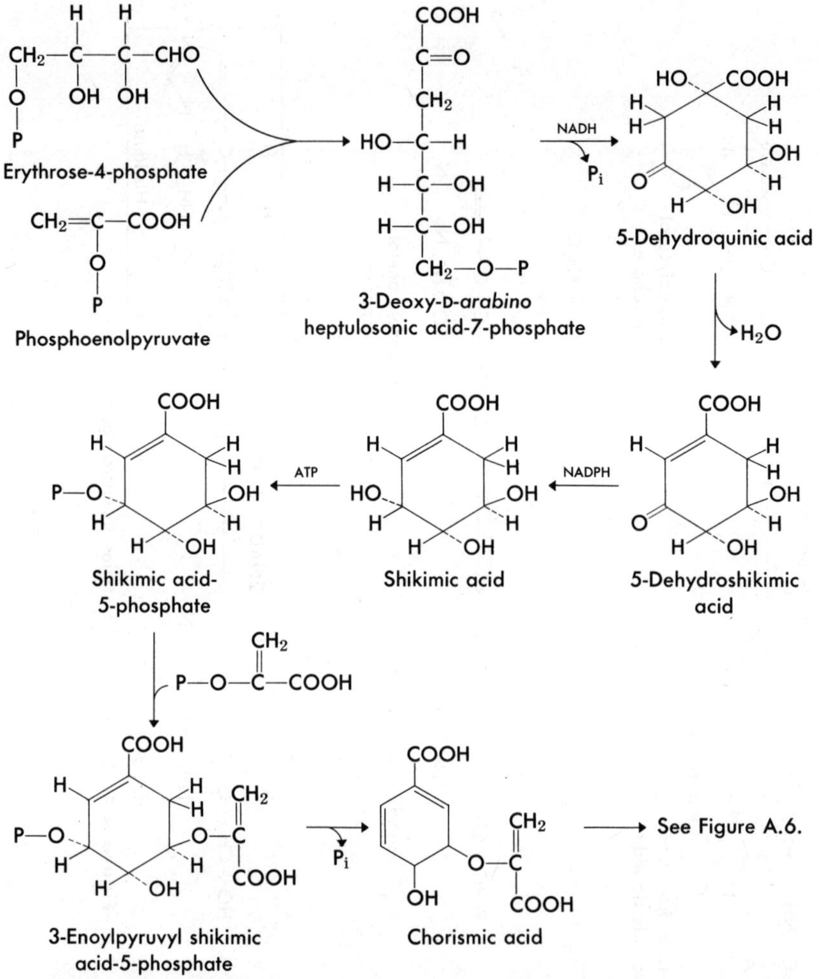

Erythrose-4-phosphate

Phosphoenolpyruvate

3-Deoxy-D-*arabino* heptulosonic acid-7-phosphate

NADH

P_i

5-Dehydroquinic acid

H_2O

5-Dehydroshikimic acid

NADPH

Shikimic acid

ATP

Shikimic acid-5-phosphate

3-Enoylpyruvyl shikimic acid-5-phosphate

P_i

Chorismic acid

See Figure A.6.

Figure A.5. Biosynthesis of aromatic amino acids in microorganisms (part I).

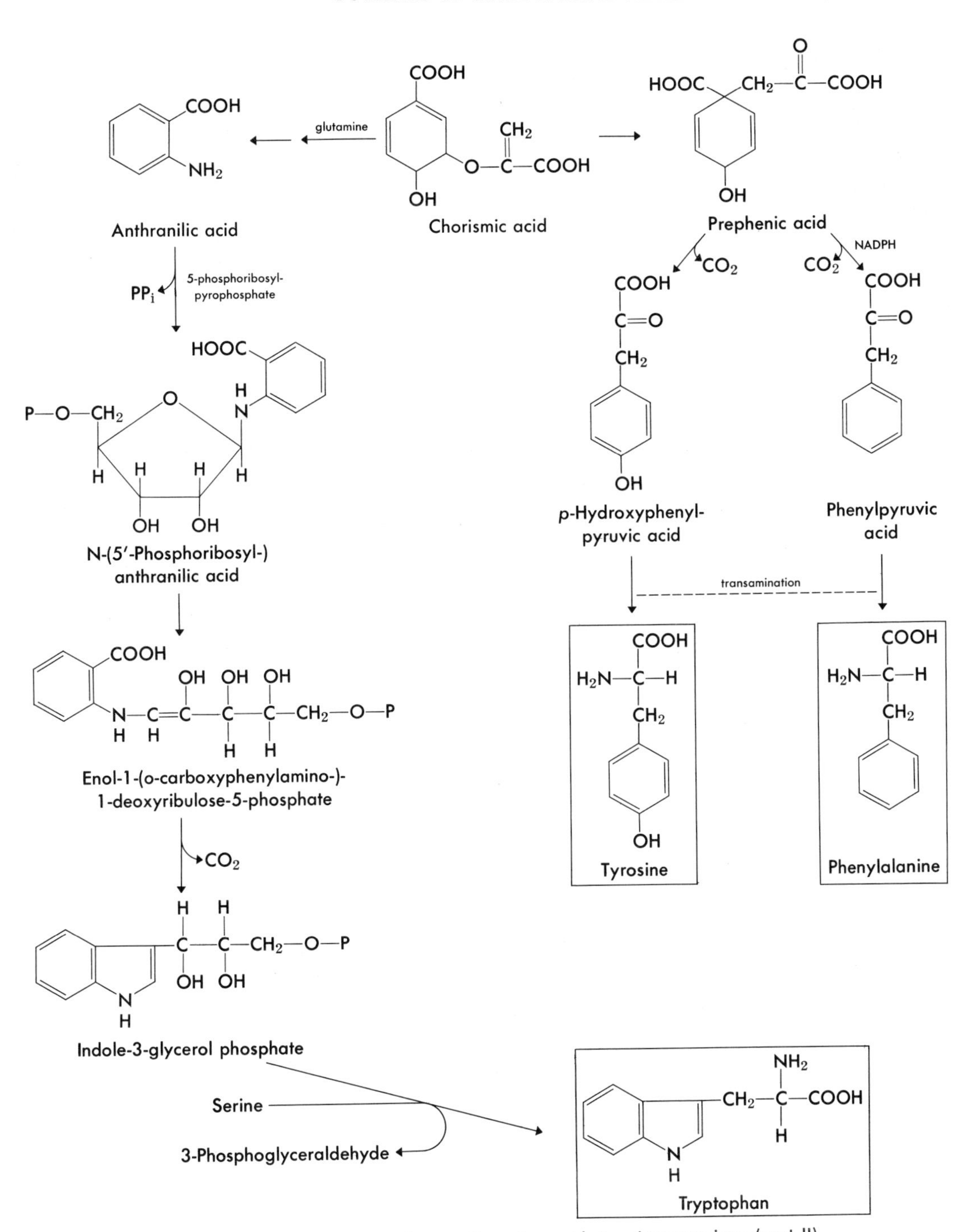

Figure A.6. Biosynthesis of aromatic amino acids in microorganisms (part II).

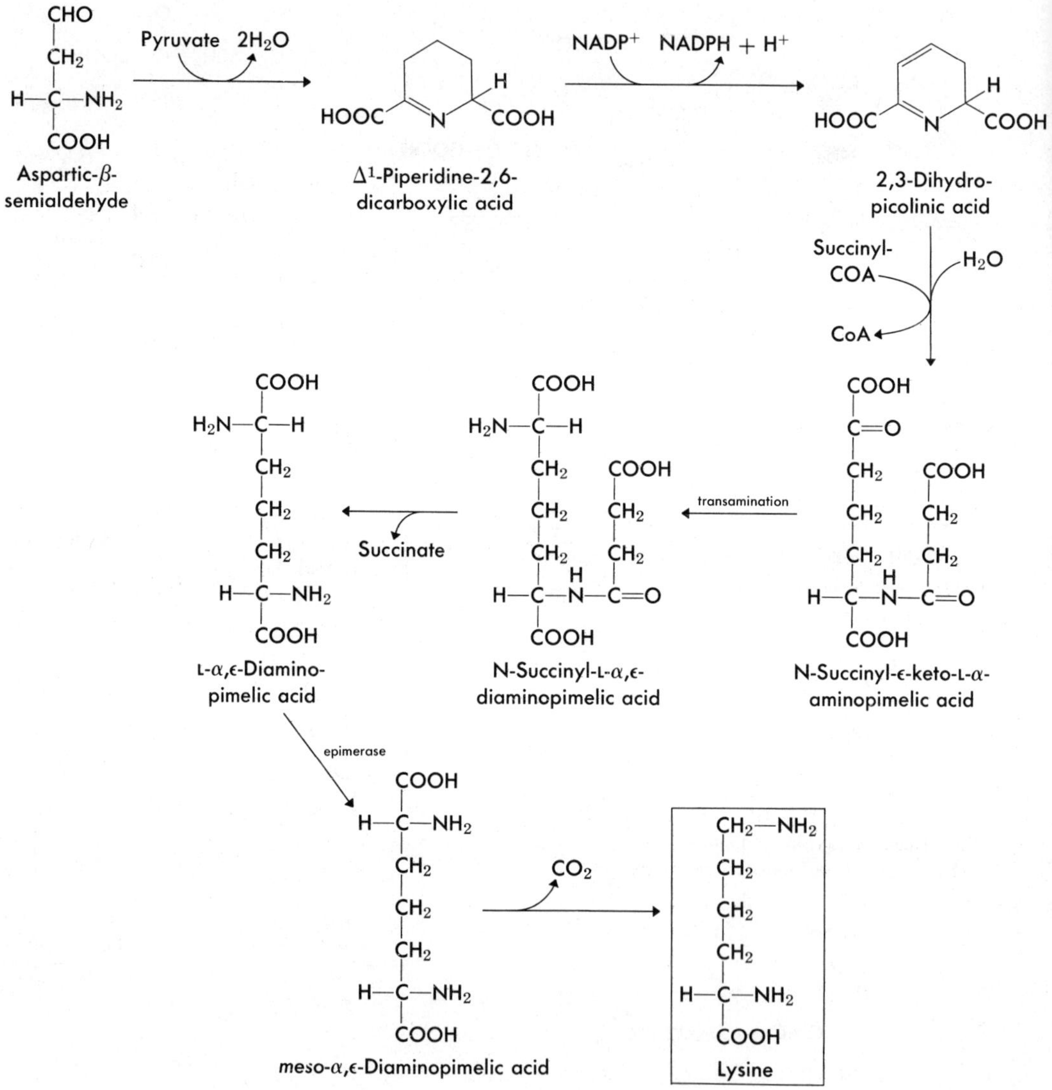

CHO
|
CH_2
|
H—C—NH_2
|
COOH

Aspartic-β-semialdehyde

Pyruvate $2H_2O$

HOOC N H COOH

Δ^1-Piperidine-2,6-dicarboxylic acid

$NADP^+$ $NADPH + H^+$

HOOC N H COOH

2,3-Dihydro-picolinic acid

Succinyl-COA

H_2O

CoA

COOH
|
C=O
|
CH_2 COOH
| |
CH_2 CH_2
| |
CH_2 CH_2
| H |
H—C—N—C=O
|
COOH

N-Succinyl-ε-keto-L-α-aminopimelic acid

transamination

COOH
|
H_2N—C—H
|
CH_2 COOH
| |
CH_2 CH_2
| |
CH_2 CH_2
| H |
H—C—N—C=O
|
COOH

N-Succinyl-L-α,ε-diaminopimelic acid

Succinate

COOH
|
H_2N—C—H
|
CH_2
|
CH_2
|
CH_2
|
H—C—NH_2
|
COOH

L-α,ε-Diamino-pimelic acid

epimerase

COOH
|
H—C—NH_2
|
CH_2
|
CH_2
|
CH_2
|
H—C—NH_2
|
COOH

meso-α,ε-Diaminopimelic acid

CO_2

CH_2—NH_2
|
CH_2
|
CH_2
|
CH_2
|
H—C—NH_2
|
COOH

Lysine

Figure A.7. Biosynthesis of lysine in bacteria and higher plants.

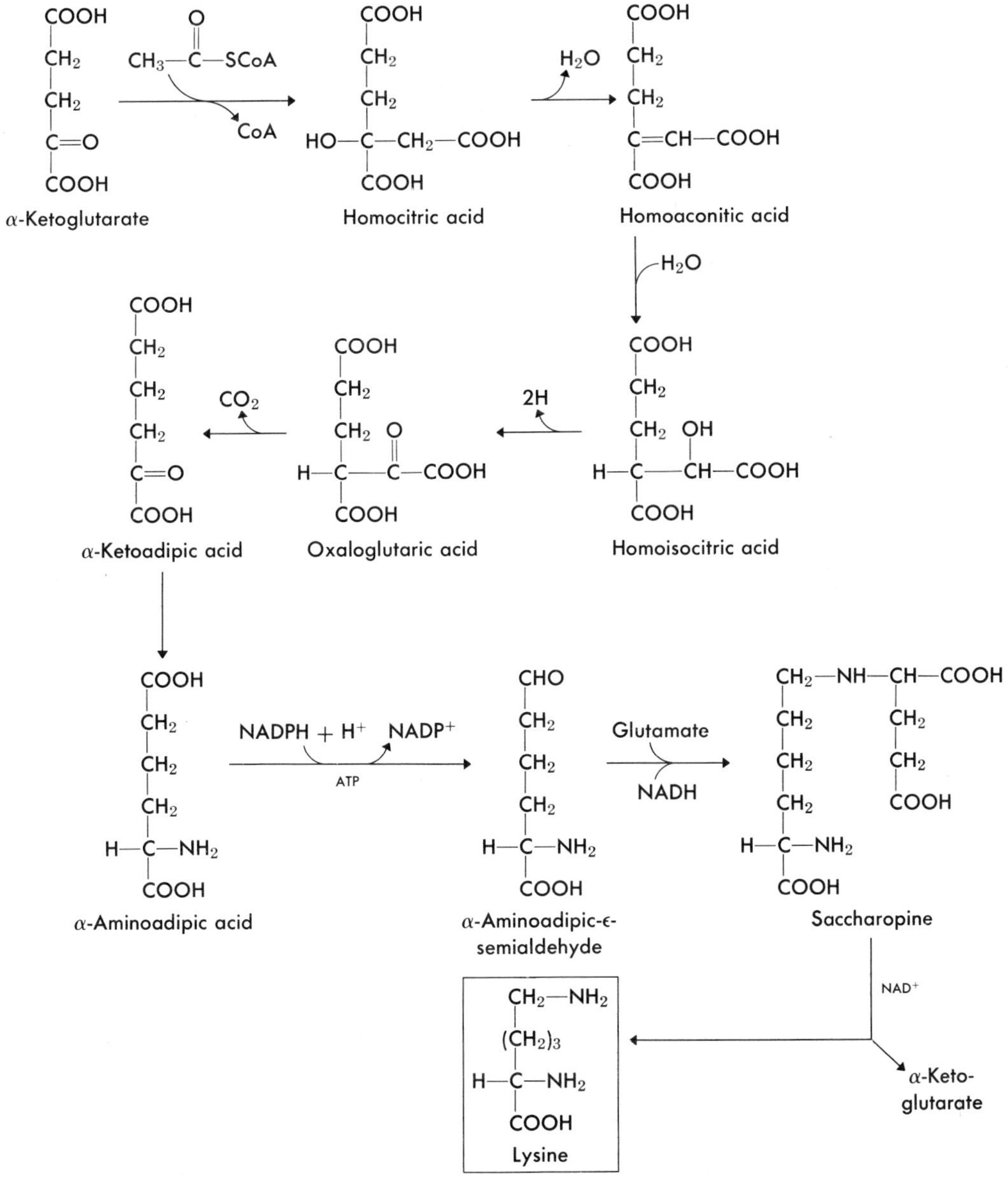

COOH–CH$_2$–CH$_2$–C=O–COOH
α-Ketoglutarate

CH_3–C(=O)–SCoA → CoA

COOH–CH$_2$–CH$_2$–C(OH)(–CH$_2$–COOH)–COOH
Homocitric acid

→ H_2O

COOH–CH$_2$–CH$_2$–C(=CH–COOH)–COOH
Homoaconitic acid

H_2O ↓

COOH–CH$_2$–CH$_2$–H–C(–CH(OH)–COOH)–COOH
Homoisocitric acid

← 2H

COOH–CH$_2$–CH$_2$–H–C(–C(=O)–COOH)–COOH
Oxaloglutaric acid

← CO_2

COOH–CH$_2$–CH$_2$–CH$_2$–C=O–COOH
α-Ketoadipic acid

↓

COOH–CH$_2$–CH$_2$–CH$_2$–H–C–NH$_2$–COOH
α-Aminoadipic acid

NADPH + H^+ → NADP$^+$ (ATP)

CHO–CH$_2$–CH$_2$–CH$_2$–H–C–NH$_2$–COOH
α-Aminoadipic-ϵ-semialdehyde

Glutamate → NADH

CH$_2$–NH–CH(–CH$_2$–CH$_2$–COOH)–COOH; CH$_2$–CH$_2$–CH$_2$–H–C–NH$_2$–COOH
Saccharopine

NAD$^+$ → α-Keto-glutarate

CH$_2$–NH$_2$–(CH$_2$)$_3$–H–C–NH$_2$–COOH
Lysine

Figure A.8. Biosynthesis of lysine in fungi.

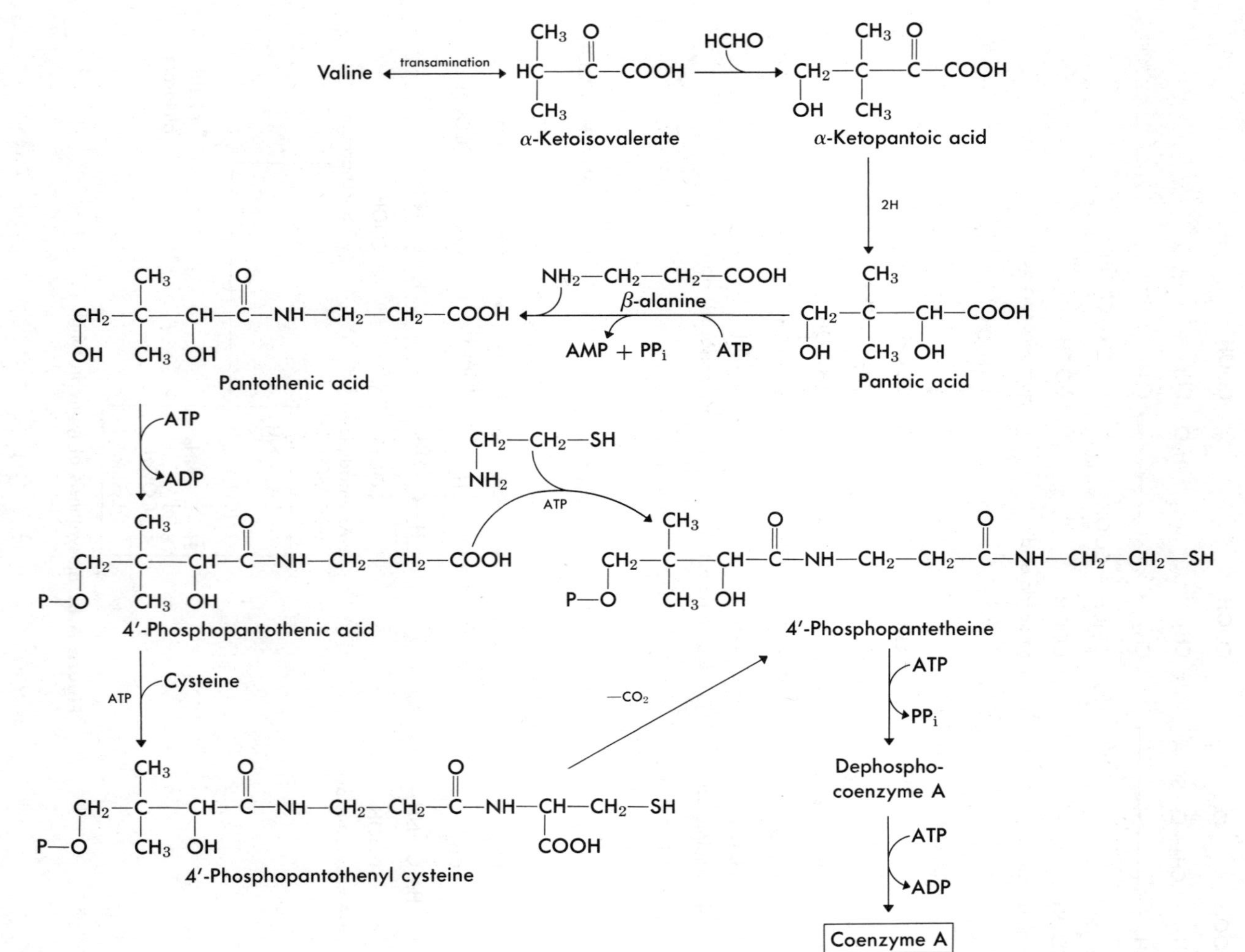

Figure A.9. Biosynthesis of coenzyme A.

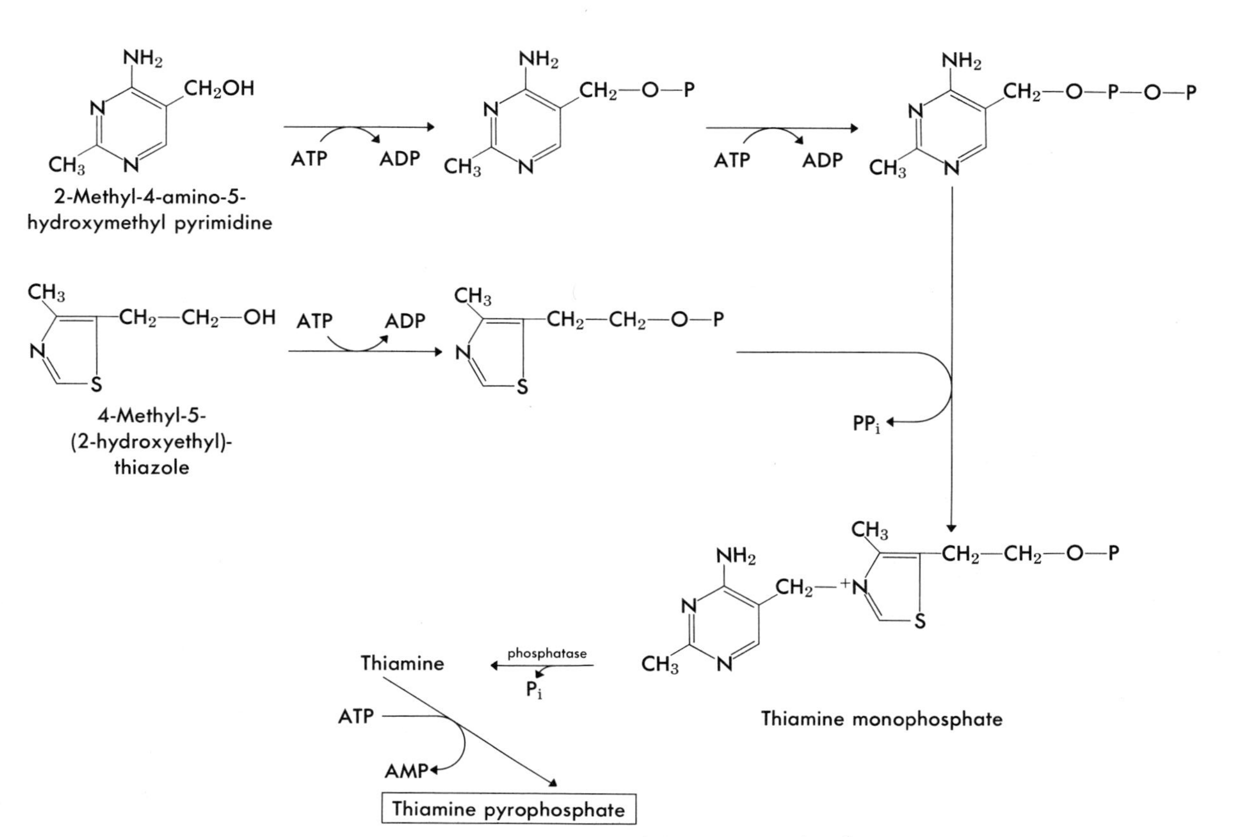

Figure A.10. Biosynthesis of thiamine pyrophosphate.

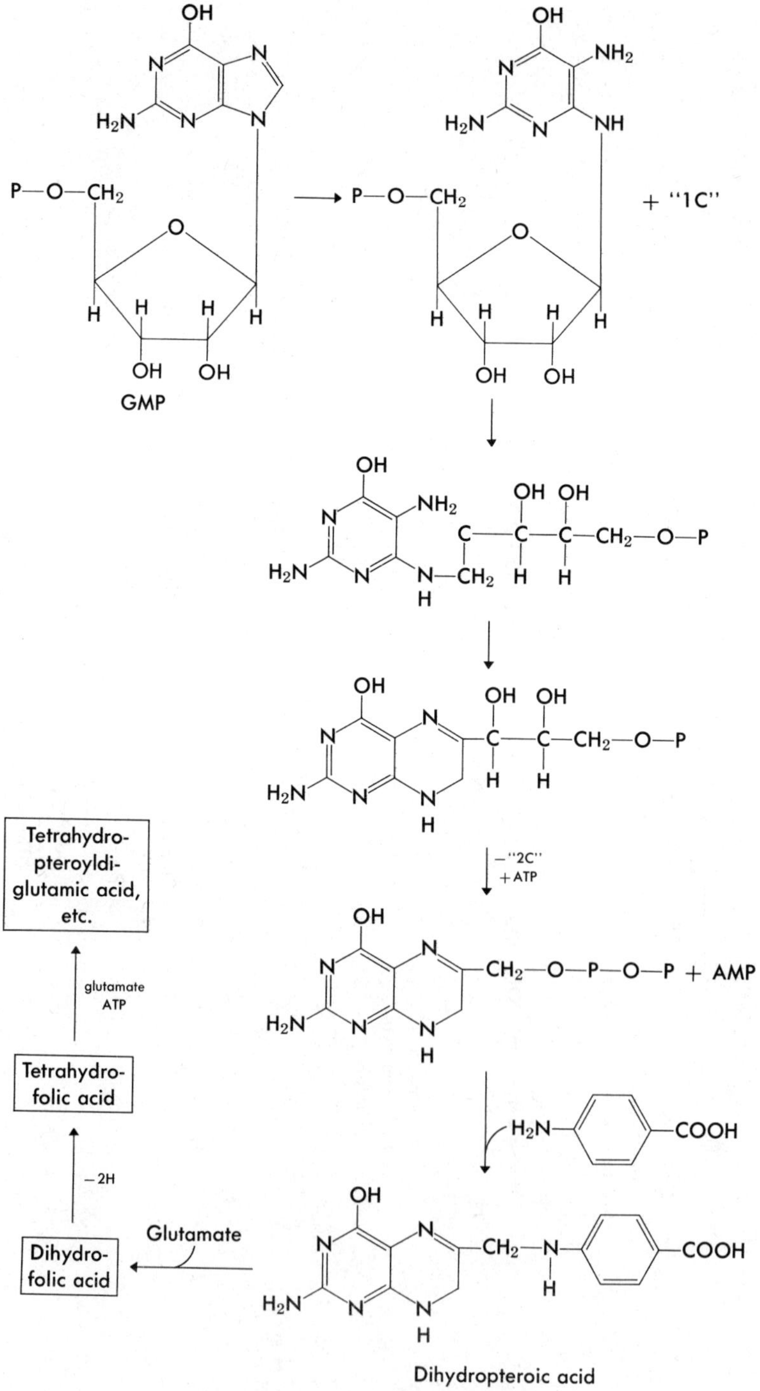

GMP

+ "1C"

−"2C"
+ATP

+ AMP

Tetrahydro-
pteroyldi-
glutamic acid,
etc.

glutamate
ATP

Tetrahydro-
folic acid

−2H

Dihydro-
folic acid

Glutamate

Dihydropteroic acid

Figure A.11. Biosynthesis of folic acid.

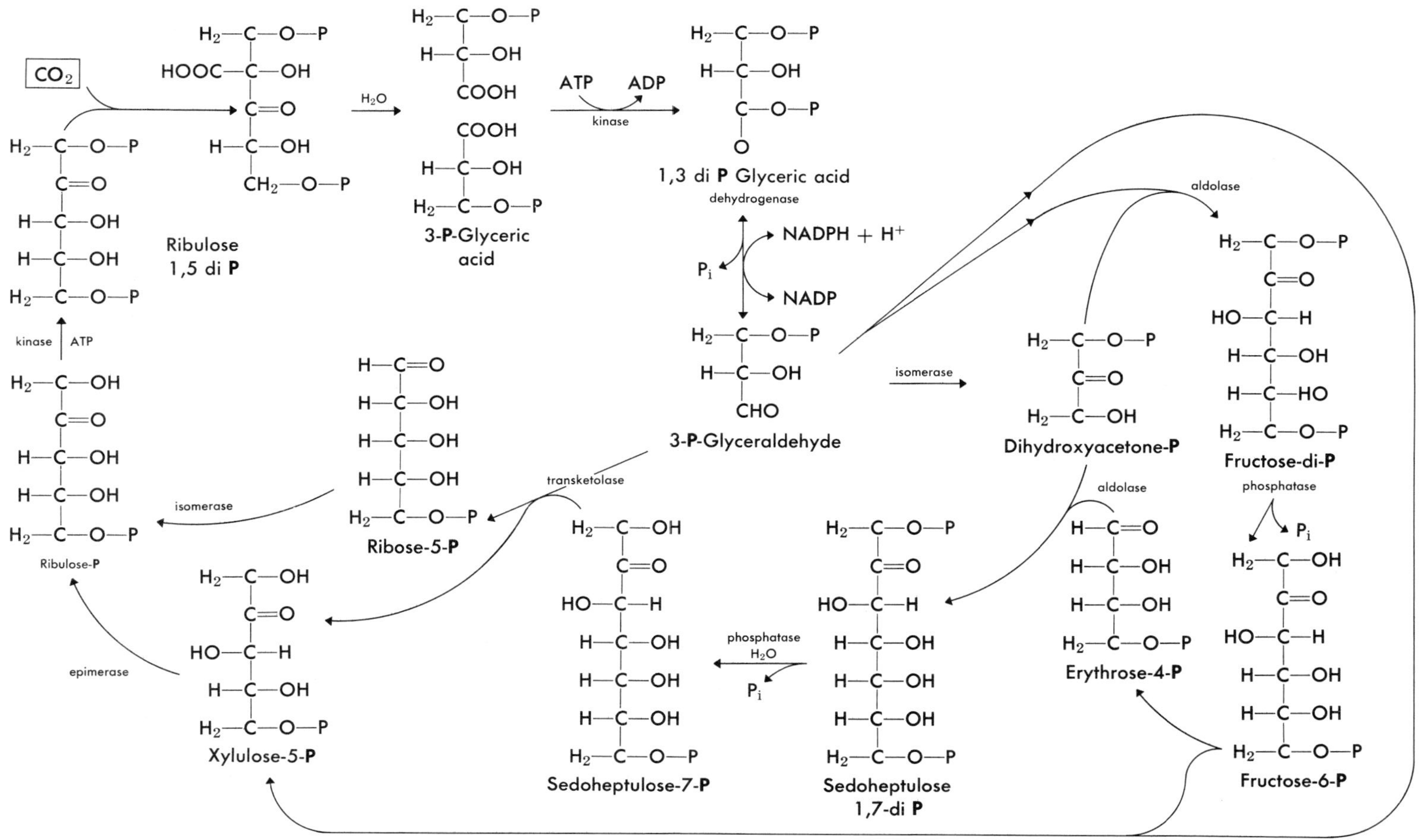

Figure A.12. Calvin photosynthetic carbon cycle.

INDEX